Project Management
Theory and Practice

Project Management Theory and Practice

Gary L. Richardson

CRC Press
Taylor & Francis Group
Boca Raton London New York

CRC Press is an imprint of the
Taylor & Francis Group, an **informa** business

AN AUERBACH BOOK

Auerbach Publications
Taylor & Francis Group
6000 Broken Sound Parkway NW, Suite 300
Boca Raton, FL 33487-2742

© 2010 by Taylor and Francis Group, LLC
Auerbach Publications is an imprint of Taylor & Francis Group, an Informa business

No claim to original U.S. Government works

Printed in the United States of America on acid-free paper
10 9 8 7 6 5 4 3 2 1

International Standard Book Number: 978-1-4398-0993-8 (Hardback)

Library of Congress Cataloging-in-Publication Data

Richardson, Gary L.
    Project management theory and practice / Gary L. Richardson.
        p. cm.
    Includes bibliographical references and index.
    ISBN 978-1-4398-0993-8
    1. Project management. I. Title.

HD69.P75R5225 2010
658.4'04--dc22   2009050315

Visit the Taylor & Francis Web site at
http://www.taylorandfrancis.com

and the Auerbach Web site at
http://www.auerbach-publications.com

Contents

PART I   *Conceptual Overview of the Project Environment*

PART III   *Defining the Triple Constraints*

PART IV   *Advanced Planning Models*

PART V   *Planning Support Processes*

PART VI   *Project Execution—Managing the Plan*

PART VII   *Monitoring and Controlling Techniques*

Part VIII   Closing the Project

PART IX   Contemporary Topics

PART X   *Professional Ethics and Responsibility*

Preface

The roots of this effort go back many years in my attempts to install standard project development methodologies into large organizations. Also, through all of those years I was involved with projects of one kind or another. Around 2003, I returned to the University of Houston to teach project management, thinking that it would be an easy subject given my previous experience. However, it soon became obvious that this subject was not well documented in a readable format and students struggled to get a real flavor of the topic. Most of the textbooks on the market were either too academic, too much IT, or too much real-world "silver bullet" quick fix advice types. Based on that assessment, the vision of correcting that oversight began to take shape. After four years of thrashing around with the topic, this text is the result. You the reader will have to decide how well it matched the goal of a readable overview of project management theory.

The academic program at the University of Houston is heavily based on the Project Management Institute's (PMI) concepts and that bias formed the foundation for the text, but not the complete final scope. The early chapters stay reasonably close to the Project Management Body of Knowledge (*PMBOK® Guide*) models, but also attempt to show how this model fits the real world. In this regard, the material in the text is viewed as a companion to the technical guide and should be of help to someone studying for various project management certifications.

There are several project-related model frameworks sponsored by PMI today, and many of these are covered in dedicated chapters in the text. Specifically, the following six topics are discussed in some detail:

- Work breakdown structures (WBS)
- Earned value management (EVM)
- Enterprise project management
- Portfolio management (PPM)
- Professional responsibility
- Project team productivity.

In addition, there are multiple chapters related to various other associated contemporary topics that are currently emerging in the industry.

Deciding on how to define what material should be covered was difficult once the basic core topics were listed. For many of the major sections, the PMI Global Accreditation curriculum learning objectives were adapted with permission of PMI and used to guide the content. Note that each of the nine text sections has a set of learning objectives stated in the header. Learning objectives for those topics not covered by PMI were developed by the author.

Even though interest in the topic of project management is growing and maturing, the subject is still in a relatively neophyte maturity stage. Many industry experts have willingly shared their work and thoughts in this effort. Their comments have been incorporated, and hopefully the resulting material did not distort their content. Based on the logic outlined above we believe the result represents a good overview of the project management environment today, but also recognize that there is more left to evolve.

The writing style used was not meant to be overly formal in the hopes that it would create a more willing reader with the belief that formal project management technical material can be similar to a dentist visit for a root canal. Also, the writing goal was to not duplicate the *PMBOK® Guide* mechanics, but stay close enough to the model structure so that one receives a complementary perspective of the common topic areas. Recognize that there are parts of the text that clearly push beyond the basic model view and extrapolate beyond current reality. Please accept these few ventures as an attempt to broaden the current perspectives and offer a potential future pathway for the overall topic. These jumps in faith were carefully taken and directionally seem appropriate. At least they should stimulate thinking beyond the pragmatic, since any professional working in this field needs to understand both the current views and be prepared to evolve with these over time. In my view the resulting document has both a theoretical model and sufficient real-world perspective combined.

Text Organization

The initial text structure starts with very basic organizational and project concepts. There is no reader background assumed in Parts I and II other than a very general knowledge of organizations and projects. Lack of any real-world exposure will make some of these topics seem abstract, while exposure will make the reader more willing to believe that projects really can be this poorly managed.

Each major section of the text focuses on what is considered a key topic goal area. Part I is intended to level set the reader with background up to a Quick Start project plan. Parts II through VII focus on major life cycle "domains," which basically translate to major project stage activities. Part VIII covers six areas that are considered to be contemporary, meaning that they exist in the real world, but the topics are in transition. Finally, Part IX deals with the ethical framework that must be understood by the professional project manager. One only has to read the daily newspaper to see the rationale for this topic being included.

Minicase Worksheets

Scattered throughout the text are sample worksheets used to demonstrate various management decision or analysis oriented tools.

I hope that you find your journey through this material interesting and worthwhile.

Gary L. Richardson
University of Houston

Acknowledgments

No effort of this scope and complexity could have been accomplished by one person in any reasonable time frame. This book is no exception to that rule. During the early incubation period several colleagues provided stimulus for this effort. First, Walter Viali, who is a 30-year professional associate, convinced me that the *PMBOK® Guide* and Project Management Institute (PMI) were the right thought leaders to provide the foundation structure for the university academic program. This has proven to be a successful core strategic decision for our university program as well as for this book. Rudy Hirschheim, Dennis Adams, and Blake Ives were instrumental in helping me move from industry to academia and then supporting me as I tried to become a professor again. Later, Michael Gibson provided the final push and support to allow me time to complete the draft material. Ron Smith (PMP and CSPM) provided several of his published worksheets and helped customize these for use as end of chapter examples. As a result of this involvement, he became a reviewer of the book and a great supporter. Charles Butler, who has been my collaborator on so many past ventures, edited much of the text and through that made the content better. Addie Tsai did much of the edit work on a very rough manuscript. Sandeep Vajjiparti and Prinitha Koya helped in various formatting and packaging chores during the early formulation stage. And thanks to Teri, who taught me about chip theory and a lot of other soft skills.

Industry gurus Watts Humphrey, Walt Lipke, Tom Mocal, Max Wideman, Frank Patrick, Lawrence Leach, and Don James contributed ideas, reviews, and material in their respective areas of expertise. Other sources such as The Standish Group, QPR, QSM, and the Software Engineering Institute shared their intellectual property.

Jerry Evans and Dan Cassler, my University of Houston office mates, continually provided an environment of friendly warmth and fun that may well be the most important support of all. Last but not least, Bob Fitzsimmons continued our 55-year friendship with frequent moral support and became the chief graphics artist along the way.

Over the past 6 years I have been blessed with 200 captive project management graduate students digging through voluminous technical sources to generate a library of raw material from which much of this text is drawn. In return for this sweat equity, a research contribution list of defined student contributors is enclosed to give some measure of credit for their work. Also, a percentage of any future royalties received will be allocated to the University of Houston for project management scholarships.

The resulting text material is a compendium of intellectual thoughts and ideas from all of the sources mentioned above, plus my own experiences. I have tried to credit all of the sources that were used, and if any were missed it was unintentional.

Finally, my wife Shawn's tolerance through all of my seemingly endless nights and weekends in the study upstairs must be recognized. Without her support this effort could not have been finished.

Author

Gary Richardson currently serves as the program coordinator for the University of Houston College of Technology graduate level project management program. This program serves both the internal and external community in regard to the theory and practice of project management. He comes from a broad professional background including industry, consulting, government, and academia.

During the early phase of his career he was an officer in the U.S. Air Force, followed by industry stints at Texas Instruments as a manufacturing engineer, and then by consulting assignments at the Defense Communications Agency, Department of Labor, and the U.S. Air Force (Pentagon) in Washington, D.C. The latter half of his career was spent with Texaco and Service Corporation International in various IT and CIO level management positions. Interspersed through these periods he was a professor at Texas A&M, the University of South Florida, and the University of Houston, and also did other adjunct professor stints at three other universities. Gary has previously published four computer-related textbooks and numerous technical articles.

Through his experiences in over 100 significantly sized projects of various types, he has observed frequently encountered issues and has been an active participant in the evolution of management techniques that have occurred over this time.

Gary received a BS in mechanical engineering from Louisiana Tech, an AFIT postgraduate program in meteorology at the University of Texas, an MS in engineering management from the University of Alaska, and a PhD in business administration from the University of North Texas. He currently teaches the PMP Prep course and other graduate-level project management courses at the University of Houston plus various continuing education courses.

Part I

Conceptual Overview of the Project Environment

LEARNING OBJECTIVES

This initial section is designed to level set the reader with various aspects of the project management field. Upon completion the following concepts should be understood:

1. Definition of a project and its general characteristics
2. Basic history of project management
3. An understanding of the typical challenges faced by project managers
4. Benefits of the project management process
5. An introductory overview of the Project Management Institute's project model
6. Some of the contemporary trends that are changing the view of project management. Basic project scope, time, and budget mechanics
7. Key project vocabulary that is needed to understand the more detailed sections that follow later in the text.

1 Introduction

The term *project* is the central theme of this text and it is a frequently used descriptor; however, there are many different perspectives regarding what the term means. A collection of keywords from various sources and individuals will typically include the following key words in their definitions:

1. Team
2. Plan
3. Resources
4. Extend capability
5. Temporary
6. Chaos
7. Unique
8. Create
9. State transition.

From these diverse views it would be difficult to construct a universal definition that neatly included all of the terms, but collectively they do say a lot about a project's composition. The Project Management Institute (PMI) defines a project as

A temporary endeavor undertaken to create a unique product, service, or result. (PMI, 2008, p. 434)

In the modern organization, the project model is used to accomplish many of their planning goals, that is, moving the organization from state A to state B (state transition). For these endeavors, resources are allocated to the target, and through a series of work activities the project team attempts to produce the defined goal. Typical goals for this type of activity involve the creation of a new product, service, process, or any other activity that requires a fixed-time resource focus.

Figure 1.1 is a visual metaphor to illustrate what a project is attempting to accomplish. The two fuzzy clouds depict an organization moving from a current state to a future state. The arrow represents the project team driving this movement. From an abstract point of view, the role of a project is to create that movement, whether that represents an organizational process, new product development, or some other desired deliverables.

Projects should be envisioned as formal undertakings, guided by explicit management charters and focused on enterprise goals. Practically speaking, this is not always the case, but given the nature of this text we need to reject projects that are not focused on improving the goal status for the organization and those that do not have the explicit support of management. Any other initiatives are not examples of a project, but rather "project chaos."

1.1 PROJECT MANAGEMENT

The management of a project consists of many interrelated management pieces and parts. PMI defines this activity as

The application of *knowledge*, *skills*, *tools*, and *techniques* to *project activities* to meet the project *requirements*. (PMI, 2008, p. 434)

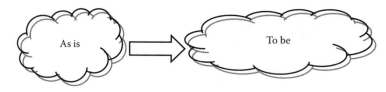

FIGURE 1.1   Project state transition process.

One of the first management issues is to define the schedule, budget, and resources required to produce the required output. These parameters are fundamental to all projects. Closely related to this set is the concept of quality, which relates to both the project target and the work processes used to achieve that target. Collectively, these items represent some of the more visible components involved in project management. Supporting this activity is another group of items related more to "how" the goal will be accomplished. This second grouping of the management focus activities involves more aspects of human resources, plus issues related to procurement, communications, and risk. During the course of the project, all of these topic areas interact with each other and therein lays the management complexity related to this topic.

1.2   ROLE OF THE PROJECT MANAGER

Essentially, the role of a project manager (PM) is to "make it happen." This does not mean that he is the best engineer, programmer, or business process technician. It does mean that he has the necessary skills to acquire, develop, and manage a team of individuals who are capable of producing the desired product. Every project has unique characteristics and therefore the roles required change accordingly. The current state of understanding for this role has defined the basic knowledge areas involved in this activity, but the operational techniques for creating productive project teams is still a fragile art form.

Many project success and failure studies have documented the basic factors leading to these conclusions. As projects have become more complex there is growing recognition that a skilled PM is the glue that brings these elements together. This involves the more mechanical management elements, but probably more important is the use of softer management skills for team motivation, conflict resolution, user communications, and general negotiation. We must not forget that project management involves humans and will never be reduced to a mechanical exercise. Nevertheless, the mechanical aspects are an important part of the overall management process in identifying what actions are required to influence changes. For example, to know that a project schedule is overrunning requires a complex set of data mechanics, but does not in itself do anything about it. Conceptualize the mechanical side of the management role as a meter—if your car's gas gauge is near empty this will stimulate the driver to find a gas station. Similarly, if the project schedule is not going according to the plan, the mechanical management process helps identify where and why. Other management action follows the meter readings.

1.3   PROJECT MANAGER SKILLS

We are tempted to say that the ideal PM skills are the ability to "leap tall buildings with a single bound, faster than a speeding bullet," but that might be a little excessive. However, it is accurate to say that this individual needs to understand how to deal with the various knowledge areas involved, with additional high skills in both personal and organizational areas. Project dynamics create an amazing array of daily issues to resolve. If one cannot organize this activity into some workable process the project will stagnate. Through all of this, it is the project manager's goal to achieve the plan. Industry project failure statistics indicate that this is more difficult than is understood by most.

At the highest level, the PM needs to bring structure and organization to his project team. One senior PM once described this problem as "putting a lot of mush in a small bucket." A significant aspect of this is formalizing the roles and relationships of the various players in regard to their functions in the life cycle.

A second PM level skill view is that he needs to be recognized as a leader of the effort. This does not mean that he is out front shouting "follow me," but he has to ensure that the team continues to move toward the required target. During early project phases, the target is not well defined, so the leadership role at that point is to bring the proper players together and help resolve various conflicts that typically emerge.

The third critical skill involves dealing with the various human resources related to the project. The most noticeable group will be the project team who ultimately will be the "builders." They collectively have the skills to execute the plan, but there are many human relationship issues that can get in the way of that effort. Project team members must be managed and nurtured through the life cycle. To properly do this, the PM needs to be an operational psychologist who understands individual and group needs. Project teams are a cauldron of human emotions. Kept at the right temperature they can produce amazing results. When allowed to boil the conflict can destroy the process. Finally, during this process, an additional role of the PM is to improve the skills of the team members and ensure that they are properly relocated at the end of the project.

In addition to the internal project team, there will be other human interactions with external groups such as users, management, and organization entities. Each of these has a different perspective regarding the project and all of their views must be dealt with. In all of these cases, the PM is never given enough formal power to edict solutions even if he knew what the solution was. The human relationships require a more open communication, motivational style with the approach being to build partnerships. Each of the human interface groups holds a piece of the project success and the PM must extract that piece from each.

1.3.1 Success Management

At this stage in the evolution of project management knowledge, there is a reasonable understanding regarding the major factors leading to success and failure (two sides of the same coin). Understanding these factors is the first step in being able to avoid the failure factors. The basic management model outlined in the text offers a reasonably clear set of processes to achieve that goal. However, the organizational environment in which a project exists may contain factors that make success unlikely. In some cases a PM is assigned Project Titanic (i.e., the ship is likely to sink). When this happens, it is important to realize that recent evidence indicates that the decisions made by the crew of the Titanic actually caused it to sink, not the initial iceberg hit. This view suggests that if the ship had been left where it was rather than moving it off the iceberg, it would likely have stayed afloat for at least a little time. So, a catastrophe could have been mitigated with the right management decisions. Here we see that bad management decisions can make a tough situation disastrous. The same scenario is valid for the project environment. A good PM might well be able to salvage the worst of projects.

So, success management requires a series of strategies. First, understand where failure comes from and mitigate as many of those factors as possible. Second, through the course of the project, influence the right set of actions to correct any deviations that threaten to become a problem. Third, when a threat surfaces, take quick action using all of the management skills at hand. Finally, if the boat is in fact sinking, you will have told all hands how to handle the situation. Management will have been informed along the way and similar warnings would have been given to others. In all of these modes the PM must be an *honest broker of information*.

If we follow all of these prescriptions, will every project be successful? Probably not. There are too many uncontrollable variables to expect that, but proper use of the tools and techniques described here should significantly improve the outcome. If we continue to look at what went wrong with the last project and try to ensure that the previous item does not recur, the next project should progress better. Experience from the Japanese quality programs has taught us how continuous improvement actions over a long period can take a country from a crude tool maker to Toyota Lexus in 50 years. We must realize that project management is not an event; it is a process. Organizations must focus on it and individuals must study it.

1.4 TEXT CONTENT AND ORGANIZATION

This text looks at the project experience from the view of a project manager. Material covered in the text has been selected from a personal database of "things I wished that I had known more about" at one point or another along the way. Also, in recent years the PMI has documented a great deal of professional project experience into the published archives on this topic and the PMI documentation is respected internationally. Over the past six years the author has been heavily involved in teaching this topic after many years in industry attempting to master it. Those two experiences lead to the amalgamation found here. The content is a mixture of the PMI model view and comparable views of practitioners. Attempts to translate this material to university and industry groups indicated that a proper source document with a reasonable dose of theory and vocabulary would help someone desiring to understand the breadth of this topic. This was the initial goal that started this effort.

The text material makes a reasonable attempt to stay consistent with the *Project Management Body of Knowledge* defining document from PMI, as well as other supporting project related standards such as OPM3, WBS, and Professional Ethics (each of these will be explained later). The collection of material contained here is a compilation of project management models, concepts, vocabulary, and trends. Through all of these elements, the goal was to make each item fit into the big picture and more importantly keep the discussion on an understandable level. If the reader wades through this material to the end, we will even share the secret PM handshake (this is probably the only joke in the text so it needs to be tagged).

Another stimulus for this effort has been the emergence of a formal educational curriculum accreditation process for project managers (PMI Accreditation, 2001). Prior to this, individuals seeking project management certification studied various reference sources and then pursued the formal certification exam hoping that they had been exposed to the right material. In an attempt to ensure that the material covered fit the PMI accreditation structure that document was used to cross reference section material content. The header for each major section will reference the accreditation goals where applicable and these will be defined as learning objectives for that section. This is intended to give the text legitimacy in regard to its topic content.

1.4.1 TEXT STRUCTURE

The text material is partitioned into nine major sections that are essentially envisioned as "peeling the onion" away from the real-world core. Each major section represents a layer of that onion and each opens up a new more complex layer related to the topic. Each of the sections builds on the previous layer and there is no assumption made as to background of the reader other than the fact they have some understanding of an organization and hopefully have seen a project in action. This material has been tested on university students during the development stage, and graduates from the program have achieved a near 100% pass rate for the Project Management Professional (PMP) Certification exam. Based on this experience, there is general confidence that the approach helps in the conceptual understanding of project management and the PMI model.

The summary below outlines the goal of each major section.

Part I. Conceptual overview: This section consists of eight chapters that collectively lay the foundation for the rest of the text. Basic vocabulary and concepts are covered here.

Part II. Projects as state change vehicles: This section is designed to sensitize the reader to the role of a project in the organization and the general mechanics required to properly initiate that effort. Basically, the project should focus on an organization goal that helps move the organization toward that goal.

Part III. Defining the triple constraints: Most projects focus on time, cost, and scope. This represents only a starting point for the project manager, but there is sufficient theoretical material in this area to justify its isolation. This topic area is isolated from other more complex concepts to help focus on these mechanics. Managing the project's triple constraints represents a core activity for the project manager.

Part IV. Planning support processes: In addition to the core management activities, the PM must also understand the role of other knowledge areas—those related to HR, Communications, Procurement, Quality, Risk, and Integration. Each of these topics represents critical management issues for the PM and they collectively have to be dealt with along with the core items in order to produce a viable project plan. Upon completion of this section, the reader has been introduced to material related to formulation and planning of the project.

Part V. Project execution: At this point, the text material has defined processes to produce a viable project plan. The management challenge in execution is to produce the planned output as defined in the earlier sections. Unfortunately, the process becomes muddier at this stage. There is more human conflict emerging as well as more change dynamics. A dose of reality enters the scene and the real management complexity is now uncovered. If these dynamics are absent, the management role in execution would be "task checker." The reality metaphor for this stage is to compare it to an airplane pilot in rough weather with various mechanical and environmental problems to deal with. Most of the material used here is still model driven, but an attempt is made to give the model a reality flavor.

Part VI. Monitoring and control techniques: There are many control-oriented aspects at this stage of the project life cycle. Rather than attempting to bundle them all together the key techniques have been separated for discussion. Each of these represents a control knowledge component that the PM needs to understand.

Part VII. Closing the project: The proper shutdown of a project is an important management activity. Rationale for this and the associated mechanics are the theme of this section.

Part VIII. Contemporary topics: It is important to emphasize that the art and science of project management are not yet matured. There are several emerging topic areas that seem destined to enter the project scene and these are bundled for discussion in this section. Since there is no recognized standard for these items, there is a certain amount of literary license taken in this section. The reader must understand that view and the fact that each of the topics will morph over time. Nevertheless, to leave this subject without recognizing the potential impact of each topic violates the design goal stated earlier.

Part IX. Professional ethics and responsibility: One only needs to read the daily news to see why this topic is worthy of inclusion. PMI has issued a code of conduct for the PM and the tenets of this code must be understood, as well as some motivational examples to show that it is a real topic.

APPENDICES

The following three additional topics exist without a clear place to slot them in the subject continuity list:

A. Financial analysis mechanics
B. Project templates
C. Document repository.

In each case, the material here is relevant to what a working PM should understand and use as part of his tool kit. Reading this text will not accomplish all that one needs to know about the subject of project management, but it does a reasonable job in a single source in covering the major topics that one should understand.

REFERENCES

PMI Accreditation, 2001. *Accreditation of Degree Programs in Project Management*. Newtown Square, PA: Project Management Institute.
PMI, 2008. *A Guide to the Project Management Body of Knowledge (PMBOK® Guide)*, Fourth Edition. Newtown Square, PA: Project Management Institute.

2 Evolution of Project Management

Project management is an increasingly important topic of discussion today because all organizations at one time or another, be they small or large, are continually involved in implementing a new business process, product, service, or other initiative. When we examine how organizations pursue changes, invariably it involves organizing a team of people with chosen skills to do the job. Management of the activities to complete this class of task is what project management is about.

We are indeed living in interesting times as regards to the topic of project management. On the one hand, it is now generally recognized that a disciplined approach to managing projects yields positive value in the resulting cost, schedule, and functionality. However, there remains great conflict over exactly what discipline is to be used in this process. In addition to this philosophical discord, technology itself continues to bring new challenges to the organization such that it is often difficult to repeat one successful approach multiple times. Manipulating the project variables the same way can produce different outcomes based on the subtle relationships inherent in the process. Also, new tools, techniques, and products continue to enter the marketplace making even 5-year-old project management strategies look very dated. So, the challenge in navigating this mine-strewn environment is to explore the subject and distill nuggets of information that have stood the test of time and then attempt to pave a pathway that can survive the next wave of technical discontinuity. In order to understand how the current situation got to its present state let us take a quick look at some of the not too distant evolutionary stages that the approach to project management has moved through. History often offers subtle insights into broad-scale phenomenon such as this. The stages outlined below are somewhat arbitrarily grouped, but are designed to highlight the more obvious driving factors that have changed the approach to managing high technology projects.

By scanning any library or book store today, you will find shelves stocked with volumes of books explaining in varying detail methods useful for successful completion of projects. Each author has their own guaranteed project management strategy designed to ensure a triumphant conclusion; yet real-world statistics still show marginal results for most projects. This section does not intend to attempt to trace all of the trodden paths related to this topic, but does attempt to look back at the people and concepts in history that have formed the foundations of project management on which modern day approaches are based.

2.1 EARLY HISTORY OF PROJECT MANAGEMENT

The basic principles related to the science of project management have evolved over many decades. The evolution of this body of knowledge mostly evolved since the early 1900s and accelerated after the 1950s. Some very early projects were quite impressive in their scale, but they did not follow what we would call the modern project management style of design. Incubation of the modern thought process can be traced to the industrial age during the latter 1880s, which provided much of the catalyst for the application of a more scientific approach to the management of project and manufacturing processes. Studies and experiments conducted by pioneers in the field during the early part of the twentieth century further paved the way for the understanding of project management as it is known today.

One has to look only at the historical structures and monuments left behind in past centuries to conclude that some form of managing a project was in place at that time. It is unfathomable to imagine that the Great Pyramid of Giza, the Great Wall of China, or any of the ancient Greek or Roman projects could have been completed without some type of project management that basically knew how to manage the variables involved. Each of these undertakings was constructed with nothing more than simple tools and manpower, often slave labor. The early PMs were technicians or engineers, generally multiskilled generalists who could deal with many situations (Kozak-Holland, 2007). The manager skill base was most likely the architect/designer of the project who understood how it needed to be constructed, and they were given the authority for allocating sufficient resources to that goal. This style of the multiskilled technical generalist overseeing projects was the norm through the nineteenth century.

2.2   APPLICATION OF ANALYTICAL SCIENCE

As organizational processes became more complex, many underlying aspects of getting work accomplished began to change. Most noticeably, the manufacturing process moved out of the craftsman's homes into formal factory settings where the products could be mass produced. This necessitated a tighter coupling of work processes and more refined versions of them. Toward the end of the nineteenth century, new technologies using electricity and internal combustion brought a further expansion of manufacturing complexity. Suddenly, employee (nonowner) managers found themselves faced with the daunting task of organizing the manual labor of thousands of workers and the manufacturing and assembly of unprecedented quantities of raw material (Sisk, n.d.). This phase basically marked the beginning point for the application of analytical science to the workplace. If one could point to a birth date for modern project management, it would likely be in the two decades leading up to the twentieth century.

2.3   FREDERICK TAYLOR AND SCIENTIFIC MANAGEMENT

Frederick Taylor is called the father of scientific management and his influence can be traced through much of the early evolution of project management thought. Taylor came from what was considered a privileged background, but entered into employment with the Midvale Steel Company of Philadelphia as a common laborer in the late nineteenth century. The prevailing wage system in place at the time was called piece work. That is, employees were compensated based on their production; more production more pay. One common practice for management was to monitor the payroll and as soon as workers began earning "too much," they would cut the piece rate to try to entice the workers to do more for less. In reaction to this, employees scaled back their output to keep the quota lower. This practice was called "soldiering" (Gabor, 1999, p. 13). Years later, this concept would be called peer pressure and added to the behavior theory of management. Taylor saw this practice and even participated. Some time later, he was promoted to gang boss at the mill and became determined to stop the soldiering. Being an engineer, his method of doing this was to find a way to define "scientifically" what a fair pay-for-performance formula would be. In order to do that, he had to do research on the best method for the job.

Thus began Taylor's application of systematic studies for various jobs and the time required to complete each task. He conducted time studies of various jobs using a stopwatch. This became a common activity in manufacturing organizations under the title Time and Motion study (Gabor, 1999, p. 17). By standardizing the work processes and understanding the needed times to complete tasks within those processes, Taylor was able to increase the output at the steel company.

In 1899, Taylor was recruited to Bethlehem Steel Works, where he conducted what is his most famous experiment, based on the loading of pig iron (NetMBA, n.d.). The impetus for the experiment was a rise in price for pig iron due an increased demand for the product. Using his knowledge of work process and time studies, Taylor set about to increase the productivity of pig iron loading. Loading the pig iron was a backbreaking labor, but over the course of time the workers with the proper skills were put in place. The average daily load of pig iron per worker was 13 tons. By conducting time and motion experiments to determine the proper timing of lifting and resting the workers could increase the production to 47.5 tons per day (NetMBA, n.d.). What is not so readily defined in history is that the workers did not adopt Taylor's method even though he showed that it was more productive. It took several more years before the concept of group behavior was better understood. As is the case with most improvements in management thought, each small step forward leaves behind other unanswered questions. In this case, why would the workers not want to produce more if they did not have to work harder?

Taylor became famous after testifying before the U.S. Congress on ways in which the U.S. railway system could be made more productive. This testimony was published in the New York Times describing his theory of saving $1 million per day through his principles. One could argue that this was the first of the management "silver bullet" ideas that represented all you needed to know to solve basically any problem. Many of the historians we examine were not afraid to tout their solutions in this way. Taylor left his mark on the industry with his 1911 publication of *The Principles of Scientific Management*. This described four of Taylor's management principles as follows (Ivancevich et al., 2008, p. 143):

1. Develop a science for each element of a man's work that replaces the old rule-of-thumb method.
2. Scientifically select and then train, teach, and develop the workman, although in the past he chose his own work and trained himself as best as he could.
3. Heartily cooperate with the men so as to ensure that all of the work is done in accordance with the principles of the science that has been developed.
4. There is almost an equal division of the work and the responsibility between management and workmen. Management takes over all work for which it is better fitted than workmen, while in the past almost all of the work and the greater part of the responsibility were placed on the workmen.

It is on these foundation concepts that others have expanded the "scientific" view of management and projects that remains in place today.

2.4 FRANK AND LILLIAN GILBRETH

The fun trivia fact about these two individuals is that they were the subject of a classic movie "Cheaper by the Dozen." Clifton Webb and Myrna Loy were parents with 12 children and this was in reality the Gilbreths. To suggest that they were experts in time and motion study would be obviously true. Frank Gilbreth and Frederick Taylor first met in 1907, which resulted in Frank becoming one of Taylor's most devoted advocates. As a result of this influence Frank and Lillian developed what the world came to know as time and motion studies (IW/SI News, 1968, p. 37). They collected numerous timing data on small human motions and cataloged the timings so that a trained analyst could construct "synthetic" time standards without having to measure an actual worker. These small time units were called "Therbligs," which is essentially Gilbreth spelled backward.

Through this pioneering research the Gilbreths contributed greatly to the knowledge of work measurement. Lillian's work illustrated concern for the worker and attempted to show how scientific management would benefit the individual worker as well as the organization (IW/SI News, 1968, p. 37). Frank utilized technology including clocks, lights, and cameras to study work processes. The effort and intensity with which the Gilbreths pursued their chosen field of scientific study is most notably showcased in their time and motion study of bricklayers. In this study the Gilbreths observed the processes needed for a group of bricklayers to complete the installation of a wall. As stated in their biography printed in the September 1968 issue of the *International Work Simplification Institute* newsletter, the Gilbreths were able to reduce the number of basic motions needed for laying a brick from 18 to 4.5 (Gilbreth Network, n.d.). This scientific method was used to show how improved processes would make workers more productive and efficient. During this period there was an expanding recognition that the workers themselves had something to do with productivity, but for now, close supervision was the solution.

2.5   HENRY GANTT

No discussion on the beginnings of project management would be complete without mentioning the contribution to the field by Henry Gantt. Gantt himself was an associate of Frederick Taylor, who first documented the idea that work could be envisioned as a series of smaller tasks. Gantt was influenced in his view through involvement in Navy ship construction during World War I. Also influenced by the research conducted by Taylor, he applied that knowledge to the construction of the ships. A concise explanation of Henry's contribution comes from the Gantt Group's document "*Who was Henry Gantt?*" (Gantt Group, 2003). It states, "he broke down all the tasks in the construction process and diagrammed them using the now familiar grid, bars and milestones." This familiar time grid is now called the Gantt chart. It remains today the most used planning and control document in industry after 100 years (Figure 2.1).

Note that the chart above defines tasks and times through the use of horizontal bars. The completed chart provides an overall view of the timeline and tasks needed to complete the project. The appearance and use of the Gantt chart has many variants, but the basic idea has changed little since its conception. We will see this chart later in the text in modern context.

2.6   MARY PARKER FOLLETT

With the increased study of work processes and methodologies to streamline productivity and increase output, industries began looking more at how to do the work than who was doing the work. Mary Parker Follett stepped out from behind scientific management theory and instead focused on the human element. She opposed Taylor's lack of specific attention to human needs and relationships in the work place (Ivancevich et al., 2008, p. 13). From this action, Follett takes the honor for spawning the behavior side of management and was one of the first management theorists to take this view.

Follett focused on the divisions between management and workers: more specifically, the role of management instructing workers on what was to be done and how it was to be done. Follett believed that each worker had something to contribute and the amount of knowledge held by workers was not being tapped. She believed that it would benefit the workplace and all of society if instead of working as individuals or separate groups that these groups or individuals worked as a whole. Here we see the beginnings of the team concept, although without an operational theory to support it.

Treating workers as something other than a means to get the task done was a concept that was counter to the Taylor school of thought outlined earlier. Gabor in her book *The Capitalist*

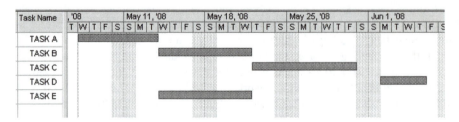

FIGURE 2.1   Sample Gantt chart.

Philosophers states that Follett's ideas came to be embraced by the most forward-looking management thinkers of her time, many of them also admirers of scientific management. Ironically, Follett's views of focusing on the worker would be accidentally validated in the future from a classic scientific management experiment.

2.7   ELTON MAYO

The evolution of scientific management principles continued into the mid-1920s following the concepts laid down by Taylor and his disciples. This area of study had attracted its share of detractors, such as Mary Parker Follett, but the visible quantification related to the scientific approach also attracted many to that school of thought. The *Taylorites* saw the factory as a complex set of processes that needed to be optimized and taught to the worker. Eldon Mayo and his research team followed that general principle in believing that one of the keys to improving productivity lay in improving the physiological environment of the worker. Looking back, we see elements from both the Taylor and Follett schools of thought in this view. At any rate, this premise led to what is known as the famous Hawthorne experiments (Gabor, 1999, p. 85).

The Hawthorne experiments were conducted by Mayo and his team from 1927 to 1932 in Cicero, Illinois at the Western Electric Hawthorne Works. These experiments were designed to examine physical and environmental influences (e.g., brightness of lights, humidity, etc.) on worker productivity. Later versions of this effort moved into the more psychological aspects to include work breaks, group pressure, working hours, and managerial leadership (Envison, n.d.). The initial studies focused on the effect that changing light intensity might have on productivity. The results of this experiment were very confusing to the cause-and-effect-oriented researchers. They observed that an increase in light intensity corresponded to an increase in worker output; however, as the lighting decreased, productivity continued to show an increase. The puzzled researchers wondered what outside variables had not been considered and set about laying out a second cause-and-effect experiment in the relay assembly process.

The relay assembly control test room was set up to measure the productivity of workers under a myriad of changing conditions. Despite varying worker environmental conditions regarding work break durations and length of the work day, output continued to rise regardless of the change. This simply did not fit the Scientific Management principles of cause and effect. Eventually, analysis of this set of experiments would open the door wide in understanding some initial concepts related to worker motivation. In these experiments, essentially none of the test variables were responsible for the worker behavior. It took more analysis before a cause-and-effect relationship was determined and this changed the field of modern management.

In the aftermath of the Hawthorne experiments, interviews were held with the test subjects. The results showed that the participants had formed their own social network that did not exist on the factory floor. Also, the test subjects felt as though they belonged to something special by being a part of the experiments. They were special and someone was paying attention to them. As a result of this new feeling, they wanted to produce like special workers should. The fact is that the group was purposefully randomly selected and was no more special than the hundreds of other workers outside of the control room. The conclusion now known as the *Hawthorne Effect* is described in the article "The Hawthorne Effect—Mayo Studies Motivation" (Envison, n.d). The results of these studies formed the basis for the foundation of what is the modern day behavior school of management.

2.8   PHASES OF PROJECT MANAGEMENT EVOLUTION

Intermingled with the basic management thought evolution was a corresponding evolution of project management thought. We somewhat arbitrarily start this history in the mid-1940s when large, time-critical projects exploded on the scene as a result of the war.

Stage I: The first major epoch of modern project management came during World War II in the mid-1940s. The atomic bomb Manhattan project and other complex military programs that followed added much insight into methods for completing this class of endeavor. In the period following World War II, these methods were translated into formalized and documented approaches. Military projects continued to push the technology envelope into the 1950s and pressure to improve technical project management continued. Many credit the activities surrounding the successful design and implementation of the highly complex contractor-developed Polaris nuclear submarine project in the 1950s as the beginning of broad nonmilitary acceptance of a model-driven approach to project management. This was reflected in the popularization of well-known network management methods such as Program Evaluation and Review Technique (PERT) and critical path method (CPM), which have proliferated into all industries since this time. These early planning and control models were initially able to be used only in large organizations because high-priced computing resources were required to manipulate the models. Further proliferation of these models had to wait until low-cost processing technology emerged in the 1970s.

Stage II: The 1970s and 1980s brought tremendous expansion in hardware and software technology offerings. Proliferation of minicomputers broke the cost barrier for operational modeling and this opened the door for improving planning and tracking of project status. General knowledge of the CPM-type network model existed, but there was still minimal understanding of the underlying management processes required to effectively utilize the model. During this period, vendors sold "canned" methodologies claiming that they would solve the project problem. They seldom did. By the 1980s, the United States was in an economic boom and the key requirement for organizations was more toward speed of delivery than efficiency. Slowing down that effort to sort through internal process improvements was low on the priority list.

A second constraining factor during this period was the organizational rethinking of the central IT department that up until now had held the keys to computing power. The period around 1970 ushered in smaller computing devices (minis and personal devices) that opened up user-based computation needed to make some of the project management tools viable. Over the next 15 years there was a deluge of software produced for this environment and then the masses started becoming computer literate. However, for one living through this era it seemed that little conceptual project management theory progress was made as organizations were moving from the highly controlled central mainframe computers controlled by a single department to a more distributed hardware environment with a "do whatever you want by yourself" mentality. The software maturity was outrunning the infrastructure necessary to support it with usable data. During this period, academic organizations and consultants published concepts, theories, and management strategies that would have moved the project discipline further along, but the general project audience was not yet convinced of the value of project management. Many looked at management software as requiring too much overhead and some feel to this day that the models developed are not appropriate for the task. The organizational model during this period was one based on speed of product delivery and purchasing software from third parties. The latter strategy was thought to take away many of the needs for project management since "the code was already written." Subsequent massive system development failures with attempts to install these "prewritten" systems uncovered another view. That is, there is more to successfully executing a project than loading code to a computer or buying some vendor's management tool. So, by the end of the 1980s, there was a new level of project management understanding. More "silver bullets" emerged, but none solved the project problem.

Stage III: As history transgresses into near present time, one often feels that the problem is understood and the answer is close. In this spirit, the 1990s are viewed as a period of maturation and proliferation of information tools, techniques, and user literacy. Small and powerful desktop devices solved the modeling capability issue and the Internet solved many of the information distribution constraints. However, neither did much to improve the organizational discipline regarding the management of projects. Desktop tools such as Microsoft's Word, Excel, and Access turned millions of computer aficionados into what looked like programmers and the project world become flooded with small pockets of disorganized data. These new capabilities improved the look of documentation, but the underlying processes were not appreciably improved.

Another evolutionary thread emerged during this period in the form of improved software packages. During this wave, major mission critical systems were being replaced by suites of integrated commercial software packages as a strategy to cut computer expenditures. Names such as SAP, Peoplesoft, Oracle, J.D. Edwards, Lawson, and others became familiar terms. In many cases this "silver bullet" solution failed to materialize and the results sometimes bordered on catastrophic for the organization. Once again the primary reason for many of these failures was not the lack of purchased product quality, but the underlying process for selection and implementation—that is, a gap in a process that could have been viewed as a project requiring project management principles. Also, because these projects represented such a large resource commitment, there was an attempt to manage them in a professional way and yet they still failed to meet expectations of budgets, schedules, and functionality. Something was clearly not working right! The one item of good news coming from these failed initiatives was that the projects involved large segments of the organization and this uncovered one of the missing issues in project—that is, communications.

As a result of these experiences, the use of formal reporting processes and metrics related to project execution began to be recognized as a requirement and senior management became more interested in this aspect. Prior to this, the prevailing lack of computer literacy by management and the lack of appropriate project status metrics allowed projects to run under the radar of management scrutiny until the project was completely out of control. There was now a growing awareness that some type of prerequisite management process must be in place prior to embracing a complex highly technical undertaking, whether that be hardware, process, or software. Organizations that failed to understand this continued to believe that project management was simply an overhead and adding cost for no benefit. By the mid-1990s, there was growing management recognition that the project environment within various organizational segments could not be allowed to go uncontrolled any longer. Also, project costs were now being captured and were recognized as a major component of the capital budget in many organizations. This stimulated the requirement for more rigorous project justification.

In parallel with the improved understanding of project success variables came an increased awareness of project activity in general. As project disaster data began to be published, it highlighted that poor project performance was a general phenomenon and not one limited to local initiatives. As an example, the popular press chronicled a $1 billion overrun for the new Denver airport basically caused by a project management error. Also, various governmental projects and the Big Dig tunnel in Boston had similar results. Once recognized, these reports became common fare for news reporting. The public was now in tune with the issue and these daily status notes opened the door to increased sensitivity regarding the need to deal with the problem.

During the 1960s through the 1990s, organizations attempted various management strategies to improve project results. Looking back on these efforts, they were lab experiments quite similar to the Hawrthorne research—looking in the wrong place for the answer. Many root causes for failure were identified, but true fixes did not emerge on a broad scale. Organizations such as the PMI became involved in searching for solution techniques and they pioneered the concept of a general project management model. Also, the Y2K (Year 2000) phenomenon became widely discussed around 1998 and prognosticators predicted doom if all computer software was not repaired by the end of the decade. For the first time, global technical projects were perceived to require completion of their mission on time and within scope. For all of these reasons, the latter 1990s ushered in a worldwide recognition of project management. The technical and maturation events described in Stage III provided a broad view of what project management is and what it can contribute. We would be naïve to suggest that the problem is now ready for solution, but it is widely documented and discussed by industry and academia. So, we enter Stage IV.

Stage IV: The key philosophical question at this point is to forecast where the current trends will take the topic and in what time period. One view is that the current trends will continue to broaden across all organizations in essentially unaltered state. That is, new products, hardware, software, and telecommunications technology all driving the same processes but using the new tools. Also, it seems reasonable to predict that the global population in general will become more literate with these tools and technologies. At the same time, the project environment will also continue down the trail of increasing complexity and its methodologies will likely become more embedded in the fabric of everyday organizational work processes.

Certainly, one exacerbating theme through this period is the dynamic nature of international organizations involvement in domestic projects. Just as the project management model began to synthesize a workable set of tools and concepts for the organization, the operational model was disrupted. Specifically, the globalization of organizational processes is now a reality and is being stimulated by the existence of cheap foreign labor sources. This in turn has led to increased outsourcing to third parties. These initial outsourcing initiatives involved relatively simple organizational processes and were marginally successful during the 1990s, but the trend continues to grow with some new approaches and a better understanding of how to manage such ventures. One of the major positive contributors to this is the increase in functionality and lower cost of international Internet telecommunications, which has in turn opened up new vendor opportunities for distributing knowledge work. This has allowed the emergence of smaller niche vendors who do one function very well. In some cases, these specialized vendors are often able to take over an entire business process at less cost than it could be operated internally. Each of these niche solutions changes the internal process of the organization and potentially changes the project management protocols related to those vendors. The number and scope of such activities is forecast to continue to increase.

Recent experience indicates that these niche vendors can be successful in the marketplace as a result of their economies of scale and specialized skill level, but once again we also see a trend that requires more complex project management techniques as the business process becomes fragmented across multiple organizational groups. As a result of these outsourcing trends, critical operational activities are being performed external to the consuming organization. One potential risk issue raised by this trend is the impact of an external vendor failure. This was a much more controllable situation in the traditional structure, but now can have a significant impact on the organization and the project. Because of the complexity and loss of internal skills, a reverse migration of these processes becomes quite difficult. For this reason, risk management in such an environment takes on increased importance. A further risk extension of the contemporary trend for outsourced vendors is that they often reside in another country—an uncommon practice prior to the late 1990s. Today, a full complement of service providers exists in locations such as India, Pacific Rim, China, South America, Russia, and others. These new technical entities continue to increasingly extract work from local U.S. organizations and basically change the project management landscape.

With the industry trends partially outlined above, there is an increasing need for more formalized and effective project management across the collaborative partner domain. Any new management processes hosted in this way must be compatible with the evolving business requirements of a global work force. As a result, planning, control, communication, and team collaboration are more critical processes in the contemporary environment. Experience has shown that the road to project success involves more than manipulating technology and tools. Success is clearly driven by proper management of the human element and the subsequent implementation of the output created by the project. In some ways, the project management tools are morphing into more of an operational management concept.

To support these contemporary organizational needs, a strong project management orientation is needed and it will have to be sensitive to producing value for the organization and not just installing new products and processes. Organizations have now become sensitive to the issue of selecting the right project to start with. That often adds another layer of management to the traditional project view. These organizations are demanding quicker cycle times to customer demands along with more complex visions. In order to gain respect, the project management function has to be seen by the enterprise management as delivering value. One threat in such an environment is for management to say "if project management can't deliver this, I'll find another way."

History has shown that time pressure can cause planning to be ignored (i.e., planning is overhead). Project management theory is built on the basic concepts of planning and control; so it is going to be up to the profession to show value in a formal planning-oriented model. The dilemma with this is how to go fast and not make costly mistakes versus going slower and accomplishing the technical goals. Management wants to know how much formalization will cost and how much value it brings. That is difficult if you do not measure the process and learn from that along the way. Resolving this conundrum remains one of the toughest challenges of this period. To support this goal, there is a growing interest in new approaches to development now going under titles such as "agile," "extreme," spiral, critical chain, and other names. These new schools of thought focus more on speed of delivery and customer satisfaction. To date, many traditional managers have resisted most of these methods because they tend to move forward before a firm vision of the deliverable along with schedule and cost is defined. The underlying belief is that lack of planning increases the future level of scope change and in turn makes the project cost more and may in fact then take longer to complete. There is insufficient proof for either side of this debate, but both appear to have positive and negative value positions. The result of this may be to conclude that each option focuses on different goal sets (e.g., schedule, customer satisfaction, cost, etc.). One of the more obvious potential benefits of the lower initial planning approach is customer satisfaction; however, the negative issues of management visibility and control are left to be dealt with. The key to future success in this arena is to find a proper blend of predefinition versus the increased customer satisfaction from a better match to their requirements. In the current traditional environment, most management groups will not approve a project without some reasonable view regarding the future project's functionality, resource commitment, cost, and schedule. This suggests at least some degree of planning to satisfy these requirements.

Collectively, all of the contemporary trends in the project environment will bring new challenges to management philosophy. Certainly, the combination of business pressure for increased speed of delivery, increased use of purchased services and commercial off the shelf (COTS) packages, outsourced service providers, and use of offshore vendors impacts many aspects of the traditional project management model. This new environment will require an improved set of tools and strategies to navigate these initiatives successfully. As a result of these dynamics, the subject of project management will remain under great conceptual stress, but will also be more recognized as a key requirement to success. Obviously, the skill requirements for this class of project will be greater more than those found in most organizations today. As a result of these trends, one should not plan on a status quo approach to this subject over time. The new generation of PMs must evolve as the organizational environment evolves to employ the new strategies.

2.9 PROJECT MANAGEMENT CHALLENGES

The upcoming challenges for PMs in the field involve how to deal with activities defined in a development life cycle and manage the associated technical resources. The following statistics provided by Successful Strategies International offer some insight into the current status of these goals (SSI, n.d.).

- Twenty-four percent of organizations now cite inconsistent approaches.
- Forty-six percent of the organizations cite resource allocation issues such as estimating and benchmarking.

On the positive front, development of project management capabilities integrated with the application of collaborative technologies has the potential ability to support a new level of globally efficient project management practices. As a result of this new capability, operational proficiency across functional management teams, organizations, and their business partners can concurrently view and interact with the same updated project information in real time, including project schedules, technical discussions, and other relevant communication activities. This capability becomes a prerequisite to move forward effectively with the underlying management processes.

2.10 PROJECT MANAGEMENT BENEFITS

2.10.1 At the Macrolevel

Higher organizational levels often have great interest and motivation to implement formal project initiatives. The typical goal of this segment of the enterprise is to ensure that the project undertakings (small or major) are delivered on time, within the approved cost budget and with the desired output. Certainly, one of the drivers for more rigorous techniques for projects is to satisfy the control objectives of management.

2.10.2 At the Microlevel

At lower levels in the organization, there is frequent resistance to the discipline inherent in a formal project management approach because it is viewed as inhibiting personal flexibility. When pressured to conform, the local PM would likely describe his role as follows.

1. Customizing the project work to fit the operational style of the project teams and respective team members.
2. Proactively informing executive management of project on a real-time basis.
3. Ensuring that project team members share accurate, meaningful, and timely project documents.
4. Ensuring that critical task deadlines are met.

The reality in accomplishing these goals is to understand that creating a mature project management environment is very difficult and even if successful does not assure project success. On the reverse side, doing nothing in this direction invites chaos.

So, what is the conclusion in regard to the value of formalized project management? Some consultants would state that good project management techniques can add up to a 25% savings for the project. This means to deliver a stated set of requirements for a lesser price. What this means exactly is open to interpretation and accuracy. Some would conclude that every such attempt they have been involved with has been a disaster. Many top managers would more likely say that they have to do more because lack of the discipline is wasting money and they are in the dark as to status. It is probably best to leave this question on the table and let the material in the text sell itself, or fail. As you go through the following chapters try to ask yourself if the technique seems reasonable compared to whatever other approaches you may be familiar with.

REFERENCES

Envison, n.d. The Hawthorne Effect—Mayo Studies Motivation. http://www.envisionsoftware.com/articles/Hawthorne_Effect.html (accessed August 14, 2008).

Gabor, A., 1999. *The Capitalist Philosophers.* New York: Times Business.

Gantt Group, 2003. Who was Henry Gantt? http://208.139.207.108/html/RGC-KO-04.htm (accessed March 28, 2008).

Gilbreth Network, n.d. Frank and Lillian Gilbreth biographies. http://gilbrethnetwork.tripod.com/bio.html (accessed August 14, 2008).

International Work Simplification Institute, Inc. (IWSI), 1968. Pioneers in improvement and our modern standard of living. *International Work Simplification Institute News, 18.*

Ivancevich, J.M., R. Konopaske, and M. Matteson, 2008. *Organizational Behavior and Management.* New York: McGraw-Hill.

Kozak-Holland, M., 2007. History of Project Management. http://www.lessons-from-History.com/Level%202/ History_of_PM_page.html (accessed March 27, 2008).

NetMBA, n.d. Fredrick Taylor and Scientific Management. http://www.netmba.com/mgmt/scientific (accessed March 27, 2008).

Sisk, T., n.d. The History of Project Management. http://www2.sims.berkeley.edu/courses/is208/s02/History-of-PM.htm (accessed March 28, 2008).

Successful Strategies International (SSI), n.d. How to Attain Project Success. http://www.ssi-learn.com/ Downloads/Presentation%20-%20Attaining%20Project%20Success.pdf (accessed August 14, 2008).

3 Project Management Body of Knowledge

As indicated in the previous chapter, the profession of project management has matured greatly over the past few years and a large part of that has been as a result of the PMI and their international organization. Through its various activities PMI has made the topic of project management much more visible to an international audience and as a result is recognized as defining the de facto description of the subject. The current version of this key definitional artifact sponsored by PMI is a document titled *A Guide to the Project Management Body of Knowledge (PMBOK® Guide)*, Fourth edition (PMI, 2008). In order to keep the material current, new editions are planned for every 4 years. This reference document models and defines the project management processes and activities that should be utilized in executing a project and it is the conceptual structure that is used in this chapter. This material follows the model structure in sufficient detail to illustrate how it works in a real-world project environment.

3.1 HIGH LEVEL OVERVIEW

The advent of a formalized view of project management has created an organized, intelligent, and analytical approach toward not only tackling large projects but the associated organizational issues as well. In pursuit of this goal, the need for an international "standards" type view was recognized and from this recognition, PMI was borne. Along with this came the creation of their most internationally recognized model guideline of which well over 2 million copies have now been published in multiple editions versions. This model is titled the *Project Management Body of Knowledge*, officially known as the *PMBOK® Guide*. Since inception, the guide has grown to become a standard that is recognized worldwide in terms of the knowledge, skills, tools, and techniques that collectively relate to the management and oversight of projects within an organization. The *PMBOK® Guide* (pronounced pimbok) is defined by PMI as:

> A Guide to the Project Management Body of Knowledge is a recognized standard for the project management profession. This standard is the formal document that describes established norms, methods, processes, and practices. As with other professions such as law, medicine, and accounting, the knowledge contained in this standard evolved from recognized good practices of project management practitioners who contributed to the development of this standard. (PMBOK®, 2008, p. 3)

This conceptual mode defines the project management processes and activities that should be utilized in executing a project and the conceptual model is used in this chapter. We will follow the model structure throughout the text in sufficient detail to illustrate how it works in a real world project environment. Those who wish to see the full PMI model processes and activity detailed descriptions can purchase the *PMBOK® Guide* through www.pmi.org, or other commercial text sources.

3.2 HISTORY OF *PMBOK® GUIDE* DEVELOPMENT

The historical foundation for the *PMBOK® Guide* model project life cycle can be traced back to Walter Shewhart's work on manufacturing system control during the 1930s. Edwards Deming updated this concept with his pioneering work in Japan after World War II and then internationally into the 1990s. The high-level structure of this model is simple and it is based on four linked components—PLAN-DO-CHECK-ACT.

The PMI was formed in 1969 and the first edition of the *PMBOK® Guide* was published in 1987 through the efforts of thousands of volunteers. As the defined processes and expertise in project management matured, a second version of the *PMBOK® Guide* was released in the late 1990s based on updates recommended by PMI members. At this point the guide became recognized as a standard by the American National Standards Institute (ANSI) and the Institute of Electrical and Electronics Engineers (IEEE). In 2008, a fourth edition was published with further updates including the structure of the document, modifications to the major processes, new terms, and increased recognition of programs and project portfolios. Through this evolutionary maturation process, the influence of the guide can now be found across the full spectrum of projects in all industries throughout the world.

3.3 STRUCTURE OF THE *PMBOK® GUIDE*

3.3.1 PROJECT DOMAINS

Figure 3.1 shows a high level view of the five major process groups defined by the *PMBOK® Guide*.

The labels shown in Figure 3.1 actually represent the historical origins of the *PMBOK® Guide* model. In modern form, the guide renames the logical groups above into more operationally recognized terms as follows:

* Initiating
* Planning
* Executing
* Monitoring and controlling
* Closing.

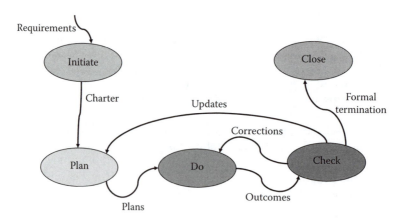

FIGURE 3.1 *PMBOK® Guide* major process groups. (From PMI, 2008. *PMBOK® Guide*, 4th ed., Project Management Institute, Newtown Square, PA. With permission.)

The flow arrows in Figure 3.1 imply that the five major phases are executed somewhat serially; however, the actual workflow is more iterative and complex than what appears from this diagram. It would be more accurate to show the initiating and closing process activities as the fixed start and stop components, while the middle of the life cycle contains the core planning, control, and execution processes. That view more appropriately reflects the dynamic nature of projects (i.e., changes during execution require replanning).

Surrounding the four core work activity groups is a formal monitoring and control shell, designed to ensure that the project goals are being met through each of the stages. Embedded within these groups are 42 defined activities that represent fundamental management processes required to execute the project.

This project life cycle model was adapted from the classic Shewhart and Deming Plan-Do-Check model of the 1930s. PMI added initiation and closing steps, and then formally redefined the five major groups as Initiating, Planning, Executing, Controlling and Monitoring, and Closing as common steps for a model project. Edwards Deming updated the Shewhart foundation conceptual model with his quality management work in Japan after World War II. The high-level structure of this model is simple and it is based on Shewhart's four linked components—PLAN-DO-CHECK-ACT. The role of each process group is summarized in the following sections:

Initiating: Outlines the activities required to define and authorize the project or a project phase.

Planning: Outlines the activities required to produce a formal project plan containing objectives, scope, budget, schedule, and other relevant information to guide the ongoing effort.

Executing: Uses the project plan as a guiding reference to integrate human and other resources in carrying out project objectives.

Monitoring and controlling: This process group of activities measures and monitors progress to identify plan variances and take appropriate corrective action.

Closing: A group of activities required to formally shut down the project and document acceptance of the result.

Within the five major process groups outlined above are nine knowledge management (KM) areas and 42 activities distributed across the various KMs. The nine Knowledge Management areas (KAs) are summarized below with a brief description for each (PMI, 2008, p. 43):

1. *Scope*—a description of the work required to be accomplished by the project team.
2. *Time*—describes the processes related to executing a timely completion of the project.
3. *Cost*—describes the processes related to estimating costs, budgeting, and project cost control.
4. *Quality*—describes the processes required to assure that the project will satisfy the operational objectives for which it was formed. This has aspects of planning, quality assurance, and quality control.
5. *Human resources*—describes the processes involved in acquiring, developing, and managing the project team.
6. *Communications*—describes the processes related to timely information distribution related to the project.
7. *Risk*—describes the processes related to managing various risk aspects of the project.
8. *Procurement*—describes the processes required to manage activities related to the procurement of products and services related to the project.
9. *Integration*—describes the activities needed to integrate all of the other KAs into a cohesive and unified plan that is supported by the project stakeholders.

Embedded in each of the KAs lower level 42 activity descriptions are the inputs, tools and techniques, and outputs that drive each process. The *PMBOK® Guide* document provides a good high-level definitional roadmap for project management and it represents an internationally accepted standard. However, it is not designed to be a tutorial to hand carry the PM through all of the somewhat abstractly defined steps. Rather it is a general knowledge model structure to provide guidance from which a specific project model can be constructed. Experience and training are required to turn this standard model view into workable project management tools and processes.

The five major *PMBOK® Guide* process domain groups serve as a model project life cycle to illustrate and define what should occur in each of these major segments. For instance, monitoring and control activities are designed to reveal variances from the approved plan, which would then trigger corrective activities to move the plan back towards its defined goal, or to trigger approved changes in scope that cause the project plan to be altered. If we could assume that the project would proceed without deviation, the major stages would evolve from initiation to planning to execution to closing and in this ideal situation, a control stage would be unnecessary. Obviously the real world is not so orderly so, variances are the norm and changes do occur. These events in turn cause interrelated decision actions to move across all of the phases and knowledge areas. The key role of the life cycle is to manage and control such dynamics and attempt to help make the overall process more stable. Failure to maintain a reasonable level of stability as the effort moves through the life cycle leaves the project environment in a state of chaos, which typically results in failure to achieve the desired objectives.

The development process implicit in the *PMBOK® Guide* requires that the proposed project be formally evaluated on its business merits, approved by management, formally initiated, and then undergo a detailed planning cycle prior to commencing execution. Within each life cycle step there is a coordinated management process designed to ensure that the project produces the planned results. Once the appropriate stakeholders have approved the project plan, the subsequent execution phase would focus on doing what the plan defines (nothing more and nothing less). Overseeing the execution phase and all other phases is an active monitoring and control process designed to periodically review actual status and take appropriate action to correct identified deviations. After all the defined project requirements are produced, the closing process finalizes all remaining project paperwork and captures relevant lessons learned that are used to improve future efforts. When examined from this high level perspective, the *PMBOK® Guide* project model is a deceptively simple structure, but we should recognize that this simple view hides significant real world challenges in executing the defined processes.

The flow arrows in Figure 3.1 imply that the five major phases are executed somewhat serially; however, the actual workflow is more iterative and complex than it appears from this diagram. It would be more accurate to show the initiating and closing process activities as the fixed start and stop components, while the middle of the life cycle contains the iterative planning, control, and execution processes. That view more appropriately reflects the dynamic nature of a project life cycle (i.e., changes recognized during execution require replanning). Each of the five major domains has a particular goal in the life cycle. An explanation of these goals is provided below.

3.3.1.1   Initiation

This process group is involved with the activities required to define and authorize the project or a project phase. One of the most important aspects of the Initiation process is the evaluation of the project vision from a business goal alignment perspective. In other words, how does the vision support organizational goals? The decision to approve a project must also consider it in competition with other such proposals based on factors such as resource constraints, risks, capabilities, and so on. After consideration of these factors by management, formal approval to move the project into a more detailed and formal planning phase is signaled by the issuance of a formal Charter. This step outlines the basic approval of the project and the constraints under which it is to be governed. The guide defines that a PM is formally named at this point to move the effort forward. Project Charters represent the formal authorization step and it formally communicates that management is behind the project.

3.3.1.2   Planning

This process group relates to the activities required to produce a formal project plan containing specified deliverable objectives, budget, schedule, and other relevant information to guide the subsequent ongoing effort. The principal goal of the Planning phase is to produce an accurate, measurable, and workable project plan that has considered the impact of all knowledge areas. This particular phase consumes the second highest amount of resources throughout the life cycle and its goal is to lay out a path for execution that can be reasonably achieved. The key output from this phase is a formal project plan outlining not only the scope, schedule, and budget for the project but also how the project will deal with integrating the other areas of Quality, Human Resources, Communications, Risks, and Procurement.

3.3.1.3   Execution

This process group uses the project plan as a guiding reference to integrate all work activities into production of the project objectives. The actual project deliverables are produced in the execution phase. During this cycle, the PM has responsibilities including coordination of resources, team development, quality assurance, and project plan oversight. Given that a project plan seldom, if ever, goes exactly according to the original vision, it will be necessary to deal with resource usage and task status variances along with new work created by change requests that are approved by the project board. Another important activity is to communicate project status to various stakeholders outlining the current state and expectations for the project. The ultimate goal then is to deliver the desired result within the planned time and budget. During this phase, the PM's mantra is to "work the plan."

3.3.1.4   Monitoring and Controlling

This process group measures and monitors progress to identify plan variances and take appropriate corrective action. Monitoring and Control transcends the full project life cycle and has the goal of proactively guiding the project towards successful completion. As unplanned changes occur to schedule, scope, quality, or cost, the PM will have to determine how to react to the observed variance and move the effort back to an appropriate strategy. Much of this activity is driven by performance reporting, issues (variances or process issues), and the formal change management process. In addition, one of the most critical aspects of this phase is the risk management process which involves monitoring various aspects of project risks including technical, quality, performance, management, organizational, and external events.

3.3.1.5   Closing

Formal project closing involves a group of activities required to formally shut down the project and document acceptance of the result. Also, this step completes the capture of lessons learned for use in future initiatives. It is widely noted that the closing phase gets the least attention; however, the guide model requires that all projects formally close out the activity, including both administrative and third party relationship elements. The basic role of this phase is to leave the project administratively "clean" and to capture important lessons learned from the effort that can be shared with other projects. In regard to third party agreements, it is necessary to view the formal contractual closing as vital. Failure to execute final vendor status for the project state can open up future liability for the organization if a supplier later makes claims for nonperformance. If this occurs, the project organization would then have to scramble to rebuild the status with old records (often poorly organized) and missing team members. Similarly, documentation of lessons learned during the project has been found to provide valuable insights for future projects.

Finally, a close-out meeting or team social event is important in order for the team to review the experience and hopefully see the positives in their experiences. Too often, a project team just walks away from the effort without receiving any feedback. This can leave the individual feeling that the effort was a waste of time and this negative attitude can carry over to the next project assignment.

3.3.2   Knowledge Areas

Knowledge areas represent a set of competency skills and processes that must be properly utilized by the PM throughout the life cycle. The guide outlines nine knowledge areas: Scope, Time, Cost, Quality, Human Resources, Communications, Risk, Procurement, and Integration. These skill and process-based items interact throughout the five life cycle Domain Groups of Initiating, Planning, Executing, Monitoring and Controlling, and Closing. Basic roles for specifics within each of the nine knowledge areas will be summarized below; however, much of their detailed content will appear in various formats throughout the book. The goal at this point is to briefly introduce them as key components of the life cycle model. These represent fundamental vocabulary concepts and are important ideas in the overall management framework. A brief description of each knowledge area role is summarized in the sections below.

3.3.2.1   Scope Management

In its simplest form, scope management involves the work efforts required to ensure that all defined requirements are properly produced based on the developed requirements statement. High level work definition is decomposed into a lower level work breakdown view. During the detailed planning phase, this activity involves translation of a formal statement of requirements, while later activities deal with control of the requirements change process and verification that the ongoing results will meet customer expectations. The primary scope output for the planning cycle is a Work Breakdown Structure (WBS) that provides subsequent guidance for various follow-up project activities.

3.3.2.2   Time Management

This process deals with the mechanics and management requirements for translating the defined scope into work unit activities and then monitoring those activities to ensure "on-time" completion. The guide defines six processes in this area: Define Activities, Sequence Activities, Estimate Activity Resources, Estimate Activity Durations, Develop Schedule Development, and Control Schedule (PMI, 2008, p. 43).

The Define Activities process translates the identified scope requirements deliverables into a set of work activities required to complete the deliverables. The Sequence Activities process involves linking the work activities into a network plan structure. Step three in the time management planning process involves estimating resource requirements for the defined activities, which in turn leads to a resource allocation process and an activity duration estimate. Once the various work units have been translated into duration estimates, they can be sequenced together to create an initial project plan. The reason for illustrating the guide processes here is to show that they are designed to provide guidance in the development of required project management activities.

In concert with the time planning activities, the project team will evaluate each of the other knowledge areas to produce a final schedule for presentation to management. During the execution phase, a schedule control process monitors the ongoing status and guides the PM through appropriate corrective action.

3.3.2.3   Cost Management

This knowledge area includes various activities and processes that guide the budget creation process, then establish a control function to guide the project through the execution process. The *PMBOK® Guide* defines three processes related to this knowledge area: Estimate Costs, Determine Budget, and Control Costs (PMI, 2008, p. 43). Basically, the process of generating a project budget involves estimates of human resources (quantity and skills) and material costs for each defined work unit. The values that are determined from this process help the project team develop a Cost Budget which includes not only the direct work cost estimates but also various other cost components needed to support the overall project activity. Cost budgeting organizes the values and estimates from the various sources and produces a cost baseline that is used to measure project performance. Cost Control deals with monitoring costs and understanding observed variances that occur as the project processes. A second part of the control process is to create a management process to handle cost changes, inappropriate charges, and cost overruns. More details on the mechanics of this process will be described in Part III of the book.

3.3.2.4 Quality Management

This knowledge area focuses on all aspects of both the product and project quality processes. The three-model defined processes are: Plan Quality, Perform Quality Assurance, and Perform Quality Control (PMI, 2008, p. 43). The Quality Planning process is designed to focus the team on organizational or industry related quality standards for the different product and project deliverables. Once the appropriate quality standards have been established and documented in the Quality Management Plan, the remaining two processes focus on the appropriate activities needed to operationally satisfy the respective quality goals. The quality assurance process reviews the state of the project from its ability to deliver the required result, while the quality control process covers the tactical procedures to measure quality of the output.

3.3.2.5 Human Resources Management

This knowledge area focuses on actions related to the human element of the project. These activities consist of four operational processes: Develop Human Resources Plan, Acquire Project Team, Develop Project Team, and Manage Project Team (PMI, 2008, p. 43). These activities include a staffing plan that outlines project staffing requirements throughout the life cycle. Staff acquisition is based on this plan. In a matrix type project organization, project resources are typically not dedicated to one particular project, but are leveraged across multiple projects and sourced from various departments in the organization. This complicates the resource allocation mechanics for the PM and makes the acquisition step more complex than one might anticipate. Once a project is underway, the process of team development starts and continues through the life cycle. This includes both individual and team training with the ultimate goal being to improve the overall skill of the team members even after the project is completed.

3.3.2.6 Communications Management

Communication problems represent one of the most causal reasons for project failure. The communications management activities are designed to support the information needs of the project stakeholders. The five model communication processes are: Identify Stakeholders, Plan Communications, Distribute Information, Manage Stakeholder Expectation, and Report Performance (PMI, 2008, p. 43). One of the key aspects involved in these processes is to identify who the targets are for communications, then explicitly plan how communications will flow during project execution. The latest version of the guide has expanded focus in this area to balance recognized deficiencies in most projects (PMI, 2008, p. 43).

3.3.2.7 Risk Management

The basic goal of risk management to is minimize the probability of negative events hurting the project and maximizing any opportunities that exist for positive events. The management of this class of activity is very difficult in that a risk event has not occurred during the planning cycle, but if does occur later, the plan will be potentially affected. Identification of such events is a complex undertaking and represents a critical success factor for the PM in that he can do an excellent job of managing the defined effort only to find that some unspecified event, internal or external to the project, wipes out the entire value of the effort. It is important for project teams to understand risks areas of vulnerability and have plans in place to deal with them. The Risk Management process involves some of the most complex project management tasks which will be discussed in greater detail later in Chapter 22.

3.3.2.8 Procurement Management

There are many situations that lead the project team to decide to procure material or human resources from third parties. The Procurement Management processes are utilized to manage the acquisition of these items. These are defined in the guide model as: Plan Procurements, Conduct Procurements, Administer Procurements, and Close Procurements (PMI, 2008, p. 43). This area of project management has historically been considered very mature, but in recent years, the increase in outsourcing has made this a troublesome area for the PM as his project resources have become scattered across wide geographic areas. The procurement planning process lays the policy groundwork to guide how project resource needs will be externally acquired and the remaining activities in this area are involved in selecting vendors and entering into contractual or formal relationships for the supply of goods or services. Upon completion of the project, contract closure reviews performance of the contract and resolves any remaining issues.

3.3.2.9 Integration Management

The guide describes Integration Management as the first knowledge area; however, it is hard to explain this process until one has a better understanding of the items that need to be integrated. We have now discussed eight other knowledge areas that must be kept in synchronization. The guide defines this role by saying that it "includes the processes and activities needed to identify, define, combine, unify, and coordinate the various processes and project management activities within the Project Management Process Groups" (PMI, 2008, p. 71). There are many examples that can be offered to illustrate this activity, but one should suffice for now. Let us assume that a major change request has been approved for the project. Obviously, this represents a change in scope. From this, it would seem reasonable to suggest that it might well bring with it changes in schedule (time) and cost. Related to this could be changes in staffing or procurement or risk and so on. The key idea of process integration is that changes in one knowledge area process often spawn changes in others. From a basic viewpoint, project management is Integration Management.

3.3.2.10 Overall Process View

As we have seen in the discussion above, each KA has multiple subordinate processes—42 in all. Some of these processes are linked directly to inputs or outputs from other knowledge areas in various segments of the life cycle, while yet others are invoked somewhat randomly at various times as necessary. As an example of linked processes, scope definition items lead to time results, which in turn lead to cost parameters. Other process linkages such as communication are not as deterministic as to life cycle points. Previous editions of the guide recognized this two-type process view, but the more recent versions dropped that designation. Regardless, it is still a good way to understand how the various items work in the overall scheme. Figure 3.2 shows a physical distribution of the 42 processes across the major life cycle process domain groups. In this view, note that each knowledge area is uniquely numerically coded to better show how the various management processes are distributed across the project life cycle. Also, each process group is segmented into either a core or a supporting role even though the guide does not specifically distinguish them this way. The core tasks are essentially performed in a somewhat defined order and are generally required activities for each project, while the supporting activities are more dependent upon the nature of the project and they may be performed in any order, iteratively, or even minimally in some cases. The process box numbers reflect the *PMBOK® Guide* KA chapter number and reference number sequence. For example, *5.3 Create WBS* would be described in the guide Chapter 5 as the third reference process.

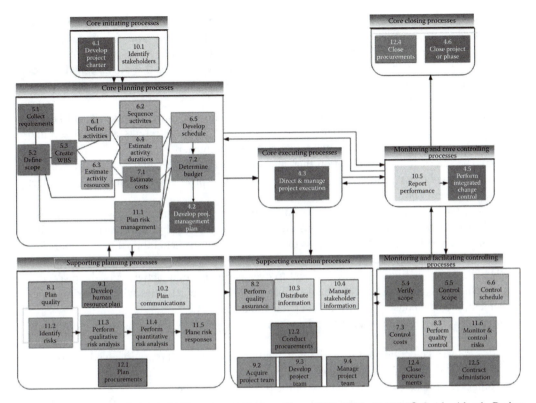

FIGURE 3.2 *PMBOK® Guide* 2008 process overview. (From PMI, 2008. *PMBOK® Guide*, 4th ed., Project Management Institute, Newtown Square, PA. With permission.)

A more detailed technical discussion related to the Process Groups and KAs described above can be found in the *PMBOK® Guide*. Embedded in each of the guide's KAs lower level, 42 process descriptions are summaries and descriptions of the inputs, tools and techniques, and outputs that drive each process. Recognize that the guide document provides a good high-level roadmap for project management and it represents an internationally accepted standard. However, it is not designed to be a project management cook book. Rather it is a general knowledge base to provide guidance from which a specific project model can be constructed. Experience and training are required to turn this standard model view into workable project management tools and processes.

3.4 KEY INTRODUCTORY VOCABULARY TERMS

One of the values of defining a standard project life cycle model is the consistency of approach and vocabulary that it brings to the organization. There are many vocabulary terms that are important in communicating project management concepts and these will be seen frequently as we navigate through the various text topics. At this stage, 10 terms have been selected to include here as being "fundamental" to the model approach. The following terms and their related concepts are summarized here and future chapters will amplify how they fit into the overall project management scheme:

Portfolio: A collection of projects or programs. This term is often used to describe the total organizational package of proposed and approved project efforts. A related management decision process should attempt to optimize the organizational value based on selection of the active portfolio members.

Progressive elaboration: This term is important to understand in that it describes how details related to project requirements can evolve in increments or steps. The planning stage can be viewed as a series of such elaborations leading to the desired final plan.

Project life cycle: Project evolution is divided into a series of phases that are designed to provide better management control of the overall project. The defined phases collectively constitute the life cycle.

Project management: This term represents the application of tools, techniques, skills, and knowledge to the project domain. Also, the term typically represents a formalized and standardized organizational methodology used to execute a project.

PMI: The international organization that sponsors the *PMBOK® Guide*, PMP certification, and various other professional initiatives designed to proliferate a more mature project management knowledge base.

Project Management Office (PMO): A contemporary organizational unit assigned responsibility to coordinate the selection and oversight of projects within the enterprise. Specific roles for a PMO function vary across organizations, but the general goal is to support formal project selection, approval, and execution.

PM: The person assigned by management to oversee the ongoing activities of the project in pursuit of its goals. This individual coordinates life cycle activities with senior management, project sponsor, users, and other stakeholders.

Sponsor: This is a senior level individual who provides general vision, guidance, and funding for the project.

Stakeholders: The collection of individuals and organizations who are involved or affected by the project. Some stakeholders exert influence over the direction of the project, while others are impacted by the outcome of the project deliverables.

User: The individuals or organizational groups that utilize the project's output product. This group is often called the "customer" for the project.

3.5   ANCILLARY MODELS

PMI continues to provide knowledge and leadership in the definition and certification of project management principles. There are various focus areas and certifications sponsored by PMI and these are internationally recognized. Further details on these initiatives can be found at www.pmi. org. All of these initiatives impinge on the project world in some form. All of the operational items from this list have text chapter references, but the reader should understand the basic significance of other programs or enhancements mentioned below. A summary of PMI sponsored items that represent an expansion to the basic *PMBOK® Guide* follows, as shown below with text references wherever applicable:

WBS: See Chapter 12.

Earned Value: See Chapter 27.

OPM3: See Chapter 31; relates to organizational processes that support the project.

Certified Associate Project Manager (CAPM) Certification: Similar to PMP certification for less experienced professionals.

Program Manager Certification (PMgP): A senior manager certification for individuals who have a high level of experience and manage collections of large projects.

Code of Ethics and Professional Responsibility: See Chapter 37; a published set of code of conduct and ethics and responsibilities for the PM.

Risk Certification: Released in late 2008, this certification is designed for those project technicians who work in the risk management arena of projects.

Team Leadership: This area is under review for some amplification in terms of techniques or certification.

3.6 SUMMARY

This chapter has provided a brief overview of the PMI project management model as described in the 2008 edition of the *PMBOK® Guide*. The concepts and philosophy of this model are reflected and supported in the structure and topics covered in this text. Individuals serious about project management as a profession should seek certification from PMI as either a PMP or a CAPM. The PMP certificate requires more work experience than the CAPM, but both are respected professional credentials and they add professional marketability for the holder.

The *PMBOK® Guide* has been approved by the ANSI, which should further expand its industry recognition. Many organizations use this certification as a prerequisite for hiring or job assignment as a PM. Further specifics on various PMI certification and standards programs can be found at www.pmi.org.

REFERENCE

PMI, 2008. *A Guide to the Project Management Body of Knowledge (PMBOK® Guide)*, 4th Ed., Project Management Institute, Newtown Square, PA.

4 Industry Trends in Project Management

4.1 STANDARDIZING PROJECT MANAGEMENT

One of the secrets of success for companies such as Disney, Nokia, Johnson & Johnson, Vodafone, and Virgin Air is that they have produced phenomenal customer products, spawned from innovative ideas and then followed that up with project/development methodologies that facilitated their innovativeness to deliver their projects to market more quickly than their competitors (Charvat, 2003). Experience has shown that the use of a standard project methodology is often an effective strategy for developing projects in the IT, energy, aeronautical, social, government, construction, financial, or consulting sectors. A project methodology contains processes designed to maximize the project's value to the organization (Charvat, 2003). In its operational form, a methodology must accommodate a company's changing focus or direction. Over time, it becomes part of the organizational culture and embedded in the way projects are executed. To be effective over a long term, a methodology must fit the perception of activities required to execute the project. This perception involves both technical and personal views. For that reason, individual managers often resist using someone else's methodology. This statement suggests that project management contains both science and art aspects.

4.2 ENTERPRISE PROJECT MANAGEMENT

Enterprise Project Management (EPM) and Portfolio Management are two of the newer management practices shaping the view of project management in action today. Project Portfolio Management (PPM) is an organizational level methodology that caters to the enterprise-wide collection of projects, while EPM is less well described but essentially deals more with the overall organizational processes utilized in executing projects. EPM enables organizations to manage projects, as a collective portfolio of activities, rather than as separate, isolated initiatives with no overlap (QAI, 2005). EPM provides a big-picture perspective for all project assets, such as enterprise goals, staff, equipment, and budget; it allows projects to be aligned across departmental boundaries and containing key strategic organizational initiatives. This approach is designed to provide visibility to all project stakeholders so that overall project initiatives remain in alignment with organizational objectives, and required status information is properly communicated.

EPM can be viewed from different viewpoints. One perspective of EPM is that it is a software tool that helps plan and implement all projects in an organization. Another view is that it is a process whereby projects are selected, planned, controlled, and implemented from a central location that assures all projects report progress to a single element. A formal definition of EPM is as follows:

> Enterprise Project Management is an enterprise view of the Project Management activities. It is a strategic decision by the organizations which aims at linking the organization's mission, vision, goal, objectives and strategies in a hierarchical fashion to ensure that the organization allocates its resources to the right projects at the right time. (Ireland, 2004, p. 1)

In operational form, EPM represents the model showing how to incorporate the art and science of project management into a new way to do business. It focuses on consolidating project principles across the entire organization with the goal of optimizing organizational value for the chosen ventures. Companies such as American Express, ABB, Citibank, and IBM are acknowledged as pioneers of businesses pursuing this approach and each has now taken substantial steps toward applying EPM principles on an enterprise level (Morris, 2000). EPM is a methodology that combines standardized project management processes and supporting tools to better meet an organization's project or program management goals (Landman, 2008).

One of the major distinctions of EPM versus project management or PPM is the implied high-level integration process throughout the organization. In the EPM organization, no longer would the project teams feel like they were on an isolated island. Rather, various formalized organizational processes would be actively involved in formal support of the effort. Figure 4.1 offers a conceptual model of an enterprise's general structure when it adopts the EPM approach to project management.

The theme of this figure is to indicate that the various components integrate the project process life cycle into the organizational goals. A brief description of each component is presented below.

Portfolio management: The normal industry designator for this function is PPM. Various other titles could be used for this activity, but its primary role is to establish some form of high-level portfolio view of the organizational project and process environments. This entails defining operational details regarding what projects are being worked on and data related to proposed projects. In some cases, the portfolio will also include the status of existing processes and products as indicating what work needs to be done to upgrade them. Out of the portfolio view would be derived engineered proposals that would collectively produce the greatest impact on organizational improvement toward its defined goals. Think of this activity as a very complex analysis of current state versus desired state. The role of a particular project would be to move that segment of the organization toward the desired state. This activity implies a strong role for senior management in the goal setting and project selection process. Section 4.2 of this chapter, along with Chapters 32 and 33 will pursue this topic in greater detail.

Project management: The role of project management involves delivering the defined results that were specified in the PPM process. One of the central themes of this text is to describe what that role entails. However, it is important to recognize that this involves more than just defining those processes. There is still great resistance by many in believing that such rigor is required to produce successful outcomes. So, an EPM organization would have to establish a mature organizational culture in this regard. There is mounting evidence that the current ad hoc practices of project management are not successful and that improved rigor does help improve the resulting outcomes (Thomas and Mullaly, 2008). Nevertheless, recognize that project management involves more than the mechanics as outlined here.

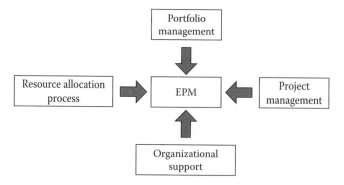

FIGURE 4.1   EPM architecture components.

Organizational support environment: The analogy often used is that projects are like seeds in an organizational flower bed. This implies that projects cannot successfully exist without the support of their host organization. This support comes in various ways:

- Senior management for general support and high-level decision making
- Users who help with requirements definition
- Functional managers who own the critical resources
- Organizational support processes that are needed by the project team
- Capital resources that must be supplied by the organization in competition with other groups.

Each of these provides valuable support to the project team in various ways throughout the project life cycle.

Resource allocation process: The typical project organization is staffed as a matrix, which means that required project team human resources are supplied by various functional organizations or external vendors through some form of planned relationships. Of all the project critical success factors, access to appropriate technical and business knowledgeable resources is the most significant variable assuming that the proper target has been selected. Failure to garner those resources in the needed time periods will yield less than desired results in time, cost, or technical outcome. Exacerbating this issue is the dynamics of the project world and the multitasking approach that is often used for the resources. Competing demands across multiple projects for scarce resources means that the resource allocation process becomes a vital component in the overall EPM process. There are other views to the EPM process and it certainly is more complex in operation than described here, but this is the essence of the model; this level discussion provides the necessary foundation for the material that follows.

4.3   EPM IN OPERATION

The various EPM functional decision processes collectively serve as a basis for selecting projects that have a high degree of both strategic and operational fit to the enterprise goals. From the top-down, the model provides both general and organizational guidance to project-specific implementation. Each of the decision process nodes are linked in a way to ensure consistency of purpose and predictable outcomes (Ireland, 2004).

4.4   IMPLEMENTATION AND ADVANTAGES OF EPM

It is the responsibility of senior management to determine whether sufficient benefits exist for moving to an enterprise level project management organization. Step one of the decision process would be an analysis of existing projects to determine whether the project selection process is adequate and whether the existing project slate is less than optimal in terms of size, risk, profitability, and strategic fit compared to the organizational goals. If a decision is made to switch to an enterprise level focus, it is necessary to design a transition strategy away from the current method. In this transition, it is important for senior management to be the leader and make sure that all stakeholders understand the rationale for the change and the associated benefits. The potential benefits for EPM would include the following:

- Greater management visibility into project status in terms of profit and alignment to organizational strategies
- A disciplined approach toward allocation of resources to the projects
- Resources allocated to right projects result in more timely delivery of the project deliverables (Ireland, 2004).

4.5   OTHER TRENDS IMPACTING PROJECT MANAGEMENT

Organizational maturity: There is a growing recognition that the maturity of the organization is related to project success. Recognition of this basic concept has spawned a growing interest in defining what a mature organization looks like and does. One aspect of maturity is to grade just how much support the organization supplies to the project, meaning that the project does not have to invent all of its necessary processes, but can simply attach itself to them and move forward with minimal process overhead. There is a great deal of evidence to support the notion that moving the organization up the maturity scale can improve project success rates and operational productivity—two of the needed items in the project manager's goal list.

The emergence of project management maturity models is a fairly recent phenomenon evolving from the classic Capability Maturity Model (CMM) developed by the Software Engineering Institute (SEI) at Carnegie Mellon (Grant and Pennypacker, 2006). Out of this classic research, the concept of process maturity has been pursued by thousands of organizations and involving billions of dollars in project expenditures.

As described earlier, the EPM model involves both organization and project level processes. The concept of maturity is embedded in both of these levels. The CMM grading scale is based on a numerical value ranging from 1 to 5, with 1 basically indicating no formal process and 5 indicating an optimal level of maturity. The basic question for this discussion is "Maturity of what?" The PMI model as generally described in Chapter 3 helps in answering this question in its definition of the nine project knowledge areas and specifications for their related processes. From these two perspectives, one could envision a two-dimensional model of five maturity grades mapped against the nine knowledge areas as a method of portraying an organizations status from the project view (Grant and Pennypacker, 2006). The intersection of these two dimensions, knowledge area and maturity level, can then be further decomposed into various lower-level knowledge area process components to produce a measure of operational maturity at a fairly low level of granularity. One could then envision this result as a three-dimensional view of the overall project maturity—maturity measure by knowledge area by knowledge process. The key point for this discussion is to recognize that both projects and the organization have operational maturity that helps support effectiveness of the project activity. This view of maturity can be assessed for organizational, departmental, and project levels. Our focus is to primarily deal with the project level, but it is hard to isolate that from the enveloping organizational components.

Contemporary trends: Samples of trends that are impacting the working environment of a project manager are as follows:

- Increased use of technology within the daily operational processes. This includes project modeling, computer-based document management, web-based tools, enterprise-level resource databases, and sophisticated collaboration tools
- Increased customer-driven projects (both internal and external to the organization)
- Geographically dispersed teams. The trend of working in a virtual environment has resulted in project management across multiple time zones where communication becomes more long distance and impacted by cultural and language diversity
- Moving from stand-alone project management to portfolio model perspectives increases the complexity of the overall process and requires the project manager to become more focused on broader organizational goals
- Implementation of PMOs adds a layer of centralized control above the project level, which has the potential to increase bureaucracy and inhibits the speed of decision making. Many project managers (and some sponsors and clients) will not like this added layer of "help" for their projects.

Project manager skills: In order to survive in this highly dynamic world of today, the project manager must develop new skills and work habits in order to keep proficient in dealing with the changing trends as outlined in this section. Basically, the modern project manager must do the following:

1. Keep abreast of current best management and technical practices
2. Develop skills in customer-centric communication
3. Learn to coexist in the more politically charged EPM-type organizational environment
4. Develop soft skills in regard to team motivation, diversity, and corporate culture
5. Improve in emotional intelligence (feeling and thinking)
6. Improve management skills in developing high-productivity teams while using less formal authority mechanisms.

4.6 PROJECT MANAGEMENT PERSPECTIVE

Project management should be viewed as a set of techniques, theories, and tools that collectively help organizations effectively execute designated projects. However, it is also important to recognize that the use of these mechanical methods does not automatically guarantee project success. Over the last 50 or so years modern project management has developed into an applied science (albeit still a pseudo one), which has proven helpful in the achievement of the project goals. However, the current set of methods and tools are quite young by management standards; hence there is every expectation that the operational mechanics and theories for managing projects will continue to evolve. Topics such as risk management, organizational maturity models, virtual team management, and project value analysis are likely to be at the forefront of this evolution. At the tactical project level, there is now growing acceptance of general management approaches to project management but resistance remains as to its value in the success of the project. In spite of this resistance, management continues to press for more success in this arena. Project management as described here is the general model being pursued by growing numbers of organizations (Barnes, 2002).

In order to improve knowledge regarding project management value, PMI chartered a three-year research study in 2005. The results of this study published in 2008 concluded that project management added value, but also showed that understanding that value was more complex than using simple financial measures (Thomas and Mullaly, 2008). The interested reader is encouraged to review this groundbreaking research to get a better idea of how projects bring value to the organization. Other studies have shown that projects do in fact produce value, but often not in the way originally predicted. Our goal in this text is to understand the fundamental mechanics of producing successful project outcomes and some of the key external organizational factors that support the process.

DISCUSSION QUESTIONS

1. Why would the existence of standardized project management procedures help achieve higher rates of project success?
2. If it makes such logical rationale for organizations to properly select their portfolios of projects, why is this activity not well practiced?
3. Using the high-level description of EPM, describe some of the implementation issues that you see for the organization pursuing this goal.

REFERENCES

Barnes, M., 2002. *A Long Term View of Project Management—Its Past and Its Likely Future.* Presented at 16th World Congress on Project Management. Berlin.

Charvat, J., 2003. *Project Management Methodologies: Selecting, Implementing, and Supporting Methodologies and Processes for Projects.* New York: Wiley.

Grant, K.P. and J.S. Pennypacker, 2006. Project management maturity: An assessment of project management capabilities among and between selected industries. *IEEE Transactions on Engineering Management.*

Ireland, L., 2004. Enterprise Project Management—A Strategic View. http://www.asapm.org/resources/a_epm_ireland.pdf (accessed January 30, 2009).

Landman, H., 2008. *Enterprise Project Management Using Microsoft Office Project Server*. Ft. Lauderdale, FL: J. Ross Publishing.

Morris, H.W., 2000. Winning in Business with Enterprise Project Management. *Project Management Journal* (June).

PMI, 2008. *A Guide to the Project Management Body of Knowledge (PMBOK® Guide)*. Newtown Square, PA: Project Management Institute.

QAI, 2005. International Conference on Project Management Leadership. http://www.qaiindia.com/Conferences/pm05/papers_15.htm (accessed May 13, 2005).

Thomas, J. and M. Mullaly, 2008. *Researching the Value of Project Management*. Newtown Square, PA: Project Management Institute.

5 Project Types

In earlier discussions we theorized that all projects have more similarity than most understand, but the question is "how much similarity?" In order to explore that idea we need a starting point. Archibald studied the question regarding where projects exist and for what purpose. His research identified projects in the following 20 industry areas (Archibald, 2004):

1. Aerospace/defense
2. Automation
3. Automotive
4. E-business
5. Environmental
6. Financial services
7. Government
8. Healthcare
9. Hospitality events
10. Information systems
11. Information
12. Technology/telecom
13. International development
14. Manufacturing
15. New product development
16. Oil/gas/petrochemical
17. Pharmaceutical
18. Retail
19. Service and outsourcing
20. Utility industry.

Obviously, each of these industries has unique technical characteristics in their projects; however, the key point is that all have a somewhat similar life cycle. We recognize that there will be differences in the underlying technology, formality of documentation/communication, risk, human skills required, and a host of other subtleties, but fundamentally there is a strongly consistent management view appropriate for all.

Archibald and Voropaev compiled project data in an attempt to classify projects into a smaller and more coherent classification grouping by characteristics. The hypothesis of this effort was that a classification scheme could begin to help in better understanding various life cycle methodology needs based on the features and processes found in a particular group. From a categorization model theory, there is a belief that each group could have more specific defined processes, tools, and methodologies that fit their needs and the potential for reusable components that would save development time and effort. Success from such a definition would be at least one level more granular than that defined in the *PMBOK® Guide*, which tends to be more a single level view. Even though this effort has yet to reach maturity, it is still worth examining. Table 5.1 shows their draft of 11 project categories. Collectively, this list provides a good overview of the international project environment and scope.

TABLE 5.1
Recommended Project Categories/Subcategories

Categories (Each Having Similar Life Cycle Phases and a Unique Project Management Process)	Examples
1. Aerospace/Defense Projects	
1.1 Defense systems	New weapon system; major system upgrade
1.2 Space	Satellite development/launch; space station model
1.3 Military operations	Task force invasion
2. Business and Organization Change Projects	
2.1 Acquisition/merger	Acquire and integrate competing company
2.2 Management process improvement	Major improvement in project management
2.3 New business venture	Form and launch new company
2.4 Organization restructuring	Consolidate divisions and downsize company
2.5 Legal proceeding	Major litigation case
3. Communication Systems Projects	
3.1 Network communications systems	Microwave communications network
3.2 Switching communications systems	Third-generation wireless communication system
4. Event Projects	
4.1 International events	2004 Summer Olympics; 2006 World Cup Match
4.2 National events	2005 U.S. Super Bowl; 2004 Political Conventions
5. Facilities Projects	
5.1 Facility decommissioning	Closure of nuclear power station
5.2 Facility demolition	Demolition of high-rise building
5.3 Facility maintenance and modification	Process plant maintenance turnaround
5.4 Facility design/procurement/construction	Conversion of plant for new products/markets
Civil	Flood control dam; highway interchange
Energy	New gas-fired power generation plant; pipeline
Environmental	Chemical waste cleanup
High rise	40 story office building
Industrial	New manufacturing plant
Commercial	New shopping center; office building
Residential	New housing subdivision
Ships	New tanker, container, or passenger ship
6. Information Systems (Software) Projects	New project management information system. (Information system hardware is considered to be in the product development category)
7. International Development Projects	
7.1 Agriculture/rural development	People and process-intensive projects in developing countries funded by The World Bank, regional development banks, US AID, UNIDO, and other UN and government agencies; and Capital/civil works intensive projects—often somewhat different from 5. Facility Projects as they may include, as part of the project, creating an organizational entity to operate and maintain the facility, and lending agencies impose their project life cycle and reporting requirements
7.2 Education	
7.3 Health	
7.4 Nutrition	

continued

TABLE 5.1   (continued)

Categories (Each Having Similar Life Cycle Phases and a Unique Project Management Process)	Examples
7.5 Population	
7.6 Small-scale enterprise	
7.7 Infrastructure: energy (oil, gas, coal, power generation, and distribution), industrial, telecommunications, transportation, urbanization, water supply and sewage, and irrigation	
8. Media and Entertainment Projects	
8.1 Motion picture	New motion picture (film or digital)
8.2 TV segment	New TV episode
8.2 Live play or music event	New opera premiere
9. Product and Service Development Projects	
9.1 Information technology hardware	New desktop computer
9.2 Industrial product/process	New earth-moving machine
9.3 Consumer product/process	New automobile, new food product
9.4 Pharmaceutical product/process	New cholesterol-lowering drug
9.5 Service (financial, other)	New life insurance/annuity offering
10. Research and Development Projects	
10.1 Environmental	Measure changes in the ozone layer
10.2 Industrial	How to reduce pollutant emission
10.3 Economic development	Determine the best crop for sub-Sahara Africa
10.4 Medical	Test new treatment for breast cancer
10.5 Scientific	Determine the possibility of life on Mars
11. Other Categories?	

Source: Archibald, R., 2004. Presentation at PMI Latin American Forum. www.pmforum.org/library/cases/globalslides.pdf (accessed June 10, 2008). With permission.

Within the standard project model, there are eight basic knowledge areas (plus integration). Each of the project types categorized in Table 5.1 represents unique management issues for these areas. The challenge for the project manager is to identify the critical success characteristics for his project category and then use that knowledge to develop a management process that best leads to success. Each project type has varying priorities for achieving their schedule, budget, technical requirements (scope), and quality. Many would say that all are required in every case, but given the uncertainties in this environment, there must be a priority order defined to facilitate future goal tradeoffs. For example, in the commercial aerospace environment, quality (safety) would be paramount. Beneath that highest level goal, there likely would need to be design tradeoffs for other project goals (i.e., fuel economy, maintenance costs, speed, etc.). The point made here is that all design objectives cannot always be met. When that situation exists, a key part of the project management process is to trade off lower priority items in order to salvage maximum value for the effort. The traditional view of this is to manage planned technical requirements, cost, and schedule based on their respective priority rankings.

To carry this theme further, some projects involve high inherent risk. In this situation, the project manager typically pursues strategies to minimize risk exposure, but in some cases this cannot be avoided based on the goal of the project. To illustrate, building a manned spaceship to Mars is a radical example of this. Even though there is extensive effort to mitigate risk, the fact is that the project goal has significant embedded risk that cannot be completely avoided. There are similar issues in more traditional projects but probably less obvious. Whenever an organization decides to pursue a new technical product and establishes a project team to develop that product, there is significant risk involved. The technology may not work; the market may not want the product as produced, along with other outcomes. In these situations an organization has to pursue a risky effort in order to gain competitive advantage. Achieving the risky venture may bring a competitive advantage to the organization, but failure is another possible outcome. So, it is important to recognize that managing high-risk projects is quite different from constructing a standard building, although there have been examples in the past where the wrong decisions have been made even here (i.e., failure to let concrete dry before removing forms causing the entire building to fall).

One memorable risk management example was observed recently from a deep-sea multibillion dollar drilling rig project where various quality tests were cut for the sake of saving time and budget. As the completed rig was being floated out in sea, it sank and was a complete loss. One could easily argue that this was a management tradeoff decision but even if the budget and schedule were met the ultimate result was a project failure.

So, it is important to recognize that during the course of a project it is common to have to deal with trading off project goals owing to the dynamics of the process. When this situation arises, the project manager must make intelligent decisions to juggle the variables in order to achieve the best result possible. In order to do this, there should be an understanding of the relative importance of each management parameter and knowledge area involved. Understanding the idiosyncrasies of the various project categories may well be the first step in making a more coherent management process part of the base theory. Until then, it is up to the skills and intellect of the project manager to wade through this minefield.

REFERENCE

Archibald, R., 2004. Presentation at PMI Latin American Forum. www.pmforum.org/library/cases/globalslides.pdf (accessed June 10, 2008).

6 Project Organization Concepts

Previous discussions have outlined the general life cycle of projects and generally described the types of activities that have to be accomplished in that life cycle. In this section, the focus is on how to allocate the technical and process skills to support getting the job accomplished. The first item to recognize is that for a project of any size, the combination of knowledge and skills will not reside in a single group through the life cycle. During the early phases, the requirement focuses more on WHAT is to be done and this evolves into HOW the various work activities will be performed. Various skills are required at different times through this process. So, we have to recognize that it will normally not be the most efficient use of human resources to allocate individuals to the project for the duration. This can be done for a core portion of the team, but not for the total resource picture. This suggests that the typical project then has an organizational problem built around moving appropriate resources into and out of the effort according to some approved plan. Also, these resources will be used in other initiatives, so that there is a global perspective to the problem.

6.1 PM ROLE

Within the structure of the team, there are other organizational issues to resolve as well. According to the guide model, a companion decision defined in the originating Charter document is the assignment of a PM. That then becomes the first organizational decision. Is this person allocated to this task full time or part of other duties? That decision has a cascading effect on other organizational issues. In the formal vernacular there are three titles that are used for the PM roles.

1. PM—a person given formal responsibility for the project and some defined level of authority regarding expenditures and other decisions. This person "occupies the center of the interactions between stakeholders and the project itself" (PMI, 2008, p. 26).
2. Project coordinator—a person who often works for the project sponsor or some other management level person who is given responsibility to keep track of project status, but has little else in the way for formal responsibility. This approach should be used only for small, low priority efforts.
3. Project expeditor—this person is assigned to help the technical team participants and collect data, but they have essentially zero responsibility for the effort. Think of them as information distributors. This form also does not fit the requirements for a project of any size.

For projects of any size, the proper approach is to formally assign a full-time PM. The key beyond this is to provide that individual with appropriate levels of authority and a clear vertical reporting relationship.

6.2 REPORTING RELATIONSHIPS

A typical project structure would report to a project sponsor who is the executive level person who *owns* the project and is in charge of its overall success. One way to accomplish the external project management role is to establish a formal project board who is formally assigned to deal with project issues that fall outside of the PM's domain. The most typical of these would be approving change requests. The sponsor might be one of the board members and the other members typically come from key user areas impacted by the project. The board should be given delegation power to oversee the project within the Charter constraints. They will help interpret requirements issues, approve scope changes, and generally aid in helping resolve any issues that cannot be handled by the PM.

One possible modification of a reporting structure would be to have a project created as part of a larger collection of related projects called a *program*. In this case, the sponsor's role would be served by the program manager and the board structure would work the same, but may be more interlocked across multiple projects. This linkage issue would be required because the various projects would have to fit together in order for the whole to work as envisioned. NASA space projects are the best-known examples of large programs with critical project interrelationships; however, this same pattern emerges as organizations begin to recognize that projects need to mesh together in order to drive their high-level goals.

6.3 TEAM RESOURCES

The issue of resource ownership is one of the most contentious management decisions in organizational design. We have already described the project requirement for needing various skills dynamically allocated through the life cycle; however, this still leaves the question of *ownership* of that resource while they are involved in the project. Can the PM select those employees he wants on the team? Does he handle their formal performance appraisal and handle HR items such as raises and bonuses? Can he fire them for lack of performance? In most organizations, the answer to all these questions is no! This clearly is one of the Achilles heels for the PM. Functional managers of the organization will maintain formal ownership of their resources and allocate them to the project hopefully on the agreed on schedule. In the classic matrix model, the PM must then deal with these resources without a lot of formal control. Low performers can be given back to their home organization, but dealing with them is left to the functional management. This description should give insight into some of the skills needed by a PM. In this situation, he must be able to negotiate with the functional manager and motivate the team without formal power. Managers who need formal authority will be frustrated by this model.

In some cases a project will be completely housed inside of a single functional organization and many of the authority issues are minimized in that case; however, the variety of skills required for a complex project often makes this an infeasible approach in that a variety of resource skills are needed from across the organization. Beyond the source and ownership aspects of the resource there are many other people-related issues that affect the project organizational structure.

6.4 TEAM PRODUCTIVITY AND SIZE

Putnam and others have performed research on team size and they found that maximum productivity occurs in team sizes in the range of 5–7 (Putnam, 1997). There are many behavioral and communication reasons why this might well be the norm, but it is a common observation. Unfortunately, some project teams need to be larger than this. From a project team organization, this implies that one team organization strategy needs to be collections of relatively small focused groups such that the overall team becomes productive. The idea is sound, but once again we run into the dynamic resource allocation issue.

One possible approach to achieving a cohesive team goal is to envision the project as a collection of work units in which the team members participate in defining. As these work units are defined, it will help the various team members see how they fit into the broader scheme. Later, as individuals are assigned responsibility for a work unit, it will be up to the PM to communicate with these individuals and basically cheer on their performance. Achieving this measure of responsibility and recognition are major factors in achieving overall team productivity.

As team size grows beyond the optimum, the management process of team coordination and work responsibilities also grows. Communication issues rise in importance and process formality becomes more of a requirement.

6.5 TEAM'S PHYSICAL LOCATION ISSUES

Physical housing locations for the team are a common issue in a matrix-type structure. In this situation, the team members permanently belong to various functional groups and have assigned space there. So why allocate duplicate space? The simple answer to this is to facilitate team communication and improve the social team culture. This is especially important in the formulation stages of the project where requirements are being confirmed and technical issues discussed. The initial core project planning team forms the foundation for the future work requirements, and mistakes made at this point are compounded if not discovered early. During this period, the core team is working to translate the Charter issued by management into a viable set of work activities. Representatives from various skill groups are required for this effort as they translate the business vision into technical work units. This formulation period is an important element for success and the key players should have full access to each other and strive to develop good working relationships. Projects formed with unresolved requirements will spawn conflict that will likely stay that way to the end. Good personal relationships become important components of a successful project. From these, the participants can best negotiate future conflicts.

One of the best methods to facilitate communication and productivity is to place the team in a colocated space. Beginning with the post-Charter initiation phase, and following later into the formal planning phase there is a great deal of decision volatility and potential conflict between organizational and technical members as the requirements are being translated into a project plan. During this period, the team should be physically close and have every opportunity to develop personal ties with other members. Relationships built during this period will be vital later in the life cycle when unanticipated issues have to be resolved and will involve tradeoffs.

Private offices are often status symbols in organizations, but that is not the right layout solution for this project phase. A large open workspace called a *war room* has been found to be conducive to more open communication and team building. In this mode, draft documents can be hung on the walls for review and discussion. There should be space for planned visitors to come in for consultation (i.e., future users, vendors, management, etc.). Once the planning cycle is complete and an approved project plan is produced, it is then more feasible for the various work teams to exist external to this environment. The fact is that the size of the total project team often makes an open space arrangement infeasible; however, there is still value in a small core team staying in close proximity for coordination purposes.

Studies performed at IBM during the 1980s compared productivity of knowledge workers in different physical environments. Two factors surfaced from this analysis. First, a closed office environment was conducive to better concentration. Second, proper tools are important. In the case of a project team, this means that some form of quiet office space is appropriate when work is isolated to a single individual. Surprisingly, many companies today use open cubicles as a layout strategy and workers often complain about noise and interruptions. Also, lack of access to an outside window view is also a common complaint. This layout format suggests that organizations are focused on the cost of housing a worker, but may not have given the same level of attention to a productive environment for that worker.

A second factor emerges as the team becomes more physically separated. Where physical closeness resolved the internal collaboration requirements during the early phase, the team is now more separated and the communication process must adapt to that. Email has long been the favored communication tool for teams, but that technology period is now passé. Modern collaboration tools are available to improve the ability of team members to communicate and have appropriate access to technical and management artifacts being created and used by the members. Actually, the process of capturing and distributing project information is important regardless of physical location. Modern document content storage and retrieval technology available through Internet links offers a viable way to accomplish this information access goal. Various topics throughout the text will describe examples of project management artifacts that need to be accessible to team stakeholders.

Since projects are often viewed as temporary organizational activities, the assignment of appropriate physical space for the team is often not given proper consideration. Normally, this creates the obvious problem with team members scattered about. However, several years ago, the author experienced a team location event that was educational. Given lack of office space, this project team was moved to the "old 1940s Quonset building out back." This building had only open space inside (no walled offices). Surprisingly, this move turned out to be a saving strategy for the project. The team was *stuck* out in back together for over a year and over time came to view the hut as their home. This isolation from other organizational units made the group very cohesive. Because of this they learned to respect each others opinions and all developed a common view to the project goal. As a result, this very abstract and political project delivered a result that most likely would not have happened if the players had stayed in their original offices and communicated only at planned meeting times. The key for the PM is to find a way to create this type environment regardless of the physical location situation for the team.

Projects represent one of the best-known organizational breeding grounds for conflict given their complex interrelationships among the organizational groups. Early conflict can occur over the interpretation of exactly what the project is meant to accomplish, the desired time schedule, or whether the stated vision is technically achievable. Later, that conflict can evolve into one of resources or technical options. Regardless of the source, one way to decrease this is to have good communication processes among the various groups. Communication theory suggests that up to 90% of conflicts are the result of misunderstanding between the parties. Much of this can be better dealt with if the individuals or groups have good relationships and can discuss the issues in an open manner. During the early project phases this can be best accomplished by recognizing the role that physical space can have in the conflict equation. The greater the separation of the members, the more difficult these interrelationships become.

6.6   VIRTUAL ORGANIZATIONS

A clear definition regarding what constitutes a virtual organization is still in the formulation stage and therefore varied. Some of the common characteristics of such organizations are that they are geographically distributed and likely linked together via some form of technology-based collaboration tools (often just email). The members in such an organization might have different employers, yet collectively their goal is to operate as a single organizational entity. Customer call centers today are an example of a virtual organization in that they are very mechanically efficient from both a process and technical perspective at handling calls (note we did not say that the customer likes this approach, but it is cost efficient). If you are the customer call organization XYZ you will get organization ABC's call center representative who will answer as "Hello this is organization XYZ." That individual might well be in another continent, but modern Internet or telephone technology makes the geography transparent. This type of organizational virtual fragmentation occurs across many skill and service lines.

As a similar example, a software development project for organization XYZ might employ software programmers in organization ABC located halfway around the world. Similar allocations could occur for a wide variety of virtual work activities needed by the project. One form of a virtual organization involves employees from the same organization scattered across wide geographic regions. A second form involves employees from multiple employers combining to accomplish a project goal. In this case, the relationship formality must be somewhat more rigorous. In these situations, formal work specification and competitive bidding-type actions are generally required. When the project team members are from the same organization but geographically dispersed, the required formality of the relationship is somewhat less, but the basic management issues are quite similar.

A third hybrid model of a virtual organization is the geographical division of work by the time zone, "chasing the work around the sun." What this means is that three geographically dispersed teams will ship their work at the end of each day to the next location to the West. So, the work might migrate every eight hours from Texas to China to Russia and back to Texas for the start of the next day. In an environment such as this, there is tremendous productivity potential since the same unit of work is being processed for 24 hours, rather than the typical eight. This is a doable model in some cases, but obviously a very difficult work coordination process. Work models that have described this have been information technology projects doing design, coding, testing, and user acceptance. Each geographic group serves one of the roles in this linkage, so small items are routed through the chain each day as the local unit closes.

Virtual organizational relationships are on the rise as companies continue to find ways to partner for common good. In each case, the management and collaboration issues resulting from these arrangements are challenging to the PM.

6.7   ORGANIZATIONAL CULTURE

Essentially, every project exists in an organization and the project work environment is impacted by that external culture. Ideally, the project team needs support from the organization. However, in many cases, the team feels that they are on a lost island alone, trying to survive. So, organizational cultures can be either supportive or restrictive to the project. It is important that the organization support the project in every way, since this initiative has been chosen to further some organizational goal. The services of the host organization should be made available as needed. Also, various stakeholder and management groups should strive to support the effort.

Most organizations are structured according to skill specialties—engineering, manufacturing, information technology, marketing, legal, and so on. This structural form is defined as a functional organization and some would say that this structure is dysfunctional for the project. In such structures, the functional roles are so inbred over time that they become micro organizations of their own and take a very independent view of their internal resources, goals, and so on. A name given to this culture is *stovepipe*, meaning that each functional entity tends to live in its own space. Another popular metaphor for this is a *fiefdom* with its *Tribal Chief*.

Project processes basically work horizontally through the organization structure in its need for a diverse set of resources and this requires the functional culture to be supportive. The most noticeable collaborative example between the function and the project is the flow of resources. Failure to achieve a reasonable allocation of quality and quantity of resources for the needed functions will sabotage the project almost certainly. Such support can either come voluntarily or through some formal authority relationship at a higher level. One of the cultural challenges is to find a way for the functional organization to view the project as a partner in the effort.

The concept of culture is also a relevant term for the overall organization. Ideally, the operational mode for top management is to be a supporter of the project and not a control overseer, but that is often a difficult tightrope to walk. If project is no longer fulfilling its need, then top management has the obligation to shut it down. However, if the project is struggling and needs help with priorities, resources, or refereeing then top management should be prepared to step in and save the project. We have previously seen in the Standish survey success criteria (www.standish.com) that lack of senior management involvement is one of the two top reasons for project failure. That role is part of the cultural need.

Organization culture is also like family. It can be harmonious or combative. The more organizational maturity that the organization has, the more likely it is to produce a positive environment for the project. Part VIII of the text discusses various contemporary strategies for producing a mature organization that is supportive of the project role.

6.8 SUMMARY

Organizations are complex collections of human beings. The project organization is housed inside of its host enterprise envelope and draws much of its energy from that mother organization. Within this structure the project's physical team organization structure is important, but the physical organization is not the entire picture. Other related factors such as formal authority processes to achieve timely decision making, conflict resolution, resource allocation, and general skill availability impact the project organizational perspective. In an ideal case, the PM wants to be able to have access to needed skills exactly when he needs them. He wants the team member to be motivated to accomplish the defined work as specified (scope, time, and budget). Similarly, he wants timely management decisions on any issue he raises. Finally, in this ideal heaven, he wants everyone to rush to his aid when Murphy comes calling (i.e., things will go wrong at the worst possible time). That, then, is the organization that we need to engineer for success.

REFERENCES

PMI. 2008. *A Guide to the Project Management Body of Knowledge (PMBOK® Guide)*, Fourth Edition. Newtown Square, PA: Project Management Institute.

Putnam, Doug. Spring, 1997. "Team Size Can Be the Key to a Successful Project," QSM. www.qsm.com (accessed June 15, 2005).

7 Project Life Cycle Models

7.1 OVERVIEW OF PROJECT METHODOLOGIES

Formalized project management methodologies used today are varied in form and discipline. In some cases, PMs view a standard methodology as impractical and bureaucratic, relying on their gut instinct when it comes to managing their projects. In these situations, use of a predefined methodology is viewed to be too complex to use in real-world projects and PMs look for their own shortcuts. For these reasons and more, there is a proliferation of varieties of approaches to lay out project work with no one technique broadly accepted as being better than another. Like so many items in the project world, individual beliefs as to the best techniques take on religious overtones with little real substance to back up ones firm beliefs.

The bias here is that the increased complexity of projects today requires much tighter integration and innovativeness than ever before. Likewise, companies require more out of their projects than ever before. One way to achieve a more efficient result is to adopt newer, swifter, and "lighter" project methodologies while recognizing that project life cycles do have a great deal of similarly. This strategy is required to facilitate sharing of information across a "virtual project enterprise." So, what is this methodology?

> A methodology is a set of guidelines or principles that can be tailored and applied to a specific situation. In a project environment, these guidelines might be a list of things to do. A methodology could also be a specific approach, templates, forms, and even checklists used over the project life cycle. (Charvat, 2003, p. 3)

A particular methodology approach can be applied to a cross section of industries, or can be customized to reflect a specific project environment within a single organization. The secret to success for any methodology is that it contains solid, repeatable processes that serve as the foundation for a successful project initiative, supported by sufficient documentation. Attributes of such processes include (1) repeatable best practices, (2) consistency of results, (3) clear work definition, and (4) a quicker path to results. By selecting a more nimble methodology, or by cutting back on customized documentation heavy requirements, PMs can speed up the planning and execution processes. It is not uncommon for an organization to decide that a purchased methodology will solve all of their project problems only later to drop the idea because it did not accomplish the desired results. Methodologies have the potential of helping guide a project through its life cycle, but they alone are not the solution to project management problems. Organizations must look at their overall company strategy before thinking of implementing a standard or custom methodology. Also, the project participants must believe that it helps them achieve their goal and not a documentation overhead to suffer through.

The significant components of a methodology include the following:

1. It defines major project phases and decision milestones used to control major activity groupings.
2. It defines techniques and variables used to measures progress.

TABLE 7.1
Benefits Offered by Project Methodologies

No.	Focus Areas	Supports PM By
1	Effective processes	Defining key processes required
2	Reusability	Using key artifacts from project to project
3	Integrated metrics support	Provides techniques to evaluate project performance
4	Quality focus	Ensures proper consideration for quality management
5	Managing complexity	Provides techniques to help sort out root cause issues
6	Project documentation	Provides templates and aid in producing required documentation
7	Standard approach	Provides a mechanism for cross-project comparisons and simplifies team training
8	Consistency	Pursues projects using a similar approach
9	Project planning	Provides project planning techniques
10	Team management	Guides the project team to completion by defined phases
11	Elimination of crises management	By establishing an improved structure, future crises situations are avoided
12	Training	Supports team training through its formal process documentation
13	Knowledge	Supports lessons learned to improve future projects

3. It defines techniques to take corrective actions based on identified variances from the approved plan.
4. It assists in the resource management process.

An effective project methodology offers a number of benefits to the PM. Their basic role is to help prevent problems from occurring. Many of these are custom tailored to the organization or a specific project type, but underneath all of them is the application of project management principles (Charvat, 2003). Table 7.1 outlines a number of benefits offered by a project methodology and in turn what these benefits allow the manager to do.

7.2 LIFE CYCLE MANAGEMENT PROCESS

This section will introduce some of the key vocabulary that is prevalent in the project environment and then summarize a general life cycle model. Throughout the text, the PMI's *PMBOK® Guide* model vocabulary and life cycle view will be generally followed (PMI, 2008).

One of the values of standard methodology is that it uses a defined vocabulary. In the absence of this, communication among the team participants is more difficult as they struggle for terms to represent various project issues and processes. The vocabulary items below are frequently used and need to be understood in their context.

7.2.1 FEASIBILITY REVIEW

Most projects have some concerns regarding technical, political, resource, or economic feasibility. Feasibility analysis is designed to consider such issues and provide an early assessment of the Business Case vision viability. In some cases, the initial vision is more of a wish list than a feasible proposal, so a follow-on step is needed to ensure that the vision can be achieved and at what cost, time, and risk. The concept of feasibility occurs in more than one step as the formal planning stage crystallizes the requirements. By the end of the planning stage, the feasibility assessment should be sufficient to make a GO/NO GO decision.

7.2.2 PROJECT PLAN

Every project should proceed under the control of a plan customized to its size and complexity. One analogy for this is that you would not consider building a house without a plan, yet multimillion dollar projects are often pursued with only rudiments of a plan. The plan must consist of more than a technical goal and more than time and budget. It must also describe how the work will be performed and other aspects of the knowledge areas (KAs). A key management question for every project is how much planning is appropriate and how detailed should the plan be. Often the pressure to complete the project moves the team into execution before a coherent plan is completed. This move creates added risk and impacts the ability to control the project. According to the PMBOK® Guide a well-formed project plan should contain subplans based on the following eight familiar process components or KAs:

1. Scope
2. Time
3. Cost
4. Quality
5. Procurement
6. Risk
7. Communication
8. HR.

7.2.3 LOGICAL VERSUS PHYSICAL DESIGN

During the early stages of the project life cycle there are various techniques to describe the future view of the product or system prior to actual construction. In the case of a physical product, prototypes or mock-up models are often used (i.e., houses, commercial real estate, airplanes, cars, refineries, etc.) In some industries, this process is very mature, whereas in many others, it is not. In the situation where the project goal is less tangible, the requirements must be translated into what is called a logical design. Logical designs are abstractions from the real product. A blueprint for a house would be a logical design. Likewise, a flowchart or data model for a computer program would be a logical design.

The role of design documentation is to allow user and technical participants to examine the project goal before investing significant resources. Regardless of the method used to document the future project goal some process of requirements evaluation should be undertaken with both technical and user stakeholders prior to moving very far into the physical design activity, or execution. The role of this review is to provide the stakeholders with some better understanding regarding what the final deliverable would be. At this early stage the concern is more on "what" is going to be done and less focus on "how" it would be accomplished.

7.2.4 QUALITY CONTROL AND ASSURANCE

These activities involve the middle ground of the life cycle and are focused on delivering the item (produce or process) that was specified. Formal review points in the life cycle are used to evaluate either the technical or functional aspects of the output goal. Prior to a physical product being delivered, the technical specialist might review the planned general approach, tool selection, or any other technical issues related to the execution process. These reviews move the focus towards "how" the problem will be solved.

As physical output becomes available, the user community should become more involved in the evaluation. The operational goal at these points is to assess how the output compares to the design specifications. In the case of software, this mid-stage evaluation would be code modules that would be compared to defined requirements. Tangible projects would produce subsystems that had similar review characteristics. Reviews related to this segment would have the characteristic of a quality examination by technicians, or a user acceptability review.

7.2.5 MONITOR AND CONTROL

This high-level activity essentially lies across all of the life cycle phases and its role is to measure status and take corrective action as needed to move the project in the proper direction.

7.2.6 PERIODIC STATUS REVIEWS

It is customary to distribute formal project status communication on a periodic basis. The format, frequency, and audience for these reviews are specified as part of the formal project communications plan. The review process can be very formal or casual and occurs at defined points throughout the life cycle (i.e., time periods, milestones, etc.). In some cases, this communication would be performed through a non-face-to-face defined reporting process.

7.2.7 MILESTONE OR STAGE GATE REVIEWS

Beyond the major phase review points outlined above, most projects have some additional requirements for a more formal face-to-face technical, user, management, or other group review. These may occur at key budget cycle points or with visiting dignitaries. One example of a milestone review would be the demonstration of some key technology or pilot prototype performance. In 2008, Boeing engineers wanted to demonstrate that their new carbon fiber wing would withstand a particular level of flexing, so they built a mock-up and actually bent the wing and measured the flex until it broke. This was meant to ensure all (including the future riders) that the design was safe. This would be called a technical review, but it was also filmed and distributed to others, so probably had a dual role.

7.2.8 PROJECT CLOSE

The model recommended by this text is that all projects should enforce a formal closing process. Appropriate stakeholders and the project team should review both ongoing and final status of the project to evaluate both good and bad results. This review is designed to help the current project and future ones gain value from the current experience. This method has been proven to be a viable approach for organizations to improve their internal processes.

7.2.9 PROJECT COMMUNICATION PROCESSES

Various project communications activities focus on some unique physical product or project artifact and involve some specific collection of stakeholders (typically project team, user, sponsor, or management). These communication processes often take of the form of face-to-face meetings that have high potential to waste the member's productive time. Effective communications and problem solving is one of the most difficult goals for the project team to accomplish. Specifically, excessive use of face-to-face group meetings for this purpose is very costly and represents one of the time and productivity robbers that have to be recognized.

During the course of project execution one of the key communication roles is to provide confidence that the project is moving forward in an appropriate manner, or at least to communicate actual status. From a management control viewpoint, it is desirable to not allow the continuation of a project that has fallen below its value threshold. A phase review is the typical time to perform such analysis and at that point either specify corrective action or shut the project down. Runaway projects consume resources that could better be spent on other options.

Note that in the various vocabulary items given above, reviews are given names like stage gate, kill points, milestone reviews, acceptance test, and so on. Certainly the terms "gate" and "kill" have obvious implications to the PM and highlight the significance of this activity. Many organizations approach these key points with the attitude that the project must justify why it should go forward, so the built-in bias is to stop it. Regardless of the local culture, the communication and review processes are vital to the life cycle management process.

7.2.10   Life Cycle Models

There are many project life cycle model designs based on the project's goal objective, complexity, and bias toward planning. Each one has some unique aspect to its structure and phasing. Small projects typically combine some of the steps indicated above and exhibit a lesser level of planning detail and control. The *PMBOK® Guide* would suggest that the various KAs (risk, HR, quality, etc.) always be reviewed before deciding to omit them from formal consideration. Large, high-technology projects such as those found in the military, NASA, or other government agencies often encompass technical state of the art issues to resolve before the project could move forward. In that type of situation there should be strong consideration to the risk and requirements related to the vision. The role of formal technical reviews related to the requirements takes on particular value. Mega-projects spend a significant amount of planning and review time in the various life cycle stages because of broad stakeholder interest and the high level of resources involved. Alternatively, in the case of a lower technology effort such as that found in traditional building construction we would see a very formal logical design review phase, but likely less formality in the risk, communication, and HR planning aspects of the project so long as standard components were defined in the design. Also, the construction and other industries profit from having well-defined quality standards that can be easily specified in their requirements and subsequently implemented. This is a lesson all industries need to learn.

7.2.11   Templates

Organizations typically develop standard project templates for various types of project and these provide a good starting point for laying out a reasonable approach to a new project. There is no universal set of templates that will fit all projects, but the use of this strategy can save a great deal of time in the various management aspects of the project (See Appendix B for further discussion on templates). *Recognize that across all projects, there is much more similarity of processes than most recognize.* The argument that a particular project is unique is often a lack of understanding of basic project management principles. This point is one of the most important items to take away from this section. Simply stated, project management requirements are not that unique even across industries. The technical sequence of tasks is certainly different, but the overall management processes are basically fundamental. That thesis is stressed throughout the text, but not always accepted by some practitioners.

7.3   KEY PROJECT MANAGEMENT ARTIFACTS

For this discussion, let us define an artifact as some formal management-oriented document that results from actions internal to the project work activity. This collection of work product is designed to support the management and technical processes as they evolve through the life cycle. The list of items described below is not the complete summary, but provides a good general overview of the basic project management artifacts and processes occurring in the life cycle. More specific examples of these will appear throughout the text in their respective discussion areas.

7.3.1 Initiating

The two key artifacts that emerge from this group are the project Charter that formally authorizes the project and a follow-up descriptive scope statement called the Preliminary Project Scope Statement. These artifacts are designed to address and document the characteristics and boundaries of the project and its associated products and services, as well as the methods of future acceptance and scope control. The scope definition will be more clearly defined during the planning stage.

7.3.2 Planning

A great deal of formal documentation is produced in the various planning activities. First, each of the nine KAs would be documented in a related management plan outlining how that aspect of the project would be managed. The most well-known examples of this would be the cost and schedule management plans. Through an iterative process, each of the KA plans would be meshed (integrated) with the others until they are compatible with each other (i.e., HR, cost, schedule, risk, procurement, quality, etc.). The formal term for this is plan integration and the resulting project planning document includes all of the respective KA views for the project. This integrated plan would then be presented to management for approval. If approved, this establishes a baseline plan used to manage the project going forward. As changes in any of the KA elements occur the related artifacts would be updated so that the project plan remains a living document through the life cycle.

7.3.3 Execution

Formal management documents produced during this activity group relate to status information regarding quality assurance, human resources, procurement, schedule and cost tracking, and formal information distribution to stakeholders. The project management mantra for execution is to "work the plan." This means to use management skill to influence a successful completion to the effort.

7.3.4 Monitoring and Control

As suggested by the title of this activity, there is a strong orientation towards control based on measurements that compare actual performance to the approved plan values. From these measurement activities, corrective actions are defined. In addition to this, there are formal activities related to scope verification from the customer viewpoint and operation of an integrated change control process designed to ensure that changes to the plan are properly approved and processed.

7.3.5 Baseline

This term generally refers to originally planned and approved baseline parameters. When more than one baseline parameter is defined, a modifier will be added (e.g., cost baseline, schedule baseline, performance measurement baseline, technical baseline, etc.) (PMI, 2008, p. 419). Baselines represent critical comparative measuring points for project performance. Management can decide to approve a new baseline value when some major event changes the overall scope of the project, or even when formal changes in scope are approved. In this context the baseline simply means the current approved value; however, some organizations require that the original baseline be maintained as a constant to highlight the amount of scope creep during the life cycle of the project. Theoretically, the baseline should change with each approved change request, but that is not the normal situation.

An important conceptual point to understand is that the various KAs interact throughout the life cycle and decisions made in any one of them can affect one or more of the others. Also, recognize that there are multiple life cycle models and these are often given names to indicate their underlying assumptions (i.e., waterfall, agile, scrum, spiral, iterative, etc.). Each new dialect will attract some level of followers who believe that particular structure is the correct one. The *PMBOK® Guide* makes an attempt to not take a position on sponsoring any particular life cycle, but it also evolves with updates approximately every 4 years, along with other subsidiary documents. This simply verifies that the art of project management is still evolving towards a pseudoscience and better understanding of the underlying principles occurs each year. This chapter is loosely structured around the tenets of the PMI model collection as we believe it is the best theoretical view and most recognized definition of this topic internationally. PMI documentation can be found at www.pmi.org.

7.4 PROJECT METHODOLOGY MODELS

When organizations describe their *methodology*, they are implying that there has been some formal thought and standardization regarding the technical and management approach to be used by their projects. This is a general term and there can be a methodology to operate a lathe in the machine shop or to execute a particular type of project. The term is prescriptive in nature and in the project management domain of this implies something akin to a standard that is formally supported by management. In fact a methodology without this characteristic is generally worthless. There used to be a common term used by computer types that said "the nice thing about standards is that we have so many to pick from." Pause for a second and think about that statement. It often applies to methodologies and unfortunately waters down some of the value of pursuing the initiative. The *PMBOK® Guide* is in fact a high-level project management methodology model sponsored by PMI (PMI, 2008). In other cases, published methodologies are most often defined by industry or consulting organizations to fit their general class of problems, organizational culture, and other such factors. For many industry types, very robust methodologies are for sale by third parties; however, buying a commercial methodology and installing it unaltered in an organization would typically result in minimal or nil value. To be successful there is a required stakeholder buy-in and customization to this type of standard that is necessary. The customization process helps supports the organizational buy-in and generally increases the odds of success. A standard model has the advantage of supplying an extensive pedagogical and documentation base to evolve from. Regardless of the evolutionary process chosen, organizations need to consider making ongoing investments in improving their standardized process methodology. Recall from the previous project success survey discussion that a standard methodology was one of the top 10 factors leading to project success. The caution here is that an organization cannot just create a methodology and send out the documentation for team members to read. This must become part of the culture, as well as a formally communicated and ongoing process supported by management with input from appropriate stakeholders. Also, it has to make sense to the project teams who will use it. A project team member once described his required methodology as "feeding a dinosaur." The implication was that it required a lot of feeding and did not bring added value. If that is the perception, it will be hard to sell the process. Project team members are under too much time stress to want to do something that does not appear to help them. This same criticism is common for project management in general and educational efforts are important to show why doing activities that appear to be wasted overhead do in fact add value. There is a fine line to be derived between good project management and excessive overhead. Only with enlightened professionals can the proper balance be achieved.

Part of the evolutionary process for successful business processes is to document past successes and continually redesign the operational processes in an attempt to optimize the next iteration. Projects have both a technical process component and an intertwined management layer. The technical components are represented by the task sequence necessary to produce a product, while the associated management tasks add management visibility and control related to those processes.

As an organization matures their methodology usage, they will tend to add processes related to the various KAs represented by the PMI model. For example, the maturation process highlights the value of formal management related to areas such as risk, communications, and quality that would have been essentially ignored previously. This statement is particularly true related to risk management activities.

One stimulus for adding additional review areas would be the negative experience for not doing them and then suffering the negative impact later (i.e., risk assessment, procurement management, etc.).

An earlier example illustrated a general set of activities relevant to the project life cycle. Many organizations have taken this type of view and customized a task list surrounding that phased approach. This model has even been given the name *waterfall* to describe how the various phases progress in more or less serial fashion. We also described how this same phased view would generally fit many other project models. Given the definition of a project, all methodologies should define a formal starting point (initiating) and a closing process regardless of how the internal work processes are connected. The middle of the life cycle will contain activities related to planning, execution, and control in some form. Logically, some level of initial planning is required and the execution phase represents the core product delivery activity for the project, so much of the technical work sequence would be defined there.

Monitoring and control strategies are more philosophical in nature in that there are varying opinions as to how much of this should be done; however, some form of this should be added to the basic workflow. Very few projects can exist without a formal status reporting activity. The *PMBOK® Guide* is very rigorous in this area, more so than most organizations follow. In addition to these broad process groupings, the concept of defined phases is fundamental to the management control process. Even in a low technology, low-risk example of a prefabricated house construction project there is a need for some defined phase management review. In this example, the phases may need to be established simply to confirm that the foundation meets specifications, or that the wiring and plumbing are installed properly before allowing the process to move forward. For more complex projects, the phase reviews become much more rigorous and include the complete gamut of project status parameters. For this more complex class of project the phase review often is dealing with the decision to kill the project, rather than just being sure that the previous step was done properly. Note in these two diverse examples the management role is to properly approve what needs to be done for successful completion and then monitor that definition and status through the life cycle. In the house construction example, the primary issue was to execute the technical sequence using industry standards, while in the high-technology effort all aspects may be of both technical and management concern. Thus we recognize that a single standard methodology would not fit both of these project models, even though the general life cycle sequence of them is somewhat comparable.

There is broad strategic organizational value in defining a standard methodology. The most obvious value is the savings related to having to reinvent a basic technical and management sequence for each project. It is much easier to tweak a standard tested model and use most of it than to think through a complete project plan each time. The standard model will have been tested for validity and the workforce should be knowledgeable regarding its processes and vocabulary. Also, management will have approved the related control aspects. Personal experience suggests that this form of project execution can provide significant productivity benefits. All of underlying rationale related to the role of methodology standardization is important to the PM and he needs to understand its value and limitations. The challenge is to motivate the project team and organization to follow something akin to a standardized approach; however, if the organization views the PM to be a maverick with wild and nonproductive ideas, he will be judged as a nonteam player. This is not the attribute that one wants to exhibit in a PM role. Realize that this is not an engineering mechanics task, but involves the additional complexities related to the associated organizational and human culture side.

7.5 SUMMARY POINTS

The basic operational values of a standard methodology have been outlined here. Realize that the various defined activities in the methodology create specific artifacts such as a scope or a project plan. Methodologies have the potential to increase operational productivity through a common set of vocabulary with which to communicate among the various participants. Also, added productivity can be gained through the use of a template library of tested life cycle plans, work processes, and documentation artifacts (see Appendix B). Having these items in place represents a mature infrastructure for the project. As use of the methodology matures, the associated organizational infrastructure becomes less of a technical reference manual and more of a real work process. As an example, one very mature organization had customized its methodology into three separate project types—basically small, medium, and large. In this fashion the methodology better fit their environment and their project teams had access to a customized set of life cycle activity documentation and templates useful to produce basic plans. Also, the project teams were trained to utilize these materials, and as a result these projects progressed with less stress than was typically found in other less mature organizations. The point to recognize here is that a methodology does not bring value unless it fits the environment and that may not be the case with a standard commercial package.

Commercial methodology models are offered by various vendors. A good starting place to pursue implementation of a local life cycle management process would be to review industry-specific project models from sources such as trade associations and web search engines. Also, organizations such as PMI, TenStep (www.tenstep.com), Method123 (www.method123.com), and other industry organizations have models and data related to this. In addition, several large vendors offer models and consulting services in this area. The important item to realize in regard to project methodologies is that they can dramatically impact the culture of an organization both positively and negatively; therefore, installation represents more than supplying templates and documents and expecting the participants to change behavior. In fact, changing human behavior is the most difficult part.

7.6 POTENTIAL SHORTCOMINGS OF PROJECT METHODOLOGIES

Though project methodologies offer a number of benefits, they have potential shortcomings too. It is equally important to understand where methodologies can possibly go wrong. Many would characterize traditional project methodologies as follows:

- Too abstract and high level to be translated into workable processes.
- Not addressing crucial areas such as quality, risk, communications, and so on.
- Ignoring current industry standards and best practices.
- Looking impressive but lacking real integration with the business.
- Using nonstandard project conventions and terminology.
- Not having appropriate performance metrics.
- Taking too long to execute because of embedded bureaucracy and administration.

An effective project development methodology is not just about focusing on life cycles, but also about shortening a company's strategic goal delivery life cycles. No matter how efficient a company is, it needs to adapt constantly in order to maintain a successful market share. In today's format, we see the current project strategies as representing a project management compromise in trading off risk for time. For this reason, one should not have a fixed view as to how they will need to proceed toward project success. There clearly is a point that some projects fall into "analysis paralysis." This means that the project studies the effort to death without ever moving into execution. In this case the effort consumes time and resources without delivering a usable product. If done properly, proper planning supports successful delivery, but it is also necessary to move forward. Knowing when to do this is an art form for the PM. Management and users will typically be pushing to move too quickly. Keep these tradeoff concepts in mind as you struggle with the variables.

REFERENCES

Charvat, J., 2003. *Project Management Methodologies: Selecting, Implementing, and Supporting Methodologies and Processes for Projects.* New York: Wiley.

PMI, 2008. *A Guide to the Project Management Body of Knowledge (PMBOK® Guide),* Fourth Edition. Newtown Square, PA: Project Management Institute.

8 Quick Start Example

8.1 PROJECT MANAGEMENT WORK PACKAGES

Envision for a moment a "package" of work that has been defined as part of a larger project. This could be things such as installing piping, testing software, or producing a user guide. For now, view this as a defined set of work activities associated with that of a larger project effort. The vocabulary term for this is a work package (WP). One metaphor for a WP is to view it as a box with the three dimensions representing time, cost, and scope (of work). Figure 8.1 illustrates this as a three-dimensional box with inputs and outputs.

In order to develop a plan to execute the contents of this box, we must make certain planning estimates required to produce the needed outputs. The goal of this section is to demonstrate some of the basic inputs needed to derive an initial schedule and budget for this element of our project. As a starting point, we will formally define a WP as a *deliverable* or *project work component* at the lowest level of each branch of the *work breakdown structure* …. Sometimes called a *control account* (PMI, 2008, p. 445).

At this point we have not seen how a WP fits into the full project structure, but note that it represents some deliverable component of the total project. Some organizations further specify that such WPs should represent approximately 80 hours of effort and/or 2 weeks of work; however, that view could change with different types of project. From a work definition perspective, this item becomes a management focus point. The most visible outcome of examining a defined WP is to see that it contains an estimate of resources, duration, and cost. In other words, we want to estimate how long it will take to do the work, what resources are required, and how much that would cost. Time estimates can be derived using many techniques and we will focus on more specific details on those mechanics later. For now, assume that a "time guru" comes up with the exact answer. Budgeting is related to time in that one of the major cost components is the labor associated with that time estimate, plus any material type resources involved. This now represents an incremental estimate for the entire project resource, schedule, and cost variables.

8.2 WP DICTIONARY

Effective planning requires many pieces of information related to the activity in order to assess the content of the work and integrate it into the rest of the project. Conceptually, we might capture this information into a single repository that we will call a work breakdown structure (WBS) dictionary here. Such a repository may not be visible by that name in real-world projects, but the type of data described below will need to exist somewhere in the project records. In order to create and track the project, it is necessary to define the following types of definitional parameters for each WP.

1. ID reference—a code used to identify where the WP fits into the overall scheme of work (e.g., the WBS).
2. Labor allocation for the task (e.g., number of workers).

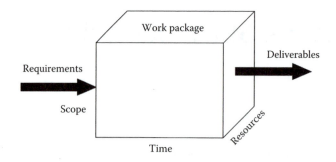

FIGURE 8.1   Project WP.

3. Estimated duration for the task given the worker allocation.
4. Materials associated with the activity and their cost.
5. Personnel responsible for managing the activity.
6. Organizational unit assigned to do the work.
7. Defined constraints (e.g., activity must be finished by …).
8. Key assumptions made in the course of the planning activity.
9. Predecessors—linkages of this activity to other project work.
10. Risk level for the activity—more work will be needed on this aspect later; for now we just need some indication of the perceived risk level (e.g., H, M, or L).
11. Work description—this may be a link to a location where such information is kept.
12. General comments—free-form statements that help understand the work required.

In addition to these core definition items, it is also necessary to have certain management oversight into the estimated values derived. In order to control these values, it is common to have additional approval fields in the WP record showing management acceptance by the performing organizational group and the project manager. With this set of information completed, we have sufficient information to produce a first cut schedule and budget for the WP. This statement is a clue that there might be more iteration later, as new information come to light.

From the dictionary estimates for work and resource allocation, WP duration is calculated. For example, if the work is estimated as 80 hours and a resource allocation is of two workers, then the number of work days (duration) would be 40. However, from this, we still do not know exactly what calendar time frame the group is working on. Workers will have scheduled time off typically on weekends and holidays; and other days allocated to vacations and nonproject company activities would extend the overall schedule beyond the raw calculation. All of this indicates more messy mechanics to deal with later and fortunately there are automated tools to handle this class of estimation from the point defined here.

Multiplying the estimated duration times with the labor rates for the defined resource types or individuals identified derives a labor budget for the WP. Material costs are added to this to produce a direct budget for the WP. Once again, there are more pieces and parts to complete an overall project budget and we will look into that later after we see the entire project elements put together.

8.3   MULTIPLE WPs

Moving one step up the food chain in our WP theory, let us see what happens when we admit that there are two WPs. For this example, we will assume that these two WPs are related, meaning that two steps are required to produce the desired outcome. Further, let us assume that the two activities are to be executed in serial form. This is called a finish–start relationship. Metaphorically, we are stacking our two boxes end to end, as shown in Figure 8.2.

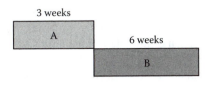

FIGURE 8.2 Multiple WPs.

For a starting place, let us assume that all workers involved have the same skill levels and can do all tasks required. They are also paid the same. These handy assumptions not only provide an easy way to duck many messy issues in the management process, but they do simplify the focus on the basic raw mechanics of scheduling and budgeting.

Given the simple two-box (WP) picture outlined above the project duration would be calculated by adding the two individual durations. So, if WP A has duration of 3 weeks and WP B has duration of 6 weeks, the calculated duration for the set would be 9 weeks. Simple, huh? One should be beware of this simplistic arithmetic however as there is a difference in duration and elapsed time. The work force typically does not work 7 days a week and with time off the elapsed time is generally always greater than the duration estimate. So, we now have three resource variables in the discussion:

1. Effort/work—the amount of resource allocated to the WP.
2. Duration—the amount of working time required to execute the work.
3. Elapsed time—the calendar time required to execute the work.

We will explore the issue of elapsed time in more detail later.

You might be saying at this point, "What can be complicated with this straightforward process?" Unfortunately, quite a bit of complexity is associated with this. Here are a few samples:

1. Skill variations and motivation of the workers may not be as anticipated.
2. Weather or environmental issues can cause disruptions.
3. Planned resources may not be available on the planned schedule (e.g., resource conflicts).
4. One of the tasks runs into unanticipated problems and overruns the duration estimate.
5. Scope definition errors often cause estimates to be wrong. For example, if task A was specified to dig a hole 10 by 10 by 10 and later it was found that the hole needed to be 11 by 11 by 11, we could see how the estimate might be wrong even if the original estimate was accurate.

The fact is that projects are fraught with such "bumps in the night." Project plans are actually very dynamic. If you ever see a plan that does not change, you can label it "eye candy" because it does nothing to help manage the project.

The discussion above has established a base set of terms and mechanics that can be used to move into the next stage.

8.4 PSYCHOLOGY OF ESTIMATING

Estimating project or WP durations is both an art and a science. It involves skill, history, and judgment to produce reasonably accurate estimates. One can approach the process from either the total project (top-down) or the WP level (bottom-up). A top-down view would normally be estimated by comparing the new venture to a previous similar project. This approach is called analogous estimating. A second approach is to define the effort with forecasting elements derived through extensive statistical studies. So long as the new venture fits the statistical model the effort can be reasonably estimated through the corresponding estimating parameters. This method is called parametric estimating. Finally, if you have ever been involved with craftsmen who do one line of work, it is common to find that they estimate through their personal experience. For example, a roofing contractor will estimate how much it would cost to install a new roof using his estimate based on the size of the roof and his homegrown multiplier. When estimating is done in this fashion, certain rules of thumb are developed through experience. These are called heuristics. One or all of these methods might be employed for a particular estimating task. Each method has its strengths and weaknesses. Conceptually, the bottom-up approach should be the most accurate because it looks into each work requirement in detail; however, this is also the most time-consuming approach.

At some point in the life cycle, a detailed WP estimate needs to be derived, whether that be just a confirmation of a previous number or a refinement. The key value of a bottom-up process is that it presses for a working level buy-in of the number, which would not have been achieved by the top-down approach. The bottom-up approach involves a detailed estimate for each WP and then adding the component parts. Done properly, the bottom-up approach should yield the most accurate result, but is recognized as the most time consuming and assumes that clear definitions of each WP are known. Since scope definition evolves through the planning process, clear definitions of all WPs will not be known early in the life cycle. For this reason, early estimates will have to come from one of the top-down methods. Early accuracy levels are probably in the ±75% range, while refined estimates at the end of the planning cycle might converge to ±10%. In any case, we recognize that these numbers are prone to variation and this becomes part of the management process to control.

8.5 PROCRASTINATION

For everyone who has ever been a student, this term has been practiced. We estimate how long it would take to do a particular assignment, then wait until the last minute (or beyond), and try to blitz the effort. What usually happens is that we find our estimates to be optimistic and crises results. This is such a common concept that it is named the student syndrome. Translated into the work place, it manifests itself slightly different. The typical scenario goes like this:

> Team leader Bill is asked to estimate the effort for his WP. He has experience in this type of work so has a good feeling for the general level of effort, however he also knows about the "bumps in the night" items. His initial duration estimate is two weeks, but he does see complexity and some unknowns, so his approach is to double the original value. He turns in the estimate as four. This will give him some security to not overrun and maybe even make him look good if he comes in early. Later, knowing that his real estimate was four time periods Mr. Procrastination comes to visit—wait until you have to get started just like you learned to do in school. Bill has a lot going on and this is not on his critical path right now. He waits for two weeks, then decides there is no more time to wait and must start work. Sure enough the task takes one week longer than originally estimated and it overruns the four-week time submitted for the plan to five weeks total.

Let us look at what just happened in this example. Time estimates are doubled to protect overruns, but procrastination causes it to happen anyway. More importantly, a 3-week estimate is then transformed into a 5-week effort—a 67% overrun caused by procrastination. At this point we will not offer a potential solution to this phenomenon, but it is too prevalent to ignore this tendency in the management view of projects.

We have summarized four basic requirements needed for a coherent estimating process. These are as follows:

1. Clear definition of work (scope)
2. Motivated and skilled team members (resources)
3. Efficient allocation of resources to WPs
4. Avoidance of the student syndrome (procrastination).

If these issues can be addressed, we have now moved pretty far down the management trail, but do be aware that each of these would be called nontrivial to solve.

8.6   DEVELOPING THE WHOLE PROJECT VIEW

Up to this point, we have looked at low-level components of the project and possibly begun to develop a working management model for that level. Unfortunately, a project consists of hundreds or thousands of such elements. The interaction of these elements threatens to blow up this simple view. It is true that this increased size and complexity clouds the management process; however, the fundamental building block elements consisting of WPs will stay essentially intact. The key additions from this point are WP interactions with each other and the impact of the environment on stability. This is this point for a moment—if all WPs were actually as described thus far, the process of developing requirements, schedules, and budgets, and the overall management process would be relatively simple. The fact that most organizations do not see project success rates much above 50% (many lower than that) suggests that other factors get introduced. Nevertheless, building a solid management foundation is the first step in beginning to deal with the vagaries of the external world. For each one of those that we identify, we need to develop a shield to minimize its negative impact.

It is always difficult to suggest that any one item is the most significant one to resolve in a project, but there is heavy evidence from industry researchers that failure to properly define project scope creates problems later and often leads to decreased success rates.

8.7   PROJECT SCOPE

The *PMBOK*® *Guide* defines this term as "The *work* that must be performed to deliver a product, service, or result with the specified features and functions" (PMI, 2008).

Note that the scope definition is based on the work to be done to accomplish the desired goal and *not* a definition of the product requirements. The product is created from the work. This is a key philosophical point, albeit subtle. In order to ensure that the new requirement set is met, both the users and the builders must have the same understanding of the deliverables. The process of gathering these dispersed specifications is called scope definition and the result of this is often translated into a WBS.

In order to start the scope definition process, we must first have a preliminary vision regarding the ultimate output goal—we would call this a preliminary scope statement. In the initial stage, this would be defined using general English verbiage describing what the desired outcome would be. As this definitional process evolves, it is common to begin trying to put some work structure to this definition—that is, how would the project team go about executing this requirement. If the desired project outcome is a tangible product, then decomposing that product into its component parts and analyzing the work to produce each part would lead to the scope definition. As an example, if we were designing an airplane, the first chore would be to agree on the technical characteristics of the device—speed, height, workload, and other technical factors. From this general description, technicians who were familiar with the requirement would then begin decomposing elements of the plane into logical groupings learned from previous experience in designing airplanes. For example, Figure 8.3 shows how a first cut scope definition for the airplane might be represented schematically.

In this large project example, more detailed WPs defined underneath each of the seven major components and likely other supporting work efforts even beyond that would be needed. We will see more of the WBS mechanics at a future point. At this point, the idea is to show how scope definition is approached and to show its role in work definition.

Scope definition does not occur in one sequential cycle. It requires multiple iterations between various team members to analyze, confirm, and digest the overall requirement. This evolutionary scoping process involves the processes of *elaboration* and *decomposition*. This simply means that the level of understanding evolves and higher-level requirements are broken down into smaller, more manageable work units. That process continues until the entire requirement set is properly defined in workable units. Clearly defined work units, WPs, are defined at the lowest level of the WBS. We will describe more about how this is done in the theory section that focuses on the WBS

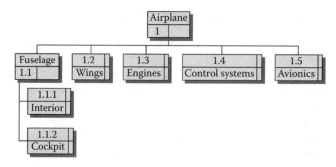

FIGURE 8.3 Airplane WBS.

(see Part 3 of this book). The essential role of scope development is to produce a working definition of the activities to be performed as part of the project. This activity lays the foundation for the project to move forward.

8.8 EXAMPLE: POOL PROJECT MECHANICS

To make the scope development process more visible, we will use a more familiar project example and illustrate how to organize the pieces into a formal schedule and budget. A pool construction project will be used to demonstrate this since everyone has a general idea about how a pool looks like and its general components. Since this is a mature and relatively small project, it is possible to elaborate the whole activity set and not have to deal with the complex interplay of defining what these might be. We will also assume that the WBS structure only requires one layer to sufficiently describe the work. As in previous examples, our goal is to simplify the explanation before opening Pandora's box of complexity later. The sample pool project is envisioned in Table 8.1 to include 11 activities.

In this case, the 11 activities would be equivalent to WPs, since the project is relatively small and there is no need for an intermediate grouping of activities. So, we can represent this project schematically in WBS format as follows:

At this point, we introduce the concept of WBS labeling in regard to the overall structure. In this case, the top label is "1" and the subordinate WPs or activities would be reflected by the decimal notation. So, excavation is 1.1, install electrical is 1.4, and so on. In this case, the labels numbers

TABLE 8.1
Pool Project Work Activities

WBS	Activity	Duration (d)	Cost ($)
1.1	Excavation	5	6000.00
1.2	Reinforcing structure	3	5900.00
1.3	Install piping	4	3800.00
1.4	Install electrical	3	3900.00
1.5	Blow pool concrete shell	5	13,500.00
1.6	Install decking	5	6000.00
1.7	Install pumps and blowers	4	4200.00
1.8	Install filtration system	2	3600.00
1.9	Install landscaping	5	6000.00
1.10	Checkout system	1	600.00
1.11	Final approval	0	0.00

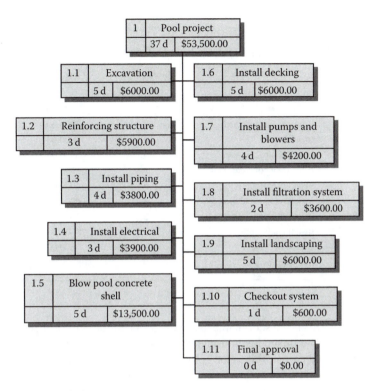

FIGURE 8.4 Pool project WBS.

range from 1.1 to 1.11. Note also that the total pool duration and cost is a summation of the work units at the level below. This shows that the project would require 37 work days and a direct cost of $53,500. In other words, all direct costs for the project are contained in the subordinate WPs. In this fashion, the WBS defines the scope of the effort. All work would be shown here and if not, it is not in the project scope. There are some additional idiosyncrasies to this idea, but this is good enough for our introductory example (Figure 8.4).

From this point, we can begin the translation of scope into a project schedule. In order to do this, we need to define the following five data elements for each of the 11 WPs:

1. Activity name
2. Activity ID (WBS code)
3. Duration estimate for the activity
4. Cost estimate for the activity (from resources allocated plus material costs)
5. Predecessor links to other activities.

The only new idea on this list is the process of defining predecessors. Basically, the values shown in this column simply relate to how the various activities link to each other. For example, a "3" indicates that the activity links to activity number 3. To illustrate this process, we restate the 11 pool project activities and complete the definition for the five scheduling attributes. This yields a project activity definition listing as shown in Table 8.2.

An explanation of the role and meaning of each column is summarized as follows:

Number: A sequential line number for each activity. This is used to identify predecessor links.
WBS: This is the code described above and is a WP reference used in various places throughout the project.

TABLE 8.2
Pool Project with Precedence Relationships Defined

No.	WBS	Activity/Task	Duration (d)	Cost ($)	Predecessor
1	1	Pool project	37		
2	1.1	Excavation	5	6000.00	
3	1.2	Reinforcing structure	3	5900.00	2
4	1.3	Install piping	4	3800.00	3
5	1.4	Install electrical	3	3900.00	3
6	1.5	Blow pool concrete shell	5	13,500.00	5, 4
7	1.6	Install decking	5	6000.00	6FS + 5d
8	1.7	Install pumps and blower	4	4200.00	7
9	1.8	Install filtration system	2	3600.00	7
10	1.9	Install landscaping	5	6000.00	9, 8
11	1.10	Checkout system	1	600.00	10
12	1.11	Final approval	0	0.00	11

Activity/task name: This is the short name used to describe the activity. Longer names will be defined elsewhere and linked here through the WBS ID code number.

Duration: As indicated earlier, this value is the estimated total activity working time, given an assumed staffing level. A zero duration signifies a milestone or review point.

Cost: This value is derived from the sum of resource costs and other materials required for the WP.

Predecessor: This code is used to link one activity to another. Note in Table 8.2 that activity number 5 cannot be started until activity number 3 is completed. Activity number 7 shows that it cannot be started until activity number 6 is completed plus 5 extra work days. This is called a lag relationship.

There are other predecessor relationship coding options that we will review later, but they are simply modifications of the ideas represented here. From a structural model view, all projects would look this same way. The WBSs would be larger and have more layers and the predecessor relationships would be more complex, but the mechanics would stay the same.

Basically, the precedence linking relationships are used to establish the technical sequence for executing the project and this requires expertise in the technology of the project. In some cases, the sequence specifications are obvious in that holes must be dug before the reinforcing structure is done; however, in some situations the sequencing may be done to optimize utilization of the team, or because it is rainy outside and the inside activities can be worked on. Regardless of the linking logic, the role of predecessor relationships is to drive the activities through the desired sequence, which in turn translates the work into a schedule.

To show that we are making some visible progress toward developing a schedule, let us translate this same data into Microsoft Project (Figure 8.5). Note that Microsoft Project understands the terms outlined above and is able to manipulate them according to its internal rules. The display shown here translates the planning items into a *Gantt bar chart* project schedule. This format is a common one used to show schedules. Some data such as the attached calendar schedule have been hidden for this example, but they are readily available within the software. This diagram describes the sequence of work from the starting point through completion.

It is important to recognize that once the five basic WP size and sequence parameters are defined, the process of generating a project plan is essentially mechanical and can be handled by software. If we can assume for the moment that the planning WP duration and cost estimates are valid, this

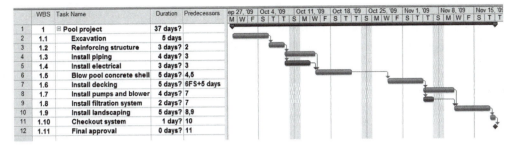

	WBS	Task Name	Duration	Predecessors
1	1	⊟ Pool project	37 days?	
2	1.1	Excavation	5 days	
3	1.2	Reinforcing structure	3 days?	2
4	1.3	Install piping	4 days?	3
5	1.4	Install electrical	3 days?	3
6	1.5	Blow pool concrete shell	5 days?	4,5
7	1.6	Install decking	5 days?	6FS+5 days
8	1.7	Install pumps and blower	4 days?	7
9	1.8	Install filtration system	2 days?	7
10	1.9	Install landscaping	5 days?	8,9
11	1.10	Checkout system	1 day?	10
12	1.11	Final approval	0 days?	11

FIGURE 8.5 Pool project Gantt chart.

process would yield a predictive schedule for the effort. However, before starting to feel too good about this nice looking plan, remember that Murphy's law says that things will go wrong and at the most inopportune time. Nevertheless, this set of mechanics is an important project management step and these planning artifacts are useful tools to verify scope and plans among the various project stakeholders. Likewise, the artifacts created to this point are all useful in communicating vision and intent.

Status tracking: After a project is approved by management as to scope, schedule, and budget, the next and more complex problem is to keep it under control. Status tracking is the basic activity that supports the monitoring and control process. There are three common project status questions that will need to be answered as a project unfolds. First, what was the initially approved project plan? Second, what is the current plan and how does it relate to that original plan? Finally, what are the forecast completion parameters for the project—generally defined by time and cost, but could also include other technical specifications?

Too often, project plans are updated weekly or monthly to show the current status view without regard for the original plan. It is a common omission for project teams to want to disregard the original planned values because keeping them visible would reveal how poorly the project has progressed to date. Another common view of status tracking is that there is none. The original plan is posted on the wall and not changed or monitored. Worse yet, it might be changed regularly with no recognition that this was not the intended plan. Good project management will monitor the ongoing status of the project as compared to the approved plan. Moreover, it will make periodic assessments of forecast completion status. The operational philosophy is that status communication must honestly reveal three status situations—original approved plan, current view, and future forecast. And it must do this in near-real time—that typically means weekly or monthly cycles as limited by the availability of actual tracking data from internal systems. There are many techniques to aid in this process but the most important step, regardless of the method, is to obtain initial scope approval from key stakeholders and then gain appropriate management approval for that plan. This process attempts to clearly define time, cost, and scope parameters for the project. Once that is done, the project is ready to move into the execution phase (*Note:* we are still ignoring some other planning activities such as risk assessment at this point). The final planning step then is to freeze the approved values for use in future status comparisons. This set of values represents the *original baseline* or *baseline.*

8.9   QUICK START WRAP-UP

The quick start overview has attempted to show a summary of key issues involved in basic project scope, time, and cost planning steps. In reviewing this process, several simplifying assumptions were made to focus on generating a complete first cut schedule and to summarize the general process for moving from scope to time to budget. More on this process will appear in various other sections of this book. These simple examples do portray the general process, but do not illustrate all of the complexities that a robust example would need to deal with. That said, we now need to start over and begin moving the puzzle pieces back into a proper structure.

At this stage, you have a reasonable mechanical understanding regarding how a project goes from the fuzzy vision stage to defined tangible activities and on into a schedule and budget. The vocabulary terms shown below represent key pieces in this discussion. If any one of these terms are not clear from the quick start examples, you should mark that term and either go back to review this section or watch for further explanation in subsequent discussions. The quick start vocabulary list is as follows:

1. Activity
2. Work package
3. WBS
4. WBS dictionary
5. Predecessor
6. Duration
7. Elapsed time
8. Baseline
9. Scope
10. WBS code
11. Gantt chart
12. Microsoft Project.

DISCUSSION QUESTIONS

1. What is the difference in effort, duration, and elapsed time? What would happen to these three parameters if the allocated resources were cut in half?
2. If a WP represents an item that has to be completed to execute the project, why would the total project schedule not be the sum of all WPs?
3. What is the difference between a WBS and the WBS dictionary?
4. Define the five basic items necessary to generate a schedule.
5. Name some reasons why a WP estimate might be wrong.
6. How does procrastination affect a schedule?
7. What does Murphy's law have to do with management problems?

REFERENCE

PMI, 2008. *A Guide to the Project Management Body of Knowledge (PMBOK® Guide)*, Fourth Edition. Newtown Square, PA: Project Management Institute.

CONCLUDING REMARKS FOR PART I

This part of the text has described many base level concepts related to the project management historical background and operational role. The historical evolution of this activity can be traced back through several hundred years; however, the mature phase of project management has occurred over approximately the past 100 years. Our brief overview of that history is used to show how the thought processes evolved and continue to evolve today.

Examples are used here to show the general order of evolution for projects as they move from initial vision on through to closing. At this point we know that projects have significant similarity, regardless of the industry or type. There is some disagreement regarding how best to move a project through its life cycle, but little disagreement about the underlying KAs involved in that process. The PMI entered this scene during the 1960s with the goal of defining best practices for project management and they are pre-eminent in this role in their publication of the Project Management Body of Knowledge, called the *PMBOK® Guide*.

Important vocabulary and concepts are introduced in this section that will be used throughout the remainder of the book. Future parts of the book will take many of these introductory items and elaborate them to greater detail. It is important that the reader should not get lost in the details of the upcoming lower-level discussions and forget what the overall goal is—to accomplish the desired project goal as approved by management and in support of organizational objectives.

Throughout the text there will be many examples of projects that failed at near catastrophic levels. The key question in each of these examples should be to examine why that occurred given the visible facts. Just realize that no textbook model will make a poorly conceived project successful. The modern PM must use every trick in his tool kit to overcome the myriad of negative issues that can exist to create failure. It is important to understand that the model view outlined here is a core knowledge requirement to begin this process. The models illustrated here have been developed by distilling the experiences of thousands of professionals. They are not pure academic models that are untested in the real world.

Part II

Projects as State Change Vehicles

LEARNING OBJECTIVES

PROJECT SELECTION GOALS

1. Understand the role that projects play in organizations
2. Understand the concept of selecting from a portfolio of project initiatives based on value alignment with enterprise goals.

PROJECT INITIATION GOALS

1. Determine and document project goals by identifying and working with project stakeholders
2. Determine, describe, and document products or service deliverables by reviewing or generating the scope of work, requirements, and/or specifications to meet stakeholder expectations
3. Identify, document, and communicate project management process outputs by selecting appropriate practices, tools, and methodologies to ensure required product/service delivery
4. Identify and document project schedule, budgetary, resource, quality, and other constraints through coordination with stakeholders
5. Identify and document project schedule, budgetary, resource and other assumptions by determining information that must be validated or situations to be controlled during the project in order to facilitate the project planning process
6. Define the project strategy by evaluating alternative approaches in order to meet stakeholder requirements, specifications, and/or expectations
7. Identify and list performance criteria by referring to product/service specifications and process standards
8. Identify, estimate, and document key resource requirements

9. Define and document an appropriate project budget and schedule by determining time and cost estimates

10. Provide comprehensive information by producing a project Charter document indicating formal approval of the project's planned scope, schedule, and budget.

Source: Initiation goals adapted from the PMI, *Handbook of Accreditation of Degree Programs in Project Management*, Project Management Institute, Newtown Square, PA, 2007, pp. 17–20. Permission for use in this format granted by PMI.

Formal projects serve an increased role in most organizations today. These transient aggregations of skilled resources are used to transform organizations from one operational state to another. This part of the book focuses on the role of projects in the organization and the initiation process.

It is now understood that projects should be spawned in support of organizational objectives. As simple as this concept sounds, it is found much less pursued in reality. Technology projects in particular are found to be difficult to conceptualize and deliver. The history of project failures is marred by organizations that blindly jump into a technology-based effort only to find that managerial trauma follows. Conceptually, a project emerges from some vision that has a positive promise for the organization. From this point, the project life cycle process moves this vision through a series of managerial and technical approval steps. Through these steps, the basic goal is to attempt to ensure that the vision does in fact help align with organizational goals and create a positive benefit compared to cost. Organizational processes are needed to support the various project activities regarding management, process owners, skilled resources, and material resources. In return, the project team is charged with delivering a working version of the defined vision.

Measuring the future value of a vision remains a challenge today and techniques to accomplish this are still somewhat arbitrary. In 2008, research sponsored by PMI revealed that project management does have value, but probably not as simply defined as previously thought. Also, this research revealed that the role of project management is an aid to this achievement. This extensive research reveals the complexity of the project environment. There is no single method to evolve a project vision and to measure its value. A PMI-sponsored research project found that the value was derived from the following five major sources (Thomas and Mullaly, 2008, p. 188):

1. Satisfaction—user-based view
2. Alignment—organizational perspective
3. Process outcomes—improved process efficiencies
4. Business outcomes—efficiency or profitability of the organization
5. Return on Investment—cost-benefit view of the project.

Evaluation of the future value is not easily performed at this point in the project life cycle. Also, experience shows that the future project value is often not as initially predicted. In some cases, it is not as good and in others it yields a higher and major competitive advantage for the organization. Clearly, the ability to accurately predict the outcome of a project is still a fragile art. This statement includes the future use of the outcome product and the cost and time of producing the output. Both ends of the project value chain have significant variability; therefore the concept of value calculation remains foggy. There is a lot to be learned from the PMI research, but still much more in how to predict these parameters during the initiation and planning phases.

The two chapters in this part will explore these selection and valuation issues in more detail. As you read this material, remember that the concept of value can have multiple dimensions and is certainly more than a simple financial metric.

REFERENCES

Thomas, J. and M. Mark, 2008. *Researching the Value of Project Management*. Newtown Square, PA: Project Management Institute.

9 Role of Projects in the Organization

Each organization will have a somewhat unique role for projects based on their business type and level of maturity. This issue can be looked at in two parts. One part involves how a project's value is defined and the second part deals with how project targets are created and pursued by the organization. Both views have undergone significant evolution over the past 20 years. Regardless of the underlying mechanics, formalization of project roles in the enterprise has been increasing over the past several years. Most organizations now look at the project model as the preferred operational approach to pursue changes in their products or processes.

9.1 PROJECT VALUATION MODELS

The process for selecting project targets within the organization remains quite diverse across various industries. Traditionally, these decisions were driven within localized business entities; however, more recent trends have moved toward a more centralized view of project selection and management (Chapters 32 and 33 will discuss these trends in more detail).

While it is obvious that projects dominate the landscape of most organizations, the rationale for creating them is not so obvious. It would seem reasonable to conclude that a project is created to pursue some credible organizational goal, but what goal? In reviewing the typical stated benefits for a project, it is common to hear such attributes as follows:

- Achieve competitive market advantage
- Cut operational costs
- Satisfy compliance requirements
- Achieve a strategic objective (that may not be financially quantified)
- Improve employee morale.

In the traditional view a project was often chartered to improve an existing process. For example, automating a payroll system might be justified by cutting labor cost, improving preparation cycle time, or cutting processing errors. In many such early cases, an attempt would be made to justify the project by showing how it produced some tangible benefit over its projected life cycle. Technology-based projects often had the characteristics shown for the payroll example—cost, cycle time, and lower processing errors. Justification for such projects was often based on a schematic cost benefit view as exemplified in Figure 9.1.

Using this valuation model, the initial costs would be represented by a down arrow as illustrated in Figure 9.1, while upward arrows would represent the future benefits. Monetary values would then be estimated for all cost and benefit flows over the projected life of the product. From this view, various financial metrics could then be produced to show a value parameter. Typical evaluation metrics used were payback, net present value, internal rate of return, or benefit-cost ratio (see Appendix A for more details). In the case of projects designed to replace human labor, this method was reasonable, but as the breadth and complexity of project goals increased, it became obvious that a valuation model was much more complex than described above.

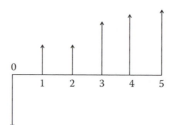

FIGURE 9.1 Project cost-benefit forecast.

A second evolution of the project valuation model added consideration of intangible benefits to the structure, although these considerations were kept separate since monetary values were difficult to derive for this category. The basic problem with intangibles is that various stakeholders might value them differently. Financial types were more sensitive to forecast actual while operational types might be more sensitive to items such as morale, ease of operations, and so on. In any case, recognition of intangibles opened the door to a more complete view of project value. Projects that would have been rejected in the first case using tangible criteria might now be accepted for less financial reasons. Use of classical financial metrics continued and some organizations refined this model to begin recognizing risk and variations in the forecast values. Initially, the level of risk would have been considered intangible in a project assessment. However, failure to consider the downside of risk wiped out many projects that otherwise showed a great tangible financial benefit. Intangibles could also be seen in customer or employee reactions to an initiative. Installing an automated customer response system could show clear financial returns in regard to cost to respond to a customer. However, if it chases the customer away, is this still considered a successful project? Cost effective for sure, but does not help the organization. Intangibles are often subtle, but clearly have to be considered in the valuation process.

A third component of project valuation perspective recognized yet another role of projects in the organization. This view hits at the core of the organizational goal structure. That is, projects should be created to support organizational goals and these goals can be very diverse in nature. Whereas the traditional view of a project was more local in scope, the third evolutionary wave moved that view toward a more top-level perspective. It now became necessary to deal more with linking organizational goals to projects than just evaluating the project itself. The fact is that, some projects might be undertaken with no tangible estimation of future value. Management might decide that some initiative should be undertaken to evaluate the potential of a technology or market. From this view, they would Charter something akin to an R&D pursuit. Some government regulatory-oriented projects have a similar lack of tangible benefit. Historically, high-tech organizations have derived values from projects in quite unexpected ways. The perceived reason to pursue the initiative was not at all where the future benefits came from. The new view of projects seeks to encompass tactical versus strategic initiatives, along with environmental, stockholder, and other perspectives. Clearly, the new generation of evaluation metrics has to include both soft and hard financial justification with organizational goal perspectives linked to the project objectives.

The concept of linking project roles to organizational goals remains a fundamental aspect of project value measurement; however, the enterprise view of value measurement continues to mature in its perspective as we will present shortly. Expanding the view of project value to include multiple attributes leaves behind the idea that some quantifiable metric can be used to compare one project with another. In reality, the valuation activity has long been political in nature and that situation has grown as the cost of projects increased and the level of review expanded to top management. Today, project overview remains extremely political, risky, and technically complex. It certainly should not be thought of as simply defining the cost versus benefits with the decision made obvious from that calculation. Defining multidimensional project valuation criteria compounds the complexity model in regard to comparing dissimilar proposals. Many different techniques are described for this in the literature. Merkhofer describes multiple types of metrics useful for allocating resources. This includes the following (Merkhofer, 2003):

- Opportunity cost analysis
- Sensitivity analysis (variable estimates)
- Multiperiod planning.

This level of model sophistication remains beyond the capability of the typical organization, but it does reveal some of the analysis issues that are relevant in the discussion. Concepts are related to missed opportunities, variation in estimates (risk), resource availability constraints, multiperiod views, project grouping, and so on. In this environment, the process of project selection has risen to a new plan of sophistication, but not the one often pursued by organizations. One thought for this is that the perceived level of complexity encompassed by this view chased away many candidates.

The fourth evolutionary wave brings the project valuation story to the current time period. During the latter 1990s, proliferation of project activity in organizations made it clear that some type of centralized control was needed for project selection and resource allocation. Multiple studies of project selection indicated that organizations were not getting full value from their allocation of resources. Merkhofer reported that only about 60% of the value was realized (Merkhofer, 2003). The remaining 40% of available value was lost due to errors in the project selection process and weaknesses in business processes. Recognition of this phenomenon spurred industry interest in two directions: better project selection techniques and better project management to deliver the defined results.

The fourth wave is described as a portfolio view of projects. No longer were projects reviewed on a stand-alone basis with local benefits. Now, projects were viewed at the enterprise level and focused on organizational goals. The analogy used was that of a capital allocation model. That is, the potential for initiating projects was well beyond the availability of resources to support that activity, so *the primary selection goal is to select and complete a slate of projects that optimize organization goal achievement.* The vocabulary term for this is *goal alignment. Unfortunately,* we are still left with the basic question regarding how to do this mechanically in terms of a single valuation metric. Chapters 32 and 33 will explore the operational project portfolio mechanics in greater detail. For the remainder of this discussion, we will focus on basic organizational strategies that are used to develop project candidates.

9.2 PROJECT SELECTION STRATEGIES

Hirschheim and Sabherwal envision a project role consistent with the business personality. In concert with this, they describe a framework consisting of three project selection strategies: Defenders, Prospectors, and Analyzers (Hirschheim and Sabherwal, 2001, p. 215). In this structure, the primary role of the project is one of the following:

Defender—defends a stable and predictable but narrow niche in its industry.
Prospector—seeks new opportunities and creates change in the market.
Analyzer—seeks to simultaneously minimize risk while maximizing opportunities for growth.

Each of the strategy types outlined above would result in different project structures as summarized in Figure 9.2, strategic alignment profiles. This figure is insightful in that it shows how an organization should approach the project selection process in order to be in philosophical alignment with what the broad company mission and goals are. This view also dictates other decision strategies such as project organization structure, resource acquisition source, and the target change level to the business.

Alignment profile	Business leadership	Partnering	Low cost delivery
Business strategy	Prospector	Analyzer	Defender
Role	Opportunistic	Comprehensive	Efficient
Sourcing	Insourcing	Selective sourcing	Outsourcing
Structure	Decentralized	Shared	Centralized

FIGURE 9.2 Strategic alignment profiles. (From Hirschheim, R. and Sabherwal, R., *California Management Review*, 44, 1, 87–108. With permission.)

A Defender-type organization tends to work toward maximum operational efficiency and the likelihood is that the organization does not have high growth objectives. On the other extreme, the Prospector organizational culture seeks out targets of opportunity and the project selection bias should follow that path. This would result in higher project expenditures and more aggressive use of new technologies. Figure 9.2 also shows associated implications in regard to strategies toward sourcing and organizational structure. Unfortunately, some organizations do not define clearly what they really are trying to be and in fact are schizophrenic in their project selection behavior. Both of these traits can cause a mismatch in the project selections and a less effective alignment with organizational goals. One example of this occurs when the organization selects a particular high-risk strategic goal, while leaving many other project selections in the Defender mode. Another mismatch situation occurs when the timing of one proposal does not match with another, even if both are appropriate technical choices. Beyond these examples, there are also many other situations that create internal misalignment and cause the selection process to be out of synchronization. In order to better identify these issues, it is necessary to have a common valuation process that matches the organizational goal structure.

The project selection process represents the demand side of the equation, while monetary and human resources are related key supply side considerations. In most cases, it is the supply side that represents the critical constraints that most impact what projects can be undertaken. Secondly, organizations are also constrained by the level of tolerance they have for levels of change created by excessive project initiatives and this represents a subtle constraint that is often overlooked. All organizations have change absorption limitations, but the operational question is that at what point do you curtail spending and where do you allocate scarce resources.

Another aspect of the resource allocation issue is benefit timing. It is much easier to envision the value for short-term opportunities and often deemed much riskier to pursue strategic targets with a long-term payback. Based on this, many organizations focus only on the short-term initiatives, which in turn will limit their market position over the longer term. In order to maintain the organization over the longer time period, some mix of tactical and strategic options must be pursued. Imagine the buggy whip company that acted in the Defender mode and continued to refine that business model by being the most efficient in the industry only to wake up one day to find out that no one needed buggy whips any longer. Clearly, some consideration should have been given to the changing market view at the expense of operational efficiency.

All organizations need to stay positioned for strategic change and their project slate should support that evolution. As an example, Apple computers saw that they were not being competitive in that industry and modified the project slate to completely change their business model with iPod and iPhones. These were nontraditional, high-risk projects that moved the company to a new market and beat the competition. Failure for the host organization to invest in these initiatives would have resulted in huge competitive loss. Imagine what this business decision must have looked like with a computer person going to senior management saying something like "boss, I want to start making a small computer-like device to play music." Most Defender organizations would have responded to this by saying "we don't do music, we do computers. You need to understand our goals." Conversely, Prospector organizations must have a management culture that is willing to look at new perspectives. Several years ago, Digital Equipment was the leading producer of minicomputers. When the proposal for producing personal computers was made to the founder, his response was that no one wants a computer in their home. Digital no longer exists in the market and that is the penalty for being dogmatic in regard to project selection criteria.

Regardless of one's belief about the correct way to allocate resources, it is clear that there needs to be a formal project decision-making mechanism that adds a degree of high-level management and control to the process, whether this is at the departmental or enterprise level. The chosen allocation slate should be consistent with the organizational goals. This suggests that the way to obtain maximum value is to centralize the allocation of these resources and focus them on management approved target areas. Some would argue that centralized planning adds a level of bureaucracy that in turn takes away the agility of the organization to react quickly to business changes. The challenge is to not let the process get bogged down in bureaucracy, yet keep an overall organization focus.

In 2003, BMC undertook a survey of approximately 240 respondents to quantify the state of alignment maturity in medium to large organizations (BMC, 2003, p. 3). The results of this survey provide a good discussion structure for the topic and some analytical data to show how organizations actually do the project selection process. The survey grouped 45 alignment practices into four broad areas: Plan, Model, Manage, and Measure. The level of operational maturity was categorized into five groups (BMC, 2003, p. 4):

- *Chaos*—no standard process.
- *Reactive*—multiple processes/procedures in place; little standardization.
- *Proactive*—standards and documentation exists; minimal compliance assurance.
- *Service*—processes are standardized and compliance is managed; some automated tools.
- *Value creation*—processes have been matured to best practices; continuous improvement and benchmarking in place.

Based on the maturity grading scale, there were six qualitative conclusions drawn that were very interesting, (BMC, 2003, p. 5):

1. There is a strong relationship between business alignment maturity level and the participants' assessment of overall efficiency and alignment.
2. Organizations that rated highly on item one also showed positive assessments in managing change.
3. Existence of a strong management organization was linked to item one.
4. Integrated metrics and scorecards by mature participants provided consistent management data across all functions.
5. Top performers in the study based on maturity scores were also viewed as top in the qualitative assessments.
6. Overall, the study population did not rate highly in the alignment area, although there were a few who were very mature in this area.

Within the model groupings, there was more process maturity observed in measurement than in modeling capability. The average score was halfway between reactive and proactive, with 70% of the respondents below the proactive state. Large organizations fared better than middle-sized ones in terms of maturity across the model groups. Quantitative studies of this type add improved understanding of the value resulting from proper goal alignment and ultimately how to accomplish the goal. Another key point derived from this study is that organizations have not yet achieved mature decision processes in the alignment arena and more management focus is needed in this area in order to improve the project selection process.

9.3   CONCLUSION

Maintaining the project's goal alignment with its host organization is a fundamental management activity that lies above the project management domain. There is a familiar adage that "garbage in results in garbage out." This is certainly true in the case where a poor project approval decision process can lead to nothing but a bad outcome, no matter how well the effort is managed. We have also seen evidence that there is much to be gained in regard to added business value and improved perceptions when this activity is properly carried out. Industry surveys and other research indicate that most organizations can use improvement in this process. In any case, selecting the right portfolio of projects to pursue is a fundamental activity in the global view of project management.

REFERENCES

BMC Software, 2003. Business Service Management Benchmarking Study Stage II. Routes to Value: An Actionable Approach to Business Service Management (A White Paper). www.bmc.com (accessed January 26, 2005).

Hirschheim, R. and Sabherwal, R., 2001. Detours in the path toward strategic information systems alignment. *California Management Review*, 44, 1, 87–108.

Merkhofer, L., 2003. *Choosing the Wrong Portfolio of Projects: And What Your Organization Can Do About it* (A Six Part Series). Published in www.maxwideman.com (accessed June 25, 2005).

10 Project Initiation

Projects can be spawned from a variety of sources, but one needs to recognize that a lot of work is left to be done in the time period from the original vision to a completed project deliverable. Elaboration of the vision into an approved project initiative is the goal of the life cycle initiation phase, which in turn is the predecessor to a more formal and detailed planning activity. The basic goal of the initiation stage is to evaluate the merits of a vague vision and from this develop a preliminary business case for pursuing this vision. The business case provides necessary financial, risk, and goal justification for approving the expenditure of resources in comparison with other initiatives that are also being considered. In all situations, the availability of resources constrains the level of activity that can be supported by the enterprise; so the basic management goal is to select those targets that best align with the goals of the organization. Project initiatives tend to have characteristics of improving current business processes, growing the business, or transforming the business. Consideration of the project proposal involves analysis of the investment level, organizational capability to accomplish the initiative, inherent risk, competitive needs, and return on the investment, among others.

Beyond the activity of evaluating how the new vision will impact the organization, the management considerations in approving the project to move forward involve the following:

1. How well does this proposal mesh with enterprise goals?
2. How long will the effort require?
3. What is the cost of the initiative?
4. What is the related resource requirement and from which sources?

It is important to recognize that the initiation stage is "fuzzy" in regard to the accuracy of data quantification that can be produced. Resource estimates often contain errors of 100% and the estimates regarding project time and cost can be equally inaccurate. Certainly, the goal of this first-phase process is to do the best job possible in decreasing these predictive errors and from that point make the best decision possible in selecting the right projects to approve and move forward into a more rigorous planning effort.

Another management issue present in the initiation stage is that not all parties involved in the decision process have the same view of the initial vision or its value. This lack of a consistent view often breeds confusion and internal conflict among the various stakeholders; so it is necessary to attempt to homogenize the perspective and get key players on the same page as to the project goals. Without some agreement as to the future target, it is impossible to produce a valid decision document. If this conflict cannot be resolved, the initiation phase begins to take on a more of an analysis of alternatives view, which multiples the work required in producing multiple decision options and parameters.

Once a decision has been reached as to the scope target and its associated time and cost parameters, there are two remaining important initiation management steps (PMI, 2008, p. 45).

1. Create a project Charter based on the approved parameters for the effort (i.e., deliverables, schedule, and budget).

2. Develop a preliminary scope statement that describes the stakeholder's needs and expectations.

For large projects, this same process could be repeated for each stage.

From a theoretical view, a project should be initiated to help in moving the organization from one operational stage to a second more desirable one. Think of this as one enterprise maturity point to another. For example, the organizational planning process might define the need to improve its customer relations or financial reporting capability. From that initial goal vision the roots of a project are spawned. One approach to the next step is to create a formal business case, which contains a narrative description outlining what the project is to accomplish and some perception of what its value would be. Associated with this first definitional step is the emergence of a management level sponsor, who is charged with reviewing the situation and producing a business case to support this vision being selected in competition with other project proposals. The sponsor would now assign some level of technical resource to analyze the proposal in greater detail. This analysis could be assigned to knowledgeable business process technicians, called Subject Matter Experts (SMEs). They would review the target topic and lay out proposed alternative approaches for the vision—one standard alternative is to do nothing. Financial, operational, and business value of other alternative solution scenarios would be defined and a rough cost and schedule estimate developed for each feasible alternative. From this list of options, the review team will produce a preferred solution. The final business case format would describe the primary alternatives and present recommendations to the sponsor. From this point, there would be a formal management review, either as part of the larger portfolio reviews or simply as a stand-alone project. The level of formal management review would vary depending on the size and impact of the proposed initiative. In many organizations, the defined alternatives would be translated into a standardized presentation format that would then be reviewed by a central screening committee. The project screening authority would rank the various project proposals and rank the list of project options. Some decision level authority would then make the final decision as to disposition of the proposal. If approved, a Charter would be issued to show that the project was formally accepted and a PM is assigned. From this official starting point, the detailed planning cycle commences.

A second possible path for project spawning comes from internal ideas generated within the organization. These grass roots level ideas can be submitted through a formal system or through some other less formalized collection method. The key difference in these two approaches lies primarily in the organizational level of origin. The grass roots method starts within the organization business levels and from this point one must find some level of management support sufficient to develop a formal business case for that idea. In this scenario, the initial idea driver is often a mid-level manager who is seeking to solve a local problem. In order to match these proposals with all others, a business case should be developed and follow the same approval path as the one described earlier. The challenge inherent in this method lies in the fact that the ideas tend to be local and not necessarily good fits for high-level goals.

Ideally, the total slate of proposed project candidates should be viewed as a project "portfolio," with each proposal having an estimated resource cost and a forecast business value. Out of the portfolio set, some projects would be approved to move forward, some would be deferred until later, and some would be rejected. This process is similar in concept to a set of possible stock market investments being considered with a limited budget. In order to optimize the portfolio return, one would rank the choices, possibly using multiple criteria and then allocate limited resources to those projects (investments) selected. The stock and project selection models have quite similar underlying concepts, except that the project value equation appears to be more multivariate. To carry the stock selection analogy one step further, even after a choice is made and funds committed the situation can change such that the original decision is recognized to be invalid. In this case, the item (project or stock) moves to a rank level where it is not worth continuing. In this situation, the decision should be made to stop further resource expenditures (e.g., halt the project or sell the stock). Too often this ongoing review process is missing and bad projects continue to consume resources long after their value has disappeared.

Recognize that the project value can change during execution simply by cost or schedule overrun or by changes in the business environment that lower the value of the effort. So, project evaluation is not just a front-end process. It should continue through the life cycle to ensure that the current plan is valid. The job of a PM is not to keep the project running toward some predefined goal if the effort is no longer appropriate. His job is to be an honest broker of information and be sure that appropriate status metrics are properly transmitted to management. The ultimate interpretation of the value comes from management's view of the project in context with dynamic organizational goals. Keeping a dead project alive to consume resources after the value has disappeared is considered an unethical PM activity, yet one that is often hard to live by. The decision to terminate a project becomes almost like losing a family member (sorry for the morbid analogy). Many long hours of work will have been spent trying to nurture the project back to health and admitting defeat is very hard. Most PMs are very goal focused and walking away is not something native to their DNA.

The role of formal project initiation is important for many reasons. Ad hoc projects spawned throughout the organization drain away critical resources that would be better spent focused on higher-level organizational goals. Mature organizations have well-developed processes to handle the project initiation process, and a project does not emerge in isolation without a formal management review and approval. Remember, a project has both a value and a cost. These two metrics need to be analyzed by comparison with other initiatives. The goal is to pursue initiatives that optimize the allocation of assets to the betterment of the enterprise. As indicated above, the ideal approach is to link all project activities to tactical and strategic plans. In order for the formulation process to function as described, the complete slate of proposed and existing projects should be analyzed on a consistent basis using some standardized presentation review format (e.g., the business case template for new initiatives or a standard status reporting format for existing projects). There is great debate in the industry today as to whether approval decisions should be made solely on the basis of a quantitatively calculated Return on Investment (ROI), or some mixture of financial and other criteria. It is the author's belief that project approval decisions must be made on the basis of both financial and other criteria. Whatever the approval process used, the key for project selection is that it should be made through a formal management review activity. Ultimately, senior management is responsible for the use of enterprise resources and the results arising from that allocation.

Project Charter: The important concept behind issuance of a Charter document is that management explicitly authorizes the existence of a project. One subtle impact of the formal chartering process is to limit ad hoc project resource consumption. A formal chartering process is a critical management activity that is designed to focus resource expenditures on those visions that have been judged to best align with the organization goal structure.

The Charter document contains various authorizations and information designed to help with subsequent activities. The following list adapted from the *PMBOK® Guide* identifies typical items included in the document (PMI, 2008, p. 351):

1. The Charter is approved through a formal management process and signed by the project sponsor who is normally responsible for the budget associated with the effort
2. The PM is named and given some defined level of authority to allocate initial resources for the effort
3. A definition of business need is stated, including the project description and general product deliverable requirements
4. Project purpose or justification is summarized
5. Milestone events and a high-level schedule are documented and approved
6. Role of various support organizations is defined
7. Key stakeholders impacted by or involved with the project are identified

8. Summary budget information is defined
9. The business case used to justify the project is included.

Following this step, a second companion document to the Project Charter is also created to help clarify some of the vagueness inherent in the initial goal statements.

Preliminary scope statement: It may seem redundant to create another specification document in the initiation stage immediately after the Charter is approved; however, there is a valid explanation for doing just that. Imagine that users and management are dealing with the original vision and the subsequent Charter. During this process, their view is primarily on the desired output of the vision and not so much on the mechanics or feasibility of achieving that output. As we move the process into a more detailed planning step, it is important to clarify the requirements in terms more amenable to deliverables and work definition. Also, there are some future project management implications that need to be resolved at this stage. The list below represents typical areas where clarification is needed. These items will be expanded from the Charter language and restated as necessary. This elaboration process is captured in a document titled a Preliminary Scope Statement and it includes the following (PMI, 2008, p. 351):

1. Project and product objectives
2. Product or service requirements and characteristics
3. Product acceptance criteria
4. Project boundaries
5. Project requirements and deliverables
6. Project constraints
7. Project assumptions
8. Initial project organization
9. Initial defined risks
10. Schedule milestones
11. Initial WBS
12. Order of magnitude cost estimate
13. Project configuration management requirements.

For the items listed above, the preliminary scope statement's goal is essentially a review and edit of the original specification into terms that are more understandable by the project team. A classic example of this would be to translate the requirement "to leap tall buildings with a single bound" into "must be able to jump a four story building in a single attempt 95% of the time." The original specification requirement would obviously have to be technically resolved later and would remain a point of confusion for the technical implementer. So, the preliminary scope document attempts to clear up as many such definitional issues as possible while the initial requirement goals are fresh. The goal should be to write in clear language what the project output requirement consistent with the terms of the Charter. Clearing up such requirements early helps resolve future confusion and wasted efforts.

In addition to the technical cleanup process, the preliminary scope document should also include some specification regarding various management activities for the new project. Typical of these would be as follows:

1. Initial project organization
2. Project board—management steering process
3. Project configuration management requirement
4. Change control process to be used.

Upon completion of the approval and specification documents, the project is ready to move into the formal planning process.

An approved project Charter and follow-on Preliminary Scope Statement represents the two key formal initiation documents that will be used to guide future planning decisions. Given these two artifacts, a high-level vision of the project objective and its value is documented, along with rough estimates for budget, schedule, and other related decision factors. The subsequent challenge for the PM is to refine these still crude statements into a detailed formal project plan that will once again need to be approved by appropriate management. Often times a project will be approved subject to refinement or rationalization of the initial project parameter estimates (i.e., cost, schedule, and functionality). Seldom is management willing to approve the entire project based on the business case unless many similar projects have been handled prior to this.

Once a project is underway, it is important to recognize that the Charter needs to be re-evaluated minimally at each formal review point to reaffirm the ongoing project scope and direction of effort. As changes are approved in the project life cycle, the plan will need to be adjusted and reissued as necessary; however, the Charter document can only be changed at the source of approval—typically at a senior management level. The project plan's role is to be a dynamic blueprint for the project showing a current view for the effort.

10.1 ENVIRONMENTAL FACTORS TO CONSIDER

Often missing in the project initiation process is a consideration of environmental issues beyond just high-level descriptions of scope, cost, and schedule. An equally important view is to assess items in the environment that can impact the project's success. There are a variety of well-known reasons why projects fail and many of these are external to the project domain itself. The Standish organization has been tracking criteria for IT project success and failure for several years, and these surveys offer great insight into issues that are important to success (Standish, 2003). Since their inception during the 1990s, these annual surveys have surprisingly varied very little in regard to the top 10 items linked to failure. Given this relatively static perspective, it seems prudent to review these and other similar factors as a first look at the project environmental factors. Table 10.1 contains the top 10 2003 Standish success factors and the frequency by which each factor was identified.

Four of the top five Standish items have consistently appeared in the survey results over the years, and for this reason one must assume that these are major management issues to deal with. It is also interesting to note that technology issues do not show up high on the list. One reasonable conclusion from this is that project failure is more often than not a management problem and not some

TABLE 10.1
Factors Leading to Project Success

Success Criteria	%
1. User involvement	17
2. Executive management support	15
3. Experienced project management	14
4. Clear business objectives	14
5. Minimized scope	12
6. Agile requirements process	7
7. Standard infrastructure	6
8. Formal methodology	5
9. Reliable estimates	5
10. Skilled staff	5
Total	100

Source: Standish, 2003. *Chaos Study.* The Standish Group, Boston, MA. www.standishgroup.com. With permission.

technology issue gone awry. Basically, every project needs to be reviewed with these characteristics in mind. If one cannot be assured that a particular success factor is present, then assume that there is an increased risk level for failure. From a global environmental level, organizations should try to evaluate the status of these factors for internal project assessment. The sections below will offer some further discussion on selected items in the Standish list.

10.1.1 User Involvement

In many cases, project scope is initially outlined by a business entity and approved by the project sponsor. This group often believes that their involvement is complete. History suggests that such is not the case. In order to be successful, a project team must build an ongoing close partnership with the business sponsor, key SMEs, and the project team who will execute the requirements. The confusion here is that users often feel that they do not understand the technical underpinnings of the effort, therefore are not needed. The fact is that there needs to be recurring communication of the requirement as translated from plain language into technical form. Only the users can effectively judge this translation, but that may not be until a visible product results much later. Omission of close user support leaves the technical team to interpret requirements and innovate to fill the gaps. This can move the project in the wrong direction and create excessive scope creep once the problem is recognized. Continuous communication is required to ensure that the right business problem is being solved, even if it is not the one originally understood.

To counter the lack of user involvement, many organizations believe that the project team should be managed by a business person in order to ensure close participation between the business and technical sides. There are pros and cons for such a strategy. On the one hand, this increases the linkage to the business side of the effort, but placing a business person in charge of a technical team can lead to flawed technical decisions that in turn lead to equally bad project results. So, there is no clear answer as to the best way to deal with project team management. Regardless of the option chosen, there must be clear communication input from both the business view and the technical side. Roles of both parties need to be understood and managed. The business side should view their primary responsibility as the "what" of the project, while the technical participants should be held responsible for the detailed "how." Project team organization roles and responsibilities should be created to support this view. Regardless of the structure selected, failure to achieve user involvement through the life cycle will almost surely doom the effort to less than desirable results.

10.1.2 Executive Management Support

Following closely behind the user involvement issue is that of the senior management role; however, the project related interaction in this case is not so obvious. It would appear that a well-structured project team with appropriate business and technical participants could execute the requirements without senior management involvement. This might in fact be the case if requirements were completely and accurately defined at the initial stage, cost estimates were accurate, schedules accurately defined, and all other extraneous events insignificant to the outcome. Unfortunately, few of these situations exist in the typical project. Requirements evolve and grow as more is defined in the course of the project life cycle. Various stakeholders surface with differing views on scope and direction for the team. Unplanned technical issues often surface that lead to adverse results for the plan. To top all this, the organizational priorities may change over time in such a way as to cause corresponding changes in the project goal set. For these reasons and more, the role of senior management in these situations is to be an engaged supporter and referee.

Senior management initially approves the Charter and should then allow the project team to manage the ongoing process so long as they stay within approved boundaries. The team is obligated to communicate their status to the management layer and present periodic reviews at key decision points. Senior management must understand enough of the ongoing situation to judge how to support the effort from the enterprise viewpoint. This does not always mean to continue funding a project, but it does mean that they have a role in deciding when and how to support the effort. If senior management is not involved at a reasonable level they cannot make these difficult decisions. In the final analysis, the project is their ultimate responsibility. Within the approved Charter guidelines, the project team and business sponsor must do their best to accomplish the defined goal. But in the final analysis the senior management group owns the project and has authority to approve disposition of any out of bounds conditions. Probably stated better, the sponsor, the PM, and the team should all feel like they own the project. If the effort goes well, the reward should be left with the team. If it begins to erode, the ownership issues creep up the organizational ladder until all are left with the responsibility. This is a subtle point, but an important management concept.

10.1.3 Experienced PM

This item has increased its importance from previous surveys and the reason for this increased focus is subjective at this point. One reasonable guess as to why it is elevated over time is that projects continue to be more complex in structure and knowledgeable project mangers are better able to demonstrate their value in this environment. In concert with this, the term "experienced" really means someone who has studied the art and science of the topic and is sensitive to the type of issues being described here. Historically, a PM might be selected from the ranks of senior technical individuals in the group. This is called the *halo effect*—a good technician was thought to be a good PM. However, experience shows that selecting a good technician for this role does not necessarily translate into a good manager. Managing people and work processes is a different skill and mindset than doing technical work.

One of the subtle points in project management is that the PM acts as an agent to the senior management of the organization, the business sponsor, the stakeholder community, and his project team. This dual agency often places the PM in a conflict situation and the only way to navigate through this is to be honest to all parties by presenting timely, visible, and accurate project status for all to judge. In many situations, the PM is tempted or pressured to sugar coat status with the belief that the team will shortly resolve the current deficiency. This is not an appropriate management behavior strategy and should be resisted. Accurate project status should be timely reported to appropriate support levels in the organization for their review or resolution.

The efforts of groups such as the Standish Group, PMI, SEI, and other such research-oriented organizations have been instrumental in education and communication of factors and approach that aid in successful project execution. The author's bias suggests that individuals who are trained in this art are beginning to distinguish themselves through their positive project results. In other words, project management is a learnable skill and process that can help produce successful outcomes. That does not mean that everyone who is trained the same way will be an equally good PM, but it does say that they should be better than they were before they were trained.

Various surveys over the past few years have indicated that project results have improved over time and this quite likely is a reflection on the improved underlying project processes used. It is also interesting to point out here that the role of the PM is ranked higher in the survey's success criteria than the availability of appropriate technical skills. One interpretation of this ranking is that successful execution of a project is clearly more than allocating skilled technicians into a group and awaiting the desired outcome.

10.1.4 Communications

Studies conducted on PM activities show that they should spend approximately 90% of their time communicating with various stakeholder groups (Mulcahy, 2005, Chapter 9). One of the less visible and measurable activities in project management is the communications process. This activity is vital in resolving conflicts, work issues, and coordinating the various project activities. Failure to perform this function adequately has major adverse implications. Often times, a technically oriented manager will hold the belief that his primary goal is to produce a technical product and that will make the users happy. History shows that this is not the case. This statement is not to suggest that a PM does not have to be technically literate and involved in technical activities within the team, but even that role becomes one of communicating.

10.1.5 CLEAR BUSINESS OBJECTIVES

One might argue that this factor is simply a restatement of the user involvement factors listed above. The key point to understand here is to answer how a set of project requirements gets created. In order to accomplish this goal, it is necessary to involve a larger collection of users to produce a consensus-based view—that is, user involvement on a broad scale. Conversely, one can have a sufficient quantitative level of user involvement that never is able to agree on the objectives. So, failure to accomplish a consensus will surely leave the technical team in a quandary. Given this latter view, we need to recognize that there must be a process that translates the original objectives into usable project requirements and not just a collection of user wish lists. Even the stated requirement of "leaping tall buildings with a single bound" is not a sufficient statement for the team but might sound pretty good to the user side.

Requirements generation remains one of the weak technical areas of project methodology, and the effectiveness of the approach taken in dealing with this situation affects the project outcome. Traditionally, an analyst from the project team would personally interview various business sources to document their views on the problem. The individuals who are used by the project team to define requirements are often called *domain experts*. At the end of the interview cycle, the analyst would combine these views into a requirement statement that would then be circulated to the various stakeholder entities for comment. Later in the life cycle, a similar serial process would be used to refine these statements into more specific requirements that could be used by the technical team in the system design process. One fundamental problem with this serial process is that it takes too long to schedule and go through the approval process and, second, there must be an effective give and take across the participants. Also, many argue that users are not capable of completely envisioning a requirement prior to being able to physically see some part of the result in some tangible form. The fact is that requirements are easier to define once the user sees a working component. Given these requirement traits, we must be aware that the process must be more concurrent group oriented and less serial in nature. One proven way to improve the requirements definition process is to convene the proper domain experts in a single place away from daily work diversion for a process known in the literature as Joint Application Development (JAD). JADs typically require two to five days of extensive focus on a target area. This semistructured process is lead by a trained person who guides the group through a series of steps designed to produce a draft requirements definition document. This process is capable of producing not only an improved requirements document, but it also achieves a high level of consensus among the group members since they have the opportunity of sharing views during the discussion. The key to success for the JAD process is that the group is locked away to focus on the problem and they collectively hear what others have to say in such a time frame that they can either agree or argue a counterpoint. This type of consensus building is important early in the project since it helps to focus the team on key issues. If agreement cannot be reached as to general scope and technical strategy, the management function would be required to decide how to proceed, either by making a unilateral decision on diverse opinions and communicating that decision, canceling the effort, or by supporting another round with some guidance on the disagreement items. At any rate, the goal would be not to leave this stage without reasonable buy-in from the domain experts. The JAD format can be used in various stages of the project to quickly collect stakeholder and technical opinions on complex topics. PMs should recognize that group sessions of this type are valuable consensus forming techniques and they can address many of the technical and behavioral issues that will cause problems later if left undefined or unresolved.

TABLE 10.2
Project Duration, Team Size Affects Project Success

Project Size	People	Time (mo.)	Success Rate (%)
Less than $750 K	6	6	55
$750 K to $1.5 M	12	9	33
$1.5 to $3 M	25	12	25
$3 to $6 M	40	18	15
$6 to $10 M	+250	+24	8
Over $10 M	+500	+36	0

Source: Standish, 1999. Chaos: A recipe for success, Boston, MA. www.standishgroup.com. With permission.

10.1.6   Minimized Scope

As we traverse the list of key project success factors, various cross relationships emerge. In the case of project scope, size is a factor linked to several other issues. For example, small projects have less complexity, less communication requirements, less user involvement needed, and so on. Also, a minimized scope touches on the ability to produce clear objectives. Once again the Standish group offers some insight into the impact of scope size on project success. Table 10.2 shows the relationship of team size and cost on project success.

One point clearly visible from this data is that a project needs to be of reasonably small size if success is of prime concern. Of course, some projects by their nature cannot be structured into small increments; therefore in these situations the role of project management needs to be recognized as more complex. When a large-scale effort is dictated, issues such as risk management, scope control, communications, HR, and others become of increased importance and must be dealt with more formally. Given the Standish data shown earlier, it seems reasonable to suggest that the initiation goal should be to formulate all projects as moderate size, or to attempt to decompose them into aggregates of related smaller projects with well-defined relationships. Clearly, increased size is a significant negative factor in project success.

10.1.7   Other Success Factors

The remaining factors out of the Standish top 10 list are of lesser importance to success, but nevertheless are worthy of a brief comment.

10.1.7.1   Agile Development Approaches

There is a growing school of thought among IT system development technicians that requirements cannot be adequately defined at the outset; therefore they espouse an alternative requirements definition strategy. That is, quickly formulate a basic plan and then proceed to develop a series of small iterative solutions that the user can evaluate. In this fashion, the project would be executed by managing creation of these "chunks" of capability. This methodology school of thought is most often called Agile or Extreme (among other names). In using this approach, the user would work closely with the technical team to produce quick and small instances (chunks) of the solution. In some cases, these would be called iterative prototypes. The goal of this approach is to quickly generate user value and to quickly validate requirements. One clear advantage of the iterative methodology is that project deliverables are seen by the user earlier than would be found in the traditional "big bang" serial life cycle implementation approach. Conversely, some would argue that this could cause extensive rework and therefore additional cost. Scope of this discussion does not allow us more time to wrestle with these advantages and disadvantages, but the contemporary PM should be versed in both approaches and customize his project to fit the situation.

User Requirements?

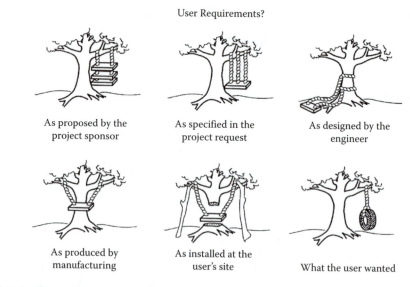

As proposed by the   As specified in the   As designed by the
  project sponsor   project request   engineer

As produced by   As installed at the
 manufacturing   user's site   What the user wanted

FIGURE 10.1   How requirements can go wrong.

Regardless of one's conceptual methodology, there is general agreement that defining accurate requirements is fundamental. Our belief is that some level of early definition of the business objective is needed. From that point, either the iterative or the traditional JAD (or both together) approach could be pursued effectively to serve the requirements definition role. Cartoons similar to the one shown in Figure 10.1 have draped walls in almost every system analyst area. When first seeing this example, people laugh. After being in the job of a few years, all understand its meaning. Communicating accurate requirements is hard work and often not done well. This classic cartoon describes a not too far out scenario showing how requirements get garbled through the organizational layers.

10.1.7.2   Existence of a Standard Process Infrastructure

This item implies that each project should not have to reinvent the organizational process environment. By utilizing standard processes, the new project can simply plug into a defined and well-managed foundation without having to deal with reinvention of equivalent supporting functions. Too often, a new project will have to create its own tools and processes that have to be engineered as part of the project work. These support tasks are essentially nonproductive to the end requirement and consume valuable resources. If excessive, such activity will negatively impact the project capability to deliver. Use of standard management components and processes should be stressed wherever possible for this reason. When new tools and technologies are chosen, recognize the added risk created by the creation of these untested items.

10.1.7.3   Use of a Standard Methodology

Previous discussion has described the role and structure of a methodology. If properly applied, this form of standardization can be productive because of the underlying processes that are familiar to the various stakeholders. From past efforts the methodology should have been tested and the result will be usable components such as templates, operational processes, lessons learned, and a common vocabulary. There is research evidence from William Ibbs and others that organizational maturity has a positive impact on project success (Ibbs and Kwak, 2000). The sensitive issue here is to ensure that the methodology is flexible enough to handle different project types and sizes. One approach does not fit all situations. The issue of increased risk management has come up in other discussions. In this case, the risk element is that management oversight and documentation specified in a methodology will in effect sabotage team productivity. More control and oversight should be applied to larger and mission critical projects; less for smaller and noncritical projects. One size methodology is not the correct approach. Project teams left to their own devices often choose to do little documentation or status communication. Obviously, this is not a desirable outcome for most efforts, so it is necessary to think through the control and management activities for each effort and obtain agreement regarding how much is appropriate and then specify this as part of the formal project plan.

10.1.7.4 Reliable Time Estimates

There are many contributing issues that can lead to false expectations from the stakeholders. When schedule, budget, or technical estimates are bad, the project is often viewed as a failure regardless of other factors. Basically, the original project estimates influenced the decision process to approve the project and they create associated user expectations. If one or more of these estimates is flawed, there may be a significant impact on the value of the project. Sometimes, estimating errors are the result of inadequate requirements definition, whereas in other cases, it is the result of inadequate skills or erroneous assumptions. Simply stated, a bad estimate yields a bad result in the stakeholder's eyes.

10.1.7.5 Availability of Appropriate Skills

Failure to provide timely quantities of skilled resources is an important and visible attribute for project success. The fact that this obviously critical item is last on the top 10 list likely means that it is widely recognized and already focused on as part of project management. It surely does not mean that it is number 10 in importance. Every manager recognizes that skill levels are variable across different workers, so their impact of budget and schedule are clear in this regard. Also, there may be some aspects of the project that can only be accomplished by a high skilled individual. We will see implications resulting from skill variability in many future discussions.

10.1.8 INDUSTRY AND ORGANIZATIONAL CULTURE

Beyond the internal organizational factors described above, there are also various differences observed by industry and organization. As an example of this, the Standish surveys document that the retail sector experienced the highest project success rate, while government projects tend to exhibit the lowest (18%) (Standish, 1999). Much of this difference in project performance can be explained by the types and sizes of projects typically pursued in these industries. Certainly, the government often pursues projects that are of megasize scale and we have seen that these tend to have less success than other smaller initiatives. Another factor that does not yet appear on the Standish survey is that of individual organizational culture. It is difficult to describe this term, but it basically relates to the internal support environment for the project and the general maturity of the organization. A well-conceived project that is appropriately managed will have a higher probability of success than one that does not have these characteristics. Also, an available pool of appropriately trained resources affects success rates. A third aspect is related to issues inherent in a global organization. For instance, a strong departmental organizational form with a territorial view is not conducive to complex intradepartmental types of projects. There are also other cultural factors that can affect project performance and when initiating a project this class of issues must be recognized in their role of either nurturing the effort or creating roadblocks and constraints. We have used the analogy previously that a project is like a flower seed planted in a flowerbed. The organization is the flowerbed and if it does not nurture the project seed, it will die. This is the impact that culture contributes to the project and its ultimate success.

10.2 PREDICTING A PROJECT'S SUCCESS

Countless studies over the last three decades point to similar findings: Approximately 50% of projects undertaken have been late and/or exceed budget, and approximately 25% will be cancelled

TABLE 10.3
Project Trends

	1998 (%)	2000 (%)	2002 (%)	2004 (%)	2006 (%)
Succeeded	26	28	34	29	35
Failed	28	23	15	18	19
Challenged	46	49	51	53	46

Source: Standish, 2008. *The Trends in IT Value.* The Standish Group, Boston, MA. www.standishgroup.com. With permission.

before completion. The success of a project most typically depends on people—not technology or tools. It crosses all industries, from manufacturing to financial services.

Considering how much and how rapidly technology evolves, one would think that the ability to implement successful enterprise projects would have improved at a comparable pace. But it has not, according to a 2008 report by the Standish Group, a Boston-based research firm. The study found that IT project delays and cost overruns have not improved much since 2000: 47% of projects are over budget, and 72% do not meet projected timelines, compared in 2000 with 45% and 63%, respectively. Table 10.3 categorizes these trends over time by measures of success, failure, and challenge.

Because technology advances so rapidly—with significant dynamics occurring every 6 to 12 months—projects tend to have high degrees of uncertainty. Methodologies change, skilled personnel turnover can be high, and there are currently no standard benchmarks for evaluating abilities and skill levels. A project's success ultimately depends on people, not technology or tools. This statement holds true across all industries, from manufacturing to financial services.

Table 10.4 provides a sample technique to illustrate how to perform a quick success assessment of a high-technology project. Local experience will be required to supply appropriate results in your environment and this customization will require tweaking of individual parameter weights to achieve better outcome prediction. The sample worksheet is intended to illustrate how the evaluation of certain key success parameters can lead to a quantitative measure for project success. In this model, any score less than 75% suggests that the project could be headed for trouble. Worksheets of this type can be customized to fit local conditions by adding appropriate items and weights to the model. Also, experience will supply information to allow tweaking individual weights to better predict outcomes.

10.3   FORECASTING THE SUCCESS OF TECHNOLOGY PROJECTS

Instructions: Use the worksheet in Table 10.4 to fill in your own numbers under the "Your Project" column and perform the indicated calculations.

To illustrate the variation in evaluation criteria, a second worksheet is shown in Table 10.5. In this version, the project type is higher technology, which raises the risk. These two variations show how weighted criteria can be part of the initial valuation process.

Instructions:

1. Fill in your project assessment values (0–5) under "RISK SCALE." The higher the number, the greater the risk.
2. Multiply your risk number by its corresponding "SUCCESS VALUE" and enter the "SCORE" results for each item.
3. Add up the scores to get your project's TOTAL SCORE.
4. Values at the bottom of the table translate risk level for this project. Future experience can be used to adjust these values accordingly for local data.

TABLE 10.4
Project Success Worksheet

	A	B	C	D	E
	Success Weight[a]	EXAMPLE		YOUR PROJECT	
QUESTIONS TO ASK ABOUT YOUR PROJECT:	High = 5 Medium = 3 Low = 1	Answer Yes = 1 No = 0	Success Value A × B	Answer Yes = 1 No = 0	Success Value A × D
1. Project is part of the execution of the business strategy	1	1	1		
2. There is a project sponsor	3	1	3		
3. You have full backing of project sponsor	3	1	3		
4. Project does not have multiple sponsors	3	1	3		
5. Has real requirements	5	1	5		
6. Has realistic deadlines	5	0	0		
7. Uses current technology	5	0	0		
8. Everyone knows the "big" picture	1	0	0		
9. Project processes (e.g., solution design/ delivery) are defined	5	1	5		
10. Project processes are understood	3	1	3		
11. Project processes are accepted	1	1	1		
12. Project plan is defined by participants and affected people	5	1	5		
13. Doing some small experiments to validate project	3	0	0		
14. Overdue project plan tasks elicit immediate response	3	0	0		
15. Issues are converted to tasks in project plan	3	0	0		
16. You do not have a great dependency on a few key resources	3	0	0		
17. Enforced change control process	5	1	5		
18. Risks are correctly quantified	3	0	0		
19. There is project discipline	5	1	5		
20. PM is not the final arbiter of disputes	3	1	3		
21. There is a communications plan	5	1	5		
22. There are weekly status meetings	5	1	5		
23. There is a one-page project-tracking dashboard	3	1	3		
24. All project documentation is stored on shared database	1	0	0		
25. Lessons learned are adopted to improve project processes	3	0	0		
Totals = sum of 1 through 25	85	15	55		
PROBABILITY OF SUCCESS[b] =	C total ÷ A total = 64.7%			E total ÷ A total =	

Source: This worksheet is adapted from a similar version published in *Baseline*, May 2006. The version shown here is approved by the author Ron Smith. With permission.

[a] Assumes weight is constant.

[b] Probability of success less than 75% suggests you need action plan to change key "no" answers to "yes."

TABLE 10.5
Project Success—Technology Projects

TOOL: Project Success Assessment Table	SUCCESS VALUE	EXAMPLE: Oracle Upgrade		YOUR PROJECT	
		RISK SCALE	SCORE	RISK SCALE	SCORE
TOP 10 REASONS PROJECTS FAIL	Total – 100%	0 – Minor 5 – Major	Value × Risk	0 – Minor 5 – Major	Value × Risk
1. Incomplete and/or changing requirements	25	5	125		
2. Low end-user involvement	15	3	45		
3. Low resource availability	10	1	10		
4. Unrealistic expectations	10	2	20		
5. Little executive support	10	1	10		
6. Little IT management support	5	3	15		
7. Lack of planning	5	4	20		
8. System/application no longer needed	5	1	5		
9. Bleeding-edge technology	5	2	10		
10. Other/miscellaneous	10	0	0		
TOTAL SCORE			260		

WHAT YOUR TOTAL SCORE MEANS: 0–125 = high probability of success. Review week areas if any; 126–250 = low probability of success. Work one week areas; 251–500 = cancel project it will most likely fail.

Source: This worksheet is adapted from a similar version published in *Baseline*, June 2008. The version shown here is approved for use by the author Ron Smith. With permission.

10.4   CONCLUSION

In this section, we have looked at the general process by which a project moves from the fuzzy vision stage into a management approved stage. The project Charter authorizes the effort to continue and it represents a formal approval from management. This document should assign the PM and formally delegate to him sufficient authority to move forward with the initiative. The next step in the process is to take these vague requirements and produce a formal project plan that considers in greater depth the issues confronting the effort. Part III of the book will take a more detailed view of the steps following this step.

REFERENCES

Ibbs, C.W. and H.K. Young, 2000. Calculating project management return on investment. *Project Management Journal.*

Mulcahy, R., 2005. *PMP Exam Prep*, Fifth Edition. Minnetonka: RMC Publications, Inc.

PMI, 2008. *A Guide to the Project Management Body of Knowledge (PMBOK® Guide)*, Fourth Edition. Newtown Square, PA: Project Management Institute.

Standish, 1999. Chaos: A recipe for success, Boston, MA. www.standishgroup.com

Standish, 2003. *Chaos Study*. The Standish Group, Boston, MA. www.standishgroup.com

Standish, 2008. *The Trends in IT Value*. The Standish Group, Boston, MA. www.standishgroup.com

CONCLUDING REMARKS FOR PART II

Part II has described many base-level concepts related to the project management historical background and operational role. The historical evolution of this activity can be traced through several hundred years; however, the mature phase of project management has occurred over approximately the past 100 years. A brief overview of that history is used to show how the thought processes evolved and continue to evolve today.

Examples are used here to show the general order of evolution for projects as they move from initial vision on through to closing. At this point, we know that projects have significant similarity regardless of the industry or type. There are some disagreements regarding how best to move a project through its life cycle, but little disagreement about the underlying KAs involved in that process. The PMI entered this scene during the 1960s with the goal of defining best practices for project management and they are pre-eminent in this role in their publication of the Project Management Body of Knowledge, called the *PMBOK® Guide*.

Important vocabulary and concepts are introduced in this part that will be used throughout the remainder of the book. Future parts of the book will take many of these introductory items and elaborate them to greater detail. It is important that the reader should not get lost in the details of the upcoming lower level discussions and forget what the overall goal is—that being to accomplish the desired project goal. Throughout the book there will be many examples of projects that failed at near catastrophic levels. The key question in each of these examples should be to examine why that occurred, given the visible facts. Just realize that no textbook model will make a poorly conceived project successful. The modern PM must use every trick in his tool kit to overcome the myriad of negative issues that can exist. Certainly, it is important to understand that the model view outlined here is a core requirement to begin this process.

Part III

Defining the Triple Constraints

LEARNING OBJECTIVES

Upon completion of this part, the following concepts should be understood as to why they are required as part of the core planning process:

1. Identify, analyze, refine, and document project requirements, assumptions, and constraints through communication with stakeholders and/or by reviewing project documents to baseline the scope of work and enable development of the plan to produce the desired result.
2. Develop the WBS using the Scope Statement, Statement of Work (SOW), and other project specification documents. From this definition, decomposition techniques are used to facilitate detailed project planning, which is then used in the executing, monitoring and controlling, and closing processes.
3. Analyze and refine project time and cost estimates by applying estimating tools and techniques to all WBS tasks in order to determine and define the project baseline, schedule, and budget.
4. Understand how to develop the resource management plan required to complete all project activities and then match planned resources to those available from internal, external, and procured sources.

Source: Initiation goals adapted from PMI, *Handbook of Accreditation of Degree Programs in Project Management*, Project Management Institute, Newtown Square, PA, 2007, pp. 17–20. Permission for use granted by PMI.

What is a project plan? Basically, it is a formal forecast of the future project state. It will define the deliverables, work process, and other information relevant to the project. The core variables dealt with in the plan are scope, time, and cost; however, other supporting variables will also be included. In the project management vernacular these core attributes are called the *holy trilogy*. In ctuality, there is a lot more to project planning than this but these elements are the most visible plan output parameters.

There is wide diversity of views regarding what constitutes project planning. The real world simply defines this as publishing a budget and schedule. Other groups might suggest that project planning involves a broad scope statement along with the budget and schedule. At the other end of the maturity spectrum, the *PMBOK® Guide*'s view is more robust in its consideration of other decision variables. It defines project planning as "… documenting the actions necessary to define, prepare, integrate, and coordinate all subsidiary plans into a project management plan" (PMBOK 2008, p. 48).

In this model-based view, the mention of "subsidiary plans" implies that other considerations are required in the planning process beyond the elementary view of scope, time, and cost mentioned above. Therefore, the guide implies that planning also involves the other six project KAs described earlier as encompassing:

- Quality
- Procurement
- Risk
- Communication
- HR
- Integration.

This expanded view means that the project plan should deal with all nine KAs in an integrated manner. One way to look at this is to recognize that scope, time, cost, and quality fundamentally represent the output goal attributes of the project, whereas procurement, risk, communications, and HR represent underlying decisions within the project to achieve those output goals. All of these supporting decision areas then have to be integrated into a coherent whole in order to achieve a viable project plan.

This part will focus primarily on the output triple constraint basics of scope, time, and cost development. Figure III.1 shows this in relation to the other KAs. Part IV will expand on this basic view by describing how decisions from the other six areas are melded into the final plan and how decisions in these areas affect that plan. The reader must recognize that many real-world PMs will take many shortcuts to develop a plan and personal experience suggests that few real-world projects follow the full rigor outlined in Parts III and IV. Nevertheless, when shortcuts are taken the PM must recognize what negatives may emerge as a result.

There is now strong evidence to support the idea that modern techniques in project management do in fact improve success probability (Thomas, 2008, p. 180). In some cases, a project failure is

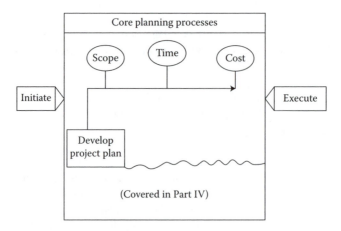

FIGURE III.1   Holy trilogy planning model.

not one of deficient planning, but getting the organizational participants to follow the plan. Just recognize that a good plan is not the only condition for success. It is a road map, but there is always unexpected environmental roadblocks found along the way.

To make the initial planning discussion more coherent and understandable, some simplifying assumptions are used in this part. These assumptions help to keep a more narrow focus on the basic mechanics of scope, time, and cost. In this first view, Chapters 12 through 14 specifically, the planning process will be assumed essentially constraint-free. In this assumption, there are no resource limitations, so the derived schedule is technically based only on the requirements and if the defined resources are provided the work will be completed as indicated.

This assumption set will be relaxed in Part IV, but in Part II we will concentrate on the basic core mechanics and steps for generating a scope, time, and cost plan.

Given the risk- and error-free assumptions the Part III segment of the planning process can be traced through the following eight basic steps:

1. Define the technical project specifications
2. Activity definition
3. Activity sequencing
4. Resource estimating
5. Activity duration estimating
6. Schedule development
7. Cost estimating
8. Cost budgeting.

These steps constitute the core activities of the project planning process. However, it is important to recognize that other planning activities are equally important to deal with, but this starting base is the best foundation onto which the other supporting activities can be added and discussed in a more coherent manner. Part IV will have that as its goal.

One way to partition the project planning topic is to view Part III of the book as the holy trilogy process planning mechanics and the upcoming Part IV as dealing with the remaining six supporting KAs: quality, HR, communications, risk, procurement, and integration. Logic for this topic organization is that the elements described in Part III are more mechanically orderly in their interrelationships than found in the remaining areas. The six KAs for Part IV have the attribute of being more iterative and less process structured.

Figure III.2 illustrates the process linkage between the three initial holy trilogy KAs.

The sequential flow illustrated in Figure III.1 is important in that a formal scope definition provides the foundation for schedule development and the resulting schedule then becomes the base artifact for project cost and budget definition. Many projects start off with these assumptions reversed. Schedule and cost can in fact be hard constraints, but the sequence defined here is that the flow of plan development should start with scope and let that dictate cost and schedule.

So long as our earlier assumptions remain valid, this stage one process is very orderly and represents one of the fundamental activities of project management. Specifically, Chapters 12 through 14 will outline what would be considered fundamental scope, time, and cost planning concepts and mechanics. Following this basic overview, a Technical Appendix supplement to this part offers

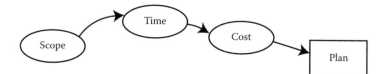

FIGURE III.2 The project plan triple constraints.

some additional advanced views related to this topic area. The appendix contains the following additional views:

Chapter 15—Deals with variable time estimates

Chapter 16—Reviews the concept of a project schedule as a chain of events as opposed to the traditional time-box approach

Chapter 17—Provides an illustration of operational techniques to simulate project performance.

Upon completion of this part, the reader will have a reasonable foundation of basic scope, time, and cost planning mechanics.

REFERENCES

PMI, 2008. *A Guide to the Project Management Body of Knowledge (PMBOK® Guide)*, Fourth Edition. Newtown Square, PA: Project Management Institute.

Thomas, J. and Mark M., 2008. *Researching the Value of Project Management*, Newtown Square, PA: Project Management Institute.

11 Project Plan Development

Approval of the project Charter moves the project process into a more formal plan development stage. During the earlier concept-visioning period the project was likely supported by a small number of individuals who philosophically believed in the endeavor and may have even been passionate about it. The planning process now moves the issue beyond this sponsor emotional point into steps designed to translate the original vision into a workable technical approach to achieve that vision. The goal now becomes one of defining the path and work elements required to achieve the vision.

In order to achieve future project success, the appropriate planning process should open up a more logical and technically driven analysis of the required work effort. Planning stage participants now include a wider variety of organizational and technical skill backgrounds and these new participants tend to view the proposed effort with less bias than the original visionaries. The primary goal of this second-stage activity is to collectively resolve the ambiguities remaining in the original definition and produce a work plan that will guide the project to completion. First, the process has to define in greater detail the WHAT (requirements) and then work toward architecting HOW the effort can be accomplished. This evolutionary activity will uncover a breadth of diverse opinions regarding the merits and technical approach required. This activity will likely produce political, organizational, and technical conflicts that the PM will be challenged to resolve. Through all of this, it is the role of the PM to work toward a common positive team spirit regarding successful completion of the project as compromises are sought. The planning participants must look at this stage as an attempt to find a workable integrated solution to the problem, while at the same time attempting to deal with the various internal and external issues that can negatively impact the outcome.

It should be recognized that the real-world PM will often be required to take many shortcuts in developing the project plan, and personal experience suggests that few real-world projects follow the complete rigor as described in Parts III and IV of the text. Many project teams do not do it because they fail to recognize the value in this level of analysis, while others do not do it because they believe that their environment is too dynamic and uncertain to plan.

In past years, organizations have purchased various models and computer programs that promised to make the project experience more success and many of these ventures failed to produce the desired result. There are multiple reasons to explain why these past quick-fix solutions failed. Certainly, one common scenario is the tendency for senior management and key users to prod the project to move on into execution before a planning cycle is complete. In many of these cases formal planning is not respected as having value. This rationale is often justified by saying that something different would happen anyway, so why waste time creating a document that does not map to reality. Regardless of the reason for taking planning shortcuts, realize that the views regarding project planning are controversial in organizations. It is also important to realize that many projects fail because these issues were not properly considered.

One example to illustrate the operational nature of a project plan involves work elements dealing with fragile technology. It is true that the impact of a yet unknown risk event cannot be included in the base plan because it has not happened yet; however, we must recognize that it might occur and find a way to manage the potential events in this category. The project plan defines work units as they are expected to occur, but when the risk events actually occur, the operational plan will have to be adjusted accordingly. A final project plan is a documented view of the anticipated project, hence changes during the planning or execution stages need to be continually updated in the operational plan.

Another characteristic of the planning process is its iterative nature. One should not expect to plow through the process in sequential work unit order and end up with an approved result. A more realistic vision is to view planning like "peeling an onion." As one layer is uncovered another layer of issues becomes visible and better defined. Maybe the vision of fear should be kept with this analogy as this is another emotion one might have as the project complexity is recognized. One example of this layering view is that uncovering project scope details will help understand the related time and cost requirement. A less obvious situation is that estimating work unit times will eventually define resource capacity issues, which then may require replanning to match the available resources. So, not only is the idea of a dynamic project expectation part of this evolution, but the iterative nature of the process is also part of the understanding. Once project constraints (time, cost, resources, etc.) are encountered, it will be necessary to replan what was already planned to deal with those issues. Failure to do this will invalidate the previous work. All too often the constraints are not taken into account, only to find later that the project will not work as defined. Ideally, a plan should be a future roadmap for the project and it should work just as a roadmap works for an auto trip. It should recognize where detours might exist and define ways to deal with such events. This is a risk-oriented view of the process. One does not just view the future path as clear roads and sunny skies. If you have a good roadmap and then receive information that the road ahead is under construction it will be possible to look at the map and take an alternative route. The project plan should have these same characteristics. So, when you think about planning keep the roadmap vision in mind. Also, understand our conceptual planning keywords: layering, iteration, contingency, and status information.

11.1 ARGUMENTS FOR PLANNING

All project teams are typically pressed to move on into execution quicker than they would prefer. Management and other stakeholder groups push the team to move forward and start getting visible items accomplished. Users feel like they have defined the requirements with the original business case or vision statement and the planning process is not adding anything to the process other than time.

Planning should not be viewed as a waste of time if you can produce a reasonably well defined target and believe that the technical steps necessary to achieve that target are known. The flaw in this logic is that this scenario is seldom the case. User requirements are typically not well defined or understood even by the user much less the project team. Also, project team members often have diverse views regarding how to produce the item. Given these fuzzy issues, there are perspectives that favor the development of a less thorough initial planning process with some form of iteration in developing the output. In other words, let the users see something working and then define the next step, iteratively marching toward some completion point. From a management viewpoint, the lack of a defined schedule, budget, and some definition of the final goal makes this approach hard to sell. Based on this view and the belief that iteration can waste resources, many organizations opt for a more planned initial stage prior to beginning execution. The sections below summarize a few of the common supporting reasons for this.

11.1.1 PROJECT MONITORING AND CONTROL

One of the classic rules of management is that you can only control what has been planned. This means that the omission of a coherent plan also means that you are giving up a significant ability to control the project. From a management perspective, the plan becomes the baseline on which to measure performance status through the life cycle.

11.1.2 Conflicting Expectations

Given that a project vision often is spawned in one segment of the organization, it is common to focus the solution on that segment. As the scope and impact of the project is better understood, there will likely be conflicting views regarding how best to orient the requirements for the benefit of the overall organization. What might be very productive for one segment of the organization could well create chaos elsewhere. Failure to define and review the overall requirements with a broader stakeholder group will often leave issues to be uncovered later when they will be costly to correct. Even when the issues are relatively minor, a stakeholder's frustrations are often caused by his lack of understanding the project directions—how to best execute the requirement, supporting processes, resource issues, and so on. The classic example of this would be to produce a costly product successfully according to the original requirements only to have the user population say "we can't use this," or to find a better product already in the marketplace.

One of the major purposes of the planning process is to evaluate the various views of the vision and work toward one that the overall organization understands and agrees to support. Everyone may not agree with the result, but they should agree to work positively toward that agreed upon goal and to understand why that particular choice is either appropriate or approved. Failure to go through a planning process would omit this negotiation process and buy-in result.

11.1.3 Overlooking the Real Solution

Often times, a new technology looks promising and brings the hope of some breakthrough solution to a perceived problem. Racing to achieve the use of a new technology can result in similar negative outcomes as described above. In this case, the new technology could require changes in organizational processes, structure, governance, or reward systems (Henry, 2004). In other situations, a narrow perspective of the project goal might completely miss the proper target. A valid solution at one point in time could be absolutely wrong, given dynamics in the organization or the external marketplace.

11.1.4 Competing Solutions

In the organizational environment, multiple project proposals and activities in place are likely at any one point in time. As efforts continue to define the technical approach or scope direction for a particular project, different solutions can emerge elsewhere. Experience indicates that the initial approach is often not the best strategic approach. Unfortunately, the zeal of the various parties can turn the selection process into a battleground of egos and parochial positions. It is up to the project governance structure to ensure that reasonable options are viewed and explored without turning the process into an "analysis paralysis" activity with no direction. Certainly, one of the reverse side risks of planning is over planning. Finding the right balance between too much time spent and not enough is a critical skill for the management team.

11.1.5 Misaligned Goals

We have discussed in previous sections the requirement that a project align its direction with organizational goals. As obvious as this point might be, it is one that is often difficult to achieve in the operational environment.

11.1.6 Quality Solutions

Even in the situation where the target project vision is proper and all are in synch with it, there is still a need to a produce a quality output. In order to achieve that goal the resulting plan must find an appropriate balance between the vision (scope), cost, schedule, risk, and quality. Very few projects are worth pursuing without regard to the associated cost, time, or level of risk. Likewise, few projects can afford to produce the highest possible quality. So, the planning function must find the correct balance between these competing goals. The only way to estimate these parameters is to carry out a reasonable level of planning.

In order to create an effective plan that properly deals with the issues outlined above, there are five items that must be resolved.

1. A combination of technical, resource, and process considerations must be dealt with
2. Diverse stakeholder expectations must be negotiated and documented
3. A proper balance between tactical and strategic needs must be reviewed and resolved
4. A solid review of the business case must be completed and matched with the subsequent requirements developed during planning
5. The organizational goal alignment requirement needs to be verified.

11.2   PLAN PROCESS AND COMPONENTS

There is wide diversity in organizational views regarding what constitutes a project plan. Certainly, the real world often defines this simply as a budget and a schedule with high-level statements of objectives. In order to be the type of roadmap outlined here, it must be recognized that a viable plan consists of many interrelated components involving the nine KAs (risk, communications, HR, etc.). In an overall planning model-based view, there are essentially primary and supporting sets of activities required to create a plan (the author's view and not specifically reflected in the guide). In this view, the final plan will contain a collection of related subcomponents. The primary plan components are often summarized for presentation to senior management and this view summarizes various key issues that are relevant for the project (i.e., cost, schedule, risk, etc.). Some project plans need to emphasize risk and time, whereas others might focus heavily on quality, time, or availability of critical resources. In any case, the total project plan should produce a broad vision of expectations and issues that are relevant to the activity and its stakeholders. The KAs form the organizational structure to view plan components.

11.3   INITIAL PLANNING VIEW

The following chapters in this section will view the planning process by making some simplifying assumptions in order to focus on the basic mechanics of scope, time, and cost. Figure 11.1 provides an overview of that simplified planning process view.

Basically, this view assumes that the requirements are well defined with essentially no uncertainties. From that viewpoint, all project resources are the same and available when we need them. By making these simplifying assumptions, it is possible to deal with some core planning concepts and avoid some of the more onerous sides of planning. If we admit at the outset that all nine KAs are in play, at the same time these core concepts would be harder to sort out. This restricted view is rationalized by the broad overall concerns over time, cost, and scope. These are the most visible output variables in the base structure of the project plan.

In this initial discussion, it is important to understand the linkage between scope, time, and cost. As simple as this linkage is, it is often misunderstood by many project participants and management. This lack of understanding causes many projects to be chartered with a fixed schedule and budget prior to a clear definition of requirements. This inverted approach has been named *Project Titanic* in the industry because it typically leads to a sunken project. The model sequence of the next three chapters as described in Figure 11.1 is basically sequential through scope, time, and cost. In other words, scope leads to time, which then leads to cost.

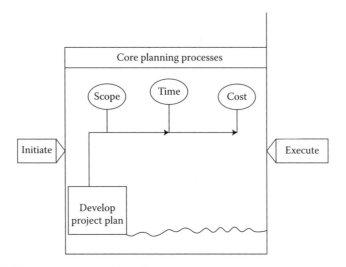

FIGURE 11.1 Model planning process (phase I).

From a project management perspective, it is imperative to approach the planning process by first performing a careful definition of project scope (deliverables). From this view, a preliminary schedule can be constructed and from that point a preliminary project cost can be derived. This triple set of linked processes is a cardinal component of project management. The plan resulting from these three initial steps is called the *First Cut Plan*, meaning that other processes are yet to be performed. Part IV of the text deals with the additional KA steps required to complete the planning process and produces an approved project baseline plan.

11.4 PLAN ARTIFACTS

It is important to recognize that the planning process is initially triggered by an approved project Charter signed by appropriate management. This formal initiation step includes a preliminary scope definition blessed by the sponsoring source. Both of these early artifacts are preliminary in nature in that they do not provide sufficient guidance to complete the project or the required visibility to measure its general technical or organizational viability. A second-stage basic planning question involves resolution of these broader views. There are many possible decision support artifacts that need to be produced as part of the planning process. The following list provides typical samples of planning artifacts that have high probability of being included in the final project plan documentation:

1. Approved scope definition—an expanded human language statement outlining the required project deliverables. This list will be reviewed by a broader group of stakeholders than the preliminary version.
2. A WBS that decomposes the project into work units necessary to produce the defined output.
3. Assumptions and constraints made as part of the planning process.
4. Work unit time and material estimates.
5. Work unit relationships (sequence).
6. Time-phased resource allocation plan (human and capital).
7. Major review points—technical and management.
8. Documentation requirements for subsequent phases.
9. Testing and user acceptance plans.

10. Project team training plan—skill requirement definition.
11. Status reporting metrics and delivery process.
12. Communications plan.
13. Risk management plan.

Items one through seven will be described in Chapters 12 through 14, while the remaining five items will be generally dealt with in other parts of the text. This summary list of artifacts is not meant to be comprehensive, but more to show that planning involves a broader perspective than is understood by most.

The degree to which the various output artifacts are utilized in a particular project depends on many factors. For example, large or costly projects often have sufficient impact on the organization to require a thorough plan in both scope and depth. Projects involving new technology need a heavy focus on risk management and work definition. Projects involving a new-type target or skill should be approached carefully with adequate planning. In general, smaller commodity-type projects can be pursued with less complex planning. Projects involving a common theme might require much less detail in their plans, or might be able to use templates from earlier similar efforts. The point of each example type is that projects have a lot of similar structural characteristics, but also have significant differences in their internal emphasis characteristics. It is important for the PM to decide on appropriate levels of detail for each KA in each project. Some organizations require a mandatory core set of activities and a supplementary collection of optional items.

11.5 REAL-WORLD PLANNING PROCESS

It would not be appropriate to leave this section without a comment regarding the real-world view of project planning. Do not be surprised to find that few real-world projects follow the rigorous planning definition described in these sections. Many do not do it because they do not recognize the value in this level of analysis. Many others do not do it because they do not believe that you can plan accurately, therefore a waste of time. So, the effort would be wasted in documenting something that will not come to pass. Regardless of the reason, realize that the views regarding project planning are controversial. It is also important to realize that many projects fail because these issues were not properly considered. As an example, risk management is one of the new hot topics in project management and many are now recognizing that this area can create significant issues if not dealt with properly in the planning phase. Our mission here is not to take sides on the ideal level of initial planning, but rather to show why each of the KAs has an important planning perspective that must be considered in the management process.

11.6 CONCLUSION

The key point to recognize in this overview is that the project plan should attempt to define the impact of each KA and deal with those issues to the degree required for a particular project. In some cases, a KA is not of critical concern and can be minimized in the plan. However, the nine KAs have

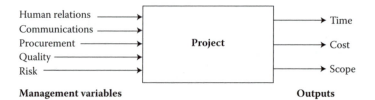

FIGURE 11.2 Planning variables schematic view.

been identified by very seasoned PMs and are defined in the formal model of project management. From this, one might suspect that each of them is relevant to some degree in every project plan.

Figure 11.2 illustrates a schematic way of looking at the planning process. In this view, the basic output is represented by the three core holy trilogy components, while five management related variables are needed to generate those outputs. Internal to the box is the concept of integrating all eight of these into a consistent set of objectives and plans. This is not a perfect characterization but conceptually puts the right focus on the management view.

In the final plan, all eight KAs are manipulated (integrated) to a state whereby each is viable and consistent to support the resulting plan outputs.

REFERENCES

GSAM, 2008. Guidelines for Successful Acquisition and Management of Software-Intensive Systems. Version 4.0, U.S. Air Force Technology Support. www.stsc.hill.af.mil (accessed August 28, 2008).

Henry, G. 2004. Best-Laid Plans. www.Projects@Work.com/content/articles/221772 (accessed August 31, 2008).

12 Scope Management

An initial scope view was created for the initial project approval process. We now need to realize that this preliminary attempt to document the project objectives only serves as a starting point for the project team. The initial view typically focuses on the logical requirement, but little in regard to the technical work requirements to achieve those goals. In order to structure the project work it is now necessary to translate the output requirements into work units required to produce those deliverables. Also, the initial verbiage outlining the project requirements was not rigorous enough to support a detailed planning process or to give other decision makers enough information to approve the initiative. The goal that we are looking for in this activity is to identify "the sum of the products, services, and results to be provided as a project" (PMI, 2008, p. 440).

In order to ensure that these criteria are met, both the users and the technical builders must have the same understanding of the required deliverables. The process of gathering these dispersed specifications is called scope definition and the result of this is often translated into a WBS.

The formal project-planning phase commences with various scope definitional activities designed to produce a clearer understanding of the project work units, deliverables, assumptions, and constraints. This activity leads to the development of a detailed WBS that represents the primary scope definition artifact for the project. Essentially, scope analysis must provide definition for the following:

1. Definition of stakeholder *needs, wants, and expectations* translated into prioritized requirements.
2. Definition of *project boundaries*, outlining what is included in the project scope and what is not included.
3. Definition of the project *deliverables* including not only the primary product or service, but all interim results as well. This includes items such as documentation and management artifacts.
4. Definition of the *acceptance process* to be used in accepting the products produced.
5. Lists and defines project *constraints* that must be observed by the project.
6. Lists and defines project *assumptions* and the impact on the project if those assumptions are not met during the course of the life cycle (PMI, 2008, p. 116).

Italicized items above highlight the key resolution points that are essential elements needed for a clear specification. The *PMBOK® Guide* outlines the following three sequential steps for scope determination:

- *Scope planning:* This initial step is designed to formulate the approach for definition and management of scope through the life cycle. This is essentially a level of detail decision point.
- *Scope definition:* This activity is an extension of the preliminary scope statement produced during the initiation phase. The earlier effort was designed to provide a general view of scope, while this second iteration will be more rigorous in its structure and will include inputs from a broader group of stakeholders. The goal for this stage is to produce refined

definitional statements in the six areas summarized above. The resulting set of specifications is documented in the updated Scope Management Plan.

- *WBS creation:* The WBS is a hierarchical representation outlining the structure of work for the project. It decomposes the project into layers of smaller and smaller groups of work until the lowest level represents manageable work packages (WPs). The WBS is a fundamental core document for the team and drives many of the subsequent phase activities (PMI, 2008, p. 117).

12.1   DEFINING PROJECT WORK UNITS

Envision for a moment a "package" of work has been defined as part of a larger project. This could involve tasks such as installing piping, testing software, or producing a user guide. For now, view this simply as a defined set of work activities associated with a larger project effort. The vocabulary for the lower-level work units is a WP. One metaphor for a WP is to view it as a box, such as the one shown in Figure 12.1, with the three dimensions representing time, cost, and scope (of work). The sequencing of WPs will be discussed in Chapter 13.

In order to develop a project plan to execute the contents of this box we must first make certain technical decisions regarding the work necessary to produce the desired product. This section will discuss some of the basic processes required, derive an initial schedule and budget for this the project. As a starting point let us define a WP as "a deliverable or project work component at the lowest level of each branch of the work breakdown structure ... Sometimes called a control account" (PMI, 2008, p. 422).

At this point, we have not illustrated how a WP fits into the full project structure as yet, but the key point for now is to recognize that it represents some deliverable component of the overall project that requires work to accomplish. The full collection of these would then represent the complete project scope definition.

One common rule of thumb is that a WP should be defined and sized to represent approximately 80 h of effort and/or two weeks of work; however, that definition might not be appropriate for some project types. The key point is that this work unit definition becomes a management focus point for the entire life cycle of the project. A WP will be defined and managed by its estimated requirements for human and material resources, support equipment, and other parameters related to the work. From these input parameters the WP will produce some defined output. In other words, we basically want to estimate how long it will take to do that work, what resources are required, and how much the effort would cost. In essence, this is the concept of a project plan.

12.2   WP PLANNING VARIABLES

Effective project WP planning requires many items of information related to the activity in order to assess the content of the work and integrate it into the rest of the project. Conceptually, it is

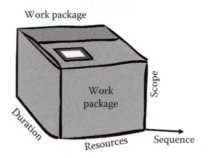

FIGURE 12.1   WP dimensions.

desirable to capture this information into a single data repository that we will call a WBS Dictionary. Such a repository may not be visible by that name in real-world projects, but the type of data described below will need to exist somewhere in the project records. In order to estimate and track the project activities, it is necessary to define the following types of definitional parameters for each WP:

1. ID reference—this is a code used to identify where the WP fits into the overall scheme of work (e.g., the WBS)
2. Labor allocation for the task (e.g., number and skills of workers)
3. Estimated duration for the task, given the planned worker allocation
4. Materials associated with the activity and their cost
5. Name of individual responsible for managing the activity
6. Organizational unit assigned to do the work
7. Defined constraints (e.g., activity must be finished by …)
8. Key assumptions made in the course of the planning activity
9. Predecessors—linkages of this activity to other project work units
10. Risk level for the activity—more work will be needed on this aspect later; for now we just need some indication of the perceived risk level (e.g., H, M, or L)
11. Work description—this may be a reference link to a location where more detailed information is kept
12. General comments—free-form statements that help understand the technical aspects of the work required.

In addition to these core definition items, it is also necessary to have certain management oversight into the estimated values derived. In order to control these values, it is common to have additional approval fields in the WP record showing acceptance by the performing organizational group and the PM. With this set of information completed, we have sufficient information to produce a first cut schedule and budget for the WP. This statement is a clue that there might be more iterations later as new information arises.

From the planned work and resource allocation data, an initial schedule can be calculated using the WP duration values. For example, if the work effort was estimated as 80 h and a resource allocation of two workers, then the number of work days would be 40. However, from this we still do not know exactly what calendar time frame the group is working on since workers will have scheduled days off typically on weekends and holidays, plus other days allocated to vacations and nonproject company activities. It needs to be recognized that workers are typically not allocated 100% of the time to an activity, which in turn affects the actual observed time to execute the task. All of this represents more messy scheduling mechanics to deal with later.

12.3 MULTIPLE WPs

Moving one step up the food chain in our WP theory, let us review the example first discussed in Chapter 8, "Quick Start." This example assumes that two WPs are related, meaning that two sequential steps are required to produce some desired outcome. This view is technically called a Finish–Start (FS) relationship. Metaphorically, we are stacking our two boxes end to end as shown in Figure 12.2.

For this example, let us assume that all workers involved have the same skill levels and can do all tasks required. They are also paid the same. These handy assumptions allow us to avoid many messy issues in the calculation process, but they help simplify the focus on basic raw mechanics of scheduling and budgeting. Review the discussion in Chapter 8 to refresh the vocabulary terms effort/work, duration, and elapsed time. We will see more of these terms in Chapter 13 and beyond.

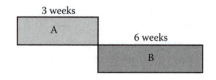

FIGURE 12.2   Multiple work packages in sequence.

12.4   DEVELOPING THE PROJECT VIEW

Up to this point, we have looked at low-level components of the project and begun to develop a working management model for that level. Unfortunately, a project consists of hundreds or thousands of such elements. In a large project, the interaction of these elements threatens to cloud this simple view. It is true that increased project size and complexity can make the management process look foreboding; however, the fundamental building block elements consisting of WPs will stay essentially intact. The key conceptual additions from this point are WP interactions with each other and the impact of the project environmental issues on the original planned view.

If all WPs could actually be accurately estimated as described, thus far the process of developing requirements, schedules, budgets, and the overall management process would be relatively simple. However, the fact that most organizations do not experience project success rates much above 50% (many lower than that) suggests that other factors somehow get introduced. Nevertheless, building a solid WP level management foundation is a key step in dealing with the vagaries of the external world. Regardless of where or how we start the planning process, the goal is to converge to this level of granularity.

12.5   DEVELOPING PROJECT WBS

It is always difficult to suggest that any one activity is the most significant one to support the management and work control process, but there is heavy evidence from industry researchers that indicates failure to properly define project scope creates problems later and often leads to decreased success rates. One proven technique that has been found to be of broad value in the project planning and control process is the WBS. Our bias is that a properly created WBS is the most important planning and control artifact and it has broad impact across the various project management processes. The fact is that a well-developed WBS is the best tool available to define and communicate various aspects of scope and status for the project. In this section, we will review some of the key concepts and mechanics related to this activity.

As the project definition moves from a high-level vision-oriented definition of requirements toward the technical "what's" there is a need to translate the English verbiage into something more akin to a structured technical work definition. This translation process is the fundamental role of the WBS. The value of this approach is that it serves as a good communication tool between user plain language requirements and the technical work required to produce those requirements. Both technical and nontechnical groups can understand the result, and it provides a communications bridge to confirm that the stated requirements are being produced by the defined project work units.

Before jumping too deeply into this topic let us illustrate a WBS with a simple example. The simple structure involving construction of a house is shown in Figure 12.3. The nine subcomponents are meant to identify major skill and work areas necessary to complete the work. The role of each component is to deliver a defined portion of the overall project requirement. Collectively, these combined boxes represent the total scope of the project. Note that a box dedicated to project management is included in the scope definition since it is a required work activity and consumes resources.

Various sources describe general rules of thumb regarding the WBS construction process. As an example, PMI's *Practice Standard for Work Breakdown Structures* provides guidance and sample templates for various types of projects (PMI WBS, 2001). The accepted approach shows the top

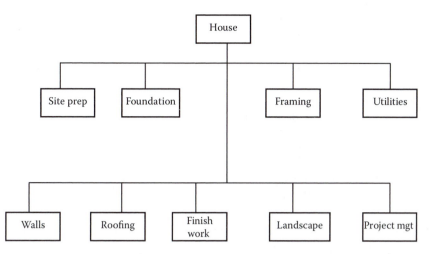

FIGURE 12.3 Basic WBS structure.

WBS box indicating the short name of the project or program that is being defined. The second level would contain names of the project, project phase, or major subsystem depending on the scope and type of project plan. The third level continues the scope decomposition and by this point, the structure should begin to reflect the major work groups and/or key deliverables. According to the Department of Energy (DOE) methodology, the first three levels of a WBS are:

Level 1: Shows major project-end objectives.
Level 2: Contains major product segments or subsections of the end objective. Major segments are often defined by location or by the purpose served.
Level 3: Shows definable components, subsystems or subsets, of the Level 2 major segments (DOE, 1997).

Even though standard construction rules have logical value for future project comparison and potential reusability, there is no one right answer for constructing a WBS. Its primary role in the process is to describe the work structure from the eyes of those responsible for delivering the output, that is, variable depending on project size, technical approach to the effort, and the organizations doing the work. There are several ways to construct a proper structure. Samples of these are the following:

1. Using standard templates from similar recurring project types
2. Modifying the structure used from a similar effort
3. Defining the major work organization groups and then decomposing the structure from the top-down
4. Start with lower-level defined work elements and aggregating them upward into a logically hierarchical structure
5. Packaging the structure using the required deliverables as guidance (Schwalbe, 2006, p. 163).

Regardless of the method used or resulting structure, the final WBS view should eventually contain a set of reasonably small WPs at the lowest level. These collections of defined work effort represent the total scope of the project. In other words, project requirements should be mapped to specific WPs to ensure that all user-approved requirements have been defined in the technical work structure. Envision a WP as a defined and managed unit of work. As another metaphor, these are the molecules in the chemical compound—we might need to expand this analogy to include the view that the human and material resources assigned to these molecules are then the atoms.

As mentioned previously the general sizing rule of thumb for a WP is that it should be approximately 80 h of work and two weeks in duration. For large projects this might be too restrictive and for small projects it might be excessive, but it gives a general sizing metric. For operational reasons the WP should normally be linked to a single organizational work group, or at least have a single manager assigned responsibility for the effort. What is not obvious at this point is that the WP will be the basic item used for detailed planning, execution, and control. In order to construct the project plan, it is necessary to define the related work content, resource requirement, schedule, and budget, plus other requirements relevant to the individual WPs. As the project progresses, actual status will be collected for these items. All WBS summary aggregations above the WP level are simply groupings of their lower level items. Be sure that you have a clear understanding of the WP concepts introduced thus far. These items are the basic technical and management building blocks for the project and they are the items that must be in place to help drive the project to successful completion.

In addition to the visible project activities related to product delivery, there are other supporting activities that should be reflected in the WBS. A sample of these follows:

1. Developing product approval processes with the future user
2. Project planned meetings (team and external)
3. Team management/customer interfaces activities
4. Quality inspections and defect repair processes
5. Training activities (team and users)
6. Project management activities
7. Project formal communication requirements (status reporting and presentation preparation)
8. Additional project related management processes that need to be developed (i.e., change control, quality assurance, etc.)
9. Project startup activities
10. Planning for deployment of the project output and ongoing support
11. Creation of future operational service level agreements with outside support groups
12. Project closeout.

Many of these activities represent key life cycle management decisions more than work unit specification; however, if issues such as this are not unaccounted for, the required staffing level will be inaccurate. What this means is that the WBS must also include environmental work activities that on the surface do not look like requirements. These are requirements that are put in place to improve the probability of success.

The WBS structure is a great tool for showing the basic work organization of the effort, but it is weak on task definition. To supply needed details to this structure a companion document called the WBS Dictionary is used. The purpose of this document is to provide needed descriptive detail for each WBS component. Specifically, the following details should be recorded for each WP element shown in the structure:

1. Statement of work (SOW) description
2. Codes to support tracking of organizational resources and project financial details (i.e., WBS or organizational accounting codes)
3. Deliverables
4. Acceptance criteria
5. Associated activities/tasks (predecessors and successors)
6. Milestones
7. Responsible organization for the work

8. Resource estimates
9. Start and stop schedules (this may be kept in the Project plan)
10. Quality requirements and metrics
11. Technical reference
12. Contract information
13. Constraints and assumptions (PMI, 2008, pp. 121–122).

Because of the data intensive nature of the dictionary items defined above, many organizations skip lightly through this process. It is possible to avoid having a single data source dedicated to document many of these items, but for future ease of access this class of data needs to reside in a formally defined location. If the project plan supporting data is documented in a formal computer-based repository, the internal project team can have ready access to needed work items. Searching for data can be a very wasteful activity for team members. Details related to items such as schedule, budget details, resources, organizational assignments, and task relationships should all be contained in the formal project repository.

Each WP requires a SOW that can be used by the assigned resources to execute the requirement. In addition, quality and related technical reference items should also be kept in the WBS Dictionary. The combination of a WBS, its dictionary, and the corresponding project plan provides a formal communication source for the project stakeholders. Regardless of the data capture strategy, appropriate specification of these data elements is fundamental to the project management activity.

Development of a good WBS structure is not an easy task and the guiding principle is that it helps both the project team and external stakeholders understand how the project is structured. In order to be of maximum value, the project plan structure should map to the WBS, therefore one approach to developing the WBS structure would be to align it around the development methodology being used. In many cases, there are supplemental activities that are external to a standard product development methodology. In these cases the project plan becomes a hybrid of the basic technical methodology with supplemental management and support work efforts attached.

Recognize that a detailed WBS will not always be fully decomposed to the WP level during the planning phase. At some point in the planning process, the decision could be made that the requirements are sufficiently defined and then leave further elaboration for the execution phase. This means that some segments of the WBS would contain work units at a higher level of definition that described thus far. This also means that there is a higher potential error in the planned values. These higher-level work units are called *planning packages*. They have the same general definitional needs as a WP and would be mechanically dealt with as schedules and budgets are derived. Regardless of the work unit level reflected in the planning phase, each box in the structure would be linked to some specified organizational entity. Prior to actual work being performed on a particular unit, a more detailed work unit plan should be completed. This form of plan evolution is called the *rolling wave* approach.

12.6 WBS MECHANICS

To develop a WBS, you must first identify each of the major tasks or goals required to fulfill the project objectives. The top level of the structure should represent the highest view of the project. This could be a program in which the project is one component. Alternatively, it could be major phases of the project, or major deliverables. The list below offers some basic decomposition steps to consider:

1. Identify the top-level view that represents how the project will be defined within the program, phase, or component structure.
2. Identify the goals of the *entire* project. Consider each primary objective as a possible top-level element in the WBS hierarchy. Review the SOW and project scope documents to aid in this decision.

3. Identify each phase or component needed to deliver the objectives in step one. This will become second-level elements in the WBS.
4. Break down each phase into the component activities necessary to deliver the above levels. This will become third-level WBS elements.
5. Continue to break down the activities in step three—these tasks may require further decomposition. This process should continue downward until the work units are identified to a single organizational unit owner. Where possible, the defined work units fit into the size definition supported by the organization, with the standard rule of thumb being two weeks duration and 80 h of work. If this is not feasible at this stage the unit should be labeled a planning package and marked accordingly. It is a management decision regarding further decomposition of the larger packages, but scheduling accuracy and future control granularity will be lessened if large work unit packages remain.

The theory and concepts for a WBS are easy to relate to; however, a basic question remains regarding how to identify the correct structure for the WBS? The list below contains items that may help decide on the proper packaging structure:

1. Are there logical partitions in the project? What phases make sense?
2. Are there milestones that could represent phases?
3. Are there business cycles that need to be considered (e.g., tax period, production, downtime schedule, etc.)?
4. Are there financial constraints that might dictate phases?
5. What external company life cycles might impact the project?
6. What development methodology process will be used?
7. Are there risk areas that need to be recognized (e.g., technical, organizational, political, ethical, user, legal, etc.)?

Project team involvement is imperative for moving the WBS through the "yellow sticky note" stage. Using this technique, team members discuss proposed views and draft possible package names on sticky notes that are then arranged on a wall into the draft WBS hierarchy tree structure, as shown in Figure 12.4.

Through subsequent discussions various sticky notes will be moved around in the WBS until a particular work organization is agreed upon. This process should normally work from a top-down view of the project scope. Once a particular WBS level is defined, it may be possible for a portion of the structure to be allocated to a subgroup for more detailed discussion and decomposition. Later, each subgroup would make their presentation to the whole team to ensure that other "across-the-tree" editing is not needed. This should generally occur on a level-by-level basis through at least the planning package definitions. As the process moves downward the level of interaction across the structure should

FIGURE 12.4 Developing the WBS with sticky notes.

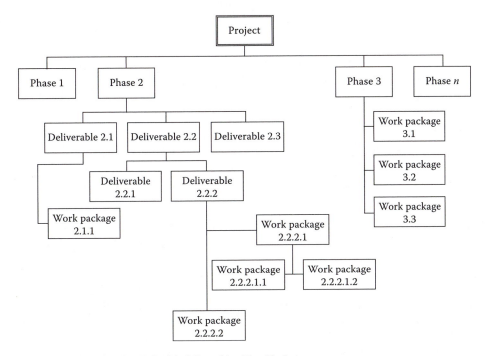

FIGURE 12.5 Sample draft WBS with deliverables identified.

decline as the work begins to align itself into specific organizational units. When the draft process is complete down to at least planning package sizes, a final project team review of the overall structure is required. At this point, the team must buy into the structure and its represented work. This process benefits the project team, as they will gain a deeper understanding of the project scope, roles and responsibilities, potential risk areas, and critical assumptions. In addition, this overview stimulates team communication and a spirit of collaboration. Sessions of this type can complete a draft WBS structure in fairly short time periods, assuming that the project scope is familiar. However, if the project involves high complexity or a new-type venture, then this process might be iterative and require multiple sessions to resolve. Regardless, the process of decomposition and review remains the same.

Figure 12.5 illustrates a sample draft WBS. In this model some boxes have been labeled as a deliverable rather than a summary package. The focus point of this view is that defined deliverables were part of the discussion and the team wished to ensure that the structure contained the specified requirements. In any case, it is important to review the final structure to ensure that the work defined will produce the required deliverables and that should be included in the quality management aspects of the scope development. In other words, if work unit 2.1 is to produce a specified deliverable, that should be part of a future quality check on completion of that activity.

Keep in mind that the number of levels in a WBS structure depends on the size and complexity of a project. Another stylistic consideration is to use nouns as titles in the structure, rather than verbs. This helps focus on what is to be done and not how it will be produced. *The focus of the WBS structure is on the work required to produce the project deliverables!*

12.6.1 WBS NUMBERING SCHEME

There are various schemes used to label the boxes in the WBS structure; however, in most of these a decimal point approach is used to reflect the hierarchical layers. Mature project management organizations such as the Departments of Energy and Defense have standardized schemes for numbering and the higher-level structures. A less elegant method is to label the top box "1," then the

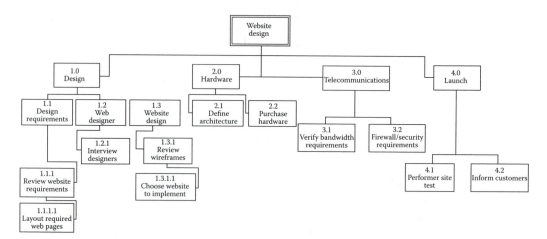

FIGURE 12.6   Sample WBS numbering scheme.

layer below would be 1.1, 1.2, 1.3, and so on. If the top layer is labeled "0," then the layer below would be defined as 1, 2, 3, and 4. Regardless of the scheme, the layers below generally attach a decimal-type notation. The sample WBSs shown in this section illustrate typical numbering notation (see Figure 12.6).

It is also advisable to leave some sequence numbers open for future changes and additions to the scope. This could be done by using increments of 10 in the numbering sequence. In any case, recognize that future changes will add work units to the structure, so the numbering can get messy and disorganized without space to add. As a final point on the schematic mechanics, realize that these numbers will be linked to the other project processes for cost and schedule tracking, risk, communications, HR, contracting, and others. This coding system provides an excellent method to communicate understanding of the overall process.

12.6.2   WBS Dictionary

The companion artifact to the WBS schematic is the WBS Dictionary. Its goal is to add content to the structure. Table 12.1 illustrates typical data elements that are included in this definition.

TABLE 12.1
Sample Data Dictionary Template

Project	WBS Task No.		Person Responsible	
Task Work Description (What work is authorized)				
Task Deliverable				
Acceptance Criteria				
Duration	Cost $ Total		Deliverables	
Due Date	Preceding Activity	Team Member Assigned	Succeeding Activity	Team Member Assigned
Resource Assigned Purchasing				
Approved by Project Manager			Date	

Note that each box in the WBS structure would require this type of information in order to support the planning process.

12.6.3   OTHER WBS VIEWS

The concept of structuring work has many operational benefits to the project management process. In order to support different control or analysis needs, it is typical to sort the basic WBS data into other views. The most common restructuring occurs in the case of third-party vendors being involved in some subset of the project. In this situation, the element of the WBS that is contracted will generally not be under the control of the internal project team. Assuming that is the case, then the contracted "branch" of the structure would essentially be extracted from view and allocated to the contractor for management. In some cases the contractor would need little added specification for their activity, while in more complex situations they might have to be fully involved in the decomposition process to the WP level just as though they were internal members of the team. If the contractual arrangement is fixed price, that part of the plan might be shown on the WBS as a single summary box. However, other reporting needs or contractual options might necessitate further WBS detail be displayed, but most likely not to the vendor's internal WP level. This substructure is called the Contract WBS (CWBS). It is connected to the master WBS, but generally stays under control of the third-party vendor to manage without specific details being visible to the buyer.

A second presentation form for the WBS is organizational centric, meaning that it is sorted into collections of work by organizational group. This view is called an Organizational WBS (OBS). Similar to this view is a time-phased sort by resource categories used to show total skill requirements represented by the WBS. This view is called a Resource WBS (RBS). So, if we were attempting to capture product costs or levels by resource type, it is possible to define that level of detail in the WP and sort the data accordingly.

There is yet one more sorted view of the WBS based on risk categories or time phasing of risk events. It is confusing to recognize that this is also labeled an RBS as well. In this case, the concept is a risk breakdown structure. So, the warning here is to read this acronym in context to sort out which view is being described.

As projects evolve, changes invariably creep into the requirements set. These often occur because more is now understood about the system and if the changes are properly managed this can be a reasonable process. As part of the change control process management has to agree that the change value is positive and a desirable choice. This typically means that the budget will increase because something extra has been added. When this occurs, it is possible to re-establish the project plan baseline. This would be called the *revised baseline* and the process is called *rebaslining*. It is important to understand what each of these baseline points represents. As an example, it is not appropriate for the PM to be blamed for a budget overrun if management has approved a 20% increment in the project requirements through various changes. In this case, a 20% growth in actual costs over the original baseline may be reasonable. On the other hand, comparing end of project costs to the original baseline does show how much the scope expanded during the project. Analysis of these comparisons can provide important lessons learned for future projects, and for that reason the role of various baselines needs to be understood.

12.6.4   TRACKING STATUS OF THE WP

The status tracking process will be based off an approved plan. Typically, the approved plan is frozen through the life cycle, so that actual comparisons can be made to it. This frozen view is called the project *baseline*. Most organizations have a formal financial system designed to collect actual project cost information and in most situations the formal enterprise system will be the system of record for project accounting. Other project information is captured in various operational repositories, so it is then necessary to link between the formal accounting system and the project

9000	Project X
9000.10.1	Subgroup A
9000.10.10	WP 1
9000.10.11	WP 2
9000.10.12	WP 3
9000.20.1	Subgroup B
9000.10.10.1	WP 4
9000.10.2	WP 5

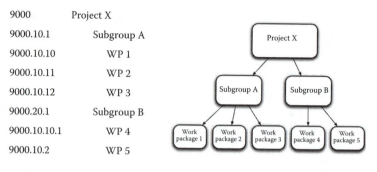

FIGURE 12.7 WBS coding scheme.

systems. In order to compare planned time or cost data between these two sources, it is necessary to establish a linkage mechanism. Developing a translation key that maps accounting codes to the project WBS typically does this. Here, the financial system will need to be mapped to boxes in the WBS structure for purposes of actual cost collection. These common mapping points are called *Control Account Package* (CAPs) and they can link to low-level work, planning, or summary level activities. The level of visibility into actual resource usage desired dictates location for CAPs. It is theoretically desirable to have actual cost data captured at the WP level, but in many cases the extra administrative work to do this is not justified. So, location of the various CAPs in the WBS hierarchy is a management decision. In operational mode, actual costs would be collected for the appropriate CAP, which in turn would allow for plan versus actual financial comparisons to be made at that level.

The example provides a simple cost collection approach using WBS-type coding. This can be accomplished in several ways, but the key is to use a coding scheme that allows different hierarchical WBS levels of detail to be aggregated. Also, note the companion indented format WBS structure and a corresponding equivalent schematic version. For this example, the assigned project code is 9000. This number would be prefixed throughout the WBS structure as illustrated in Figure 12.7.

By using this mapping structure, it is possible to track planned and actual costs status at the desired CAP. In the sample case, if we wanted to review the budget and current costs for Subgroup A (9000.10), we would simply add up the 9000.10 account hierarchy (e.g., WPs 1, 2, and 3). Likewise, project × total cost could be derived by adding up all 9000 subgroup accounts. It is important to review this rollup logic until you understand how costs are aggregated in the WBS structure.

The key to obtaining status from hierarchical coding schemes of this type is to define an appropriate hierarchical account code logic in the WBS so that the lower level activities roll up properly. In some cases, there can be multiple account requirements within the project (e.g., tracking tactical and capital costs for each account). In such situations some dual coding or allocation scheme would be required. In addition to project cost collection, there is potentially a need to summarize costs back to organizational groups. These additional requirements further complicate the coding process, but the same type code logic would support these needs. Also, recognize that there are other many other coding approaches that can be used, but the sample one shown here illustrates the idea reasonably well and any other schemes should work along the same lines. For the examples used in the rest of the book simple coding schemes will be used, but keep in mind that large organizations would have to deal with a more complex code structure, or a mapping algorithm to link the simple internal scheme to the enterprise financial system.

12.7   WBS CONSTRUCTION MECHANICS

At this stage in the project planning process, the goal is to translate the logical requirements into a technical structure that will produce the defined deliverables. The six steps outlined below are intended to provide some guidance in regard to how to look at the requirements and translate this into a technical work structure for the project. Also, keep in mind that there is no book solution to this. The guiding principle is that the resulting structure will map to the way in which the project will be envisioned by the organizations executing the processes.

1. *Identify the WBS components.* Group this level using one of the following structuring approaches:
   a. Major project phases (i.e., phase I, phase II, etc.)
   b. Major projects under a larger program
   c. Methodology life cycle phases
   d. Major deliverables
   e. Organizational responsibility
   f. Geographical location
   g. Process sequence
   h. A hybrid of the above structures.

   Regardless of the WBS structure chosen, it should reflect the way in which the project will be managed and executed.

2. *Decompose the lower tiers into activities.* Decompose the next layer to an appropriate level of detail. At this level in the structure, the activities identified should have the attributes outlined earlier for the WBS Dictionary. The minimal definition items are as follows:
   a. A defined management owner who is responsible for the work
   b. Clearly defined and measurable outputs
   c. Defined estimates of resource, cost, and duration
   d. Quality definitions that can be monitored through performance criteria associated with each output
   e. WP sizes should be close to guidelines (40 h and approximately two weeks in duration). Review any deviations.

   This decomposition process continues downward through the structure until a sufficient scope definition is achieved. This final planning level can be to the WP level or to some higher summary activity (planning package), if it is judged to be sufficiently clear for planning purposes.

3. *Develop WBS ID code scheme.* Using a defined coding structure, assign a unique identifier to each WBS activity based on its hierarchical level (Figure 12.8). The coding structure will be used to reference WPs and track actual status of the work. In developing the number scheme consider how future changes to the requirements might be accommodated. As mentioned earlier, incrementing the codes by say ten rather than one might help with this.

4. *Assign owners to activities.* Each WBS box entity requires assignment of a responsible manager who is charged with overseeing that aspect of the work. Activity owners assist

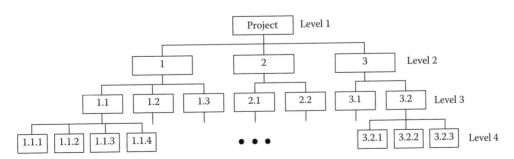

FIGURE 12.8 WBS planning structure.

in planning their activities and are later responsible for ensuring that the work gets done to specifications and within agreed schedule and resource constraints. When no owner is explicitly assigned to a work unit the ownership role defaults to the responsibility of the PM.

5. *Define completion criteria.* Documentation for project WBS activities is an important planning requirement. The purpose of this is to provide a measurable mechanism to judge future status of the activity. These requirements should provide guidelines to evaluate completeness. A sample work unit completion criteria requirement is shown below:

   Example completion criteria—Complete six error-free runs of the approved test script against the integrated code produced by the WP and document the results in the project test database.

   Activity owners identified in the WBS structure are responsible for the maintenance of this data; however, both the owner and the PM should approve the contents of these documents to ensure that they are clear and appropriate for the task.

6. *Support area review and further analysis.* At this point, we have completed the initial planning scope definition steps that translated the initial requirements into work units. The next step involves reviewing these work units in consideration of the various support activities required to execute the project. These support activities come from various KAs that need to be recognized before the scope definition can be complete. Example considerations must be reviewed in the following areas:

   a. Human relations (skill availability, capacity, and organization)
   b. Quality plans
   c. Risk assessment and response plans
   d. Communication plans
   e. Project organization and staffing plans
   f. Procurement plans (third-party human resources and material)
   g. Project management.

   Each of the support areas outlined above can impact the project work scope. For example, the decision to outsource a portion of the scope would impact defined work units related to that activity. In similar fashion, decisions on risk mitigation or transfer would likely impact other aspects of the project. Decisions from these related areas must be integrated into the WBS plan since they impact the resulting scope, schedule, and/or budget. Once these various decisions have been incorporated into the WBS structure and recorded in the WBS Dictionary, the next step is to move forward to complete resource estimates for each WBS activity. From a theory viewpoint, this would mean that the decisions would be at the WP level, but from a practical viewpoint it might stop at higher-level aggregations when it is decided to move forward using some planning level activities. When taking this short cut, there is a risk that lesser knowledge of the work required will contain higher estimating error rates. Nevertheless, it is a management decision to decide when it is time to move the plan forward and leave further definition to a later time.

7. *Management approval of the plan.* Before the scope definitional process can be considered complete, it is necessary to review it with regard to various stakeholders' view of the project. The WBS serves as a good communication tool to discuss the technical and management approach envisioned for the project. From this view, it also offers a somewhat refined high-level view of time and cost. This level of review should be considered as a mid-stage plan assessment. From this point, further work is needed to produce a viable schedule and budget before seeking formal approval of the plan; however, if a confusion over requirements surfaces at this stage that are counter to the original business case, these points should be presented to management for resolution.

12.8 REQUIREMENTS "IBILITIES"

On the surface, the requirements definition concept seems pretty simple—identify what the customer wants and then figure out technically how to construct it. The general scope discussion to this point has had that flavor. However, there are a set of not so obvious issues that fall into the crack of the requirements definition. We call these the basic nine "ibilities":

1. Traceability
2. Affordability
3. Feasibility
4. Usability
5. Producibility
6. Maintainability
7. Simplicity
8. Operability
9. Sustainability.

Each "ibility" represents a work unit attribute to be considered in the requirements definition evolution from concept through design. We must recognize that the project goal is not just trying to produce a stated deliverable. It must also consider a broader look at the resulting attributes of the result. In order to do this it is necessary to review the approach taken and adjust the scope statement according to each of the nine *ibility* attributes to ensure that the approach chosen appropriately matches the real requirement. In many cases, a particular solution will involve a trade-off of one or more of these attributes based on time, quality, functionality, or cost constraints. These decision alternatives will present themselves along the following general lines:

1. Present versus future time aspects
2. Ease of use versus cost or time
3. Quality versus time or cost
4. Risk of approach
5. Use of new strategic technology versus a working tactical approach, etc.

As the project moves through its life cycle processes of scope definition, physical design, and execution each of these considerations should be reviewed. All too often one or more of the *ibilities* is ignored or overlooked and the result is downstream frustration by someone in the chain of users or supporters of the item. The section below will offer a brief definition and consideration review for each of the *ibility* items:

Traceability is the ability to follow a requirement's life span, in a forward and backward direction (i.e., from origin, development, and specification, its subsequent deployment and use, and periods of ongoing refinement and iteration in any of these phases).

Affordability relates to a match of the design approach to the budget. There is always pressure to cut costs through the design, but in many of those decisions cause some adverse impact on another of the *ibilities*.

Feasibility can wear many hats in the project environment. The most obvious of these is the technical feasibility of the approach. Often times, stretching to achieve some performance goal will go beyond the existing technical capabilities and create additional risk. In similar fashion, the lack of critical skills availability can adversely affect the outcome. Think of feasibility as anything that can get in the way of success, whether that be technical, organizational, political, resource, or otherwise.

Usability is similar in concept to operability, except that in this case it more involves the resulting value generated by the output. It is what the process or product does in the hands of the future user. This can be either reality or perception based, but is certainly a concern for the project team to deal with.

Producibility is an attribute associated how the actual item will be created. In many cases, there is a gap between the designer and the builder, so the key at this stage is to be sure that the building community is represented in the design and probably even in the initial requirements process. Think of this as a "chain" of events that need to be linked together and not just thrown over the wall to the next group. Each component in the life cycle needs to consider this attribute.

Maintainability deals with the item in production. The consideration here is how much effort is involved in keeping the device ready for operations. In the case of high-performance devices there is often a significant downtime for maintenance. Having a device capable of "jumping tall buildings with a single bound" sounds good, but what about if it can only do that about 10% of time, with the remaining period being down for some type of maintenance. There is clearly a trade-off consideration here. Maybe the design trade-off in this case is to design a way to perform the maintenance quicker, but certainly the best choice is not to ignore the issue.

Simplicity is an overarching concept. Complex is the natural state. The goal needs to be finding ways to achieve the required output as simply as possible. This is a motherhood statement, but a real requirement to keep in perspective.

Operability involves the future user's ability to easily and safely use the product or device. Many years ago, aircraft designers found that the location of gauges, switches, and knobs had a lot to do with the safe operation of the airplane. Every device has characteristics similar to this. Think of this attribute as not changing the requirement, but rather making the functionality easier to use and safer. Automobile designers in recent years have found this to be an issue with some of the new functions being installed in the modern car.

Sustainability is likely the least understood of the ibilities. This goes beyond all of the other attributes in that it evaluates the ability of the process or product to exist long term. Will the underlying technology survive? Will the design last as long as required? In high-technology projects this can be one of the most difficult factors to deal with given low predictably of the next new technology. Maybe "predictability" is in fact the tenth *ibility*. If the project team had an accurate view of the technical and organizational future this goal could be better achieved. All too often, an underlying technology is used in the design only to find in a much too soon that some better technology has been introduced to make the current approach obsolete.

The final word on this set of concepts is that they are important to both short- and long-term success of the project. One of the keys in both requirements and design reviews is to go through the nine *ibilities* list and resolve the trade-offs outlined here. This process may well be equally important to getting the user requirements correct because if the correct choices are not made here, then the user will still feel that the requirements were not met.

12.9   MOVING FORWARD

This chapter has illustrated how scope definition is the functional jumping off point for project planning. At this stage the important point to understand is that planning decisions made in multiple other areas can impact other aspects of the plan including scope. We have touched on the integrative characteristic that KAs can have on each other. A more detailed discussion related to that idea will be deferred for now.

The next chapter will focus on time management as the core planning activity that follows from scope definition. Essentially, this involves an evolution of scope definition and the time required to produce defined work units. This data, in turn, will be used to develop a translation into the project schedule and corresponding budget. Traditionally, the basic project output planning parameters are time, cost, and functionality—classically called the *triple constraints* or *holy trilogy* of project management. In order to successfully generate these desired outputs several supporting management activities are required in both the planning and execution aspects of the project. We will see various discussions on these issues throughout the book.

REFERENCES

Department of Energy, 1997. *Cost Codes and the Work Breakdown Structure*. Department of Energy. www.directives.doe.gov/pdfs/doe/doetext/neword/430/g4301-1chp5.pdf (accessed October 15, 2008).

PMI WBS, 2001. *Practice Standard for Work Breakdown Structures*. Newtown Square, PA: Project Management Institute.

PMI, 2008. *A Guide to the Project Management Body of Knowledge (PMBOK® Guide)*, Fourth Edition. Newtown Square: Project Management Institute.

Schwalbe, Kathy, 2004. *Information Technology Project Management*. Boston, MA: Course Technology.

13 Time Management

The process of creating a project schedule follows directly from the scope definition process and builds from the WBS. The interface point for this is at the activity level in that the goal is to create a sequence and cycle time (duration) for each of the activities in the project, then meld these together to create a schedule. The sections that follow will construct the schedule in the following steps:

1. *Defining project activities:* This segment will translate the WBS work and planning units into entities for time estimation. Each of the items will be constructed as having a scope, time, and resource component. The activities below will produce the time variable.
2. *Resource estimating:* This activity reviews the general availability of various resource quantities and skills needed to complete the project.
3. *Duration estimating:* This step focuses on defining the amount of work required to accomplish the activity, then allocating specific resources to each activity in order to estimate the working time for each. This estimate may either be a single deterministic time unit or a probabilistic (three-point) one. For this segment of the discussion we will assume a single time estimate.
4. *Activity sequencing:* This step establishes the desired order for the activities to be processed. This is called defining the predecessor–successor relationships.
5. *Schedule development:* Creates the final schedule and involves dealing with various constraints and other support activities that will impact the base scope-driven schedule (i.e., risk, resource availability, vendor issues, etc.).

The five time-related processes outlined in Figure 13.1 indicate a slight sequence difference from the one described here. These differences seem minimal if one recognizes that the key issues in basic time management are to first define the work units, then create a duration estimate, and sequence those work units to produce a first cut plan. We will actually see that there is much more to deal with in this process, but the fundamental mechanics must be understood first.

13.1 DEFINING PROJECT WORK ACTIVITIES

The earlier discussion on scope management outlined the process by which a tangible project scope was created from vague visions into a more technical work centric structure as defined by the WBS and its companion WBS Dictionary. This section will move from that level of definition to show how a first cut project schedule can be derived from this scope view (i.e., WPs and the larger non-decomposed planning packages). Some related vocabulary for this process is described below.

Activities represent defined work efforts that will be represented in the project schedule. Each of these will consist of resource definitions and duration time frames for execution for each work or planning package. The project schedule is then composed of these two basic building blocks

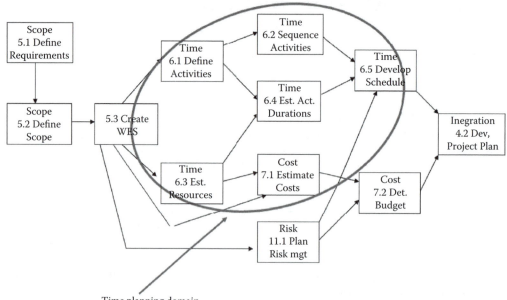

Time management domain

Time planning domain

FIGURE 13.1 Time management processes. *Note*: The reference numbers in the diagram relate to *PMBOK®* *Guide* chapter and process (6.4 is Chapter 6 and process number 4). (Adapted from PMI, *PMBOK® Guide*, 4th ed., Project Management Institute, Newtown Square, PA, 2008.)

optionally packaged into sets of summary activities. From a project management viewpoint, this represents three granular levels of activities:

- Basic WBS defined WPs with detailed planning data
- Planning WPs still requiring elaboration and containing planning level information
- Summary activities composed of groupings of work and planning packages.

In some cases, it may be desirable to schedule a specific individual task that is subordinate to its host WP, but this decision is left to the WP owner and will not be considered in this discussion. Such lower-level task allocations would be done primarily to aid internal team coordination. Regardless, the first three items are present in all project plans. There is frequent confusion in the use of the terms *activity* and *task*. The typical industry vernacular is for an activity or lower-level task to be called a task. In theory, an activity generally involves more than a single individual and is owned by a defined manager. As a conceptual view, think of an activity as a unit of work normally performed by one or more individuals in a single organization. Remember that the WP rule of thumb is that it is sized to approximately two weeks duration and 80 h of effort (work). Real-world distinction of these concepts is not nearly so clear, so do not be too concerned about the label. The key time management concept is to define the relevant parameters for such work units and to ensure that they have a specified individual responsible for its execution.

One common practice is to focus primary time management attention on the earlier project phases and detail later phases at the higher planning package level. Elaboration of the planning packages would then "roll" through the life cycle as the project moves from phase to phase. When executed in this fashion, the planning process is called "rolling wave" and this also fits the concept defined earlier called "progressive elaboration." Another option is to fully define the entire WBS structure before leaving the planning phase; however, one might question whether this adds real value to the overall process, given the uncertainty of the effort at this point. It is a management decision to dictate the level of work specification prior to approving the effort to move forward into execution. This choice is viewed as a risk versus accuracy question. The trade-off is higher risk in moving quickly versus potentially improved estimating accuracy to perform the full WP decomposition prior to leaving the planning step.

13.1.1 ACTIVITY DEFINITIONAL DATA ELEMENTS

As described in the previous WBS discussion, defined work units minimally need the following planning data defined in order to support further project plan creation:

1. Work unit name
2. Time estimate
3. Cost estimate
4. Individual(s) responsible to execute the task
5. Material resources required to execute the task
6. Any operational constraints (e.g., required completion, start, etc.)
7. Associated technical details (predecessors, assumptions, etc.).

Information related to the work unit details should be kept in a WBS Dictionary, or an equivalent formal repository for future reference purposes. One of the key roles of activity definition are first to make sure that the items described in the WBS structure have been properly translate. Secondly, resource estimates for each of the defined work or planning entities need to be estimated. At this point in the project, an increased level of control is needed over who can load or change this data in the project repository (i.e., WBS Dictionary or other repository). The manager in charge of the individual work unit is primarily responsible for the estimate, but the PM should have final control over the number being used in the resulting plan. Importantly, no changes should be allowed without mutual permission of these two individuals.

In this section, we will use the term "activities" for all work units. This term can represent WPs, planning packages, summary activities, milestones, or tasks. The key issue for consideration is how these are to be sequenced through the project. Previous high-level decisions have described how the project life cycle process is viewed, including major phases and key milestones. The next question now becomes how to define the sequence of activities in the project plan.

13.2 ACTIVITY SEQUENCING

There are two important management considerations for sequencing. First, there is a concept called *technical sequence*, which is based on the technology of the project. For example, the technical sequence required to build a building would say that the foundation would have to be in place before constructing the walls. Likewise, a software system would have to be constructed before it could be tested. All project efforts have some sequence that is dictated by their underlying technology.

A second consideration of sequencing falls beyond the technical issues. That is, many activities can be performed at various times. In these cases, the PM may sequence them to accommodate availability of resources, weather, or other factors. In both the technical and arbitrary sequencing situations, it is incumbent on the skills of the project team to make the appropriate activity sequencing decision. However, it is important to recognize that these decisions do impact the schedule results, so sequencing considerations are often an iterative activity until the plan is fixed.

Regardless of how the sequencing decision is made, a schedule will result by stacking the WPs in the order prescribed. This is the essence of the sequencing process.

13.3 DURATION ESTIMATING

One of the most difficult time management activities involves estimating the time required to produce a defined work unit or the entire project. There are many techniques and factors involved in producing such estimates. Creating a duration estimate is not only a science but also an art. It is a science because the estimator is often utilizing historical data, mathematical formulae, and statistics to determine the estimate for a work unit. Also, it involves art because each situation is somewhat different and the ability to customize the value requires skill that is obtained through study, observation, and experience with projects (Baca, 2007, p. 135). This section will briefly explore different estimating techniques. These techniques are most often used to determine the resources that are required and from this to compute the resulting duration and cost for the activity or project. These estimates then become building basic blocks to create the project schedule and budget.

The estimating techniques reviewed here include expert judgment, analogous, heuristic, Delphi, parametric, phased, effort distribution (top-down), bottom-up, and Monte Carlo techniques. No one of these is optimal for every case and in most situations multiple approaches will be used to confirm the derived value.

13.4 TIPS FOR ACCURATE ESTIMATING

One of the goals of work estimation is to create the most accurate duration, budget, and resource requirement possible. If an estimate is too low, the project will likely not finish on plan, exceed its budget, and suffer from inadequate resources. As a result of this, the project will have a higher level of cancellation as senior management loses confidence in the project. Conversely, if estimates are too high then excessive budgets and human resources will be dedicated to a project where they are not needed. Also, the projected value of the project will be defined at a lower point than it deserves, which then may make the initiative look undesirable. Recognize that estimates are used to evaluate a project's merit. Errors in these create issues in both directions. Low estimates will cause a negative view later when the plan starts to go awry, whereas high estimates may make the project look nonviable. Both situations are critical from a management viewpoint.

There are several potential reasons why it is difficult to make accurate estimations regarding the cost, duration, and resources of a project. One basic reason lies in the fact that every project has variables that are difficult to anticipate. It is impossible to incorporate every possible scenario into the estimate. The more complex the project, the tougher it will be to make accurate estimates. Size and newness of the project target are the most difficult factors to deal with. Project team members often pad estimates to ensure that they do not overrun the target. In other cases, the result is to tell management what they want to hear or simply to be overly optimistic (Soomers, n.d.). Estimates can also be inaccurate because the specifications for the project are poorly defined (Verzuh, 2005, p. 168). The worst case scenario is to use optimistic estimates in order to get the project approved. As indicated above, both situations lead to future management issues as reality surfaces.

While it is impossible to have estimates that are 100% accurate all of the time, there are tips that can be followed before, during, and after estimating to ensure that the estimates are as accurate as possible. These tips include the following (Soomers, n.d.):

1. Create and maintain a database that records the actual time, cost, and resources spent on each task in your project. This data can then be used to make estimations on future projects and to identify the historically accurate buffer time needed to realistically complete the project.
2. Create standard planning documents such as specifications and project plans that are used consistently for all projects.
3. Carry out a detailed requirements analysis of the project's work requirements.
4. Compare the new estimate to a former project to determine if it is more or less complex, then adjust based on size, technology, the number of groups involved, or other metrics.

5. Apply multiple estimating techniques to arrive at the final estimate. Recognize the potential variability of the estimate and decide if a single time estimate is prudent.
6. When making an estimate, identify the assumptions, constraints, and caveats that were used to produce the value. Monitor these factors throughout the project to ensure that the environment has not changed.
7. In situations where the estimated budget or duration does not appear adequate to complete the project, propose an upward or downward adjustment to the design criteria. These criteria can include factors such as quality, features, schedule, and cost.
8. When planning the project, consider simpler and more efficient ways to do the work. The simpler the tasks, the easier it is to make an estimate.
9. To avoid a chaotic scramble of the project rollout at the end, start planning and estimating the project rollout from the very beginning.
10. When estimating a project characterized by limited information, consider a phase-based approach. In the first phase, the focus should be on refining scope.
11. Categorize the project's deliverables into the "must-have" and the "nice-to-haves." This will help the PM to create contingency plans.
12. Create a lessons-learned database for assistance in future projects. Use this database from past projects to create best practices to apply to the estimating technique being utilized.
13. Make sure that the estimator has experience with the type of work being reviewed and that they understand the estimating technique. The estimator should also consider skill of the individuals who will be performing the work (Verzuh, 2005, p. 169).

13.4.1 TYPES OF ESTIMATES

There are essentially four levels of accuracy in estimates: ballpark, order of magnitude, definitive, and budget. Generally speaking, this is determined by the time available to make the estimate and the level of detail known about the project. If someone were to ask for a quick estimate with little knowledge, the local expert would respond with a ballpark number. This type of estimate is used for rough evaluation and often contains error rates of 100% or more. These estimates are really only useful in deciding whether or not it is worthwhile to spend more time refining the number (Verzuh, 2005, p. 171). Industry participants call this a WAG—politely stated, this is a Wild Guess.

The next level of accuracy occurs in the early planning period and is similar to the ballpark number, except some more level of detail exists at this point. Order of magnitude estimates are considered calculated guesses based on high-level knowledge of the target. In this case, the estimator reviews the major factors that influence the project's size, then uses these factors to make an estimate based on these comparisons or some other high-level estimating tool. Order of magnitude estimates are often called SWAGs—a more sophisticated WAG (Baca, 2007, p. 132).

A third level of estimate is the definitive estimate. These are typically derived by reviewing each work unit in the WBS and adding the results, whereas the former two options looked at the issue top-down. Because of the way they are produced, a definitive estimate is often described a bottom-up. It is always dangerous to project just how accurate any particular estimate methodology might be, but a bottom-up review of work should be accurate within 25% or higher assuming that the project scope is stable. If an organization has long experience in the particular project technology and type, then the error rate should be closer to 10%. These types of estimates are considered an accurate measure for a specific project element because they are based on a thorough understanding of the requirements and the availability of resources (Baca, 2007, p. 133). In order to perform this level of estimate, the scope definition needs to be complete and the WBS work units would be used to create the value.

A work unit estimate contains multiple views. The first stage involves an estimate of the amount of work required to complete the defined deliverable. Within this process are assumptions related to the skill of the workers. That value is the *work* related to the activity. From this, a decision is made as to the level of resource to be allocated the activity, which in turn will impact how long it will take.

This process yields the *duration* of the activity. It is also necessary to make assumptions as to the effective time that the workers have for the task (Baca, 2007, p. 134). It would not be proper to believe that a work task requiring 80 h of work could be completed by a 40 h per week skilled worker in two weeks, or in one week with two workers. It is common to find that a worker is only 70% productive because of other activities not related to the direct task at hand and this insight is one place where art and experience are required. Given this set of factors, the 80-hour work task with one worker assigned would now have duration of approximately 80/0.70 or 2.85 weeks rather than 2.0. The final type of definitive estimate is a calendar or elapsed time estimate. This is simply the number of calendar days needed for task duration. For example, duration of 2.85 weeks would consume in excess of three calendar weeks because of time off for weekends, holidays, or other nonwork periods. Verzuh states that "A detailed estimate includes all schedule and resource information and a time-phased forecast of the project budget, cash, and resources. This represents the estimate that will be used to manage the project and evaluate its success" (Verzuh, 2005, p. 172).

The fourth estimate type is called a budget estimate. This term could be related to either the time or the cost for the project or work unit. Additions to the base definitive estimates could include such items as contingency funds, fees, or other nondirect factors (Baca, 2007, p. 134).

13.4.2 ESTIMATING TECHNIQUES

13.4.2.1 Expert Judgment

Expert judgment is a very popular technique for making both high and work unit level estimations. According to a software industry study, 62% of cost estimators in this industry use the expert judgment technique (Snell, 1997a). An estimator using this approach relies on his expertise and is guided by historical information and experience with similar projects. For improved accuracy, expert judgment is often used in combination with other estimating techniques. As an example, imagine that you need to get an estimate on how much it would cost to fix the transmission in an automobile. You could either say that further evaluation will cost so much and then you would give a definitive estimate, or based on history with this brand you might be willing to at least give a range. Lack of experience with that brand product would take away this option.

There are three main advantages for using expert judgment as an estimating technique. First, it requires little data collection and simply uses experience from past projects. Secondly, it has sufficient flexibility to be adapted to the conditions of the current project. Finally, it provides a quick assessment of representativeness because the expert will have a large knowledge set from which to derive the estimate (Snell, 1997a).

There are also drawbacks to consider when using the expert judgment technique. First, the estimate provided by the expert will not be any better than the objectivity and expertise of the expert (Snell, 1997a). An accurate estimate requires an expert that has extensive experience dealing with the type of task or project being estimated. Secondly, estimates made by experts can be biased, which produce a number with no easy way to verify the logic of creation. Finally, since the expert is often basing his or her estimate from personal memory, the expert may not have all relevant information which could make this estimate inaccurate.

13.4.2.2 Analogous Estimating

Analogous estimating has similar traits to expert estimating in that it is based on prior experiences. However, in this case, the comparison is based more on data from the previous project. This process would typically produce an order of magnitude estimate unless the new project is very similar to the comparison one. This estimating technique often uses measures of scale such as size, weight, and complexity from a past task in order to make the estimate (Callahan, n.d.). When using analogous estimating, the estimator needs to factor in any differences between the new work being estimated and the previous task being used for comparison. Examples of complexity factors include any new technology that is now being utilized or any changes in the complexity of the task or project (Callahan, n.d.).

Analogous estimating is best used in the early phases of a project before significant details are visible. This method provides quick and easy estimates for projects or tasks that are not very complicated; however, the main drawback is that the results are often not very accurate (Baca, 2007, p. 136). To make sure the estimate is as accurate as possible, use past projects that are very similar in fact and not just in appearance, and that these match the expertise of the estimator (PMI WBS, 2001).

A simple example of an analogous estimate would be to determine how long it would take to unload a ship and move it back to its home port. Past experience with this same size ship and cargo offers the first view. From this, a measure of resource availability, weather, and any other differences would need to be considered. Based on this type of analysis the estimator should be able to derive a reasonable work or elapsed time estimate.

13.4.2.3   Heuristic Estimating

When using heuristic estimating, the estimate is based on a "rule of thumb." These "rules of thumb" estimates are based on parameters derived from past experiences. In that sense, heuristic estimating is based on expert judgment, except in this case the estimate is translated into a mathematical type expression. Also, expert judgment estimating has to be performed by the expert, whereas heuristics estimating can be transferred to others who are capable of manipulating the defined relationships (Mind Tools, 2008). Imagine the situation where you need an estimate for a new roof on your house. The person that comes to provide this has never seen your house or possibly even done roofing. What he has is the heuristic estimating formula. Assuming that the roof type is specified and relevant to the model the estimate is derived by plugging in the parameters. In this case, it may be as simple as square footage. An expansion to the formula could result if the roof has a sharp slope. This class of estimating would typically be found in situations where the same type project is performed repeatedly such that the relationships are well established—roofing, suburban house construction, installation of a water heater, and so on. Whether the vendor is willing to make this a firm fixed price or not is based on the potential variability of the task. The house construction bid might have several caveats and be only an order of magnitude estimate, whereas the roofer might feel comfortable making his bid a fixed price.

The benefit of using heuristics is that it is easy to use and does not require a lot of new detailed research. It is important to formalize the heuristic estimating parameters so that they can be used with others. In many situations, the heuristic estimate is not expected to be highly accurate and they should only be used in situations where the inherent risks are acceptable (Mind Tools, 2008). If we were estimating that a car can still drive 100 miles on one-fourth tank of gas that is one thing, but an airplane pilot could not use this same technique to estimate fuel requirements.

13.4.2.4   Delphi Technique

The Delphi technique is another variety of expert judgment technique. It is most effective when making top-down estimates in the early stages of projects where there are many unknowns and a single expert is not sufficient because of the breadth or uncertainly of the goal. A radical example of this is "How long will it take us to get to the moon?" In complex situations, the Delphi approach gathers the estimates from a group of experts with the goal of combining their estimates to eventually reach an unbiased estimate (Snell, 1997b). The Delphi technique involves several steps including the following:

1. Experts are given the specifications of the project and an estimation form
2. The group meets to discuss any estimation or product issues
3. Each expert provides his or her individual estimate without collaboration with others
4. Estimates are returned indicating the median group estimate and the individual's estimate
5. The group meets again to discuss the results and individual logic for their estimate
6. Each expert again provides his or her individual estimate
7. Steps 3–6 will be repeated until the group of experts reaches a consensus (Snell, 1997).

TABLE 13.1
Delphi Technique Estimation Form

Job Name: Gary's New Mansion

Date: 6/30/2009			Est. Units: Days				
WBS	**Task Name**	**Est. 0**	**Δ1**	**Δ2**	**Δ3**	**...**	**Key Notes**
1	Foundation	10	+3	+2			Question soil composition
2	Walls	15	+3	+2			Drawings not clear
3	Plumbing	5	+1	+1			
4	Electrical	5	+1	+1			
5	Landscape	9	+6	+5			
6	Ext. paving	12	+4	+4			Scope not defined well
	Est. Delta		**+18**	**+16**			
	Total	**56**	**74**	**72**			

Table 13.1 shows a sample estimation form that can be used by the members of the Delphi panel to record their estimations. In this example, an estimate is given for each of the six tasks in the project and totaled at the bottom. Once the first iteration is complete, the panel meets to discuss the results; then estimators review their estimations in the Delta column. The Deltas are totaled and changes to the total estimation are made at the bottom. This process will continue until a consensus is reached.

Normally, the Delphi technique is used for more highly complex estimates for which there is little historical background; however, the concept works in situations where a nonbiased estimate is needed. In the example illustrated in Table 13.1 the original time estimate was 56 days. After two iterations, the Delphi experts refined that as shown by Δ1 and Δ2. The first round was summarized as having the average "expert" estimates at +18 above the original estimate. The second round shared the logic of Δ1 estimates and that total variance shows as +16. From this, a consensus estimate would be made as 72 days.

Because this technique seeks estimates from multiple participants, it tends to remove bias and politics that can occur when an estimate is based on only one expert's judgment. Group meetings also allow experts to discuss any issues or assumptions that may impact the estimate. One of the main drawbacks for this type of technique is the amount of time it can take for the panel of experts to reach a consensus. Larger numbers of experts will increase the number of iterations required and therefore the time it takes to reach a final estimate. Another potential issue is the experience level of the panel. If the panel is made up of individuals that are not very experienced, the estimate will not be so accurate. It is also important to make sure that strong facilitators are available to guide the group and keep the group focused on the topic. The central idea of the Delphi is to anonymously share the estimates made by others along with their basic rationale. The next iteration would allow the group to think about other views and possibly adjust their estimates. This is essentially a consensus building process.

13.4.2.5 Parametric Estimating

According to the *PMBOK® Guide*, parametric estimating is a technique that uses statistical relationships between historical data and resulting work levels (PMI, 2008, p. 150). Parameters based on size, footage, or other scope-related values can be used to produce time and cost estimates for the related work. Given the mathematical sophistication of the parametric estimating model, these are usually computer-based techniques. The process of creating a formal parametric model is to collect data from thousands of past projects and from this produce a regression-type model and considers the impact of each relevant variable. Parametric techniques are most useful in the early stages of a project when only aggregate data is available. These estimates are considered order of magnitude in accuracy because they may lack requirements precision (Kwak and Watson, 2005).

Parametric techniques were first developed by the U.S. Department of Defense (DoD) during World War II to estimate high technology projects such as weapons and space exploration. The technique is most commonly used now in the construction industry. For example, Eastman Kodak created the EST1 Estimating System in order to estimate the construction cost for building additions (Kwak and Watson, 2005). Other applications include, but are not limited to, determining the conversion costs associated with new technologies in electronics manufacturing or estimating the cost of developing intellectual property such as engineering designs.

The technique can be applied to any situation in which sufficient historical data are available. From this data Cost Estimating Relationships (CERs) can be derived statistically to correlate with resulting work estimates. Project characteristics can include functions, physical attributes, and performance specifications. CERs are based on two different types of variable relationships. One type is based on an independent variable being used to predict the cost of the dependent variable. This type is called a cost-to-cost relationship. An example is using the cost of labor hours for one component to estimate the cost of labor hours for another component. The second type is based on when the number of outputs can be used to estimate a variable such as labor hours. This is called a cost-to-noncost relationship. Relationships can vary from one-on-one to very complex algorithms. The majority of CER relationships are linear, which would mean that a single value of the independent variable would be associated with the cost of a dependent variable.

The parametric estimating technique needs to take into consideration the type of development methodology that was used for each project because it could impact the total project cost. Each method has unique characteristics that will affect cost. Some of the methodologies include the waterfall method, incremental development, spiral development, and prototyping. For example, when developing software using the waterfall method, the cost of documenting the requirements prior to coding needs to be considered because this is a common practice in this method.

The main advantage of parametric estimating is that it provides the estimator with a quick estimate that can be based on a limited amount of data. It also provides the estimator with an understanding of the major cost drivers of the project. This is because the CER relationships from historical projects reflect the impact of design changes, schedule changes, and cost growth. A final advantage is that by taking the development methodology into consideration when creating the estimate, the estimate is less likely to contain errors and random variations. The estimate will also be less subjective.

The main drawback of the parametric estimating technique is the accuracy of the estimate it provides. According to one researcher, estimates that are based on the final project design will be roughly in the range of +10% to −5% accurate. If the estimate is based on project designs that are from 1% to 15% complete, the estimates accuracy will range from +30% to −25% (Kwak and Watson, 2005). Accuracy can also be affected by such factors as the experience of the individual making the estimate, changes in scope or design specifications, and other incorrect assumptions. Accuracy is especially problematic for construction companies that use parametric estimating to produce bids for projects. They will not win the project if the bid is too high, and the project will not be as profitable if the bid is too low. So, high accuracy ranges make the provider vulnerable to adverse results. With the increased amount of historical information available in databases, innovations in statistical applications, and the availability of expert data over the Internet, the accuracy of parametric estimating techniques should improve over time.

13.4.2.6 Phased Estimating

Phased estimating is based on a commonsense rule of thumb, which involves avoiding time commitments when there is not enough supporting information. Instead, the estimating process is broken up into a series of sequential decisions. At each defined decision point, which is called a phase gate or kill point, a management review process will determine whether it is appropriate to continue with the project, redefine its scope, or to terminate the project. According to Verzuh, the performance baseline should be considered reset at each phase gate and the project will be re-evaluated at each phase gate. In terms of using phased estimating to create an estimate for cost and schedule, there are a series of steps to follow when making estimates. These steps are as follows (Verzuh, 2005, p. 174):

1. Break the product development life cycle into phases. Each of these phases will be considered a subproject.
2. For the first phase of the life cycle, detailed estimates are made for the cost and duration for this phase. At this point, the estimator should also create an order of magnitude planning estimate for the entire product development life cycle. With this, the project is approved to move to the next phase.
3. When the first phase is completed, a decision must be made regarding whether to continue with the next phase. The review process must make specific decisions to re-evaluate the project direction and make changes to its scope or product requirements and specifications, or whether to cancel the project altogether.
4. If the project is approved to continue to the next gate, a detailed estimate is created for that phase and the order of magnitude estimate for all subsequent phases is updated based on current information.
5. The cycle of phased estimating continues until the project is either completed or terminated.

When using the phased estimating or *rolling wave* technique, it is important to understand what is required at each phase gate. There are three main components that each active phase gate should contain. First, the phase must specify its required deliverables. Each gate will have a different deliverable or set of deliverables. Secondly, each gate should define a set of success criteria. These criteria are used to determine whether the project should proceed to the next phase or be terminated. The final component is the specific outputs. These outputs answer the question "What is the purpose of the gate?" The number of phase gates that will be required depends on the size and complexity of the effort. When the project scope is well defined, fewer phase gates are required. Project teams should try to consistently use the same gates at consistent points for similar projects as this will help to improve the review process.

The main benefit of using the rolling wave estimating approach is that it allows the effort to move forward quicker and recognizes that early planning in an uncertain world may be a waste of time. However, this approach requires much more monitoring from management and the level of predictability is low in regard to future budget and resource requirements. This approach does not help greatly in comparing two options in that the decision metrics are known to have high ranges.

In this model, the amount of uncertainty regarding the project will be decreasing as the project progresses, so each subsequent phase estimates will be more accurate (Verzuh, 2005, p. 174). Another subtle issue in this approach is that the project team has more potential to meet the short-term phase goals, whereas the longer-term full estimate approach could lead to situations where the team had no chance to be successful.

One of the main criticisms for the use of phased estimating is that many experts feel that the project team loses accountability. Their argument is that the team cannot be accountable when the baseline is reset after each phase gate. One method to hold the team more accountable is to require justification for the cost and schedule estimates. When estimates change, the project team should be able to justify these changes with evidence of changes in difficulty or scope (Verzuh, 2004, p. 4).

Another criticism is that many organizations mistake the phases of the development life cycle for phase gates. These organizations think that they are applying phased estimating because they have multiple phases in their development life cycle. Verzuh offers the following criteria to test whether phased estimating is being used (Verzuh, 2004, p. 5):

1. The phases are not necessarily linked to formal stages in the organizational life cycle model
2. The business cases for the project are updated with each phase
3. Cost and schedule baselines are formally changed at each phase
4. Some project's scope is changed because of results from the previous stage
5. It is not uncommon to cancel a project at a phase because of previous phase results
6. Some projects receive higher priority as a result of increased value in their business case compared to when the project began.

Organizations whose phase reviews perform in this matter are utilizing phased estimating in their estimation methodology. Those organizations that do not meet these criteria are more typically using milestone reviews instead of phase gates. This latter approach tends to hold the original baseline plan intact through the life cycle and measure variances from that point.

13.4.2.7 Effort Distribution Estimating (Top-Down)

This method basically looks at the project as a whole and apportions the total estimate into high-level groupings. An approach such as this is necessary in the early stages of planning when sufficient details for lower level approaches are not yet available. Once the high-level value is derived a high-level WBS is used to assign percentage values to lower levels in the structure. Normally, this type allocation would only go to one or two levels below the top of the structure. Historical data for similar past projects are typically used to determine the percentages for each phase and/or the summary activities within a phase (Horine, 2005).

This technique works reasonably well in organizations using a common methodology for similar type projects along with good historical data. This method also works well with the phased estimating technique in that the historical data provide reasonable estimates for future phases without extensive definition. For example, when a phase gate is completed, the actual amounts from the previous stage can be used to project future values.

One hazard in using the top-down approach is that an aggregate estimating error for the project is proliferated through all segments by the apportioning method. Another drawback is that the technique uses historical data to define the apportioning formula (Verzuh, 2005, p. 176). If the projects are not technically similar, these values can provide erroneous results that are once again proliferated through the life cycle. For example, if the organization's projects do not have the same number of phases or the phases are different for every project, it will be difficult to apply a consistent apportioning formula to new projects.

To illustrate a top-down approach to estimating, imagine that a project goal is to design and deploy a new car model. The organization has defined five phases for this effort and an estimated overall budget of $250,000. These defined phases include initiation, planning, design, construction, and deployment. The construction phase is further decomposed into three activities—frame, exterior, and interior. The past projects indicate the following resource breakdown across the phases:

Initiation = 10%
Planning = 15%
Design = 15%
Construction = 40%
Deployment = 20%

Activities within the construction phase typically consume resources in the ratio:

Frame = 14%
Exterior = 13%
Interior = 13% (note that these add to the 40% value of the higher level).

Using these values, the construction phase would receive 40% of the estimated budget that would compute to $100,000 (i.e., 0.40 × $250,000). Within the construction phase, the frame work activity would be estimated to cost $35,000 (0.14 × $250,000). Assuming that the effort distribution technique was being applied along with the phased estimating technique, these estimates could be updated at each phase review point. The original allocation estimate for the initiation phase was $25,000 (0.10 × $250,000), but the actual cost was $30,000, which is 20% higher than in the plan. From this result, the overall cost estimate would then be revised by 20% to $300,000, so the new estimate for the construction phase will increase from $100,000 to $120,000, and the estimate for building the frame will increase from $35,000 to $42,000. The key issue in these comparison methods is whether the project team can modify the planned values dynamically. Some organizations will allow this and others will hold the original value constant and pressure the project team to take corrective action to stay on the plan. These are management philosophy issues beyond the estimating methods.

13.4.2.8   Bottom-Up Estimating

This technique is considered by experts to be the most accurate of all the techniques described here. The mechanics for the process are based on the WBS work planning packages. Based on this view, the total project consists of relatively small work units that can be estimated reasonably well by the performing work groups. In most cases, the estimating method used would be expert judgment, given that the individuals who will actually be doing the work would provide the values. Once the full collection of individual WP estimates is complete, they are rolled up in the WBS structure to generate higher-level aggregations. From a statistical point of view, each of these estimates would be subject to typical estimating errors; however, over the full WBS range the errors should compensate. So, the resulting estimate should be based on real views of the work to be done and the errors made in estimating less than other methods reviewed. The total project estimate is then simply the aggregation the lower-level work units.

The main advantage of using the bottom-up technique is the accuracy that it should provide. It also better involves the individuals who will be performing the work, so actual content review is better along with an increased commitment to the resulting values. It is not hard to imagine the feeling that a work group might have when an estimator brings a value to the group and says "this is how long it should take you," compared to the work group coming up with the same value. The concept of commitment is clearly on the side of the latter.

The main disadvantage of this method is that it takes a considerable amount of time to create and involves larger organizational involvement. Another subtle problem that often comes with this process is excessive padding of the value by the work groups. External estimators do not have this motive, but internal groups do not want to overrun estimates, so they may add excessive times. One reason that the estimates are often padded is to take into consideration risks that are perceived for that activity (Johnson, 2007, p. 224). The goal of work unit estimating is to derive estimates that do not have excessive padding, but represent reasonable values for the activity. In order to ensure this, some cross-checking by the PM is required. Readers interested in the psychology of this type estimating should review the Theory of Constraints (Critical Chain) discussion in Chapter 17.

The bottom-up technique is not the best method to use if a quick estimate is needed in order to make a decision regarding whether to initiate a project. First, there is not enough information available to develop detailed estimates early in the product development life cycle, and second, the required time to go through this process even at a planning work unit level is not yet justified. The other disadvantage of this technique is that estimators will often pad the estimates for each WP.

13.4.2.9 Monte Carlo Simulation

This technique got its name because of the similarities between its mechanics and roulette popularized at Monte Carlo casinos. First, work estimates are recognized as having a range of values—no other technique discussed above specifically dealt with this. To simulate a variable work activity, time numbers similar to the roulette wheel were used to represent the estimated time. From this, the project was "simulated" hundreds of times to see what the completion probability distribution looked like. Because of the volume of calculations related to this method, it is typically a computer-driven activity and there are models available to support this type of analysis. Chapter 18 will discuss the role of simulation in greater detail.

In order to drive a simulation model, it is necessary to supply a probability distribution assumption for the activity and a defining set of estimates. For example, it is common to suggest that the probable distribution is triangular and estimates for the conditions surrounding the variable are what determine the probability distribution. If the estimator has estimates for the minimum, most likely, maximum values are given the computer model will select random values and execute the plan, say 1000 times. Each of these passes will generate a discrete value and the total will present a histogram-type presentation of likely outcomes. For this type of situation, we estimate the most likely completion date as well as potential ranges. In many project situations, we want to generate a plan that has high probability of not exceeding some number, even though the planned date is earlier. Chapter 16 will illustrate this method further, but recognize that the type of analysis derived from simulation can be very important given the uncertainty related to all estimating techniques.

The Monte Carlo technique provides two main advantages when used for estimations. First, the technique provides a better understanding of the potential range of actual outcomes for the project than most other techniques. A second advantage involves the capability of running "what if" analysis to evaluate various assumptions. In this mode, multiple complex scenarios can be defined and outcomes generated. Finally, simulation is very useful in analyzing complex project situations. The main drawback of using simulation is the amount of time that it can take to set up and run the various scenarios and the cost associated with the activity. Another potential drawback is that it is easy for individuals who do not have experience with the technique to misuse the simulation because they are unfamiliar with the assumptions and restrictions of the technique (Johnson, 2007, p. 224).

Work activity estimation is a vital part of project management. It is the fundamental building block to create schedules and budgets for the project. Poor estimating plagues a project in many ways. The ideal situation would be to produce an accurate estimate, have management approve that value, and then deliver a project on planned time and budget. This is the goal of both the PM and his supporting management.

It is important to note that estimating is not only a science, but also an art. Formulae and statistics play an important role in estimation, but experience and historical data are also part of the process.

13.5 ACTIVITY SEQUENCING

There are basically two ways to define and review the sequence of activities in preparation for a more analytical view. We can model the process using symbols to represent activities and sequence, or by using arrows or boxes to show activities. Let us look at these two classical sequencing options.

13.5.1 ARROWS AND BOXES MODELS

The Activity on Arrow (AOA) approach simply constructs the project plan as a set of nodes connected by arrows to show how the various activities are sequenced. This depiction can be high level or very detailed depending on the individual bias related to project control. This diagram structure shows an orderly, systematic series of actions that must be completed in the defined order (as directed by the arrows) to reach a specific definable objective. There are two primary elements of the AOA

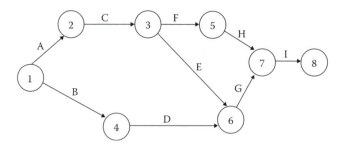

FIGURE 13.2 AOA network.

network diagram: arrows and events. Events are usually shown as circles or rectangles and are interconnected by a series of arrows indicating an activity. The activity is symbolic of a defined work unit, while an event is representative of a point in time that represents a completion of an activity and the start of the next one. In this view, it is important to understand that an activity (arrow) consumes time (Stires and Murphy, 1962). Without exception, every activity except for those that start and stop the project must have both a predecessor and successor event. Each event must have both predecessor and successor activities with the exception of the first event and last event. Events can be classified into two parts: predecessors and successors. Predecessor events are events that must occur before the following one in the network. This is called the predecessor/successor sequence. These paired relationships provide the specifications necessary to draw the network diagram. The importance to gain from this knowledge is to better draw the diagram. Once it is drawn, the pertinent information is available to be able to begin schedule calculations.

The diagram shown in Figure 13.2 illustrates the AOA model. In this example, the activities are indicated by arrows and letters to identify the activity along with an optional time estimate for each activity. In this example, activity A has an estimated duration of three time units. Node values represent terminators for the activities and indicate how the project activities would be constrained. At this point, we have established the base for calculating project duration using one of the two sequencing models—AOA. Activity on Node (AON) is the alternative option.

13.5.2 AON MODEL

As the concept of activity sequencing was in formulation stage during the mid-1950s, it was observed that the AOA structure required the use of "dummy activities" or dotted arrows to accurately portray some precedence relationships and they felt that this increased the complexity of the network (Kelley and Walker, 1989). Their solution for this was to use a rectangle to depict the activity and use arrows between the boxes to show relationship. This type of network is called AON. A sample AON network is shown in Figure 13.3.

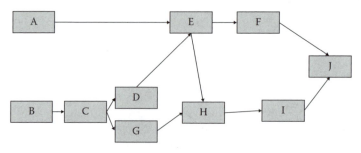

FIGURE 13.3 Activity-on-node network.

Note in this example that activity C must be complete before activities D or G can begin. As in the AOA model, this structure is now set up for calculation of project times and other schedule parameters.

13.6 TIME CALCULATION

We can develop a time dimension to the work effort using either the AOA or AON structure. Envision each of the activities as having a time dimension. Given the defined sequencing relationship implicit in each, it is reasonable to see that these model structures in fact represents the corresponding schedule of the project. So, once the work activities have been sequenced and duration estimated, a project schedule can be derived. This same data can be used to load the raw data into a software product such as Microsoft Project for schedule generation.

Recognize that if all work unit activity estimates were perfect, all resource skills available and equally productive, and all defined materials available on schedule, a project schedule is deterministic and as accurate as the duration estimates. At a theoretical model level, it is important to have this view in order to best understand the basic concepts of time management; however, it is also important to recognize that all of these simplifying assumptions are suspect. The fact is that for real world projects, the activity sequencing issue is complex and the accuracy of activity duration estimates is equally so. Thus, the job of project time management becomes one of dynamically tweaking the variables involved to maintain the plan target schedules.

Once we understand the basic time management theory, it is important to view this stage as a *first cut view*. What this means is that there will be many other issues to consider in refining this view into a final project plan. For example, factors related to risk assessment, outsourcing, resource availability, and other such factors can either expand the required time or take time away. Also, as the level of work definition improves, it is quite common to find that the original estimates contained sufficient errors to require a major rework of the initial plan—both duration and sequencing. In addition to this, management factors can require that the initial plan be rejected as being too long. All of these issues require that the time planning process be viewed as an iterative one until a final version is approved by management. During this refinement process we must keep in mind the trade-off potential between scope, time, and cost. If the current view of the plan exceeds the desired cost or time, it may be necessary to cut scope. Likewise, if scope expands with new insight into the effort, then schedule would likely also expand. Dynamics such as these dictate changes to the WBS structure and a recycle of the time planning process.

The technical sequence defines the constraints that affect how the activities can be executed. Without this, the project would simply consist of the longest task and everything would be done in parallel. The idea behind a technical sequence is that some events rely on previous tasks to be complete. For example, just as a house wall cannot be built until the foundation is in place, so too are there sequencing restrictions in all projects. Specification of these restrictions is critical to producing a viable schedule.

13.7 NETWORK MECHANICS

This section will outline the basic mechanics required to generate a project schedule. There are 10 basic parameters involved in the input and output parameters of a project plan.

1. Activity title—a short description of the work [Input]
2. Activity duration—number of working periods (days) [Input]
3. Predecessor—defining the order of execution [Input]
4. ES or $T(E)$—early start; earliest time the activity can commence [Output]
5. LS or $T(L)$—late finish; latest time the activity can start and not affect project schedule [Output]

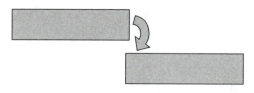

FIGURE 13.4   Finish–start relationship.

6. Slack—measure of idle time in the node or activity [Output]
7. EF—earliest time that an activity can finish
8. LF—latest time the activity can finish and not affect project schedule [Output]
9. Total float—the difference between the *earliest* date that the activity can start and the *latest* date the activity can start before delaying the completion date (LS – ES) [Output]
10. Free float—the amount of time an activity can be delayed before any successor activity will be delayed (EF – ES) [Output].

For the sample exercise, we will assume that all activity relationships are *Finish–Start* (Figure 13.4), that is, each predecessor activity has to completely finish before the successor activity can commence. A schematic view of this type of relationship is shown in Figure 13.4.

13.8   ESTABLISHING THE PROJECT ACTIVITY SEQUENCE

There are other relationship types that can be defined, but for calculation simplification the Finish–Start option will be demonstrated here. In order to generate the base project schedule, three input items must be defined.

1. Work activities in the project (WPs and planning packages)
2. Duration estimates for the work and planning packages
3. Sequence relationships (predecessors).

Table 13.2 defines a sample project activity list that these three items defined.

The sample project sequence definition specifies that activities A and B can start at any time. All other activities are constrained by some predecessor activity—that is, activity C cannot start until activity A is complete. From this duration and sequencing description, the project schedule model can be sequenced using either an AOA or an AON network model format. The schematic shown in

---

TABLE 13.2
Sample Project Definition

Activity	Duration	Predecessor
A	3	—
B	4	—
C	4	A
D	6	B
E	5	C
F	2	C
G	4	D, E
H	3	F
I	5	G, H

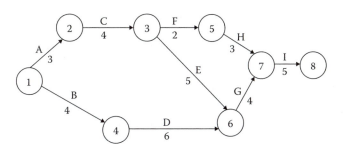

FIGURE 13.5 Sample project AOA network.

Figure 13.5 translates the project data using an AOA format. This model obeys all of the sequencing constraints identified in the project specification. Review this result to ensure that you understand what the sequencing step involves and how it is constructed. One common misunderstanding at this stage is the calculation of total project duration. Some would simply add the duration times in the table and define the project to be 36 time units or the sum of all durations. Since some activities occur in parallel, this is not a correct view as shown in Figure 13.5.

13.9 FORWARD PASS CALCULATION

From the initial setup, step two of the mechanical plan formulation is to perform what is called a *forward pass*. This means that we need to calculate the time through the network while obeying the predecessors. These early start times are recorded on the nodes in the format of "T(E)/." Figure 13.6 shows the calculated values for each of the nodes. The example calculations are quite straightforward except for two—nodes 6 and 7. For node 6, two activity paths have to be considered for the parallel activities D and E. Node 4 has a T(E) value of 4 and a following activity D time of 6; therefore the resulting value for node 6 would be 10. However, the other parallel path coming into node 6 starts at node 3 with a value of 7 and an activity E time of 5, so that the path value is 12. For forward path calculations, the highest value is selected for such parallel paths, thus we assign the value 12 for T(E) at node 6. This means that the earliest we can claim completion of project activities at this point is 12. The same type logic applies to node 7 where a value for T(E) of 16 is calculated. The key mechanic for the forward pass is to remember to take the *highest* value for any multiple input paths. T(E) values at each node represent the earliest time that these points in the project can be reached. This calculation also shows that the final node (8) can be reached at time period 21. This defines that the project duration is 21, given this set of parameters. At this point, we have a crude schedule (Figure 13.6) resulting from the duration and sequencing parameters.

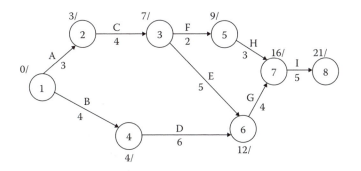

FIGURE 13.6 Forward pass calculations.

13.10   BACKWARD PASS CALCULATION

Step three of the process involves performing a backward pass on the network model. The rationale for doing this is not clear as yet, but let us first describe the mechanics of the calculations before attempting to describe the function of these values. The backward pass generates a variable defined as T(L) and it is formatted on the network node as "T(E)/T(L)." Note that the nodes in Figure 13.7 now show two such numeric values above each node. These variables are formatted as T(E)/T(L). For this basic network, the assumption is made that the planned time at termination is equal to the forward pass calculated time shown (e.g., 21 time units). Stated another way, we are saying that the forward plan specified that the project would take 21 time units and we are accepting that value for the backward pass.

In order to calculate T(L) values, we start at the completion node (8) and insert the same time as calculated for the forward pass—21 in the sample case. From this value, we work backward to the front of the network, one node at a time. For node 7, the T(L) value would be 21 − 5 (for activity I) yielding 16. Node 6 would be calculated as 16 − 4 (for activity G) or 12. As seen in the forward pass mechanics, the T(L) calculations are straightforward until you encounter a node that has multiple paths coming back into it, such as node 3. In this case, the T(L) calculations would compare the two paths for activities F and E yielding a 13 − 2 or 11 for activity F, versus 12 − 5 or 7 for activity E. The calculation rule for the backward pass is to take the *lower* value of the two paths and record that as the nodal T(L) time. This same process would be required for node 1 and at that point the value should be zero since we started with 21/21 at node 8. Review the values shown in Figure 13.7 to be sure that you understand the basic idea of T(E) and T(L) mechanical values.

Once we have the values for T(E) and T(L), it is possible to analyze two important time management factors—the longest path through the network and slack time details. The simplest item to calculate is nodal slack time. This is simply the difference in T(L) and T(E). So, node 3 would have a slack value of zero (7 − 7), whereas node 5 has a slack value of 4 time units (13 − 9). Each of these values defines the amount of time that this node can stay idle without affecting project completion date. Also, note that the computed slack for the start and finish of the network will be zero using the assumption of 21/21—that is, we are happy with the original computed project duration. You might be thinking about what you would do if this was not the case.

Three other slack-type status parameters can be derived from this view. These are total float, free float, and late finish. Each of these relates to activity views rather than node calculations, but can be derived from the node values. Total float relates to the amount of activity slack before project completion is impacted, while free float deals with the same view only for the successor activity. Late finish describes the latest time that the activity can be completed without impacting the schedule. Since activities represent the project work the PM must translate the slack mechanics to that view.

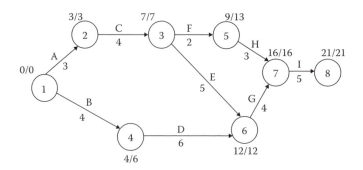

FIGURE 13.7   Backward pass calculations.

13.11   DEFINING CRITICAL PATH

Using the calculation assumptions outlined above, zero nodal slack time can be used to define the critical path. In mechanical terms, this means that the earliest time to reach a node and the latest time to leave the node are the same—that is, no idle time. Zero slack nodes constrain the network time and thus represent *the longest path through the project plan.* Any activities that cause the zero slack at these nodes then cannot be delayed without delaying the project completion date.

In order to find specific critical path activities, the key mechanic involves checking any activity bounded between the zero slack nodes to see if in fact that path is a critical activity. In some more complex networks parallel paths can both appear to be on the critical path, whereas only one actually is. In this sample case, the critical path definition is straight forward and is outlined in gray in Figure 13.8. Note that it passes through nodes 1-2-3-6-7-8 and the critical path activities are A-C-E-G-I. Specifically the critical path should be thought of as a vector containing both the activity list and the total duration, so it would be more proper to state the critical path as A-C-E-G-I, and 21 time units. To reiterate, this is the longest path through the network and represents a critical management issue for the PM (Brown, 2002).

Management importance of the critical path lies in the fact that any delay in these activities will delay project completion. Thus, it is important to focus management attention on this set of activities. Other activities that have slack generally require less rigorous monitoring since a slippage in these will not affect the completion date so long as slack remains. As an example, node calculations for the start of activity H indicate that it could start as early as period 9, or as late as period 13 without affecting the completion schedule. In similar fashion, activity F could start as early as period 7 or wait as late as period 11 with no adverse impact (this calculation is a little trickier, so it is worth looking at to ensure that you understand the meaning of the node values). This latter calculation is not as obvious as the previous one, but it is found by noting that activity H could wait until period 13 (late start) and activity F only takes two time periods. Note that activity E must start at time period 7 in order to keep the project on schedule.

Understanding slack concepts is vital in time management and requires that the PM keep these parameters in mind as the project unfolds. Keep in mind that errors in duration estimate can cause these values to change during the project, so they are not static variables by any stretch. Understanding activity slack or float gives the PM flexibility in scheduling the start of a particular activity within the early start and late start time ranges (Uher, 2003). This also allows PMs to establish proper priorities for resources across the project plan. For example, if there are two activities occurring within the same time range, one critical and one noncritical, which are competing for the same resource, the PM can allocate the resource to the critical activity first, then use the float time to delay the start of the noncritical activity.

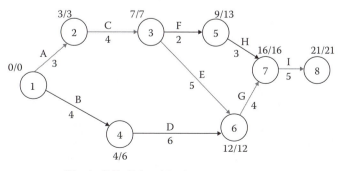

CP = A–C–E–G–I; and 21 time units

FIGURE 13.8   Defining the critical path.

13.12   MANIPULATING THE SCHEDULE

If the project begins to fall behind schedule, the PM often undertakes various initiatives to recover the planned schedule. One approach to performing this is to allocate more resources to critical path activities in order to shorten the path and thereby move the project back toward the planned schedule. This option typically increases the project budget, but can shorten the resulting schedule. Alternatively, the predecessor relationships can often be changed in such a way as to shorten the schedule. In this option, no additional resources are allocated, but this may increase risk or cost if we assume that the original plan was already optimum. The vocabulary terms for these two management options are as follows:

>   *Crashing*—allocating extra resources to the critical path with the goal of shortening the schedule.
>   *Fast tracking*—rearranging the predecessor relationships to shorten the schedule.

In order to decide which of these options to pursue, it is necessary to evaluate the value of recovering time, versus the cost. Also, consideration must be given to risk and other factors involved in the chosen approach.

13.12.1   AUTOMATED CALCULATION TOOLS

Fortunately, there is an increasingly available set of calculation tools available to handle the mechanics outlined above. To illustrate this, Figure 13.9 duplicates the sample project data in a Microsoft Project view format. Note that the software converts the manual AOA model into a Gantt bar format; however, it also recognizes that the results are equivalent to those shown for the manual example above. The only two new items shown in the Gantt view are a summary bar for the entire project and the use of line numbers to specify the predecessor rather than its name.

13.13   FORMATTING ACTIVITY RESULTS

As stated earlier, there are different biases regarding which network model sequencing format is to be used—AOA or AON. The sample data here focused on the AOA view, which contains less information on the diagram than a corresponding AON model. If an AON model is used, the activities are represented by the nodes rather than an arrow. Standard formatting for this approach would require that more of the slack and float values be computed and shown on the model. Figure 13.10 shows a typical format for displaying the activity parameters in an AON box structure. In this example, task name is used instead of activity, but recall the earlier warning that these terms are not standardized in industry. Either terminology is the same in the model mechanics.

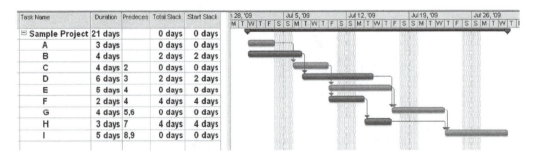

Task Name	Duration	Predeces	Total Slack	Start Slack
⊟ Sample Project	21 days		0 days	0 days
A	3 days		0 days	0 days
B	4 days		2 days	2 days
C	4 days	2	0 days	0 days
D	6 days	3	2 days	2 days
E	5 days	4	0 days	0 days
F	2 days	4	4 days	4 days
G	4 days	5,6	0 days	0 days
H	3 days	7	4 days	4 days
I	5 days	8,9	0 days	0 days

FIGURE 13.9   Microsoft Project output format for the sample problem.

Early start	Duration	Early finish
	Task name	
Late start	Slack	Late finish

FIGURE 13.10 AON parameter format.

Figure 13.11 shows what this formatting would look like in a full network with only 10 activities. Personal bias says that this format tends to be more overwhelming than the AOA view and the multitude of data shown makes the analysis more difficult. Fortunately, use of automated software such as Microsoft Project can handle these calculations and present the data in any format desired. The contemporary environment seems to favor a Gantt format as the preferred method to display schedule results. Regardless of the approach used to create a project schedule, the key issue is to recognize that a PM must understand these concepts. Different audiences will have varying needs for levels of nodal detail and the various network management software packages can format their output to fit these needs.

13.14   WHICH DIAGRAM FORMAT WINS?

In industry terms the AOA model is typically used with the PERT model assumptions and the AON with CPM assumptions (underlying mechanics of these two options will be discussed in a later chapter). It is instructive to browse the Internet using the search words "PERT" or "CPM." What you will find from this is that the outside world does not distinguish between these two terms or how they are diagrammed. Over the years, the original definitions that were outlined in this chapter have been scrambled. Networks today are drawn both ways and called by either name. The one

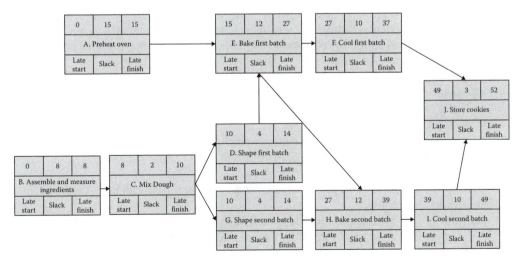

FIGURE 13.11 AON data formatting.

remaining idea is that a network diagram is a good tool to reflect activity sequencing. However, in more recent times, the emergence of easy-to-use PC-based tools such as Microsoft Project has made the manual drawing of networks passè. The fact is that the modern tool generates their results in the Gantt chart format with connectors between the bars to make the result fit the network model. So, both of these approaches have lost out to the classical format first described by Henry Gantt in the early 1900s. But do recognize that the underlying computer calculations do obey the network theories described here.

13.15 SUMMARY

This chapter describes how time-related aspects of the project evolve from scope definition and are then translated into a first cut schedule for the project. There are many more management issues to deal with related to the production of a final schedule, but for now take this deterministic model view as the conceptual process by which scope is translated into an activity list, and then that view evolved into duration and sequence specifications.

The network models illustrated here are not perfect for showing how the project will be executed, but they represent the underlying scheduling model and best illustrate the method to introduce the fundamental aspects of time management. Embedded in the network model is a quantification of the scope, work activities, duration estimates, sequencing, resource allocations, and budget. For this reason, network models need to be understood as a basic tool in the PM tool kit, even though modern computer software will handle the arithmetic calculations to generate this result.

At this stage, we focused on fixed time estimates and dealt little with the reality that time estimates in fact should be viewed as a probability distribution with ranges. Subsequent chapters will open up these limitations for further review. For now, let us move forward to the next chapter and see how project cost emerges from the schedule model.

13.16 ESTIMATING CHECKLIST

The following checklist represents a set of questions that should be reviewed as part of the work estimating process. This example illustrates how a formal list of items can help ensure a more standardized approach across various activities throughout the life cycle. In many cases, such items would have a formal signoff signature requirement for each to verify compliance.

1. Have you established a formal, documented data collection process for the project?
2. Do you have a complete and detailed WBS for the project, including management areas?
3. Do you have historical information, including costs, from previous similar projects?
4. Have you identified all sources of costs to your project (i.e., different types of labor, materials, supplies, equipment, etc.)?
5. Do you have justifiable reasons for selecting your estimating methods, models, guides, and software?
6. Have you considered risk issues in your plan?
7. Do your estimates cover all tasks in the WBS?
8. Do you understand your project's funding profile, that is, how much funding will be provided and at what intervals?
9. How sure is the funding?
10. Have you developed a viable project baseline that is synchronized with the planned schedule and funding profiles?
11. Have you established adequate schedule flexibility in the baseline?
12. Do you have a plan/process for dealing with variances between actual performance and the baseline?
13. Do you have a process for keeping records of your project activity for future efforts?

PROBLEMS

1. Use the activity specifications outlined in the table below to generate a first cut project schedule.

WBS	ID	Activity	Duration	Predecessor
1	1	Project summary task		
1.1	2	Design office complex		
1.1.1	3	Determine budget	13 days	
1.1.2	4	Determine the three best potential architects	4 days	3
1.1.3	5	Interview architects		4
1.1.3.1	6	Architect 1	1 day	
1.1.3.2	7	Architect 2	1 day	6
1.1.3.3	8	Architect 3	1 day	7
1.1.3.4	9	Select an architect	1 day	8
1.1.4	10	Prepare first draft plan	1 day	5, 9
1.1.5	11	Review the plan	0.5 days	10
1.1.6	12	Finalize the plan	0.5 days	11
1.1.7	13	Obtain construction permit	0 days	12
1.2	14	Plan office layout		2
1.2.1	15	Prepare layout plan	2 days	
1.2.2	16	Estimate costs	2 days	15
1.3	17	Buy materials		14
1.3.1	18	Rent tools and equipment	1 day	16
1.3.2	19	Purchase materials	1 day	16
1.4	20	Prepare the site		17
1.4.1	21	Excavate for foundation	8 days	
1.4.2	22	Build the foundation	12 days	21
1.5	23	Begin construction		20
1.5.1	24	Build the pillars	10 days	
1.5.2	25	Lay the roof	8 days	24
1.5.3	26	Build the walls	14 days	25
1.5.4	27	Flooring	14 days	25
1.6	28	Miscellaneous		27
1.6.1	29	Install plumbing fixtures	4 days	23
1.6.2	30	Install wires and cables	3 days	29
1.6.3	31	Plastering	4 days	30
1.6.4	32	Woodwork for doors and windows	3 days	31
1.6.5	33	Furnishing	2 days	32
1.6.6	34	Project close	2 days	33
1.7	35	Project completion	0 days	34

From these specifications, develop a network schedule and answer the following questions.

   a.   What is the project duration?

   b.   What type of activity is 1.6?

   c.   What type of activity is 1.7?

REFERENCES

Baca, C.M., 2007. *Project Management for Mere Mortals*. Boston: Addison-Wesley.

Brown, K.L., 2002. Program Evaluation and Review Technique and Critical Path Method—Background. Reference for Business: http://www.referenceforbusiness.com/management/Pr-Sa/Program-Evaluation-and-Review-Technique-and-Critical-Path-Method.html (accessed March 17, 2008).

Callahan, S. Project Estimating—fact or fiction? http://www.performanceweb.org/CENTERS/PM/media/project-estimating.html (accessed April 9, 2008).

Horine, G., 2005. *Absolute Beginner's Guide to Project Management*. Toronto, Ontario, Canada: Que Publishing.

Johnson, T., 2007. *PMP Exam Success Series: Certification Exam Manual*. Carrollton: Crosswind Project Management, Inc.

Kelley, J.E. and M.R. Walker, 1989. The origins of CPM—a personal history. *PM Network*, 17.

Kwak, Y. and R. Watson, 2005. Conceptual Estimating Tool for Technology-Driven Projects: Exploring Parametric Estimating Technique. http://home.gwu.edu/~kwak/Para_Est_Kwak_Watson.pdf (accessed April 10, 2008).

Mind Tools, 2008. Heuristic Methods: Using Rules of Thumb. http://www.mindtools.com/pages/article/newTMC_79.htm (accessed April 12, 2008).

PMI WBS, 2001. *Project Management Institute Practice Standard for Work Breakdown Structures*. Newtown Square, PA: Project Management Institute.

PMI, 2008. *A Guide to the Project Management Body of Knowledge (PMBOK® Guide)*, Fourth Edition. Newtown Square, PA: Project Management Institute.

Snell, D., 1997a. Expert Judgment. http://www.ecfc.u-net.com/cost/expert.htm (accessed April 10, 2008).

Snell, D., 1997b. Wideband Delphi Technique. http://www.ecfc.u-net.com/cost/delph.htm (accessed April 10, 2008).

Soomers, A. 12 Tips for Accurate Project Estimating. http://www.projectsmart.co.uk/12-tips-for-accurate-project-estimating.html (accessed April 12, 2008).

Stires, D.M. and M.M. Murphy, 1962. *PERT/CPM*. Boston: Materials Management Institute.

Uher, T.E., 2003. *Programming and Scheduling Techniques*. Sidney: University of New South Wales.

Verzuh, E., 2004. Phase Gate Development for Project Management—Part IV. http://findarticles.com/p/articles/mi_m0OBA/is_1_22/ai_n6134862/pg_1 (accessed April 11, 2008).

Verzuh, E., 2005. *The Fast Forward MBA in Project Management*. Hoboken: Wiley.

14 Cost Management

The goal of this chapter is to describe the activities related to project cost management in the planning phase. In other words, the focus is on how to produce a viable project budget that can be approved by management. Part VI of the book will focus on various aspects of the Monitoring and Control processes that are also part of the cost management processes.

When one thinks about project cost it is most often the view of how much money in their particular currency—dollars, euros, pounds, etc. For this view, we could say that the project would cost $100,000 and that would satisfy their needs. Others might want to know how this was divided into various resource types—$65,000 for personnel and $35,000 for material. We could translate this type of information from the WBS Dictionary data in somewhat the same fashion that was used to generate the time schedule. Previous discussions have laid out the techniques to define and schedule work units so that the dollarization process should be reasonably straight forward, right? Then the local accounting type drops by with a dazzling array of questions related to things called assets, overhead rates, depreciation schedules, chart of account data, and so on. We also recognize that the future product will need support and these representatives as they ask about various cost status aspects of the project.

As in so many of our project management discussions, the answer to the cost question is both simple and complex. Yes, we can calculate the resource costs for the project using reasonably straightforward techniques, but that simple view is not adequate for all concerned, or even the project itself. For example, accountants look at money for the enterprise in much more complex ways than the layperson. The PM must understand these views as well as those needed to manage the project. Likewise, he has to deal with various other cost-related considerations in order to fulfill his role in the organization. Previous discussions have taken a *peal the onion* view to the management process and that seems even more appropriate here. Let us take a simple approach to understand the basics of resource management, and then work on expanding into a more reasonable view after that. For the first segment, we will focus on defining the direct cost of work and planning packages.

14.1 PROJECT COST PLANNING BASICS

There are two basic project planning cost-related activities. These are (PMI, 2008, p. 165):

Estimate costs—the process related to approximating the monetary resources needed to complete the defined project activities.

Determine budget—the process of aggregating the estimated costs in order to produce an authorized baseline.

This is the time when previous definition work begins to pay off. Recall that the WBS Dictionary has defined data related to all work units—both work and planning packages. The key items required in generating a raw cost for those work units would be the level of work estimated for the unit by specific resource types. So, if the WP is estimated to require 100 hours of work by a particular skill group, the direct cost for that effort can be calculated. At this point resource cost data from the HR function is required. Since we do not know at this point who exactly will do the work and therefore do not know exactly the rate of pay, the initial cost estimate will have to use *generic rates*. This same assumption was present for the estimator who derived the work time for the task and will remain one of the potential error sources for time and cost estimates. Nevertheless, the generic labor cost estimate would be computed by multiplying work unit effort by a generic rate for that skill type. If a work unit was estimated to have 100 hours of effort and the generic rate was $25 per hour, the direct labor estimate for the activity would be $2500. Additionally, if the material estimate for this work unit were $1000, then the total estimated work unit cost would be $3500. Now, assume that we do this for all WBS work units and add all of those costs. Is that not the aggregate project cost? Why not? Certainly it is a rough estimate of the direct project costs as defined by the WBS scope definition, but it is not a cost number that represents the total actual project budget.

14.2   COST PLANNING

From a raw mechanics viewpoint, if there were no future changes anticipated to the project, we might take the calculated direct cost number and using the network plan lay out a time-phased view to show how those costs would occur over time. Figure 14.1 illustrates how dollars and time can be integrated. The typical assumption is that the planned cost is spread linearly over the activity time. This data would yield both a project direct cost and a time-phased distribution of those costs.

In addition to the direct cost items, assumptions are necessary as to how other cost items will impact the schedule. For instance, material items are often purchased in advance of the work unit and this would actually create a budget flow earlier than indicated from the direct cost calculation. Activity variability will also affect the magnitude of the actual resource flow and this can create cash forecasting issues as well. Finally, other project dynamics related to changes, risk, and other unknown events make the process of creating an accurate budget complex. The sections that follow from this point will each describe a unique characteristic that impacts the budget process. Collectively, all these have to be incorporated into the final budget view.

14.3   COST ACCURACY

Project budget accuracy is an often misunderstood concept. Too often an early initiation phase rough estimate gets carried into an approved project budget before detailed planning takes place. In any case, budgeting cost accuracy changes over the life cycle as more specifics regarding the project are known. Mature organizations realize this and make their decisions accordingly. For instance, it needs to be recognized that budget estimates at initiation can easily be ±100% for a large, complex project (maybe higher). For smaller repetitive projects, this range would be much smaller. Regardless, in general terms, the cost accuracy estimates evolve through three basic stages:

1. Rough order of magnitude (ROM): An estimate based on general knowledge of the requirement, but little knowledge available regarding specific detailed requirements (±100%).
2. Definitive: An estimate based on reasonable requirements resulting from an approved WBS, but incomplete analysis related to such areas as risk and resource capacity (±25%).

Cost Example	15 days	$65,000.00
A	5 days	$10,000.00
B	7 days	$20,000.00
C	5 days	$20,000.00
D	5 days	$15,000.00

FIGURE 14.1   Basic cost plan.

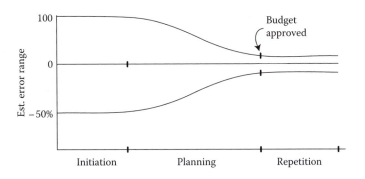

FIGURE 14.2 Planning cost accuracy.

3. Budget: The normal goal for a project budget formulated at the completion of a formal planning phase should strive for an accuracy level within 5–10%.

Figure 14.2 shows this in schematic format. In a mature organization, this notion would be more the norm and is an important concept to stress. For example, one very mature organization color coded their estimation documents. So, an estimate presented on red paper fits the ROM category, while a yellow one would be assumed in the definitive range. Finally, the formal budget would be called a green estimate. This approach let everyone know how to adjust their view of the data. More typically, only one value will be visible at any one point and it is often not clear at what stage of analysis the data represents. Project budget discussions need to have this management view.

14.4 ORGANIZATIONAL OVERHEAD

The term *overhead* is often used with a negative connotation. It is common to have certain overhead charges allocated to a project on a percentage basis with no real explanation. For example, 100% of direct costs might be added to all direct labor estimates, 5% to all material purchases, and another 50% to the total project for corporate overhead. Obviously, numbers like this will significantly increase the cost of a project, yet they are organizational reality that must be dealt with along with the direct cost items. Often times the PM does not know how these allocations are derived, nor does he have any real control over their inclusion in his budget. In many cases the only defense is to finish the project as quickly and efficiently as possible since some administrative overhead fees are time based. The necessary management strategy for this cost category is to understand the various allocations and be sure that they are included in the planning budget. If not included, the actual allocations coming later would represent a budget overrun in that the value would not have been in the plan.

14.5 SCOPE, TIME, AND COST ALIGNMENT

Before the planning cycle is over work, it is not uncommon to find that the project requirements cannot be produced in the time and cost approved in the project Charter. At this point the project is unworkable. One choice is to go back to the management approval authority to ask for more time or budget and possibly the project is so worthy that neither time nor cost are of great concern. Good luck on that one! The more typical situation is to attempt to cut scope in such a way that the schedule and budget are in line with constraints. In any case, the final planning view must balance these three variables.

14.5.1 SCOPE REPLANNING

During the course of scope development, we described prioritizing requirements as must have, needed, and some as nice to have. The nice to have category often become the first to go in the scope replanning process. Hopefully, this tactic would resolve the problem and the required schedule and budget constraints would now be met. Failure to do that necessitates the second alignment strategy.

14.5.2 FAST TRACKING

If the alignment problem is more time than cost related, it is possible to look at fast tracking the schedule. This was mentioned previously in Chapter 13. Basically, we look for ways to resequence the plan in order to make the activities fit into the required time frame. This strategy may in fact have other adverse implications in cost or risk, but if done carefully can decrease the schedule without significant increase in cost.

14.5.3 SCHEDULE CRASHING

This strategy emerges in the situation where the required schedule is not met, but more budget funds are available—a time-constrained schedule with money available. This process involves more complexity than the previous two options. Crashing involves the trade-off of budget resources for time. There are various situations in which this option is relevant to the PM. If budget is available and scope is required, then budget can be traded for time through the crashing process. In similar fashion, a contractor may want to crash a project when there are financial incentives to finish the project ahead of schedule. In order to decide whether to crash an activity, one must compare the cost of crashing the activity with the value gained by the reduction in schedule. Another less obvious reason to crash the project is to decrease indirect costs that are more time based. Direct costs are considered to be labor, materials, and equipment associated with a particular project, while indirect or overhead costs cannot be identified and charged specifically to one project. Examples of indirect costs are facilities, utilities, basic infrastructure items, level of effort activities for the project, and so on (Uher, 2003). Some of these are charged to the project monthly, regardless of status, and their effect on the project budget is tangible. Shortening the project time would decrease this class of cost.

In order to perform the crashing process two time and cost estimates are required for each activity in the network—*normal* and *crash*. These cost estimates only use the direct cost component since that is the only real component to minimize. The normal estimate is considered to be the optimum time and cost for that work unit. A crash estimate is considered to be the "absolute minimum time required for the job and the cost per each time unit from that optimum point" (Stires and Murphy, 1962). Primary concern in the crash estimate is the incremental cost to decrease the activity time. The crash relationship for activity is shown in Figure 14.3 as a time–cost trade-off curve.

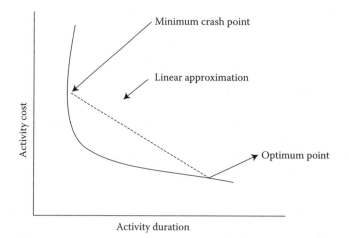

FIGURE 14.3   Time–cost trade-off curve.

A liner approximation of the incremental unit time (per day) crash cost can be determined by the formula:

$$\text{Cost to expedite} = \frac{\text{Crash cost} - \text{Normal cost}}{\text{Normal time} - \text{Crash time}}$$

This formula assumes a linear relationship through the range, which once again introduces an estimating error (Figure 14.3). However, the management benefits obtained usually justify the short-cut. If necessary, a more rigorous estimate can be derived for each time. The graph shows that each increment of time becomes more costly and eventually crashing will result in diminishing returns. At that point it does not make sense to try to shorten the project any further. Once this data are derived for each activity, we have the fundamental parameter necessary to execute the process. In order to arrive at the optimum total project investment curve, the following five-step crashing plan is required (Stires and Murphy, 1962):

1. Develop a first cut schedule using normal time and cost estimates for each activity.
2. Develop crashing information for each activity. This should include lowest crash point and crash cost per time unit.
3. Select the lowest crash cost on the critical path and reduce that activity time by one time unit. Recompute the critical path for the new time.
4. Repeat steps 3 and 4 until the desired project duration is reached, or available budget is depleted.

Data to illustrate the mechanics for crashing are provided by Figures 14.4 and 14.5 and the corresponding crashing data are presented in Table 14.1. These three items represent the starting point for the process. The goal for this example is to illustrate the crashing steps with a reasonable-sized activity set.

The first step in the crashing process is to identify the lowest crash cost on the critical path. Inspection of Table 14.1 data shows this to be activity 1.1.1 with a cost of $2000 per day. Figure 14.4 verifies that this is on the critical path, so investing these funds will shorten the project to 104 days. No more days can be taken out of this activity since it is now at the minimum duration (e.g., Minimum Crash Time).

For the second iteration, Table 14.1 identifies four activities 1.1.2, 1.2.2, 1.3.3, and 1.6.3 at the next lowest crash cost ($3000). Figure 14.5 shows that activities 1.2.2 and 1.3.3 are not on the critical path; therefore shortening these two activities would not shorten the project. So, activity 1.1.2 is selected and the duration shrinks one time unit to 100 days. The activity cannot be shortened further. We can see from inspection of the Gantt chart that the critical path will stay the same.

Step 3 selects critical path activity 1.6.3 from the options shown above. This will add another $3000 per day to the project budget; however, in this case, the activity can be crashed three time periods for $3000 each day. Taking that option adds $9000 to the budget and reduces the project duration to 97 days.

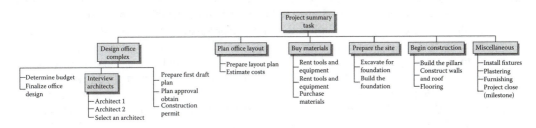

FIGURE 14.4 Example WBS for crashing.

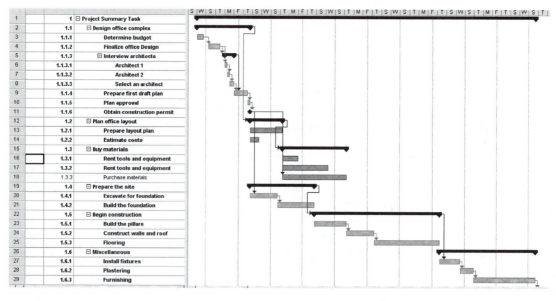

FIGURE 14.5   Project critical path (normal view).

It will be left as a reader exercise to crash this plan further. As each subsequent step is executed the incremental crash cost will increase. Also, in larger plans, it is common for multiple critical paths to emerge. The mechanics as described here will work for any network, but obviously the complexity of the setup increases. Also, it will be necessary in these situations to actually recompute the critical path for each iteration. Crashing is often not a simple process, but an important tool for the PM.

14.6   INDIRECT COSTS

Many project cost components occur through allocations or time-based charges. Facilities overhead, level of effort support charges, and other cost categories contribute to increased cost regardless of the actual project work activity. These charges can occur at both the activity and the project level. Figure 14.6 illustrates how indirect costs help justify activity crashing.

Figure 14.7 shows the impact of indirect costs on total project costs trade-offs. An understanding of these views can help establish an optimal strategy for project duration. Once again we see logic for understanding the fast tracking and crashing logic.

14.7   RESOURCE ALIGNMENT

It is one thing to tweak the plan until it fits certain scope, time, or cost constraints, but it is quite another to match this against resource availability. The initial cost estimates were based on a generic cost for each skill. We are now recognizing that the quantity or quality of those required skills may not be available at the time specified by the plan. Figure 14.8 shows this situation. In this view, the y-axis can represent either the number of resources or the cost. The situation shown indicates that the plan requires more than available, so we have a resource capacity issue. Basically, the plan is of no value in this situation because it will indicate a schedule that cannot be met and possibly a false budget as well. In order to have a viable plan, the required resources must match the allocation shown in the plan. Experience indicates that this situation is a common management issue and the cause of many projects failing to meet their plans, even with all other factors under control.

There are two basic options for dealing with lack of timely resources to fit the plan.

TABLE 14.1
Crashing Parameters

WBS	Activity	Duration	Predecessor	Minimum Crash (days)	Crash $/D ($000s)
1	Project summary task	105 d		—	—
1.1	Design office complex	17 d		—	—
1.1.1	Determine budget	3 d		3	2
1.1.2	Finalize office Design	5 d	3	4	3
1.1.3	Interview architects	4 d	4	—	—
1.1.3.1	Architect 1	1 d	4	—	—
1.1.3.2	Architect 2	1 d	6	—	—
1.1.3.3	Select an architect	2 d	7	—	—
1.1.4	Prepare first draft plan	4 d	5, 8	3	4
1.1.5	Plan approval	1 d	9	—	—
1.1.6	Obtain construction permit	0 d	10	—	—
1.2	Plan office layout	10 d	2	—	—
1.2.1	Prepare layout plan	10 d		4	5
1.2.2	Estimate costs	2 d	13	1	3
1.3	Buy materials	20 d	12	—	—
1.3.1	Rent tools and equipment	5 d		11	
1.3.2	Rent tools and equipment	14 d		12	4
1.3.3	Purchase materials	20 d	11	2	3
1.4	Prepare the site	20 d		—	—
1.4.1	Excavate for foundation	8 d	18	6	10
1.4.2	Build the foundation	12 d	20	9	20
1.5	Begin construction	38 d	19	—	—
1.5.1	Build the pillars	10 d		9	19
1.5.2	Construct walls and roof	8 d	23	7	18
1.5.3	Flooring	20 d	24	13	17
1.6	Miscellaneous	30 d		—	—
1.6.1	Install fixtures	7 d	22	—	—
1.6.2	Plastering	4 d	27	—	—
1.6.3	Furnishing	19 d	28	10	3
1.6.4	Project close (milestone)	0 d	29	—	—

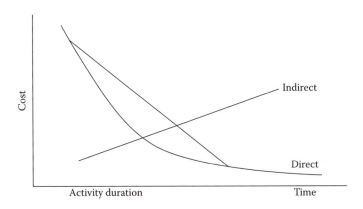

FIGURE 14.6   Indirect cost curve.

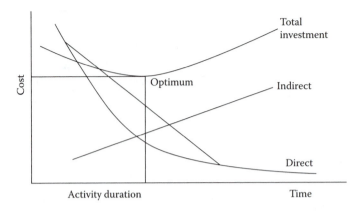

FIGURE 14.7   Total project investment curve.

1. Increase the capacity of resources available to meet the plan through reallocation, hiring, or outsourcing.
2. Lag the plan work units until the capacity matches the requirement. Increasing capacity can be achieved by reallocating internal resources from other projects to this one, hiring new employees, or contracting needed resources from third parties. Technically, each of these choices could work if properly managed.

Either decreasing the allocation of resources to various work units that will cause the activity to increase in duration or simply moving the activity to a later time when resources are available can accomplish a decrease in the resource requirement. If this is done to slack activities the action will not affect the critical path; however, moving critical path activities will also move the completion date.

Recognize that this class of problem is not solved overnight since it often involves multiple-level organizational decisions. For this reason, resource capacity issues need to be defined as far in advance as possible to allow various solution options to be resolved. In order to do this, a formal project resource plan created along the lines outlined in Part III is a prerequisite. Beyond the project plan view there must be a workable organization level resource management system that can identify aggregate capacity and then link available capacity to the project. During the planning phase we do not have to know that Joe Smith is going to be allocated to our project for some planned period, but we must know that someone with the appropriate skills will be. Failure to accomplish this basic linkage means that the project schedule slips and corresponding budgets likely slip as well. If there is one operational Achilles heel in most organizations, it is the one described here. Multiple departmental groups typically staff matrix format organization projects. The various internal resource suppliers

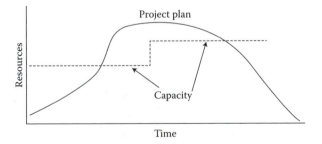

FIGURE 14.8   Resource capacity leveling.

must be able to have visibility into the portfolio of projects planned in the organization, and in order to do this, enterprise level information systems are needed to support this process.

14.8 BUDGET RESERVES

Up to this point we have attempted to view the budget environment using basically deterministic assumptions—that is, nothing will happen outside of the original planned scope and time estimates. As a modern day Wizard of Oz might say "Dude, this ain't Kansas." The fact is that a project exists in a very dynamic environment with things changing all the time. Any deviation from the plan has the potential to impact scope, time, or cost estimates. Some feel that these dynamics are so radical that planning is a waste of time. This is not the case, but clearly the plan has to recognize the dynamics. The key question is how to incorporate unplanned contingencies in a reasonable manner. Figure 14.9 shows the three major areas of dynamics that have to be dealt with in the final plan.

14.8.1 APPROVED CHANGES

The most obvious dynamic is unplanned work resulting from requested and approved changes to the original project scope. Each of these actions will occupy the project team's attention and resources, thereby taking away productive time from the current plan. When a new work unit is approved, the WBS structure needs to be modified to reflect the new scope and all planning related to how the new work fits into the current plan represents additional project work activity. It is hard to estimate how much time this activity actually takes, but clearly it is a visible amount of time that should be recognized in the project budget. These work activities have an impact on schedule, budget, and required resources for the project. One possible method to show this in the plan is to allocate a level of effort work unit with some attached budget. All change requests work could be charged to this work unit and minor approved changes could be absorbed through this.

A key management question would be how to physically incorporate an approved change request into the project work plan. If it results in additional work or new WBS WPs, the basic question is, do you allocate budget from the change pool into those packages or leave them empty from a budget point of view. This process can be handled via either the pool approach or by allocating the pool funds into defined WPs. Realize that leaving the funds in a pool will complicate the ability to track WP costs later. A more significant management issue emerges when the change request is large. The first issue is to define when a request is large. For those designated in this category, the pool approach is not appropriate. For these situations, the approved plan needs to reflect the change in some manner. This could be handled by allocating some segment of the change pool for large changes and moving the funds out of this group into the project-approved plan just as though it were there in the beginning. Regardless of the funding process selected for this class of dynamic it is not fair to the project team to ignore these events and assume that they can be absorbed into the original budget. In order to do this, the project team will pad their estimates and hide the activity, which camouflages that activity. The management goal needs to be one of tracking what is going on and not hiding behind padded estimates put in place to keep the budget intact.

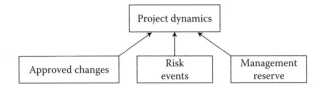

FIGURE 14.9   Major sources for project dynamics.

14.8.2 Risk Events

A second dynamic occurs from risk-related events that occur through the life cycle. A more detailed discussion of the mechanics for dealing with risk will be discussed later in Chapter 24, but for now we need to recognize that some allowance for this class of activity is needed. A risk event is called a "known/unknown." This means that certain things that the project team is aware of may not go well, but these potential events may or may not occur. The project plan assumes that they will not occur, but must cover the recognition that some will. The traditional method for handling such events is to set aside a contingency allocation that has been estimated to cover anticipated events. This allocation will be held external to the working budget and used as a particular event occurs. Think of this as a phantom work unit that is not in the plan. When the event occurs, the related corrective work unit is moved into the plan and funds are taken out of the risk contingency pool to cover it. The goal is to have a contingency fund that just covers the future needs for this class of dynamic. As in most project estimating situations, this is a difficult number to derive, but the concept is to recognize that risk events fall outside of the operational work units. We want the budget to reflect the work as planned, but have mechanisms to invoke for "bumps in the night." Scope changes and risk are two such items.

14.8.3 Management Reserve

A third dynamic is a class of issues fall into the category of "unknown unknowns." These are issues that occur, but were not planned or anticipated. This class of contingency is called a management reserve and it is designed to deal with unplanned, but required, changes that occur during the course of the project. Management practices to handle such events are not a standard industry practice item. As in other such dynamics, it is often buried under padded estimates, which violates our management approach. Ideally, the PM would have a defined management reserve contingency set aside for these events and the process would simply be to record the unplanned event and allocate necessary funds from this source to cover the item. Management reserves are typically allocated to some management entity to allocate to the project and are not part of the approved budget.

Technically a management reserve event is any work-related activity that does not meet the plan. In theory, this can be the result of any work unit that exceeded its planned amount, or any unanticipated work requirement. Project management theory says that fund requirements and fund allocations for all such events are under the control of external project management. However, as a practical matter, this is not a reasonable operational approach. From a more practical viewpoint, this level of external control for overruns can be onerous and the PM needs to have some level of flexibility for minor additional funding events. For example, if a WP overruns by 5% should that overrun be taken from a general overrun buffer or a larger externally controlled management reserve account? A PM-controlled buffer approach would seem more logical. Likewise, the requirement to pay a minor bill to repair a piece of equipment should fall into the same category. However, if a major event occurs or the project overall budget exceeds 10% that process would seem worthy of more external oversight. In this second case, a higher level of funds is also required and allocation of these will most likely not be delegated to the PM. The budget management question in this case is about the proper source and control process for additional fund allocation. An allocation for larger-size management reserve funds needs to be recognized in the overall budget, but not necessarily shown in the public budget and not under the control of the PM. One way to handle this could be to have a management reserve contingency fund similar to the risk fund, but under formal control of a group such as the project steering committee or the sponsor. In this review process, the management question would involve not only the fund allocation, but other correction actions such as whether the project should continue.

So, the recommendation for management reserve budgeting is to allow a small amount of funds to be left in the project-controlled portion of the budget to handle this class of dynamics. The main segment of management reserve would in a contingency fund held and controlled external to the project direct work. This fund may or may not be visible as part of the public budget, but will be part of the overall consideration for the project in terms of financial justification.

In this section, we have described three dynamic budgeting issues that have the potential to consume budget funds that were not defined in the original base plan. Each of these variance situations will occur to some degree during the project life cycle and some arrangement for handling each category needs to be part of the plan and somehow reflected in an overall budget view.

14.9 MONEY CAN HAVE DIFFERENT FORMS

As stated earlier one view of a budget is simply the number of dollars or euros required in executing the project. This view can become muddled as the PM begins to deal with various groups. Budget monies are often allocated to many different categories and organizational unit groups. This means that the project budget has to be structured to fit these required categories. In some cases, budget funds may come from different sources at different time frames. So, managing the budget is not as simple as just keeping a single actual total spending level within the planned value. In many cases, it is keeping various cost groups within each of their bounds. Our goal here is not to understand the accounting theory related to this issue, but it is necessary to understand and be sensitive to the fact that the budget will have various categories and sources. Some of these groupings must be considered as part of the estimating process, but others will just require that the source be marked and managed as part of the control process. Industry professionals call this bean counting.

14.9.1 BUDGET EXPENSE CATEGORIES

The following list of project cost elements provides an example to show how a budget may contain multiple fund categories within what would be called "the budget." The major categories of budget expenses are as follows (Lane, 2003):

Personnel
  Salaries and benefits (including hiring fees and bonuses)
  Training and education
  Travel
  Morale
  Staff-related depreciation
  Temporary help/consultants
  Miscellaneous (space, telecom, etc.)
Hardware
  Depreciation
  Maintenance
  Repairs
  Leases
Software
  Depreciation
  Maintenance
  Customer support
  Updates
  Repairs
  Leases
Services
  Leased lines

>　Outsourced network services
>　Security services
>　Third-party service providers
>　Miscellaneous (transport, courier, periodicals, etc.)
> *Other*
>　Etc.

Beyond the categorization aspect of the funds another example is found in the depreciation item. This is an accounting entry based on the anticipated useful life of an item. In this case the item might cost $100,000 with a life of 5 years and the amount allocated to the project might be only $20,000 per year. This is the capital funds category. Without carrying this point further, realize that a budget will have to obey the organizational financial rules and your support financial person will be needed to ensure that these are obeyed. In order to comply with this level of cost granularity, WP estimating detail will be more complex than outlined thus far. Organizational financial systems will dictate how the cost is categorized and ultimately reported in the budget format.

14.9.2 ASSETS VERSUS EXPENSES

Expenses are cash budgetary outlays for project goods and services that are consumed during the course of the project life cycle. A capital expenditure is an accounting entry in cash units, but it is not equivalent to actual dollars as reported. Simply put, capital expenditures create assets that are then depreciated over scheduled periods of time, and these depreciation expenses will be charged according to some set of rules established by the financial organization which attempts to match service value to expenditures, so that an asset cost is recognized over its useful life rather than as the money is paid for its purchase. The basic decision regarding whether to consider a purchase to be expense versus asset is its cost and useful life. Guidance on these issues is provided by the financial function and governmental regulations tend to be the guiding policy.

A capital expenditure is identified on the organization's asset schedule and some form of depreciation schedule is established for it. From this base a periodic depreciation value is entered on the project budget. This has the impact of initially showing the asset at a lesser value than the actual cost. This impact on the expense budget is important to recognize, as it can be significant. In some cases there are favorable tax credits for some investment types, which has a further impact regarding how the organization views the decision.

As a side note, it is even becoming common to view large technology-based projects as enterprise capital assets. The implication of this would be that the total cost of the project is depreciated over time, so a $1 million project that is depreciated over 5 years (straight line) would show on the organizational financial statement as costing $200,000 per year even though the organization would have expended more funds than this. Depreciation of this class of project has the impact of improving the short-term accounting profitability of the organization, but diminishes that view in future years as the depreciation expense is recorded.

The PM needs to work carefully with the financial organization to deal with proper methods for handling all budget expenditures. Some of the decisions related to asset versus cash categorization are flexible based on internal situations. It is always possible to expense an asset and take the cost penalty up front. This has the impact of showing the organization making less profit in that time period, but possibly saving current taxes.

14.9.3 BUDGET COST COMPONENTS

Table 14.2 illustrates how budget cost components might be specified for a project. It may also be necessary to define in greater detail the types of skills required if this data would be needed by

---

TABLE 14.2
Project Cost Components

Cost Category		Cost
Direct labor	Hours	$
Indirect labor	Hours	$
Total labor costs		$
Hardware acquisition		$
Materials acquisition		
Software acquisition		$
Total acquisition costs		$
Consulting or subcontracting		$
Travel or other employee or contractor expenses		$
Financing costs, such as interest on project capital funding		$
Total other costs		$
Contingency costs		$
Total project costs		$

---

internal groups. Beyond this view it is common to array cost and resource date into a time-phased plan. From that view the data could be divided into direct and indirect categories as shown in Table 14.2.

14.10 MANAGEMENT APPROVAL AND BASELINES

As the various budgeting iterations are resolved, a total project cost view becomes what the final proposed project budget. As we have seen in this chapter, there are many embedded financial considerations to consider as part of this process and each project has somewhat different cost issues and characteristics. However, recognize that more organization segments are interested in budgets than any other project activity. In any case, the assumptions that have been made throughout the planning phase are now packaged into a formal cost document with funds categorized as dictated by the financial organization. From this, enterprise management will now make a go-no-go decision on the project based on their review of this document, along with the other planning artifacts.

Once management has approved the project budget data, a formal cost baseline is set. Basically, this involves "stamping" that version of the overall plan for use in future status comparative measurement. Regardless of what occurs after this fact, the original baseline data will be kept. Recognize that the term *baseline* can apply to more than just schedule and cost. There can be a performance baseline, a quality baseline, a staffing baseline, or any other measurable parameter that management wishes to monitor.

The cost baseline is a time-phased presentation by budget category that is used as a basis against what to measure, monitor, and control for the project. This view is created by summing estimated costs by category and period, then displaying them in either tabular or graphic form. Figure 14.10 shows a graphical view in which the approved baseline is set higher than the anticipated plan. In this example, the graph is showing the cost baseline versus planned cash expenditures and funding constraints. This presentation is a key item in the project management plan. Many projects, especially large ones, have multiple cost or resource baselines, as well as baselines related to consumable production items (e.g., cubic yards of dirt moved per day). In each case, these are used to measure different aspects of project performance. These various baseline comparisons may also be included in regular project status measurement.

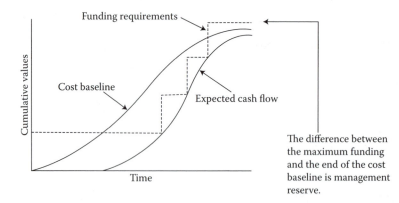

FIGURE 14.10 Cost baseline graph. (From PMI, 2008. *PMBOK® Guide*, 4th ed., Project Management Institute, Newtown Square, PA, p. 148. With permission.)

14.11 SUMMARY

The project budget is one of the most important planning artifacts and it has wide exposure across the stakeholder community. Basically, budgeting represents the cost component of the holy trilogy and is a major management consideration for any organization. It represents the primary metric benchmark for project monitoring and control.

A second theme of this chapter was to highlight the multidimensional view of funds that a budget has with various organizational stakeholders. In order for the PM to have internal organizational credibility, one must understand the formal budgeting rules and regulations.

REFERENCES

Lane, D., ed., 2003. *CIO Wisdom: Best Practices from Silicon Valley*. Boston, MA: Pearson.
PMI, 2008. *A Guide to the Project Management Body of Knowledge (PMBOK® Guide)*, 4th ed., Project Management Institute, Newtown Square, PA, p. 148.
Stires, D.M. and M.M. Murphy, 1962. *PERT/CPM*. Boston: Materials Management Institute.
Uher, T.E., 2003. *Programming and Scheduling Techniques*. Sydney: University of New South Wales.

CONCLUDING REMARKS FOR PART III

At this initial stage of the model planning process, we have focused on developing detailed information on five key planning processes:

1. Mechanics for work unit level details on the activities necessary to produce the defined outputs (scope)
2. Creation of a first cut project schedule using deterministic work and planning unit time estimates
3. Three advanced approaches on time and cost management in the Technical Appendix supplement (Chapters 15 through 17)
4. Creation of an aggregate project resource plan necessary to accomplish the defined work. This includes time-phased quantities for labor skills and materials
5. Creation of a basic project budget for the direct project work elements.

These activities do not complete all management aspects of producing a project plan, but collectively they cover what would be considered the planning basics. Part IV will incorporate the implications of various support knowledge area processes into this basic structure. Each of these future implications will potentially result in a modified view of the basic plan.

Part IV

Advanced Planning Models

15 Analyzing Variable Time Estimates

One of the restrictive assumptions made in previous chapters was that estimated activity time was deterministic and accurate. Obviously, this assumption is not valid in a real-world project, yet relaxing the assumption introduces a nondeterministic answer regarding when the project will be finished. This means that management and user stakeholders are going to have to deal with a different kind of completion discussion. The proper answer will no longer be to say that the project will be finished on June 1. Instead, the answer will be couched in statistical terms such as "we have a 50% chance of finishing by June 1 and a 90% chance of completing by August 15." It is this characteristic that keeps variable time schedules from being popular. Stakeholders seem to want a single answer even if it is wrong (which it will be most of the time). Figure 15.1 shows what this result might look like with a histogram of probabilistic completion dates. This chapter will explore a classical technique to accomplish this type of analysis.

Note from the completion diagram represented in Figure 15.1 that there is a 50% chance of project completion by February 28, the estimated project range is from February 19 to March 7. As activity uncertainty is higher, the corresponding calculated completion dates would also expand. This type of analysis is a valid way of looking at project schedules and should be pursued whenever possible. This chapter will describe the mechanics of variable time analysis and then show how the activity estimates can be translated into a meaningful view of project completion.

15.1 HISTORY OF VARIABLE TIME ESTIMATES

Variable time estimating techniques were initially formulated in the mid-1950s in high-technology military projects, most notably the Navy's Polaris project. The technique was contained in the network management model named PERT (Project Evaluation and Review Technique). Up until this point, project schedules had been created using vague scheduling tools such as the Gantt chart. The complexity of this new breed of project forced a more advanced view of the scheduling process (Verzuh, 2005). This class of project focused on time management more than cost, so the focus of the original PERT model was heavily time oriented.

Since time was a primary issue, PERT modeling used statistical techniques focused on assessing the probability of finishing the project within a given time range (Gale, 2006). The PERT model was based on an AOA network format with arrows depicting activities (see details of AOA in Chapter 13). Basically, the initial goal was to define project sequence and develop a quantitative technique to predict probable duration time ranges for each activity. From this, various statistical techniques were used to develop a probabilistic project completion forecast. The technical and schedule success of the Polaris project burst PERT into the project management limelight as a near magical tool.

Reiss later discussed the actual impact that PERT had on the Polaris project. He implies that it was never determined whether PERT helped, hindered, or played no significant role in the Polaris project, but that everyone seemed to think PERT had helped a great deal (Reiss, 1996). Nevertheless, publication of the Polaris project results stimulated PERT's emergence on the project management

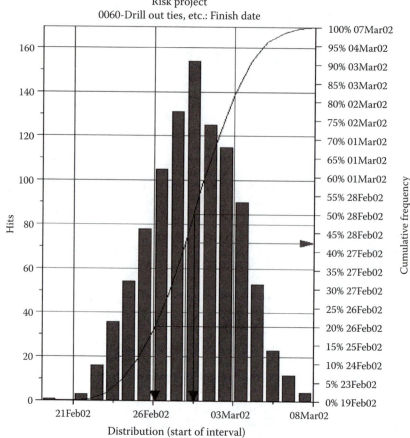

FIGURE 15.1 Probabilistic completion diagram (Pertmaster).

scene to amazing heights. During this period, a speaker at an American Institute of Management conference said, "Anyone not using PERT on their project ought not to be managing their projects at all" (Reiss, 1996).

As other projects began to use the PERT model, they began to recognize that it was mathematically complex and did not help with cost management. Also, computer software for manipulating this model was expensive to use at that time and available to only large organizations. Many felt that PERT did not work very well and there were some notable project failures attributed to its use (or misuse). The general attitude emerging at this point was that the Polaris project success was circumstantial and may not have had anything to do with PERT (Reiss, 1996). Russell Archibald, one of the PMs on the Polaris Missile project stated:

> If you followed the developments at the time, you will remember that PERT was given a lot of favorable press, with the Navy's encouragement and grand claims were made that PERT enabled the Navy to complete the program some years earlier than it would have otherwise. I do not believe that these claims are entirely true, based on my experience. PERT probably did some good as far as planning and scheduling is concerned, but both the Navy and Lockheed, as the missile system integrator, failed to recognize the area of greatest payoff: integrating the schedules of many contractors (Archibald, 1987).

Based on Archibald's testimony, it can be deduced that, realistically speaking, there were other operational flaws with PERT during its early years. As a result of these negative factors, the theoretical aspects of PERT declined and are still not used in the majority of projects today even though the historical rationale mentioned is no longer valid. At approximately the same time a second network based tool emerged. This tool was called critical path method or CPM. It utilized deterministic time estimates but dealt with cost issues better than PERT. CPM survived the early period, while the PERT model stayed in reduced form in the governmental sector.

As a result of these machinations during the early period, use of the terms PERT or CPM became muddled. Today, even when organizations talk about PERT, they generally use only a single activity time estimate for the activities and that essentially makes the PERT model usage equivalent to CPM. As the situation stands today, only a small minority of PMs can distinguish differences between the classical PERT and CPM assumptions (see Chapter 13 for a refresher on this).

So, the fundamental question here is why go back in time and resurrect something that has essentially died? The answer to that question is that the PERT model addresses a piece of project reality that needs to be resurrected to improve project management. Also, common spreadsheet computation tools can be used to perform the needed analysis. The fundamental value that the PERT model has is its ability to show the PM a probabilistic estimate for project completion. The rest of this chapter will distil that idea into what we hope is a simplistic view.

15.2 MODIFYING PERT FOR COMMERCIAL PROJECTS

PERT relies heavily on empirical mathematical averages and probabilistic distributions in order to produce its output. The two underlying empirical formulae have been translated into relatively simplistic formats, leaving only one somewhat complicated statistical relationship to deal with. For projects of 10–20 activities, the formulae are manageable, straightforward, and only require a moderate amount of work; however, when a project has 1500 activities, the mechanics require more robust utilities to manipulate (Reiss, 1996). In the mid-1950s, computer software was not readily available for this type of analysis activity and project management maturity was not sufficient to see the need for this. The reason why PERT may have initially worked for the military and subsequently on the Polaris project was a higher availability of adequate computing resources.

Most project management experts do not rank network modeling with variable time estimating as their most important activity, but it actually provides an improved insight into how the project might progress. Certainly, personal credibility is lost when a fixed date is given and the project overruns that date with little understandable explanation. The variable time concept represents reality, and with a mature management, organization can build credibility for the PM. The fact is that the model represents critical project organization mechanisms and it supports an improved level of communications (Stires and Murphy, 1962). Also, the model helps to analyze activity relationships inherent to the work elements of a project, and as a result, creates a mechanism for ranking problem areas by criticality.

Stires and Murphy summarize the use of this tool as follows:

> There is, however, the danger that a foolish manager will use PERT not as a tool, but as a weapon, a clobbering instrument. The system has that potential; but the degradation would of course result in the utter destruction of faith in the system, a hardening of the natural resistance to change, and deprive the manager of all the benefits that PERT offers when used intelligently (Stires and Murphy, 1962).

It can be deduced from this statement that their main concerns are based on an abuse of power in the assignment of duties while using PERT. As with any tool used for managing groups of people, this tool is no better than the one using it. This old adage says it well—"A fool with a tool is still a fool."

15.3 DEFINING VARIABLE TIME ESTIMATES

If you are not reasonably certain about the time it takes to execute a work unit, how do you express this? Descriptive statistics offers a traditional technique through the use of a probability distribution.

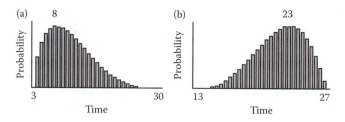

FIGURE 15.2 Skewed probability distributions: (a) positive skew and (b) negative skew.

The question then arises as to what kind of distribution to use. Early definers of the PERT model found that they needed a distribution that had the ability to be skewed in either direction. This means that we need to be able to define a task that might finish very early, but probably will not. But we also need a method to express the opposite situation or any combination between these two extremes. The two distributions shown in Figure 15.2a and b illustrate these two extreme situations.

The distribution on the left is positively skewed, while the one on the right is negatively skewed. Let us see if we can interpret these in an estimating format. In both figures, time is the *x*-axis variable and probability of occurrence is the *y*-axis variable. As an example, the negatively skewed diagram could say that the time to perform this activity was optimistically 3, most likely 8, and pessimistically 30 time units. In other words, we are not at all confident that this particular task will be any one particular value. Obviously, this example shows a wide degree of uncertainty regarding the estimate. In the case where the estimate had less variability, the shape of the curve would become more symmetrical. In this case, estimate range values of 8, 10, and 12 would suggest a more confident time estimate. This shape is an example of a situation where you have some degree of confidence, but also recognize some degree of a risk factor. Figure 15.2b shows the opposite scenario. This type activity might complete as early as 13 time units, but more likely will require times in the 23–27 range.

Project activities behave in the manner as modeled above, so the probability distribution selected has to be able to fit these shape characteristics. Based on these considerations, the PERT originators decided to use a beta distribution based on its shape flexibility. It has the flexibility of being able to be skewed in either direction, or symmetrical. The next design parameter requirement involved how to specify the shape of an activity. It was decided to use the following three estimating values to define the distribution shape characteristics:

a—an optimistic time estimate
m—the most likely time estimate for the activity
b—a pessimistic time.

There are two time-oriented parameters that need to be extracted from this. First, we compute a time that is equally likely to occur, given the defined shape. In descriptive statistics terms, this would be called the median of the distribution. PERT mathematicians then went about developing an easy to use empirical formula for this value and the result was

$$T_e = \frac{a + 4m + b}{6}$$

where T_e is the median time estimate for the activity.

A second parameter that must be defined is a measure of dispersion for the estimate. In statistical terms, this can be defined by the standard deviation. Once again, an empirical formula was developed to compute activity standard deviation as

$$\sigma = \frac{b - a}{6}$$

where sigma (σ) is the common Greek symbol used to denote this term.

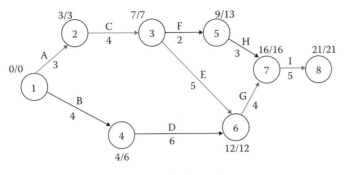

CP = A–C–E–G–I; and 21 time units

FIGURE 13.8 Defined critical path. (From Chapter 13 of this book.)

The final shape issue is to define activity variance. Once again, this is a standard descriptive statistics parameter and is simply the square of the standard deviation. So, variance could be computed by squaring the formula above, which is

$$\text{Variance} = \sigma^2.$$

Now the question is what to do with all of these statistical estimates. You will note shortly that this process will converge back to a point very close to the traditional project network. The first thing to be done is to compute T_e for each activity. Once computed, this value will be used in the network plan just as though it were a deterministic value. The network originally shown in Figure 13.8 is reproduced to illustrate this.

Assuming that the calculated T_e values were as shown on the diagram, we would still say that the most likely time to complete the project would be 21 time units, but the variance data introduce an opportunity to say more about the characteristics of this project. Before we move to that interpretation, we have to introduce one more statistical concept (are you beginning to see now why PERT was initially ignored?).

15.4 CENTRAL LIMIT THEOREM

We have shown that the variable time activities in the project can be skewed in all directions, so how can we decipher what the resulting project distribution would look like? The central limit theorem comes to the rescue for this. This theorem basically states that as the sample (activity) size becomes large, the sampling distribution of the mean becomes approximately normal regardless of the distribution of the various individual activities (Brown, 2002). Assuming that a sufficiently large set of activities exists (say 30 on the critical path), this theorem would justify projecting the resulting project distribution to be normal or symmetrical.

For example, let us now suppose that the sample network is large enough and for ease of calculation let us say that each activity on the critical path has a computed standard deviation value $[(b - a)/6]$ of 2. As we saw in the earlier formula the variance for each would then be 4.

The mechanism that affects the project duration is the critical path and the variability of that path will determine the variability of the project. Variability is determined by summing the variances on the critical path. This fact and the central limit theorem are the two tricks to this method. Reviewing Figure 13.8 above, we see five activities on the critical path, each with an assumed variance of four, so the calculated variance of the project schedule would be 5 times 4 or 20. The square root of this would yield the measure of dispersion (standard deviation) for the project as 4.47 (let us say 4.5 since

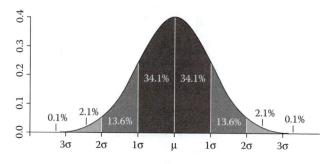

FIGURE 15.3   Normal distribution areas under the curve.

this data are not accurate enough to justify two decimal digit accuracy). From the parameters above, we know that the project has an expected completion time of 21 time periods and a standard deviation of 4.5. Using these two shape parameters, we can now describe our confidence level of project completion. Figure 15.3 provides what would be called the typical view of this in terms of a distribution with its sigma values. Also, the symbol μ represents the critical path value (21 time units) in the example case. As can be seen from the areas under the curve, values for unit standard deviations are

$$1\sigma = 68.2\%$$

$$2\sigma = 95.4\%$$

$$3\sigma = 99.7\%.$$

Looking at this data in a slightly different way, we could estimate the probability that this project will finish later than some percent. Let us now examine the probability that the project will finish after one standard deviation. Examining Figure 15.3, we see that the approximate area to the right of this point is approximately 34.1%. This then means that there is a 34.1% probability that this project will finish after time period 25.5, in spite of the fact that the original deterministic calculation said that it would complete at time period 21. This provides a good example of the power of this technique.

15.5   TRIANGULAR DISTRIBUTIONS

One of the complicating factors that have inhibited broader PERT adoption is the general level of statistics literacy in organizations. Beta distributions and the central limit theorem offer too many Greek symbols for many to digest. One simplifying view has been to use a triangular distribution. Figure 15.4 illustrates what this distribution looks like.

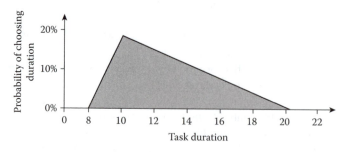

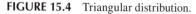

FIGURE 15.4   Triangular distribution.

TABLE 15.1

Comparison of Beta SD and Triangular Formula

Activity	a	m	b	PERT σ	σ_Δ	% Error
A	1	2	4	0.50	0.62	19.8
B	1	2	6	0.83	1.08	22.8
C	1	2	3	0.33	0.41	18.4
D	1	2	2	0.17	0.24	29.3
E	1	2	4	0.50	0.62	19.8
F	2	3	6	0.67	0.85	21.6
G	2	3	4	0.33	0.41	18.4
H	2	3	4	0.33	0.41	18.4
I	3	4	8	0.83	1.08	22.8
J	3	4	7	0.67	0.85	21.6
K	6	7	9	0.50	0.62	19.8
L	7	8	12	0.83	1.08	22.8
M	7	8	11	0.67	0.85	21.6
N	9	10	14	0.83	1.08	22.8
O	11	14	20	1.50	1.87	19.8

The three shape parameters (a, m, and b) introduced for the beta distribution can also be used to describe this distribution. For that, T_e can be estimated using either the sum of the three parameters divided by three or with the PERT model formula. Differences for the values in Table 15.1 averaged approximately 5% with the PERT model always estimating the lower value. These two formulae would be

$$T_e = \frac{a + 4m + b}{6} \quad \text{PERT beta model.}$$

$$T_e = \frac{a + 4m + b}{3} \quad \text{Triangular model.}$$

There is some disagreement regarding how to compute the standard deviation for a triangular distribution. Most organizations still use the PERT empirical beta distribution formula because it is supported in the literature; however, the University of Virginia has published the following more accurate but very complex formula for a triangular distribution standard deviation (σ_Δ) (University of Virginia, 2008):

$$\sigma_\Delta = \left\{ \frac{[(b - a)^2 - (m - a)(b - a) + (m - a)^2]}{18} \right\}^{0.5}.$$

Normally, a formula as complex as this would be avoided because of its complexity; however, desktop spreadsheets make this somewhat more reasonable. Table 15.1 shows a comparison of the results in using the σ_Δ formula against the classic beta version [$(b - a)/6$].

Note that the simplified PERT formula generates results for the sample data that average about 21% lower in comparison with the complex formula. Users will have to decide whether this error level is significant enough to warrant using the more complex version. In either case, the mechanics are the same, but clearly it must be recognized that these calculations are analytical estimates more than highly accurate numerical results. The goal is to derive a measure of project variability more than computation accuracy.

15.6 CALCULATING PROBABILITY OF COMPLETION

We have now seen the basics of variable time completion estimates. This last section is simply a way to add some flexibility to the analysis. Up to this point, the examples have used unit standard deviation interpretation (i.e., 1, 2, or 3). Once the basic shape parameters, μ and σ, have been calculated, there are other questions that can be answered:

- At what point can we be 90% confident that the project will be completed?
- If the model says that the project is expected to be completed by time period 50, what is the probability that the project will in fact be completed by time period 54?

The process to answer each of these questions involves finding areas under the normal curve that are not at the standard unit points described earlier. One easy way to accomplish a more general computation model is to use one of two Excel® built-in functions.

Normdist: Computes the probability of finishing at a specified time.
Norminv: Computes the time for some specified confidence level.

For this calculation example, we will use μ to be 50 (expected project completion time) and σ to be 4 (computed standard deviation of the critical path).

Scenario 1: What is the probability of the project finishing by time period 52, given the expected time of 50 and standard deviation of 4? This would be entered into the Excel function as Normdist (52,50,4,TRUE). The results indicate that this project would have a 69% probability of completion by that time.

Scenario 2: At what time period do we have a 90% confidence of completing this project? This would be entered as Norminv (0.90,50,4). The results returned indicate that this occurs at time period 55.

The two functions above provide a great deal of analytical flexibility to answer completion probability questions. A little practice with them will help understand their dual roles. Figure 15.5 illustrates graphically the two shape factors necessary for this analysis. In this figure, $\bar{X}$ is equivalent to μ (the expected time for the project).

15.7 SUMMARY

This chapter has demonstrated sample basic mechanics for analyzing project plans for which multiple time estimates for activities are available. The list below summarizes the eight computational steps required to perform variable time analysis using the classical model:

1. Collect *a*, *m*, and *b* parameters for each activity (optimistic, most likely, and pessimistic).
2. Compute the project schedule using values calculated for the network.
3. Calculate the critical path using deterministic rules from Chapter 13. Select the activities on the critical path for probability analysis. The calculated critical path duration becomes μ for future analysis.
4. Calculate standard deviations (σ) for each critical path activity using an appropriate formula.
5. Square the individual standard deviation values to produce a variance estimate for each activity as discussed in this chapter.
6. Sum the computed variances for the critical path activities.
7. Take the square root of the summarized variances from the step above. This will be the σ parameter for future analysis (see Figure 15.5).
8. Use μ and σ parameters to perform the desired analysis.

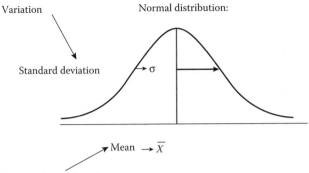

Variation

Normal distribution:

Standard deviation → σ

Mean → $\overline{X}$

Measure of central tendency

FIGURE 15.5   Shape factors for project time variable analysis.

Mature organizations need to become familiar with this class of analysis since it is a more accurate representation of a future project completion than a single estimate. In some cases, a schedule overrun can doom the project. This technique offers a way to measure the probability of an overrun value. For these reasons, the ability to model a project schedule with variable activity times is a worthy addition to the PM's toolkit.

PROBLEMS

1. Calculate the project variable time plan using the activity data outlined below and then answer the following questions:
   a.   What is the expected completion date for the project?
   b.   Calculate the probability that this project will complete by duration 155?
   c.   At what duration can we offer the management a 95% probability of being finished?

Task ID	Predecessor	a	m	b
A	...	10	15	20
B	...	14	23	50
C	B	40	60	30
D	A	40	45	50
E	D,C	30	35	42
F	A	5	10	12
G	B	3	16	25
H	E,F,G	10	15	13
I	B	7	12	13

2. Assume that the calculated shape parameters for a project are: $\mu = 50$, $\sigma = 4$.
   Use an Excel function to calculation completion parameters for the following situations.
   a.   What is the probability of completing the project by time period 56?
   b.   What is the point in time for which we have a 75% confidence level in completion?

REFERENCES

Archibald, R.D., 1987. Archibald associates. *Project Management Journal: Key Milestones in the Early PERT/ CPM/PDM Days*, 28: 29–33.

Brown, K.L., 2002. Program Evaluation and Review Technique and Critical Path Method—Background. Reference for Business. http://www.referenceforbusiness.com/management/Pr-Sa/Program-Evaluation-and-Review-Technique-and-Critical-Path-Method.html (accessed March 17, 2008).

Gale, T., 2006. Program Evaluation and Review Technique and Critical Path Method Background. Reference for Business. http://www.referenceforbusiness.com/management/Pr-Sa/Program-Evaluation-and-Review-Technique-and-Critical-Path-Method.html (accessed March 23, 2008).

Reiss, G., 1996. *Project Management Demystified: Today's Tools and Techniques*. New York: Spon Press.

Stires, D.M. and M.M. Murphy, 1962. *PERT/CPM*. Boston: Materials Management Institute.

University of Virginia, 2008. A Brief Primer on Probability Distributions, UVA-QA-0517SSRN. Darden Business Publishing. http://papers.ssrn.com/sol3/papers.cfm?abstract_id=480689 (accessed July 12, 2008).

Verzuh, E., 2005. *The Fast Forward MBA in Project Management*. Hoboken: Wiley.

16 Project Simulation

There are many reasons that can lead to project failure, but it can be argued that one of the significant and visible outcomes of performance gone awry is variance of time estimates from plan. Regardless of the reasons for this, it is sometimes easier to estimate time variance than to guess at all of the possible reasons it might occur. Traditional network models are the most frequently used tools for time analysis, yet they suffer from many limitations in modeling real-world projects. The next section will describe some of the issues that occur in projects but are not reflected in standard network models. The previous chapter described the classic PERT statistical approach to estimating variable completion times. This chapter will use the same network setup parameters, but utilize computer simulation technology to generate a similar and more contemporary view.

16.1 TRADITIONAL TIME MODELING TOOLS

The main questions that traditional network analysis seeks to answer are the following:

- How long is the project going to take?
- What is the critical path for the project (i.e., deterministic critical path)?
- How variable is the project duration (PERT only)?

Four traditional network-modeling shortcomings are summarized below. In each of these cases, a simulation model offers improved analytical capabilities for dealing with the defined scenario.

16.1.1 NEAR CRITICAL PATH ACTIVITIES

Variability of parallel activity durations can allow other near critical paths to become critical as the durations change, thereby confusing a static analysis process. The increasingly parallel nature of activities in today's complex projects makes this topic an important consideration (Williams, 2002). Envision a network with 10 essentially parallel and independent activities. Assume that all have the same duration estimate of 10 days, with each having a time probability distribution. Traditional PERT network calculations for this situation will determine an estimated time for the task, and a project critical path will be computed from that value. However, during the project execution, other critical paths may well occur as a result of unplanned time variation. So, recognize that failure to understand near critical paths is one key piece of analysis missing with the traditional network tool. Simulation modeling adds that capability to the analysis process.

16.1.2 TASK EXISTENCE RISK MODELING

A second traditional model shortcoming occurs in the situation where the actual existence of a task is probabilistic. As an example, let us say that the project involves building a bridge. Historical data from this class of project indicates that there is a 20% chance of finding archaeological remains during the excavation process, and if that were to occur this probabilistic activity could create an additional 20 extra days due to the required extra labor and handling. There is no reasonably accurate method to model this type of zero to X type task in the traditional model. Either adding the activity with some value or ignoring the event are both misleading. The ideal situation would be to evaluate the probability impact of the two events, but that is not within the traditional capability.

16.1.3 CONDITIONAL ACTIVITY BRANCHING

A slight modification of the conditional activity is a multibranch conditional option for an activity. Traditional project networks define an activity relationship structure as fixed; however, in practice, there can be situations where one path or another is taken depending on some probabilistic conditional (Williams, 2002). Modifying the archaeological example above, we could envision three possible branches as follows:

- Archaeological remains that need expert removal are found (Probability 5%, 7 days)
- Archaeological remains that can be discarded are found (Probability 20%, 5 days)
- No archaeological remains are discovered (Probability 75%, 3 days).

In this example, the three events defined would be mutually exclusive—one of them will occur but not all three.

16.1.4 CORRELATION BETWEEN TASK DURATIONS

The third nontraditional example involves the situation where tasks are interrelated across the network in a probabilistic way, but not in a work related manner. For example, there are two tasks in a project and it has been shown that one task that overruns will likely cause the second to also overrun. In other words, their time probabilities are not independent of each other. These would be called correlated tasks. By correlating the duration estimate of the two tasks, it assures that when one task is finished quickly the second will follow in the same manner. Correlations can be either positive or negative, that is, their relationship will follow in the same direction or opposite. These situations cannot be modeled in the traditional structure.

16.2 SIMULATION IN RISK MANAGEMENT

One tool that has the capability to match these real-world project complexities is simulation modeling. In such models, the network parameters are represented by probability distributions that represent each one's behavior. From this definition, a time or cost for that event is randomly created and the model is executed with this set of values. In order to test variability, the model will be run hundreds of times and each iteration creates a unique project estimate. The simulation results are collected and these summaries are used to provide analytical insights into the performance of the project. Because of the "roulette"-type randomness of the event selection, this is often called Monte Carlo simulation after the famous Monte Carlo casino. An overview of this process is represented by Figure 16.1.

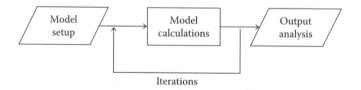

FIGURE 16.1   Simulation analysis of networks.

The simulation model will collect computed project status after each iteration for the following:

- Project completion date
- Status of any defined milestone
- Critical path activities.

Summary information from the iterations is collected and displayed in both graphs and tables.

By running hundreds or even thousands of iterations, a picture of the projects' risk profile can be built up (Khan, 2007). The profile can then be used to achieve a better understanding of the potential time or cost outcome of the project. From this understanding, project resources can be better used and risk consequences can be better understood.

Simulation models have the potential to provide much more flexibility in the project assessment process beyond what traditional network techniques can offer. However, two warnings come with this added capability:

- There is an added learning curve involved in setting up this model. Results obtained will only be as good as the model design.
- Analysis of the results is more complex than previous methods. New concepts are introduced and making sense of the new parameters will take time and experience.

Considering the limitations outlined for traditional analysis, simulation seems to be an obvious candidate. New generations of easy to use simulation utilities further enhance this opportunity. These new tools simply add into existing computer software network utilities such as Microsoft Project and Primavera. Several of these bring ease of use, which should help popularize the technique more.

As PMs learn more about situations not handled by traditional networks, more pressure will be brought on the traditional models to conform. A few of these missing items were summarized in the section above; however, there are more similar to this. Simulation tools offer one possible mechanism to bridge these gaps and may in fact be the preferred strategic method.

The fact that every project is prone to variable duration times makes a strong case for favoring simulation as a key technique. Use of probability distributions to show project completion is a much more realistic presentation than a deterministic Gantt view that is erroneous. Maturation of computer and hardware and software are also favoring proliferation of this approach. In order for this method to be successful and used more frequently, the individuals involved with project management (i.e., team, PM, users, and senior management) must be accustomed to thinking probabilistically about potential outcomes and forecasts (Raftery, 2003). This means that probabilistic estimates resulting from the planning stage would now be a probability distribution defined for each individual activity.

A summary of the rationale and potential benefits for moving to a simulation view of project analysis follows:

- Improved project understanding in regard to duration and cost
- Improved understanding of the actual critical path under different scenarios
- Better assignment of contingencies and resources on high risk activities
- Improved chances of selecting more profitable new projects based on improved risk assessment
- Better management of project expectations.

The following sections of this chapter will illustrate some of the output results from a simulation model approach. Pertmaster will be used to illustrate this, but other similar utilities exist in the marketplace.

16.3 PERTMASTER MODELING

The Pertmaster utility is now under the Oracle corporate umbrella (www.primavera.com). A vendor description of the utility follows (Pertmaster):

> Pertmaster is a full lifecycle risk analytics solution integrating cost and schedule risk management. Pertmaster provides a comprehensive means of determining confidence levels for project success together with quick and easy techniques for determining contingency and risk response plans.

Our goal here is not to go through the mechanics of using this tool, but to review the types of outputs commonly generated for this class of analysis.

As we have done before, we will use the same basic network that originally appeared in Chapter 15 (Figure 13.8). In order to make the analysis more visible, each of the activity durations will be multiplied by 10, so activity A will be 30 time units instead of three as shown. That modification just spreads out the network, but does not change the calculated critical path. The interested reader can download a demo copy of this tool with sample programs, but we will just concentrate on the sample outputs generated here.

The first step in the translation is to see what a traditional view of this would look like. Figure 16.2 shows how the project utility translates the network view into a Gantt bar structure. The critical path is identified. If we had supplied multiple PERT-like time estimates, this could have been handled by the utility, but the results would look the same as shown here. Also, if we wanted to do activity variance analysis as described in Chapter 15, that would have to be done manually or with a computer spreadsheet (no automated analysis). Note that the critical path is shown in gray as path A–C–E–G–I and 210 time units (remember, we multiplied all activities by 10). The simulation results will evaluate the probability of project completion at specified time periods by showing a probability distribution indicating dates and percentages.

Pertmaster can import the definitional data directly from Microsoft Project. A Pertmaster equivalent to Figure 16.2 is shown in Figure 16.3.

The Pertmaster Gantt display is a little different in format from Microsoft Project, but not significantly so. Note that the display of the critical path and activity relationships is very similar. The one major difference shown here is that Pertmaster works from "Rem (remaining) Duration," rather than (total) Duration. At the beginning of the project these are the same values, but in execution the remaining duration values would decline toward zero as the activity is completed. From a simulation standpoint Pertmaster would require specification for the probability distribution for each activity, whereas project has a very limited view of this type definition. One way to get a quick view of variability is simply define each activity as having some percentage variability and a singular probability format (i.e., normal, triangular, beta, etc.). Obviously, this is quicker and less accurate than

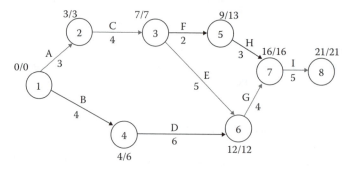

CP = A–C–E–G–I; and 21 time units

FIGURE 13.8 Defining the critical path. (From Chapter 13 of this book.)

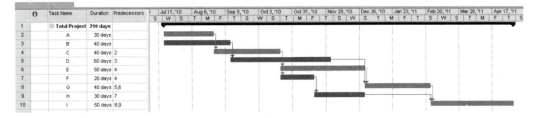

FIGURE 16.2   Microsoft Project view of the example program.

the more detailed by activity specification, but may suffice for the initial analysis. Recall that this method required more definitional work than traditional modeling, and this is one example of that.

For this example, a quick simulation is performed by instructing the model to use a triangular distribution and let each activity time vary over the range of ±10%. We then instruct the model to run 500 iterations and capture the results. Figure 16.4 shows the histogram of the time results recorded. One interesting conclusion is that the project completion could range from as early as 6th April to as late as 28th May just from a 10% error is estimating. If management or users wished to talk about completion dates this type presentation would show them how logically the completion date could have variability. For many, this would be a new approach to schedule definition, but one that needs to be added to the organizational culture.

A second analytical chart shows an output called the "Tornado chart." This is titled based on its typical look and is illustrated in Figure 16.5.

The duration sensitivity ranking shows the activities that are most influencing project duration. Variability in these tasks will have the most impact on project completion. The ranking of these represents key drivers causing the schedule, and this format can be produced for cost and risk as well.

16.4   OTHER PERTMASTER METRICS

There are a host of other analytical metrics generated by Pertmaster and each of these brings insights regarding the behavior of the project beyond what is available in the traditional PERT or CPM models. A brief list of other table and graphic analytical options are shown below. Basically, Pertmaster can describe various project situations through the production of the following outputs:

1. Tornado chart showing activities that have the greatest impact to schedule, duration, cost, or risk

FIGURE 16.3   Pertmaster view of the example program.

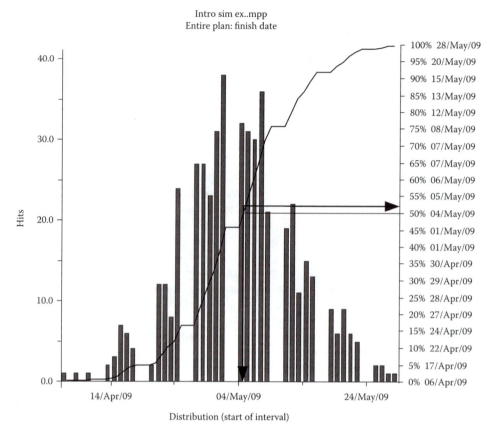

FIGURE 16.4 Pertmaster project time distribution.

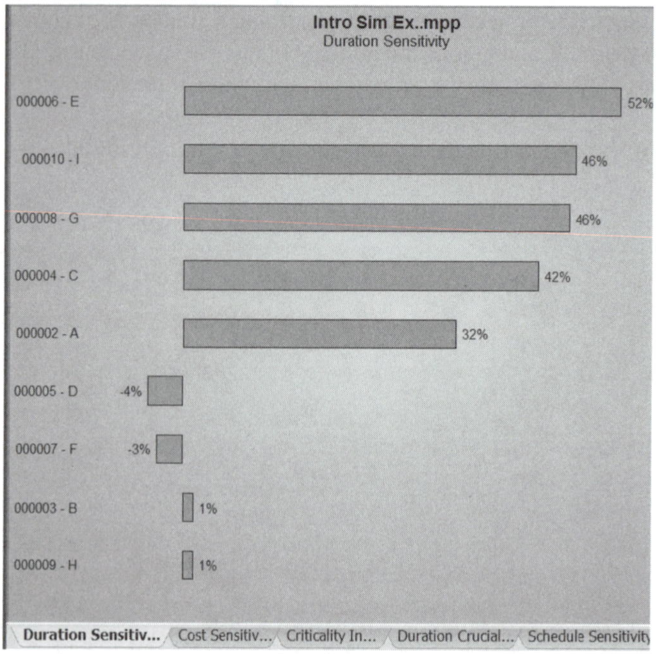

FIGURE 16.5 Critical path sensitivity.

2. Near critical path analysis to warn of activities that have potential of becoming critical as time estimates vary
3. Scatter diagram showing cost versus schedule potential outcomes
4. Time-based S-curve showing potential project schedule and cost information over time and by various project stages.

Hopefully, the examples used in this section are sufficient to illustrate the added value that a simulation model can bring in understand project issues. This class of analysis is the most complex of the examples described in this section of the text, but certainly one to be considered.

16.5 SUMMARY

Simulation is a sophisticated quantitative project risk analysis methodology. Pertmaster is the tool used to describe this methodology; however, there are a growing number of vendors offering specialized models for this purpose. Availability of cheap and powerful desktop computers provides a cost-effective method for this class of analysis.

Before simulation becomes the preferred method of project analysis, a broader recognition of the underlying reasons for its emergence must occur. The most obvious of these is that quoting a deterministic project due date is patently lying to your stakeholders. If the PM is truly going to become an honest broker of information then he must begin to educate his constituency into the variability issues of a project. In the beginning, this can simply be variability of time and cost without delving into the underlying issue too much. Later, more specifics in regard to risk management need to be introduced.

A second view of simulation in the project is one of internal analytics. All stakeholders do not necessarily need to understand all of the implications of these concepts but the project team members do. If we go back to the traditional network models introduced in Chapter 13, it is clear that those early models introduced in the mid-1950s now show their age. Clearly, deterministic views of projects need to be replaced by methods that deal with variability. Classical basic time variability was discussed in Chapter 15 and the more sophisticated Theory of Constraints will be described in Chapter 17. Collectively, these highlight various management issues in the way work is scheduled. The PERT model previously illustrated the value of variability analysis, while this chapter expanded that analytical view. Both concepts are supportive in that regard.

A simulation model is often described as a "what if" view. This means that the simulation mirrors whatever characteristics are designed in the model. It does not say that the model is accurate, nor does it say what should be done. It simply describes the scenario as defined. This has a lot of power, but the warning is that the model builder must also validate that the model represents the appropriate relationships. Said succinctly, *it describes not prescribes*. This is an important management consideration.

Simulation modeling has great potential to improve project analysis. The following advanced analytical capabilities are inherent in its process:

1. Display scheduling characteristics of projects to include time, cost, and risk variability
2. Model complex relationships between activities
3. Model complex characteristics of an activity (i.e., conditional occurrence)
4. Aid in analyzing resource allocation issues resulting from variable time results
5. Aid in identifying the real constraints of the project—activity variability, resources, and (constraint management)
6. Allow evaluation of alternative scenarios with complex assumptions
7. Show multiproject interactions.

For these reasons, the use of simulation modeling needs to be pursued by the PM. This will take some time to incorporate in the organization management process, since the new views will be quite different and the learning will require time to absorb. Clearly, project analytics are a key planning and control skill set that needs to be developed.

Traditional network models portray the project as a static series of linked activities, often with padded duration times. This view is sometimes called a "train schedule," that is, it does not represent what could be done, but what we are scheduling. In the meantime, the players often sit around and wait for the "train." In order to deal with inefficient time management practices, PMs have to break this paradigm and find more effective methods of planning and analyzing project performance. Simulation is one candidate for this.

REFERENCES

Chapman, C. and W. Stephen, 1997. *Project Risk Management: Processes, Techniques, and Insights.* New York: Wiley.

Dibiasio, L., 2007. Predicting Risk in an Uncertain World, *Primavera Magazine*, Spring, 20–21.

Dibiasio, L., 2007. Understanding Risk within the Portfolio. *Primavera Magazine*, Summer, 20–23.

Khan, S., 2007. Dealing With Risk the Intelligent Way. *World Pipelines*, 7, 29–30.

Khan, S., 2007. Taking Control of Risks in Project Schedules and Portfolios. *Project Manager Today*.

Primavera, P., 2007. Pertmaster Help Manual.

Raftery, J., 2003. *Risk Analysis in Project Management.* New York: E & FN Spon Press.

Williams, T., 2002. *Modeling Complex Projects.* West Sussex: Wiley.

17 Critical Chain Management Model

17.1 INTRODUCTION

This chapter examines one of the contemporary project planning models based on Critical Chain (CC) theory. CC models are based on the application of Dr. Eliyahu Goldratt's *Theory of Constraints (TOC)* (Goldratt, 1997). This approach to project planning and execution requires PMs to abandon traditional estimating and project control practices. Management of the CC elements is handled by the use of resource alerts and buffer management for the chain. Implementing these concepts will require a cultural change throughout the organization, beginning with a radical shift of focus from the near-term (task completion dates) to the long-term (final delivery dates). Topics described here are drawn both from published material authored by Dr. Goldratt and from other published sources by successful implementation practitioners, most notably Larry Leach and Frank Patrick.

Implementation of Critical Chain Management (CCM) looks at projects in a new light, changing the way projects are estimated, scheduled, executed, and controlled. In an ever-intensifying global competitive market, the management of projects, particularly product development efforts, is increasingly one of the factors that can produce a sustained competitive advantage. Firms that can bring products to market faster than their competitors can extract higher initial market share and margins. The underlying theme of this model is to complete prioritized projects faster and to make more efficient use of critical resources. Published experience shows that this approach requires PMs to abandon traditional estimating and project control practices. Implementation of CCM requires a significant cultural change throughout the organization, beginning with a radical shift of focus from near-term task completion dates to long-term delivery dates. Topics described here are drawn from Goldratt's ideas and other successful implementation practitioners. Industry practitioners have labeled the project management model based on TOC concepts as CCM. Patrick, Leach, Mannion, Ehrke, and others have each published particularly notable work in this area and their collective contributions have been used extensively in this chapter. These individuals translated a theoretical approach into workable techniques that can be successfully applied in the project structure. Other translations of this concept have been implemented in operational environments such as manufacturing.

The CC method accomplishes building project networks with restricted task duration, buffers to protect overruns, and disciplined resource management both within and across the projects. This discussion includes a general explanation of the TOC and how its principles are applied to project management to construct the CCM model. Also, the implementation complexity of the model and some of the challenges faced will be described.

17.2 CC CONCEPTS

The key focus on CCM is in increasing speed of project completion, but it is clear that speed must be achieved without compromising other aspects of the project deliverables such as service, features, or flexibility. As the demand for shorter project cycle times and more deliverable flexibility grows, so too does the frustration levels of both PMs and their team members. Cutthroat competition both at home and abroad and the need for process improvement encompass virtually every aspect of modern business. The popularity of downsizing, rightsizing, and re-engineering attests to the need for change. These trends will continue into the foreseeable future. Any new approach must support significant leaps in performance, but it will also have to be sensitive to the current organizational culture. The CCM model is basically a process improvement methodology resulting from a different manner of looking at tasks and their associated resources.

According to the CCM model, a project's failure to deliver as planned is tightly linked to the failure to recognize and manage it as a network of dependent events with statistical fluctuations. The CCM consists of a number of commonsense tools and processes that address this aspect of the project. In operation, these tools focus efforts on a few selected areas of the project plan, called "constraints," which restrict the project work flow. These constraints are the leverage points toward which successful improvement efforts are directed.

Project management can be improved based on utilization of the following three perspectives (Leach, 2005c):

1. *PMBOK® Guide* management model—guidance on the overall methodology of planning.
2. Total Quality Management (TQM) and six sigma concepts (covered in Chapter 21)—represents a methodology outlining how to improve organizational processes.
3. TOC concepts—describes how to overcome typical time management issues by indentifying system constraints and increasing activity throughput.

Figure 17.1 outlines how these three aggregate improvement strategies impinge on the project.

The next level concept is to analyze the resource patterns embedded in the defined tasks. A common occurrence is to find one or more scarce resources that are inhibiting the project from

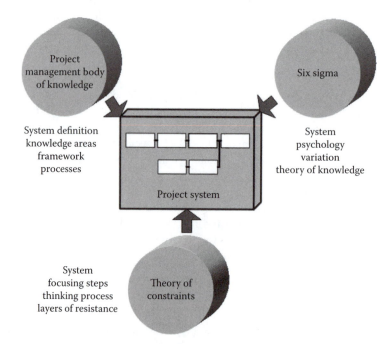

FIGURE 17.1   Three impact models on the project system. (From Leach, L.P., 2005c, *Critical Chain Project Management*, Second Edition. Norwood, MA: Artech House, p. 23. With permission.)

completing faster. Many times, multiple demands are being placed on these resources. Within these tasks, there is also recognition that the task performance process exhibits natural variation and that many of the individual tasks are interrelated (Leach, 2005c). To summarize, the following are observed:

1. Scarce resources being demanded across the global task set
2. Sequences of activities exhibit unplanned time variability.

The base state efficiency of the task set could be improved by indentifying constraints that restrict completion and deal with the scheduled time variation. Form this base set of issues CCM strives to improve the constraint situation and manage the flow issue better. The essential idea is that system outputs are limited by its constraints and the flow of work through those constraints. The management trick in implementing CCM is to identify the constraints and apply the CC logic to them. Once the constraints are decreased, improvement in work flow can occur. In this regard, Goldratt stated that (Leach, 2005c):

> Before we can deal with the improvement of any section in a system, we must first define the system's global goal; and the measurements that will enable us to judge the impact of any subsystem and any local decision, on this global goal.

The basic concept behind the CC constraint theory is best described using a physical chain analogy. The goal of a chain is to provide strength in tension. A chain's weakest link determines its overall strength, so increasing the strength of any other link other than the weakest link will have no effect on the overall strength of the chain. In similar fashion, consistently managing the weakest link in a project will improve performance of the overall project. Figure 17.2 depicts two work sequential packages chained together. The time required to execute both is then the sum of the two packages times.

FIGURE 17.2 Weakest link analogy.

17.3 CC MECHANICS

Leach outlines five steps to set up and operate the CC model. Figure 17.3 illustrates these steps and their sequence.

A typical starting point for CCM is to construct the baseline schedule using activity duration estimates based on a 50% completion confidence level. All activity milestones are eliminated and multitasking is avoided by allocating work on a priority basis. Herroelen et al. offers the following list to describe the fundamental process (Herroelen et al., 2001):

- 50% probability activity duration estimates
- No activity due dates
- No project milestones and no multitasking
- Minimize work in progress (WIP)
- Determine a precedence and recourse-feasible baseline schedule
- Identify the critical activity chain
- Aggregate uncertainty allowance into buffers
- Keep the baseline schedule and CC fixed during project execution
- Determine unbuffered projected schedule and report early completions
- Use the buffers as a proactive warning mechanism during schedule execution

To minimize WIP, a schedule is constructed by timing activities at their latest start dates based on critical path calculations. If resource conflicts occur, they are resolved by moving activities

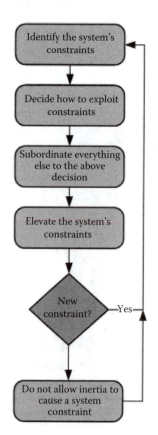

FIGURE 17.3 The five steps of CCM. (From Leach, L.P., 2005c, *Critical Chain Project Management*, Second Edition. Norwood, MA: Artech House, p. 53. With permission.)

earlier so as to not impact schedules. A CC is then defined as, "that chain of the precedence and resource dependent activities which determines the overall duration of the project" (Herroelen et al., 2001). More discussion of these mechanics is offered below, but this summary is a good overview.

In the project model view, CCM is defined as groups of dependent tasks that have potential to constrain the project schedule. In this case, the term "dependent" refers to resources and resource contention across tasks and projects, as well as the sequence and logical dependencies of the tasks themselves (Goldratt, 2007). This view differs from the classic critical path network definition as described in Chapter 13. In the traditional network critical path, only the fact that the task exists and is linked to other tasks is recognized, while the CC recognizes how related resources are used in the tasks across the project. This additional view supports the ability to observe the overall role of specific critical resources in the project or multiple projects. Resources can be managed more effectively, thereby supporting faster project completions. Faster completion leads to receiving benefits quicker and generally decreases overall project cost as a by-product.

17.4   CCM MODEL

The CCM model focuses on improving work process flow, and it represents an application of TOC principles to project management. In a project, the entire system degrades if any one element on the CC fails to deliver. In a single project, CCM focuses on the amount of time required to complete a "chain" of tasks, where as in a multiproject view, the model focuses on the collective tasks that most effect the cumulative cycle time of all the projects. The resources involved in these views are known as the organization's strategic or critical resources, also called the "Drum" resource. Operationally, CCM manages the Drum resource in such a manner as to optimize overall project performance. One metaphor for the Drum resource is "marching through the project to the beat of a drum."

There are both external and internal factors that need to be considered in setting up the CCM resource environment. A summary of these follows:

External factors

- Synchronization—task network with resources identified
- Queuing—sequencing and prioritization of tasks.

Internal factors

- Student syndrome—execution procrastination avoided
- Parkinson's law—overly conservative estimating
- Bad-multitasking—too many work tasks in play at the same time.

17.5   PRINCIPLES OF THE CC MODEL

One of the fundamental tenets of CCM is a method to deal with the way project team members are allocated throughout the project life cycle. This tenet is basically a project management discipline as applied to people and tasks. Kendall et al. (2005) offer the following summary list of concepts and mechanics used to manage the CCM process regarding people and tasks:

1. Project team members are dedicated to a single task until it is completed and to do this as quickly as possible. Period reports are required to indicate time remaining for the task. Every effort is taken to eliminate delays and work procrastination.
2. Task estimates do not have padding. They are planned at some probabilistic completion level such as 50% of what the traditional padded estimate would show.

3. Multitasking is eliminated by assigning workers to tasks in priority order and completing that task before moving on to a new task. Industry experience suggests that multitasking creates inefficiencies amounting to 30–40%.

4. Managing tasks by due date is not followed. Workers and tasks are not measured based on on-time completions. The management approach is to pass on the task to the next activity as quickly as possible.

5. By taking resource dependency and logical dependency into account, the longest sequence of dependent tasks can be seen more clearly. This longest sequence, the CC, may cross logical paths in the networks. So, the management view is to deal with resources and not with critical paths.

6. Buffers are a key mechanism to manage desired schedules. These will be defined to protect various aspects of the project including Drum resource, feeding chains, and the project itself. Project status will be monitored through buffer status and not task completion.

7. Projects are viewed similar to a relay race. When a task is close to reaching completion, the next task's resource is queued to get ready to go immediately after the preceding task is complete. In this manner, there is little delay in the chain and early completions can be used to speed the process. Every effort is made to create this culture in the project team.

8. Recognition of resource constraints requires a staggered introduction of multiple projects into the system, and resources for these efforts will be allocated on a priority basis. This method improves completion of projects in priority order, increases the outcome predictability for each project, and increases the effectiveness of critical resources by minimizing multitasking. Shorter project cycle times improve overall delivery of the project set.

17.6   BUFFER MANAGEMENT

Since every activity in the project is scheduled with no safety padding, it is expected that some will overrun their estimated times. In order to project, the project buffers are used in various places to keep the overall schedule under control. Where these buffers are located will be discussed below, but for now let us look at how a buffer works. Imagine "chains" of activities in the plan. Each activity is sized at 50% of typical estimates. This means that it is anticipated that 50% of the time the schedule will be overrun (and hopefully 50% under run). The goal of a buffer is to pad this chain in such a way that the overall planned chain length is not violated. Of course, the trick is to figure out how big such a buffer should be. For this discussion, a typical number will be used, but accurate estimation can be improved through experience with specific project types.

Figure 17.4 from Patrick provides a good schematic overview of two buffer types—feeding buffers and project buffers (Patrick, 1998, p. 6). A feeding buffer represents a set of non-CC activities that need to be protected from impacting the primary CC, whereas the project buffer is used to protect the primary CC from schedule overrun. From this example, the focus for management moves to the buffers since some tasks are expected to overrun which is not, by itself, of primary concern. If the various chains are protect, the overall project will finish on or before planned schedule. If this approach works, the CC model would be of great interest to all organizations and there is growing evidence that it, in fact, does work when properly implemented.

One of the typical ways to view buffer status is to divide them into three equally sized regions as shown in Figure 17.5. The first-third is reflected as the green zone, the second-third as the yellow zone, and the final-third as the red zone. As chain overruns occur, the buffers are "eaten." If the buffer penetration is in the green zone, no action is taken. If penetration enters the yellow zone, then the problem should be assessed for possible corrective action. Finally, if penetration enters the red zone, then careful review is required to protect the project schedule. Action plans should involve ways to finish uncompleted tasks in the chain earlier, or ways to accelerate future work in the chain to correct the buffer penetration rate (Retief, 2002).

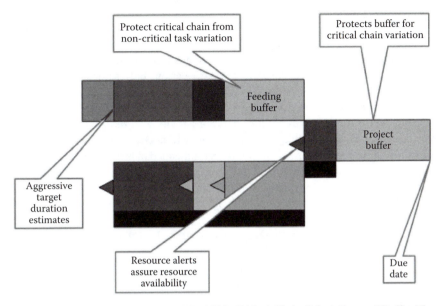

FIGURE 17.4 Buffer logic. [From Patrick, F.S., 1998. Critical Chain Scheduling and Buffer Management: Getting Out from between Parkinson's Rock and Murphy's Hard Place. p. 6. http://www.focusedperformance. com/articles/ccpm.html (accessed February 9, 2008). With permission.]

Defining project and feeding buffers will be a new skill requirement for the project team to master. Published research is available on this topic, but it will be necessary to test various methods before settling on a local approach. Leach offers detailed explanations for three methods in *Critical Chain Project Management* (Leach, 2005c). Understanding buffer sizing is a key element of project success. The first time these appear in a project plan will likely necessitate an explanation for various stakeholders, and it may even be necessary to prepare a "shadow" plan showing only key completion dates. So long as the project team meets those key dates, the lower-level view should not be required.

Conceptually, buffer sizing is a clearly defined probability and statistics-based issue. The basic methods follow similar logic similar to that described in the critical path variable time discussion in Chapter 15. However, as with most items in the project management domain, the underlying accuracy of data is the Achilles heel in accurate quantification. So, until there is a more tested local strategy, Leach (2005, p. 137) recommends utilizing the "Half the Chain" method for buffer sizing. The method was initially recommended by Goldratt as follows (Leach 2005c, p. 135):

Size the project and feeding buffers to one half of the buffered-path task length. Use only the total task length; do not count the gaps in the chain when sizing buffers.

Many organizations will likely reject such an approach because it seems too simple. In fact, it has a mathematical basis resulting from the way the time estimates were initially formed at 50%. Remember, the role of the buffer is to protect critical points in the chain from delay and approach the process with that mindset.

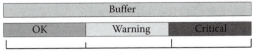

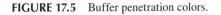

FIGURE 17.5 Buffer penetration colors.

17.6.1 Buffer Types

Discussion to this point has focused on the project and feeding buffer roles, but there are other buffer types as well. Each one serves a similar role to that described earlier—that is protecting the project from delay. This section briefly summarizes the various buffer roles.

Project buffers: As illustrated in Figure 17.3, this buffer type is located at the end of the CC to protect the project completion date from the estimate variability in the CC activities. The project is protected by this buffer because any total overrun of activities on the longest chain of dependent tasks can be protected, since the actual completion date will not overrun the planned date. As a rule of thumb, as starting strategy the project buffer is recommended to be half the size of the safety time taken out of the CC activities. The net impact of this is a project schedule that is planned to be 75% of a "traditional" project network length (Goldratt, 2007).

Feeding buffers: These buffers are placed wherever a non-CC activity chain joins the CC. Their role is to protect the CC from interference resulting from the non-CC impacting the CC schedule. Sizing of the feeding buffer is typically recommended to be half the size of the safety time taken out of the feeding chain path (Goldratt, 2007).

Resource buffers: These buffers are placed whenever a resource is needed for an activity on the CC and the previous chain activity did not require the same resource. The goal of this buffer is to ensure that the CC does not have to wait for the resource to become available.

Drum buffers: These are placed so that critical resources are available when needed. In that respect, the drum buffer is similar to a feeding buffer. They are placed in the project schedule immediately prior to the activity using the drum resource. This buffer protects the activity start date and lead time. A drum resource is considered to be a system constraint and is viewed as a potential bottleneck to the schedule. For this reason, projects are synchronized around these resource constraints, which otherwise result in schedule delays. These resources will be allocated on a priority basis and the buffer attempts to ensure that the priority project does not have to wait for that resource.

Capacity buffer: When synchronizing multiple projects around a drum resource, a capacity buffer is created between the tasks of different projects on which the drum resource is needed (Retief, 2002). Figure 17.6 shows how this buffer would be applied.

For each of the buffer types, the amount each buffer is consumed relative to project progress is the monitoring variable for status assessment. Assuming that the time variation throughout the project is uniform, the chain should consume its project buffer at the same rate tasks are completed. If the buffer sizing is perfect, it will be fully consumed as the chain is completed. By monitoring each buffer's status, it is now up to the PMs to determine the appropriate corrective actions necessary to recover buffer time as needed.

A second aspect of status tracking is to review remaining duration for all active tasks as opposed to percentage complete in the traditional mode. Resources report on tasks in progress is based on the number of days they estimate until the task will be complete. If a remaining duration value stays static or increases, the PM is warned to review that activity. This logic is caused by the fact

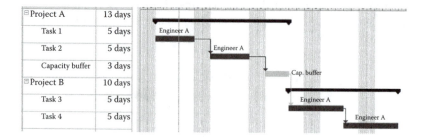

FIGURE 17.6 Drum resource capacity buffer between projects.

that no active task is static as, outlined the methods. So, no progress means that something has gone counter to the plan.

17.7 BUILDING THE CC SCHEDULE

In Figure 17.7, the letter in each block represents the resource type necessary to complete the given task. Here the assumption is that only one resource (a person or group of people) is available for each given letter and the CC duration of the task is 50% of the traditional network value and represented by the number in each block.

After the CC structure is defined, the next step is to define the size of each buffer. Preliminary rules for sizing were provided earlier as 50% of the CC length and 25% for each of the feeding chains. The published project completion date is reflected by the end of the project buffer, but not the end of the task "Q" per the original network schedule. This example shows the same project structured first as a traditional network with no buffers, then each task cut to 50%. Task durations are indicated by the number after the task ID (i.e., X40 indicates that the task is 40 days). A feeding buffer is inserted after task "N" to project the CC and a project buffer is inserted at the end of the CC to protect the project. Observed from this example, a delay in the feeding buffer chain segment will not affect the project unless that buffer is overrun. Also, note that the overall project is cut essentially in half (i.e., 128 versus 68 days).

In a CC schedule, the tasks are started as late as possible, which is convergent from the way traditional network models would schedule them. This strategy reduces WIP considerably as the work is started later and worked on until it is complete. However, additional risk is incurred by not having the security of a finished item in place waiting. Obviously, this method requires that the outlined process works effectively, also realize that this is a paradigm shift for the typical organizational culture.

In real-world CC projects, the reduction in time is found to be more often in the range of 25% less than an equivalent traditional plan with the same activities (Cook, 1998). This improvement is

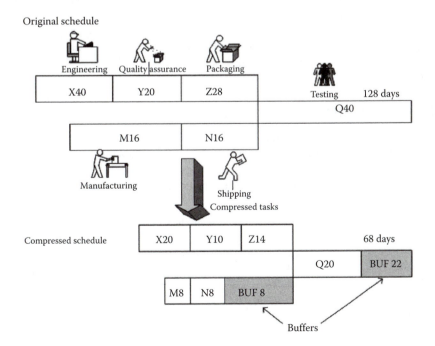

FIGURE 17.7 Translating a traditional network to critical chain.

created from aggregating the total duration protection into buffers rather than allowing each activity to have a separate buffer.

17.8   RESOURCE ALLOCATION

Once the CC structure is established, the problem turns to resource allocation and resource allocation conflicts. In Figure 17.8, resource "Y" was originally scheduled to perform two tasks at the same time so there is a conflict as to which activity to perform. In order to resolve the conflict, the feeder branch is started earlier and now becomes part of the longest chain. Also a buffer is inserted after activity "X" in order to allow time for the feeder chain to complete.

Placement of resource buffers helps the PM to focus on that particular aspect of the project. These become a communication target between the project schedule and the team. As work progresses, the resources report time estimates for task completion. When a predecessor CC resource reports having five days remaining (Goldratt recommended), the schedule process is triggered to inform the successor activity to prepare to start work on their assigned activity. In the relay race metaphor, this is analogous to getting ready to pass the baton. This mechanism represents a dynamic countdown for the successor. If the predecessor reported two days later that there was still five days remaining, this information would be passed on to the successor and a review of the activity would be performed. Note in this process that the focus of all activities is in having resources ready and prepared to execute the next task and then move as quickly as they can to completion. The "elbow" lines on Figure 17.9 represent the resource alert points for the project. Five days before each task is estimated to complete, the alert would be passed to the named resource that should then start getting ready to commence work.

The final overview step of CCM involves the manner in which the schedule is managed. Tactically, the focus in on being sure that resources are in place as required and this mechanism was explained above. Beyond that, the overall status monitoring is done through buffer management. Earlier discussion described the green, yellow, and red zones of buffer status. Project status reports would basically focus on these. More sophistication can be added by calculating "burn rate" of the buffer so that if one is decreasing faster than task completion would indicate, that would be a trigger to review what needs to be done. In this manner, the status report is more than just showing the project to be ok. It has the ability to have greater insight into trends. Unfortunately, the process of computing Earned Value as a status metric is impacted by this structure and it will have to be looked and assessed for usability.

This buffer management process highlights potential problems much earlier than they would ordinarily be discovered using the typical project a management techniques.

17.9   IMPLEMENTATION CHALLENGES

Successful implementation of the concepts outlined for the CC methodology will improve project throughput. In fact, improvement has been observed in several documented cases. However, the impact of these methods on the organization is significant and makes the conversion more complex than the method itself. There are three major areas where complex conversion will be

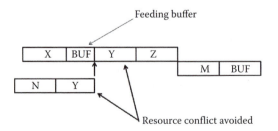

FIGURE 17.8   Resolving resource conflicts.

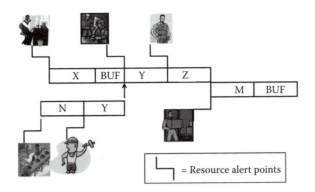

FIGURE 17.9   Resource readiness alerts.

experienced: organizational barriers, leadership challenges, and skill requirements. The sections below will describe some of the main issues in these areas.

17.9.1   ORGANIZATIONAL BARRIERS

17.9.1.1   Lack of Project Management Maturity

Use of the CC planning approach requires that the project team and other stakeholders be versed in the fundamentals of the method and that is not the traditional way of looking at a project plan. So, in order for the transition to work, there will need to be extensions made to existing planning methodology, which historically has been a difficult item to standardize. Once the new plan format is in place, status reporting and resource allocation methods would also have to change. Some stakeholders would be enticed to try the method with a promise of improved completion times. Unfortunately, there are subtleties involved that will affect deliverable quality and project priorities. Mature organizations with proven track records might be the easiest to convert, since much of the operational discipline would already be in place. The less disciplined an organization is in regard to overall project selection and resource allocation the more this change will be noticeable. PMs with a proven track record would likely be the best trend setters for this change. Stakeholders and team members would tend to follow creative leadership into new territory.

Mature supporting processes such as Integrated Change Control, risk management, issue/action management, and communication processes are essential for CCM to succeed (Leach, 2005, p. 206).

17.9.1.2   Buffer Resistance

Historically, PMs have been discouraged from visibly adding buffers into project schedules. Any real buffering at the project level was accomplished by padding individual estimates and thereby hiding them from sight. Consequently, any buffer that existed in the traditional plan was held as management reserve and was controlled by a combination of primary stakeholders or the project sponsor, not the PM. Contingency reserves are recognized in some organizations and they are like a buffer in that they are associated with project but not explicitly shown until a risk event actually occurred. So, the concept of seeing visible buffers flies against the standard operating process in most organizations. In implementing CCM, managers and stakeholders will be challenged to accept not only the existence of the chain concept and its buffers but also the acceptance of task overruns. Patrick (2001) makes the following comment regarding the nature of buffer use:

> Once developed, assessment of the full schedule, including the contribution of the buffers to project lead-time, provides a clear view into the identified potential of schedule risk for the project. In non-CC environments, when contingency is included, it is often hidden, either in management reserve or in internal and external commitments. The common practice of keeping these components off the table hides their true impact and implications. The open and explicit communication of buffers (important as

we will see in the discussion of project control) allows a clear assessment of what could happen 'in the best of all possible worlds,' versus what might happen if individual concerns accumulate to affect project performance. (Patrick, 1998, p. 44)

17.9.1.3   Milestone Tracking versus Buffer Management Reporting

CCM tracking moves the process away from milestone or activity completion into buffer status. Since task or activity estimates are at the 50% probability level (as defined here), it is necessary to anticipate schedule overruns on particular tasks. Because of this modification, the concept of "due date" loses its meaning. The due date now is as quick as you can finish the activity (Patrick, 1998, p. 5). Stakeholder agreement on this view is crucial to success. Also, securing stakeholder agreement on an acceptable project reporting methodology must be resolved during the project planning phase. Leach (2005c) and Patrick (2001) both recommend using a one-page "Red–Yellow–Green" icon approach to reporting, representing percentage of buffer penetration over time. Figure 17.10 illustrates what this status reporting format could look like.

Compared to traditional project management, the philosophical changes outlined here are a shift away from focusing on "what we've done" via reporting percent of work completed to focusing on what represents when we are going to be finished with the current task and where are the potential resource roadblocks.

17.9.1.4   Existing Processes

Organizations with established PMOs or standardized project management policies and procedures may not possess the flexibility required to implement CCM. For example, estimating at the 50% confidence level may be seen as an undisciplined approach by senior staff and management. Leach (2005c, p. 193) asserts that the final step in implementing CCPM is

> … assuring that the CCPM approach permeates all policies, procedures, and measures of the organization and is formalized into training programs to ensure that new people are properly indoctrinated into the organizational process. In the end, CCPM should not be an additional thing. It should just be "how we do business around here."

17.9.1.5   Existing Project Management Methodology

There is an old adage "If it ain't broke, don't fix it" and this can be a rationale for resisting change. Experienced project practitioners may argue that CCM is just "rearranging stuff." To counter that statement, the starting place for the method is a good critical path network that has been effectively resource leveled. Starting from that point, CC enhances the ability to optimize the schedule and set the stage for improved project monitoring and control (Kendall et al., 2005, p. 3).

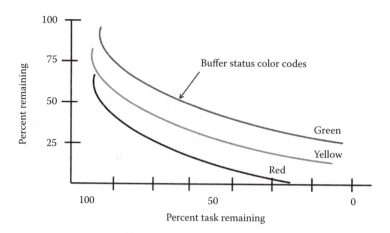

FIGURE 17.10   Buffer consumption reporting.

Suddenly faced by the challenge to estimate activities in the 50% probability range, organizations may counter with "we already use overly aggressive estimates and always work to meet or beat them" (Leach, 2005c, p. 207). Accustomed to keeping a hidden management reserve for projects, upper management may see CCM as empowering the PM to utilize a reserve (buffers) that are customarily under their control.

17.9.1.6   Existing Tools and Processes

Over time, mature organizations, may have developed project or deliverable templates with an associated set of standard defined tasks and processes. One example is an estimating technique or model that has been found to be "accurate." What this might mean is that it is seldom overrun and has significant buffering built in. The same might hold with a standard project life cycle template based on a traditional network structure. In such situations, there is likely to be resistance to changing approaches to project management that are working adequately.

17.9.1.7   Formal Tracking and Reporting Mechanisms

Organizations may either have in place a standard tracking and reporting system or may be contractually obligated to use a traditional approach for the external customer. The problem with Earned Value performance measurement calculations was mentioned earlier and the same issue could exist for other traditional reporting approaches. If milestone reporting is required for contractual reasons, Leach (2005c) recommends placing a buffer before the milestone, then treating those segments of the project like projects with specific deliverables (pp. 146–147). Christ (2001, p. 2) relates from his own experience

> There was an assumption in place that since the customer required Earned Value Management (EVM), no other method of management [approach] was authorized or acceptable. When queried, the customer indicated that factory management methods were company business, but the customer wanted data in EVM report format. This response opened the door for CCPM schedules in conjunction with EVM reporting.

More details on this approach can be found in Christ's (2001) article and EVM will be discussed in detail in chapter 18.

17.9.1.8   Resistance to Change

Classical human nature regarding change in their work process will clearly be a complicating factor in CCM implementation. Low tolerance for change is often rooted in a fear that one will not be able to develop new skills and behaviors that are required (Bolognese, 2002, p. 20). Leach (2005c) states that resistance to change is a characteristic of any stable system (p. 203). One might argue that CCM could be harder to sell in stable, mature organizations that were doing well and therefore not motivated to make what might be viewed as a radical change. Regardless of your judgment on the probable reactions in a given situation, organizational systems will naturally resist change (Leach, 2005c, p. 203).

Bolognese (2002, p. 26) comments that in certain instances employee resistance may play a positive and useful role in supporting organizational change. Insightful and well-intended debate, criticism, or disagreement do not necessarily equate to negative resistance but may be intended to produce better understanding as well as additional options and solutions.

17.9.2   Leadership Challenges

From a leadership perspective, PMs should respond to critics in a positive, constructive fashion. It is important to communicate how your project links with organizational success (Wynne, 2006). The case for CCM must also be couched in these same terms. It is not a technology initiative. Rather, it is a management technique designed to deliver defined value to the stakeholders quicker than would be done otherwise.

17.9.2.1 Project Control

CCM attempts to speed delivery of the project without sacrificing requirements. It is well known that faster completion of the project has value to the organization well out of proportion to the cost savings. Further, the longer the project takes, the greater the risk that some portion of the originally defined specifications will become obsolete. Leach (2005c) states that PMs who complain about scope creep admit to having an ineffective project change control process (p. 163). Although seemingly just a process issue, project scope control is recognized as one of the most important factors for project success. This aspect of management lies external to CCM but failure to deal with it can negatively impact any project management approach.

17.9.2.2 Planning Support

Creation of an effective project plan requires an iterative process and sufficient time (Leach, 2005c, p. 163). PMs must be allotted the time to execute the planning process and CCM may, in fact, be more rigorous and time consuming than the traditional approach since the method requires multiproject analysis of resources and buffer design on top of the traditional model (Kendall et al., 2005, p. 11).

17.9.2.3 WBS Development

In the traditional approach to scope development, the CCM team must develop an accurate WBS as a base work definition tool. A WBS is a deceptively simple appearing tool, but it contains vital information helping to organize, integrate, assign responsibility, and measure and control the project (Leach, 2005c, p. 106). The WBS is an important planning artifact to help identify work units that can be done in parallel, contracted, allocated to lower-level skills, and a host of other decisions that can help decrease the overall project time (Kendall et al., 2005, p. 9).

17.9.2.4 Schedule Development

The project team must be permitted to manipulate the WBS structured work unit schedule (early versus late starts) and establish the necessary feeding and project buffers in order for the CCM model to work. This includes a disciplined process of assigning resources to all task, and controlling those resources to work on one project task at a time. Management must support the overall process for schedule development, including resource allocation, predecessor relationships, buffer allocation, and of course the CC model itself. Patrick (1998, p. 1) reminds us that the scheduling mechanisms provided by CC scheduling require the elimination of task due dates from project plans. Christ (2001, p. 25) summarizes the following results obtained from a CC process:

> The result is a robust schedule considering resource capacity, work sequence, and task variability, with visibility provided by buffer reports. This process simplifies complex projects and provides management time to identify and address problems as they arise.

17.9.2.5 Buffer Calculations

Whether a simple or more advanced calculation is employed, buffers should always be calculated based on the length of the preceding chain. Arbitrary manipulation of buffer sizing will destroy the concept and resultant confidence of the CCM methods.

17.9.2.6 Dictated Versus Derived Scheduling

When used solely as project tracking mechanisms, a dictated "interim milestone" or "phased deliverable" defeats the purpose of using CC. Setting an arbitrary delivery date without validating the time and resource implications invalidates the plan. In some cases, an exception to this might be necessary as some external driver such as a regulatory requirement or market date may dictate such an approach. When a fixed delivery date is used, it will need to be prioritized against the existing workload to test for feasibility.

17.9.2.7 Project Prioritization

Utilizing a multiproject selection process, or some form of project prioritization and coordination at an organizational level, will help decrease the severity of resource contention in a multiproject environment. In order to minimize this issue, projects must be prioritized in terms of criticality of current commitments, value to the organization, and use of the synchronizing resource (Patrick, 2001, p. 7). Leach's view is more direct (Leach, 2005c, p. 162):

> If management does not adhere to a priority list, the multiproject system will not work. It is a simple choice, really. Behave to double throughput, or do not. Once they see the results, many management teams are able to do much better at this than many thought. After all, when the system makes more money, people's jobs are protected, and often they make more money too.

The organizational methods used to select and initiate new projects can be a source of interruption or resource contention for ongoing projects. If the new project is placed at a higher priority than some of the ongoing projects, the schedule of the ongoing projects will quickly change because of the CCM dynamic resource allocation scheme. It is necessary to keep in mind that the worst possible priority decision is not to make a priority decision, but to simply encourage everyone to do his or her best and hope the problem goes away. This inevitably causes bad multitasking and poor performance on all of the projects (Leach, 2005c, p. 161).

17.9.2.8 Resource Commitment

The practice of "pulling" an already committed resource from an ongoing project is a management challenge for any project, but can be a disaster for a CCM-driven effort. Upon receiving the advance warning that their task will start, team members must be prepared to finish up interruptible (noncritical) work and get ready to start work on the assigned critical project task. The new view has to change from "when is my task due?" to ask "when will my next task start."

The change management process must also be charged with the responsibility to protect project resources from unnecessary or significant changes in scope (Patrick, 1998, p. 4). No project can be successful when new, approved work is being added to the requirements on a daily basis. It is common for organizations to "pull" project team members away from project work in order to address some important operational issues, and sometimes this is unavoidable. Once again, failure to deliver the required resource at the right time and in the quantity will adversely affect the schedule. Management has the responsibility to make a priority call on resources, but the PM should attempt to quantify the impact of a resource reallocation and not just assume the schedule can be maintained.

17.9.2.9 Multitasking

The ability to handle multiple tasks concurrently is one of the "merit badges" of today's work environment. This is often cited in job descriptions and is true for PMs as well. This was once considered a productivity multiplier, but management and workers alike are beginning to recognize the overall negative limitations of multitasking (Anderson, 2001). Recent research is confirming those observations, pointing to multitasking as a productivity inhibitor. Rubinstein and his associates determined that for various types of tasks, subjects lost time when they had to switch from one task to another. There is an increased job complexity and time lost when workers have to switch between complicated tasks. Leach states as similar conclusion:

> Practical applications of CCPM have demonstrated the greatest gains in multi-project enterprises. The reason for this is that those environments usually require everyone to multitask much of the time. Elimination of much of the bad multitasking has the greatest impact on overall enterprise project throughput (Leach, 2005c, p. 163).

When one considers the duration-multiplying effect of multitasking, it should be clear that assigning resources to jump back and forth across project boundaries is not an efficient mechanism. Replacement of a more systemic process to multitask with synchronization, combined with the management of resources for "relay race" behavior, will go a long way to reduce this issue and will help accelerate project completions across the portfolio (Patrick, 2001, p. 8). Leach (2005c, p. 197) recommends as one of his success factors that project resources be expected to work on one project task at a time. In this same vein, removal of multitasking allocation is a stress reduction technique for team members.

17.9.2.10 Prioritization

The PM will be required to communicate task priorities to project team members. In the CC world, there are two kinds of resources; resources that perform critical tasks and those that perform non-critical tasks. The ones that have to be most carefully controlled are the CC tasks, since they most directly determine how long the project will take. The operational process must make sure that CC resources are available when the preceding task is done, without relying on a predetermined due dates. There is the underlying assumption that the work of a task will take as long as it takes, no matter what the schedule model assumes. Resources will be allocated to work on specific prioritized tasks without distraction until they are complete (Patrick, 2001, p. 6).

17.9.3 SKILL REQUIREMENTS

This view of skill relates to the management aspects rather than the technical skill to produce the product. These are skills that are not currently in place in the traditional project management methodology.

17.9.3.1 Task Time Estimation

Parkinson's law is often used to describe the behavior of workers and it particular fits the traditional estimating mindset.

> Work expands to fill [and often exceeds] the time allowed.

> **—Parkinson's law**

Goldratt (1997) based much of his original TOC model on the assertion that individuals will procrastinate when working toward a fixed due date. This is known as the "student syndrome"— Remember that term paper you had to write in school and how you procrastinated about starting it until the last minute and then went through chaos and late hours to finish? The modern project environment often works a lot like this. If CCM is to be successful, breaking out of that cycle must be a priority for the organization.

17.9.3.2 Estimating Logic

CCM requires that task durations be built with only the time required to do the work without any safety factor. The idea of estimating a task at the 50% likelihood level instead of a 90% level will be a significant philosophical shift for many. Work unit estimators will need to be coached and encouraged not to develop task estimates with built-in safety. Safety has to be consolidated into the various model buffers. Patrick (2001, p. 2) suggests using the following method for developing task estimates:

> Resources are first queried for a safe estimate—one in which there is a high level of confidence and the work unit owner would be willing to consider a commitment. This value defines the upper end of the possible time estimate. Once this upper limit is initially established, a second "aggressive but achievable" estimate is solicited—one that reflects a near "best case" situation that is "in the realm of possibility" assuming things go well in the performance of the task in question.

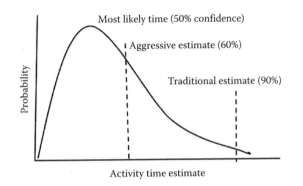

FIGURE 17.11   Estimating logic.

Figure 17.11 is a simple illustration of this concept. The traditional estimating approach is to cover time contingencies and use an estimate that has very little risk involved. This often results in the procrastination mode because the worker feels that they have plenty of time built into the estimate. In the CCM method the strategy is to push towards the 50% point and use buffers to handle overflows. This has the impact of prodding the process to not procrastinate because there is no planned idle time. The 60% time represents a slight compromise that might be used as a beginning strategy since the 50% point represents a significant cultural change for the organization.

CCM experts warn about the organizational tendency to build safety into individual tasks as protection. Leach describes the tendency of organizations to believe that all tasks are underestimated, or else they are concerned that something could go wrong to elongate the task (Leach, 2005c, p. 192). This is often the case in an organization with extensive multitasking and interruptions. In any case, there is a strong tendency to overprotect an estimate by adding significant padding to its expected duration.

17.9.3.3   Project Manager

PMs will need to avoid the temptation to pad task estimations. The stigma of time overruns will be a hard one to break for the traditional organization as they will resist the concept that 50% of every activity are planned to overrun. Team members and work unit owners will have similar biases left over from traditional management practices. This cultural item will be a leadership challenge by showing all that meeting a particular activity time estimate is no longer the criteria for success.

17.9.3.4   Team Members

Efforts to educate the team members on the principles of CCM, and the planned estimation and implementation methodology, should ideally be lead by the PM charged with the implementation. If the PM believes that bringing in instruction from an external source is necessary, then the he must also participate actively in the training effort. Ideally, CCM should be tested on a relatively small project with an experienced PM who believes in the new method. Every reasonable effort should be made to gain commitment of the team and careful metrics should be kept to evaluate the results. Team members must understand that CCM has to be fully embraced. It is difficult to isolate a small piece and still get the same value from the result.

Task estimating will likely be the first activity stressed in the implementation and the first opportunity to either commit or resist. Newbold (1998, p. 195) recommends running the team with a "Pit Crew Mentality." Everyone must understand how individual work contributes to a car or track race and attempt to translate this to the CC.

17.9.3.5 Task Status Reporting

CCM status reporting is based on team members accurately reporting task completion status. This is done not as the traditional percent complete, but focuses on the estimated time required to finish the task. Actual completion of tasks, as well as forecast for completion of tasks must be communicated in a timely manner. The traditional once a week status reporting will not work well for this requirement. Activity reporting and resource management need to be almost a real-time mechanism Leach (2005c, p. 201). Status reporting and resource management become tightly linked in the CCM operational environment since status information is used to tightly control the next step in the various project chains. Patrick (1998, p. 5) details two simple steps required to accomplish this:

Step one: Ask the responsible work unit owner how much of an advance warning they need to finish up their other work and shift to uninterruptible work so they can drop what they're currently doing and start work on their critical task.

Step two: There must be discipline in providing appropriate status updates for each active task. When the estimate to complete task A matches the advance warning needed by the task B resources, the resource management alert system will trigger and communicate this to that resource. They then need to plan on shutting down what they are doing and prepare for the upcoming task.

This advanced warning process is referred to as a "resource alert" in the CCM vernacular. Team members and their managers may resist the frequency of reporting requirement, as previous project assignments may have only required weekly or even monthly reporting, however the importance of this must be stressed. Remember, the new culture more resembles a track relay-race or race car pit crew mentality when working a CC task. The traditional model has been characterized as waiting at the train station for a scheduled train to come in. If that train were European in origin this might, but that is not the typical view of traditional project task completions. So, the resource alert process is used in CCM to manage the flow. It along with the frequent status reporting process are key processes in achieving CCM success.

17.10 BUFFER MANAGEMENT

In previous discussions we have described how observing buffer status is the new way of evaluating project status, rather than the traditional planned versus actual task status approach. The title for this activity in CCM is Buffer Management and its basic goals are to support project schedule management, avoid unnecessary delays, and focus attention on areas that need some form of recovery planning. To accomplish each of these goals the measurement process has to be more dynamic than simply publishing a once a month status report. Imagine receiving daily information on each active task and the estimated work time remaining on that task. Leach (2005c, p. 201) feels that the reason for using remaining duration rather than percentage of completion estimates is that humans tend to overestimate task percentage complete. When called upon to look forward and consider the work remaining to complete a task, the estimates tend to be much more accurate. Remaining duration is also the number needed to measure downstream status. Estimating it directly avoids the often false assumptions necessary to convert a percent complete estimate to a remaining duration estimate.

Using this reporting approach task performance, data can be entered into the network plan and from it status of each buffer can be computed. Note that as actual task values are reported on the plan the related buffer size will need to be adjusted to reflect the amount remaining. The red–yellow–green approach shown previously illustrated in Figure 17.4 can be used for high level exception reporting, but more complex status can also be derived through percent of buffer used compared to percent of chain task accomplishment type metrics. This latter view would provide a more detailed insight into how the chain is progressing. Obviously, collection and reporting tools would need to be in place to support this level of sophistication. Operating the project in this form gives it a more urgent view and that is important in changing the organizational culture to recognize that completing the project is important.

In the CCM model the project status will no longer be shown by comparing actual to fixed baselines, or milestone tracking. The usual system of task fixed time blocks and due dates is also eliminated. The only dates in a CC schedule will be launch dates for chains of tasks and final due dates associated with deliverables that are external to the project and which are protected by project buffers (Patrick, 2001, p. 6). Buffer Management is the primary status focus. Task variability impacts buffer sizes and that is the status logic. Awareness of project buffer consumption relative to the completion of the CC (and to the expected variability of the remaining work on the chain) provides an important forward-looking focal point for managing project execution.

17.11   ORGANIZATIONAL CHALLENGES OF THE CC

Scattered throughout this chapter, we have seen various new processes and concepts related to CCM. Before leaving this subject, it will be good to review the eight items summarized below as the key challenges to be faced in moving to this model. Each of these will require changes in culture, process, ego, or management. The summary items are the following:

1. *High-level management support:* As with all organization changes the support of senior management is paramount.
2. *Cultural change in managing teams and projects:* CCM changes the way in which project activity is pursued. The entire organization must understand and work with the new paradigm.
3. *Status reporting methods:* Traditional status reports will have to be replaced and all stakeholders will need to be educated in the new approaches. There will likely be the need to compromise some in regard to subproject completion reporting. There are many related status-oriented changes embedded in the CCM. It is important to remember that effective communications to the stakeholder community is a prime goal for the PM. CCM will not be a transparent change in this regard.
4. *Translate estimating techniques to 50% probability:* Taking away time padding will be a major cultural problem because of the stigma of time overrun in the traditional view.
5. *Task overruns are now the norm:* Traditional status reporting looked unfavorably at time overruns. In the CCM model, they are expected. Management and other stakeholders will have to understand this new phenomenon.
6. *Team evaluation:* In a relay race, the team wins and that is the way CCM must work.
7. *Resource allocation and project priorities:* Resource alert and formal project prioritization are required to manage the work flow process. Both of these issues require more discipline than exists in the typical organization.
8. *Multitasking avoided:* This implies that once a resource is moved to a task, they will work on that task until it is completed. No jumping around to other tasks.

17.12   CC IMPLEMENTATION STRATEGIES

There are two seemingly viable scenarios to introduce the CC model into an organization. The first scenario would be to present CCM as a challenge to an existing high-performance, recognized, and established team that is routinely assigned difficult or high-priority projects. In this case, the team would not likely be intimidated by the change in control approach and their ego could be a stimulus to move higher in respect. This would be similar in behavioral concept to the 1930s Hawthorne experiment in modern times (see Chapter 2). Offering this group the opportunity to utilize the new method to reduce their delivery time, and increase their effectiveness and value to the company, would provide the level of challenge such teams often thrive on. This strategy is similar to introducing a new tool or technology. Successful project delivery by this pilot team would serve to demonstrate the capability to the organization, thus paving the way for wider acceptance.

A second implementation scenario is to introduce the concept in a project with an experienced and respected PM, coupled with a technically proficient team, but not one who is steeped in traditional methodology. Many of the behaviors described in the CCM model would possibly appeal to such a group since they would recognize the logic behind the model. This second type environment would not be so inclined to protect an existing approach and this would avoid a defensive reaction often found in a traditional team.

The CC concepts represent an exciting new option that gives organizations the ability to increase the number of projects that can be done by the same number of resources and to reduce the average duration of projects. This approach enables an organization to confront the project completion problems that exist. Even though the processes embedded in the CCM model seem unusual, they really are more of a logical extension of current project management practices than first appears. To focus on these differences may be academically interesting, but it is counterproductive from a practical point of view.

17.13   CONCLUSION

This overview of the TOC and CC methodology concepts examined the basic mechanics and some of the operational complexities associated with the model design. It should be clear to the reader that there are potentially excellent project management ideas embedded in this model in regard to methods for improving project throughput. Certainly, the use of buffering and restricting time estimates could be implemented in some fashion even in the traditional view of projects. However, if one buys into the full CC logic described here, a significant level of organizational, management, and skill issues must be changed within the organizational culture. Specifically, CCM requires an abandonment of traditional estimating, and status reporting practices. All project stakeholders will need to buy into the new methodology in order to achieve successful implementation. The CCM view of minimizing multitasking would be a positive motivator for many team members and would help sell the approach internally, but once again this requires a change in organizational culture.

Many organizations today are searching for better ways to achieve major breakthroughs in project development cycle times in order to stay competitive. They need to complete more projects through their organization per unit of resource allocation. This goal must also often be achieved without increasing the number of people allocated to projects, or having the option of hiring additional people. Availability of skilled resources will always be a project constraint in both good and poor economic times. In healthy periods, the aggressive requirement outstrips demand and in tough economic times executives are reluctant to hire even though the demands for new projects remains.

Use of buffering and restricted time estimates could be implemented in some fashion for even the traditional view of projects. However, if one buys into the full CC logic as outlined here, a significant level of organizational, management, and resource allocation issues must be changed in the organizational culture. Notably, CCM requires an abandonment of traditional estimating, buffering, and status reporting practices.

Out of all the project management schemes proposed today, the CC logic is probably the best thought out from a conceptual point of view, but on the downside would require the greatest internal process changes to successfully implement. The uniqueness of the CC concept is that it hits at the heart of why projects take too long to execute. The best traditional project management process known can be implemented, but so long as padded estimates and multitasking remain the norm, the excessive time results will not change significantly. CC projects in mature support organizations have verified the concepts outlined here and throughput improvement in the 25–30% range have been demonstrated.

The author believes that many of these ideas and processes will be tested in various project environments broadly over the coming years. The fact is that the logic underlying the CC concepts is so compelling that it is necessary for the modern PM to understand both the power and operational complexity of this model.

Author note: A representative CC schedule is available from Frank Patrick's "Critical Chain Scheduling and Buffer Management: Getting Out from between Parkinson's Rock and Murphy's Hard Place." Reference http://www.focusedperformance.com/articles/ccpm.html.

REFERENCES

Anderson, P., 2001. Study: Multitasking is Counterproductive. http://www.umich.edu/~bcalab/articles/CNNArticle2001.pdf (accessed February 9, 2008).

Bolognese, A.F., 2002. Employee Resistance to Organizational Change. http://www.newfoundations.com/OrgTheory/Bolognese721.html (accessed April 8, 2008).

Christ, D.K., 2001. Theory of Constraints Project Management in Aircraft Assembly. http://www.cnaf.navy.mil/airspeed/content.asp?AttachmentID=57 (accessed April 18, 2008).

Cook, S.C., 1998. Applying Critical Chain to Improve the Management of Uncertainty in Projects. http://www.pqa.net/ProdServices/ccpm/ref/r9z00034.pdf (accessed April 16, 2008).

Geekie, A., 2008. Buffer sizing for the critical chain project management method. *South African Journal of Industrial Engineering.* http://findarticles.com/p/articles/mi_qa5491/is_200805/ai_n25501281 (accessed December 15, 2008).

Goldratt, E., 2007. Critical Chain. http://www.goldratt.co.uk/resources/critical_chain/index.html (accessed April 12, 2008).

Goldratt, E.M., 1997. *Critical Chain.* Great Barrington: The North River Press.

Herroelen, W., R. Leus, and E. Demeulemeester, 2002. Critical chain project scheduling—do not oversimplify. *Project Management Journal*, 33: 4.

Kendall, I., G. Pitagorsky, and D. Hulett, 2005. Integrating Critical Chain and the PMBOK® Guide. http://logmgt.nkmu.edu.tw/news/articles/criticalChain-PMBOK.pdf (accessed April 12, 2008).

Leach, L., 2004. *Critical Chain Project Management*, Second Edition. Norwood, MA: Artech House. p. 23, p. 53.

Leach, L., 2005a. EMV and Critical Chain Project Management. http://pmchallenge.gsfc.nasa.gov/Docs/Presentations2005speakers/Day%202/newideas/LarryLeach.pdf (accessed April 16, 2008).

Leach, L., 2005b. *Lean Project Management: Eight Principles for Success.* Boise: Advanced Projects, incorporated.

Leach, L.P., 2005c. *Critical Chain Project Management*, Second Edition. Norwood, MA: Artech House. p. 23, p. 53.

Newbold, R.C., 1998. *Project Management in the Fast Lane: Applying the Theory of Constraints.* Boca Raton: St. Lucie Press.

Patrick, F.S., 1998. Critical Chain Scheduling and Buffer Management: Getting Out from between Parkinson's Rock and Murphy's Hard Place. p. 6. http://www.focusedperformance.com/articles/ccpm.html (accessed February 9, 2008).

Patrick, F.S., 2001. Critical Chain and Risk Management: Protecting Project Value from Uncertainty. http://www.focusedperformance.com/articles/ccrisk.html (accessed February 9, 2008).

Retief, F., 2002. Overview of Critical Chain Project Management. http://www.hetproject.com/Francois_Retief_paper_Overview_of_Critical_Chain.pdf (accessed April 12, 2008).

Wynne, J., 2006. Running with Change. http://www.gantthead.com/content/articles/232629.cfm (accessed April 12, 2008).

Part V

Planning Support Processes

LEARNING OBJECTIVES

Upon completion of this part, the following concepts should be understood as to why they are required as part of the planning process:

1. Understand the role that various supporting process have on the project plan and its execution.
2. Understand the plan implications of integrating support processes.
3. Develop the resource management plan required to complete all project activities and then match planned resources to those available from internal, external, and procured sources.
4. Identify and implement project controls by defining the required correct processes, measures, and controls to manage project change, communications, procurement, risk, quality, and human resources (HR) to facilitate future project executing and controlling processes, and ensure compliance with generally accepted industry standards.
5. Develop a formal and comprehensive project plan by integrating and documenting project deliverables, acceptance criteria, processes, procedures, risks, and tasks to facilitate project executing, controlling, and closing processes.
6. Obtain project plan approval by reviewing the plan with the client and other stakeholders individually or in small group presentations. Negotiate open items or issues in order to confirm project baselines prior to execution of the plan.

Source: These objectives have been adapted from the PMI, *Handbook of Accreditation of Degree Programs in Project Management*, Project Management Institute, Newtown Square, PA, 2007, pp. 17–20. Permission for use in this format granted by PMI.

This part will continue the discussion of project planning, except that more recognition of real-world impact needs to be admitted. This means that all of the assumptions that were made for Part III discussion are now essentially removed. That is, we now have to deal with the following:

- Resources available are limited in both quantity and skill level.
- Work unit estimates are not accurate in many cases.

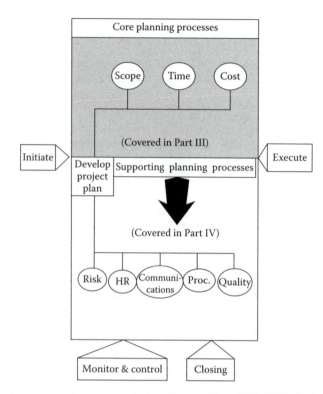

FIGURE V.1 Project management support variables. (Adapted from PMI, 2008. *PMBOK® Guide*. Newtown Square, PA: Project Management Institute.)

- All team member motivation skill capabilities are not equal.
- Conflict is a frequent work element to deal with.
- There is risk in the project.
- Expectations from external sources are difficult to keep in synch as the plan changes.

Each of these situations causes the PM to have to manipulate another series of KA management variables. Figure V.1 shows how these five new variables are manipulated to achieve the triple constraint values outlined in the previous part.

The new view of planning introduced by these additional elements expands the view of planning to show that a plan requires definition of both input requirements as well as output objectives. It also needs to incorporate these components and lay out an overall planning roadmap for achieving the defined goals. Each of the five new items represents an area of management concern in the project that needs to be dealt with as part of the planning activity. Recognize that the guide model calls all eight of these items KAs. In this discussion, it seems quite logical to characterize some KAs as generally representing outputs and some more oriented toward inputs. In each of the management variable cases, a decision made in regard to it will potentially cause a change in one of the other items, including the output trilogy. Recall that Part III approached the planning side by defining scope, and then from that time and cost were derived. In this Part IV set of topics, we are looking at the items basically surrounding the triple constraints. For example, HR represents the project team and all of its management characteristics. Communications relate to the collection and dissemination of project information. Risk events deal with items that have a probabilistic chance of occurring and impacting project performance. Procurement represents the processes required to acquire external goods and services. Finally, quality concepts are embedded in both the input and output side of the equation, but primarily deal with processes to ensure that the desired output is delivered.

So, the focus of this part is to describe how these parameters serve their role in the overall management equation.

18 Human Resource Management

HR management involves the set of processes required to plan, acquire, and manage the human complement in the project. The basic goal of HR management is to execute the project by allocating the right individuals to the correct roles at the proper time in order to complete the plan. A secondary goal of this area is to attract and maintain skilled employees and to manage them effectively (brij, 2007). This activity also includes dealing with the numerous HR issues that arise during the life cycle of a project.

The 2008 *PMBOK® Guide* defines four major processes for the HR KA (PMI, 2008, p. 217):

- *Develop HR plan*—roles, relationships, and responsibilities
- *Acquire project team*—obtain the defined HR to execute the project
- *Develop project team*—improve the "competencies team interaction, and the overall team environment to enhance project performance" (PMI, 2008, p. 229)
- *Manage project team*—"… tracking team member performance, resolving issues, coordinating changes to optimize project performance" (PMI, 2008, p. 236)

Projects primarily fail or succeed because of people. This makes the management aspect of HR a very important one—not just in the planning phase of a project where the development of a good road map for the project work is defined, but also through the other life cycle phases as the project dynamics begin to unfold. Even with an accurate plan based on the original user specifications "… requirements change so quickly these days that, unless the PM is fully aware of the issues at a business level, even a project that delivers the planned scope within time and cost may be deemed unsuccessful because its deliverables are no longer relevant to the business" (AST Group, 2001).

18.1   HR PLANNING

In spite of the dynamic nature and highly technical work product, the planning phase is charged with the development of a reasonably detailed project plan complete with descriptive work units (i.e., work units and planning packages in the WBS). Details regarding the core planning mechanics were described earlier in Chapters 11 through 14. At this early stage, the HR planning process was fairly abstract in the sense that the future project team is usually not named at this point. From this early HR view, a resource breakdown structure (RBS) for the project defines generic quantities and skills or each work unit. *Note*: Be aware that the RBS acronymn is also used in risk management to show a similar outline for risk events, so this can be confusing.

A seond key HR artifact is the *Staff Management Plan* (SMP), which is created to show details regarding how the resources will be acquired and managed. The guide defines the following items to be included in this plan (PMI, 2008, pp. 223–224):

1. Staff acquisition—internal versus external sources
2. Resource calendars
3. Location where team members will work

4. Timing information
5. Release the plan—how team members will be transitioned on and off the team
6. Training needs
7. Recognition and rewards
8. Compliance—governmental, union, financial, or other compliance requirements
9. Safety—related to work and personal risk.

If we could assume that all humans had equal skills and were available as the plan defined, the HR component of the planning process would not be overly complex. However, that is not the case.

18.2   HR IN EXECUTION

The project HR management processes that are included in the execution phase relate to the project team acquisition, development, and management. The first step to executing a successful project is assembling the proper mix of resources as defined in the project plan. This process can be tricky because of the time variability that most experience and the linkage between activities makes the output of one work process become the input of another process. Some have described this as similar to managing a relay race. The management goal is to have a planned resource standing at the finish line waiting on the previous task to complete.

18.3   ACQUIRE PROJECT TEAM

Acquiring a project team involves identifying sources for consideration, negotiating with various management units to obtain the required resources, and getting them to the project on schedule. The project team acquisition process involves obtaining the specific people needed to accomplish all phases of the project. Within this structure, each team member brings specific qualifications and capabilities to the project team.

The PM is in charge of the team selection and negotiating for these individuals from their functional managers or other sources. Key attributes for this search are required skills, work experience, availability, personal characteristics, and personal professional interests. Team members can be specifically hired for the project or acquired as contractors from outside organizations. Some project activities might require special skills or knowledge and it may be necessary to look outside of the company for this skill. Consequently, it is important to take into consideration a staff member's prior experience before assigning him to a specific activity. Personal interests and characteristics also play an important role. If the candidate is not interested in the project, it is unlikely that he or she will perform at their best. We also have to recognize that some employees do not work well with others. One solution for this issue is to separate those individuals who cannot get along well by assignments to different projects; however, in some cases, the employee who lacks people skills is the only person available who possesses a set of particular skills. In this situation, other management techniques will be required to maintain and keep team cohesiveness and performance. The final consideration is the availability of selected key team members, as the project can only function when these skills are in place.

Preassignment, negotiation, hiring, and virtual teams are the typical sourcing approaches used during the process of staffing the project team (PMI, 2008, pp. 257–258). Although a project Charter gives the PM formal authority to acquire resources, it does not give the ability to do that in any form he wishes. The project matrix resources will be owned by the varoius host functional managers and it requires negotiation with them to identify specific names who will be made available to the project. Preassignment of specific named individuals occurs when they are included in the original proposal for the project. This often occurs when a project is being performed for an external entity and the defined names were a condition for accepting the project. When staff members are allocated as part of a project proposal, they should be identified in the project Charter to minimize allocation issues later. Availability, competency levels, and personal characteristics of the staff members become key topics in the negotiation in almost the same manner that a professional athlete is drafted.

One other staffing condition that is worthy of comment involves the practice of splitting a resource into multiple pieces and allocating them to more than one project in a single time period. This is called *multitasking*. This is a common practice because technical resources are typically scarce and sometimes are not needed full time on a project. Nevertheless, the practice of splitting resources across multiple project assignments is found to create more severe scheduling issues for the PMs involved. More specifics on this multitasking issue can be found in Chapter 17. The short answer view is to avoid it if possible, because that is just one more place where allocation issues can occur.

The acquisition option implies hiring individuals or teams of people for certain project activities. This has the disadvantage of introducing new organizational members to the team and usually requires some start-up time before they can become fully productive. Only in the case of a long-term project or to resolve aggregate staff shortages would an organization hire a project.

Virtual teams are defined as project team subgroups that do not work in the same location, but share some aspect of the project goals, and have a role in the project. These can be individuals from the same organization or from third-party contracted organizations. Chapter 34 describes the virtual team environment in greater detail, but the essence of this problem is that communications are more complex and there may be cultural issues as well for teams located in other countries.

There is disagreement among project professionals regarding the type of person who should manage both the project and the subteams. One school of thought says that the leaders should be from the business area that the project impacts and the other view is that a business person makes a poor technical PM. There is probably merit in both of these views. Without attempting to take on that argument here, the essence is that whoever is put in charge needs to be sensitive to both the business and the underlying technology. This means active participation from both sides regardless of the organizational grouping. The better the team, the less important is the formal organizational structure or the skill type of the leader.

The final issue to consider before the team actually gets started is the need for training. Without proper skills, the work will not get done effectively, but the other side of this is the need to show the requirement in terms of specification by individual, along with associated adjustments for schedule and budget to accommodate the training schedule.

18.4   PROJECT ORGANIZATIONS

Project organizations can exist in a wide variety of forms. The essential element is a collection of individuals formed for a temporary time period to execute a planned initiative. Most host organizations are structured along what is called functional lines. This means that major skill groups are housed together. Names such as engineering, manufacturing, marketing, accounting/finance, legal, and others are common departmental titles. This structure is called a *functional* organization (i.e., structured around functional groups). Projects formed inside of any one of these organization departments would be under the authority of that department and are often called silo projects. The downside of this project location is its ability to obtain the various skills needed to execute the variety of skilled tasks.

A second project organizational alternative model exists for very large projects such as those found at NASA. Resource levels in this class of project are huge and the internal view consists of multiple related projects under the overall control of a program manager. In this organizational model, called a *projectized* organization, the program manager is the central authority for the resources attached to the group.

A third organizational structure is the most common for project teams. In this case, the requirements call for a variety of technical skills to be drawn from various functions within the organization and possibly even external resources. Also, the project plan calls for these resources to move in and out of the organization and only work on assigned work tasks when they are needed. The

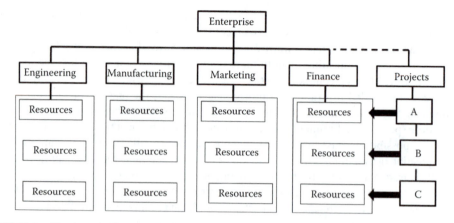

FIGURE 18.1   Matrix organizational structure.

remainder of the time these resources will be elsewhere and not charging the project for their time. For example, engineering might be heavily involved early, but less so later. Manufacturing might be the central resource for execution, but not heavily involved elsewhere. Marketing, legal, and finance requirements would be scattered through the life cycle. In this case, the resulting organization structure is called a *matrix* and is shown schematically in Figure 18.1. There are disadvantages to this structure, but the significant advantage that makes a structure somewhat like this a requirement is as follows:

- Required resources can be managed moving into and out of the project team
- Project cost is lessened because resources are only used as needed
- There is an appropriate focus on the project objectives through a single PM
- The overall enterprise utilization of resources is better handled in the mode

The paradox of the matrix structure shown in Figure 18.1 is that it has advantages for the organization, but complicates the life of the PM. The one basic issue that plagues most projects structured in this form is the weak level of control over the resources. It is up to the PM to negotiate needed resources from the owning functional departments and he has to do this without any real authority to obtain the specific resources he wants. There is an obvious conflict built-in as there will be other projects attempting to also obtain the best people. If the project schedule shifts, as it often does, the individuals originally planned may not be available. From a project view, this means that the resource allocation aspect has become much more complicated than if the resources were "owned" by the project. In addition to this, the authority delegated to the PM in a classic matrix structure is weak, meaning that the functional manager is still the team member's boss and the project team member will feel like he has two bosses as a result. This watered down authority level can be an issue if the assigned team member is not performing adequately. As a result of this dual ownership view, there is often divided loyalty of the team member between the temporary project and his permanent professional home.

18.4.1   Dotted Lines

In many ways, the matrix project team is afloat in the organization in that the project team member may not be easy to show in a formal organization chart. In some cases, this is partially resolved by having an organization structure that looks much like a functional one with a senior manager as the defined authority, but dotted lines to other organizational groups. The dotted line relationship is yet another way in which the PM has more than one boss. Nevertheless, even in this case, there are many other management stakeholders that are not formally identified by the organization chart, but nevertheless are very interested in the outcome of the project. This means that the PM has by definition multiple bosses regardless of the organization form used.

18.4.2 People Issues

Assembling team members together brings many issues that are discussed here and elsewhere in the book. Projects are breeding grounds for conflict of various types and it is not easy to keep a project team focused on a target. A.T. Kearney found in a survey that seven out of 10 teams failed to produce the desired results (Hall, 2008, pp. 23–34). The one major issue observed was in the dynamics related to moving dissimilar organizational types together and keeping their roles and relationships defined. If a distributed geography is added to this equation, the complexity of coordination grows even greater.

So, the reality of project organizations is likely matrix oriented and the second half of that reality is that this brings with it increased management challenges. Also, the issue of team collaboration (information sharing) remains one of the current focal points as team members are increasingly being dispersed.

18.5 ROLE SPECIFICATIONS

Closely related to the resource acquisition process is the idea of role assignments for the team. Once the team staffing is generally defined, the next goal is to make sure that the team members understand their roles in the project. There are several things that can be done to help the team members understand the overall project and their particular role in it. At the highest level, the WBS offers a good tool to explain the overall scope and work activity, particularly for the work units that those team members will be responsible for. Also, the project plan contains documentation on the various KAs related to the project and describes each of these. This needs to be mandatory reading for the team members.

One way to start the definition of team member roles is to convene the team for that purpose. A formal kickoff meeting is one recommended method for this. Depending on the size and scope of the project, this meeting could be a multiple day exercise and involve not only the technical and management aspects of the project, but some socialization as well. This is a good time for the team members to get to know each other and develop some camaraderie. It is important for the team to believe that this project is worth doing. No one wants to be working on a project that is viewed as not worth doing.

Once the basic technical issues are understood by the team, the indoctrination process should move to team organization and roles. Once again the size of the team will affect what level of formality will be used to define the role and responsibility structure. A general guideline for this is to keep subteam sizes in the range of seven or less. If the project goals can be decomposed into work groups of this size, there is a better chance of building productive groups and simplifying communications. A decomposed team structure of this type then requires sub team leaders who have the ability to create and manage an effective team and they become the key drivers for the project effort. Obviously, smaller projects would have only one or two such groups to deal with. Another organizational activity that should be going on at this point is bringing key stakeholders into the kickoff sessions and make sure the team knows who they are and the role they play.

The preparation of exact role specifications, usually in written form, was introduced during World War I, when staff officers compiled elaborate tables of organization for the infantry. Today, schematic organization charts are still used to show authority relationships, but they are less used than in the past primarily because of the complexity of contemporary organizations.

18.6   RESPONSIBILITY ASSIGNMENT MATRIX

A responsibility assignment matrix (RAM) is typically used in linking activities to resources in order to ensure that all work components are assigned to an individual or team. One format for this is to define four category roles in the responsibility assignment—Responsible, Accountable, Consult, and Inform. This is called a RACI chart, because it assigns those roles for each work unit. These tables can be constructed either at a high level (major project groups) or at a detailed level (WPs). The coding schemes used in a RACI chart are often used to specify who is responsible for various functions related to a work unit. In some cases, these codes can be more than the four shown here; however, a brief translation of the basic codes is as follows:

R—the organizational unit or individual assigned to do the work.
A—designates the management level person or organizational unit responsible for the work.
C—designates the person or organization responsible for functionality of the work. This can be the SME for the effort.
I—designates those individuals or organizational units who will be informed regarding status of the work unit.

Other coding schemes can be used in the matrix, but the key point here is that a RACI format chart is useful as a communications specification tool to help ensure proper coordination and information distribution at the work unit level. Tools of this type are often ignored.

A RACI table is typically constructed with activities/work units on the vertical (i.e., WBS IDs) and resources on the horizontal (i.e., by name or organizational unit) plane. Not every resource will have an entry for every activity. Table 18.1 illustrates the format for a fragment of this type chart.

A second name for this class of chart is a linear responsibility chart (LRC) and it focuses more on naming who is responsible for specified work units at the lower levels of all levels in the WBS. In this model, each row could represent a WBS ID (all of them) and each column a person or a team name. The boxes could then be completed with the letters P (Prime support), S (Support), and N (Notify). This is similar to a RACI chart, but in this case, it shows more the degree of involvement than the multiple management-type roles of the classic RACI chart.

Regardless of the coding schemes used, RAM-type charts are useful to map the work of the project as described in the WBS to the individuals or groups responsible for performing that work. For smaller projects, it is best to simply assign WBS work units to individuals; for larger projects, it may be more effective to assign the work to organizational units or teams.

18.7   RESOURCE HISTOGRAMS

A resource histogram is a bar chart that shows the number of resources required or assigned over time to a project. In showing project staffing needs the vertical bars represent the number of people needed in each skill category and by stacking the columns, the total number of resources required

TABLE 18.1
Sample RACI Chart

WBS ID	Bill	Gary	Ron	Bob	Bud	Teri
4.F	R	A	I	C	C	
6.A	I	A	C			R
2.A	R	A	I			C
2.B	R	A	I	C	C	C

for each time period can be represented. In some cases, this same view can be produced for each resource by name, type, or organizational unit. The resource histogram is a tool that is often used to produce a visual representation of resource requirements. This can show various views as well. For example, one view might be planned versus available resources to show resource capacity shortages. This is a handy format to show various stakeholder groups resource views for the project.

18.8 STAFF MANAGEMENT PLAN

"A Staff Management Plan is a subset of the human resources plan and it describes how the human resources requirements will be met" (PMI, 2008, p. 223). This document describes the types of people needed to work on the project, the numbers needed for each skill type by time period, and how these resources will be acquired, trained, rewarded, and then reassigned elsewhere after the project.

Development of a project staffing plan also involves the process for selecting and assembling a project team. This definition builds on the high-level staffing needs roughly identified first in the Initiation Stage. When developing a staff plan, the focus includes specific roles and responsibilities for team members in the project and the specific staffing profiles across the project's life. This plan describes the approach and detailed "baseline" information regarding staffing and roles. It also describes the specific roles and responsibilities as they have been tailored for the project. These are not meant to be job descriptions, but rather a summary of the responsibilities for each role. Individual responsibilities are tailored based on the life cycle phase and actual project staffing.

A SMP describes the appropriate procedures used to manage staff on the project. It discusses mentoring, cross training, primary/backup role assignments, training and development assignments, performance evaluations, performance recognitions, and promotions, as well as disciplinary actions and demotions. It states how a staff vacancy is to be handled, and what happens if it appears that the position will not be filled for a while, and what will happen if the vacancy cannot be addressed by a single person (given current skill sets available). Also, review of qualifications to ensure the replacement will be able to assume the work, addressing differences or discrepancies.

18.9 MOTIVATION THEORY

As teams are formed, the job of motivating the team members to willingly take ownership of the project goals does not just happen. An understanding of the potential triggers to spur humans into production action is a complex undertaking; however, there are a few basic motivation concepts that can help start the process. Quite apart from the moral and altruistic views for treating colleagues as human beings and respecting human dignity, research has shown that well-motivated employees tend to be more productive and creative in the workplace. Unmotivated team members tend to be less productive and supportive of the project goals. Since the job of a PM is to get things done through his team, it is important to understand some of the key triggers for energizing those individuals and the team as a whole. Accomplishing this goal is the essence of the management task.

There are many motivation theories in existence, but matching these theories to real humans remains an art form. To understand motivation one must understand human nature itself, and therein lies the problem! There are two project views for motivation. One is from the individual perspective and the second is a team (group) view. The management processes for each are somewhat different. We will first examine the individual theories and then look at how groups of humans react.

18.10 INDIVIDUAL MOTIVATION THEORIES

Since the Hawthorne experiment in the latter 1920s attempted to show the impact of light intensity on worker productivity, the behavioral school of management has been hard at work to expand knowledge about human needs and motivation. The discussion below refers to four of the classic behavioral researchers and their core set of concepts regarding human behavior and motivation. These four researchers are the following:

1. Douglas McGregor (1906–1964): Theory X and Y
2. Abraham Maslow (1908–1970): Hierarchy of needs
3. David McClelland (1917–1998): Achievement Motivation theory
4. Frederick Herzsberg (1923–2000): Motivation/Maintenance theory.

The individuals are listed in chronological order to emphasize to some degree the evolution of this research.

Douglas McGregor: His theory is based on the idea that there are two basic types of people: Theory X types who do not want to work and must be directed closely and Theory Y types who look at work as natural. The latter types will seek and accept responsibility while being self-directed. This theory opened the door for participative management rather than an authoritarian command and control form that was more prevalent at the time (*ca.* 1960).

Abraham Maslow: Dr. Maslow developed his view of humans by studying Rhesus monkeys. From this base, he formed a five-stage hierarchy of needs as it applied to humans and documented this in his 1954 book titled *Motivation and Personality* (Maslow, 1954). He illustrated this by a pyramid to show how humans move up the need hierarchy as they fulfill lower-level requirements. Figure 18.2 shows the five major need levels.

The Maslow need hierarchy model describes individuals at level one as looking for the basic items that sustain life, notably food and sleep. This would be characteristic of someone who was basically nothing and was struggling for bare survival. As level one needs are met, the individual begins to look upward to what are described as safety needs. At level two, the needs would include safety aspects of the job, financial and physical. So, a reasonable job that offered more than enough money to maintain level one needs would represent this need level. In addition, physical safety would remain a need regardless of the others items. At level three, the individual is fed and is safe; so finding social outlets takes on importance. This can be exemplified by friends, belonging to a group, and love. Level four enters with the humans wanting a sense of belonging. Terms such as self-respect and achievement are often used to define this need level. Note that these terms would seem to have the most relevance to the knowledge worker and his potential motivation triggers, since he is earning enough to satisfy the lower need levels. Level five is the summit of the need

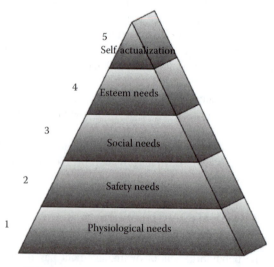

FIGURE 18.2   Maslow's need hierarchy.

hierarchy in which the humans are looking for something that allows them to reach their full potential as a person and this often is a non-work-related view of the world. Terms such as wisdom, jus-
tice, and truth are used to describe individuals at this level. Bill Gates giving away billions of dollars
is an example of an individual at this level.

Maslow's implication was that the layers of the need hierarchy were motivators for individuals at
that level, and essentially if you knew where the individual was located on the hierarchy, those items
would serve as motivators—that is, money buys food, house, cars, and so on. Even though he did
not specifically say this, the implication is that money was a motivator if it was linked to a need. A
great deal of research has gone on with this model and much of it has been refuted in its simplicity;
however, almost everyone can relate to it to some degree. We should also note that many professionals are clearly above level two and we see strong behavioral motivation correlations at levels three
and four. Occasionally, we even see somewhat successful professionals who leave good paying jobs
to go teach at universities and write textbooks. So, there is some reasonableness to the Maslow
model and it offers an interesting structure to review for a starting place.

David McClelland: This behavior model has some of the same concepts as of Maslow's, but it
begins to focus more on the motivational triggers rather than assuming that human needs lead to
motivation. McClelland hypothesized that motivation could be created by two sources: intrinsic and
extrinsic. Intrinsic motivation comes from something satisfying that one enjoys. Extrinsic motivation occurs from some external factors. One of the long-term arguments among behavioralists has
been whether money was in fact a motivator or not. McClelland classified money as both an intrinsic
and an extrinsic motivational trigger. Intrinsic because it represents success (fancy car, big house,
etc.), while the extrinsic view was that money would cause the individual to seek it out in return for
some act. The acceptance of the value of money as a motivator remains controversial to many.
McClelland believed that intrinsic forms of motivation were more effective.

In its basic form, the McClelland *Need Acquired Theory* (1965) described individuals as being
motivated by one of the three general needs (McClelland, 1965):

- Need for power—strong need to lead; increase personal status
- Need for achievement—seeks advancement, feedback, and accomplishment
- Need for affiliation—seeks relationships and human interactions.

Each individual has varying degrees of these three needs and their ratio mix determines a great
deal about their resulting style and motivation. From a work performance point of view, McClelland
was most interested in the achievement characteristics because he believed they were the type individuals who made things happen and got results; however, the downside of this was their potentially
negative impact on those around them because of their excessive demands. As in the case of Maslow,
we see these characteristics in project teams. Even if the theory is absolutely correct, we see that the
key is to find the correct balance.

Frederick Herzsberg: The Motivation/Maintenance theory was published in 1959 and it was the
first behavioral research to describe specific motivators by source (Herzsberg et al., 1959). It also
had a second mirror-type view regarding what Herzsberg called dissatisfiers. These are factors that
can cause one to be dissatisfied with the job, but do not necessarily become satisfiers or motivators
if fully supplied. The basic idea of this two-pronged approach was that individuals might stay on the
job because of certain factors, but would not be motivated unless an adequate level of satisfiers
existed. Figure 18.3 provides a high-level aggregate summary of the Herzsberg research results. The
decreasing curve denotes general strength of each satisfier (motivator) item, while the increasing
curve shows the same for the dissatisfiers.

These data have been interpreted by the author from the original Herzsberg research data and
this figure is intended to illustrate which items are generally strongest for each factor. In attempting
to quantify the relative frequency of these two parameters, the author reviewed multiple reviews of
the data from other sources. Some researchers disagreed with the methodology of the original

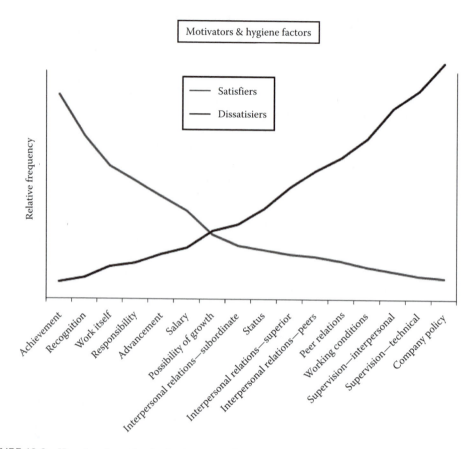

FIGURE 18.3   Herzsberg's motivation/maintenance factors.

research and refuted its conclusions, while many others discussed the conclusions. One of the issues found was that the different survey populations that tried to duplicate the Herzsberg data for engineers and accountants got different results. School teachers for example seemed to be more motivated by money than the original group. However, given the Herzsberg original survey population types the results may well match the technical project team factors reasonably well.

Attempting to quantify a human attribute with an "intensity" measure for satisfaction or dissatisfaction is not our goal here. Figure 18.3 attempts to show general relationships that have a varying probability for motivation, as well as those that have high potential for being dissatisfiers. There is general consistency with other researchers that achievement, recognition, and responsibility factors are high motivators and Herzsberg adds to this that they can also be dissatisfiers if not present in the job structure. Also, note from the figure that salary falls into the sixth position as satisfier, but is also a significant dissatisfier if not adequately supplied by the job. This translates to an interpretation that money is not a strong motivator, but can be a demotivator if it does not exist at a reasonable level. On the dissatisfier side, company policy and supervision win the potential dissatisfier awards. The PM must understand this list of factors and use the knowledge to minimize dissatisfiers and maximize the satisfiers.

18.11   TEAM MOTIVATION

Motivating a team is obviously more challenging than motivating a single individual given that teams have both individual and interpersonal aspects to deal with. A management technique that might work with a single individual can well be wrong in a team setting. Bringing a group of diverse individuals together to work on a project does not necessarily yield a functioning unit even if all members are highly motivated and have high achievement needs as defined in the behavior theories.

The first critical aspect of team motivation is to establish a clear goal direction for the project. In the team organization, individuals must execute their individual roles, but should be willing to support team goals over their own. Assuming that the team has been staffed properly, each of the individual skills is important in completing the overall goal, but no one of them alone is sufficient for success. This diversity of member personalities and skills can be a positive force if each performer can be motivated to find a way to contribute their unique capabilities when and where needed. Team organizations can be very powerful mechanisms to produce spectacular results if the team leader can provide the proper management and operational climate. This is the essence of the project team goal.

Alignment of goals, purpose, and values between stakeholders, team members, and the external organization represents a complex set of management activities for the PM. Team goal alignment and a positive internal chemistry are key attributes of a productive group. When team members are not aligned with project directions, then motivation and productivity will likely be reduced. The role of team leaders (or team members in self-directed teams) is to achieve the following five motivational goals:

1. Foster mutual respect for the individual expertise of all team members
2. Help weaker team members believe that their efforts are vital to the team
3. Support a shared belief in the cooperative capabilities of the team
4. Create individual team member accountability for their contributions to the team effort
5. Direct the individual's competitive spirit toward the team and the organization; success is defined as team success.

These protocols and attitude goals do not mean that the team sings *Kumbaya* to start each day, but it does mean that the team leader and members have certain civil behavioral protocols toward each other.

18.12   HYGIENE DISSATISFIERS

The reverse side of motivation is the existence of dissatisfiers that take away from motivation. In this regard, it is also important to be aware of any adverse environmental hygiene factors that may exist. In many cases, these cannot be resolved quickly, but sensitivity to the issues can help. The Herzsberg model offers a list of these to review, but others may also exist. One of the management keys in team motivation is to first take away as many negative influences as possible and then focus on the positive factors inherent in the project work. In addition to the organizational and physical hygiene factors outlined by the Herzsberg model, there are three environmental prerequisites needed to position the individual or team member to be motivated (AMA, 2008).

1. A clear definition and ownership of their defined tasks
2. Removal of external inhibitors that hinder the individual from accomplishing the required tasks
3. Ensure that the individual has appropriate training and skills to execute the defined tasks.

The job of a PM in a motivated and productive team is to protect his team from external negatives, whatever they may be.

Techniques to create high-productivity teams are still not sufficiently understood to the degree that a mechanical checklist can be defined. Such a list would contain much of what has been described in this section, but it still takes management skill to produce the desired outcome. More details on the creation and management of high-performance teams are presented in Chapter 35. For this discussion, we stop at this theoretical point and recognize that a collection of humans formed into a project team structure brings a wide variety of motivational trigger mechanisms that are unique for each person.

18.13   EMPLOYEE SATISFACTION

Herzsberg has provided a good starting place to understand the basic drivers for employee satisfaction. His data indicate that effective people management is about more than traditional mechanical HR practices such as work allocation and training. We are now sensitive to the value of finding these motivational triggers for achieving team member satisfaction with both their current job and growth development for the future. Part of the satisfaction dimension comes from outside the project in the way the employee sees their company as a community, where there is a social environment. Research indicates that the more successful organizations are those in which a positive socialization climate exists. Some of the organizational characteristics that foster this attitude are a feeling of concern for employee welfare, lower status barriers, good communication, high-quality training, and overall respect for employees.

Research by the Gallup organization about Great Managers and Great Workplaces also suggests that measures linked with employee satisfaction, such as turnover correlate with profitability (Gallup, 2008). Researchers noticed that organizational excellence was driven by high-performing workgroups, rather than the organization as a whole. They found that a number of situational factors such as pay and parking, which are often elements of an organizational HR strategy, do not make significant difference to the most productive workgroups in the companies studied. If they exist, they are not dissatisfiers, but neither are they motivators. What did matter were a number of other work related factors that are more firmly within the responsibility domain of team managers (i.e., interesting work, positive feedback, and friends at work). These factors also included employees having effective communication regarding what is expected of them through the setting of explicit expectations by their manager. However, individuals expressed the need to have some freedom in finding their own way of achieving these expectations if they are to feel ownership of their job.

PMs need to be able to define how to help each employee take responsibility and be able to maximize their productivity in an energized team. This does not mean working of the team for 60 hours a week. This strategy will actually yield the opposite result. The positive approach is to establish an environment for job satisfaction and assign work that makes the best use of individual talents. The new generation knowledge worker reacts very positively to recognition for good work and constructive feedback. Development of this team culture is a clear requirement for the PM.

18.14   CONFLICT MANAGEMENT

Conflict in the project team can come from various sources and involves some divergence of opinion. One conflict scenario involves two very rational team members who have a disagreement over some technical direction issue. Other scenarios are represented by disagreements over other aspects of the project—scope, schedule, budget, and so on. Each of these situations demands some resolution. In all cases, conflict must be seen as a problem to be solved rather than a war to be won. The key theme of conflict management is that it cannot be left to fester and increase. Confrontation is the PM mantra. He must be on the lookout for its existence and activate a strategy to deal with it. However, the strategies are situational, and therein lies the management art.

There are a number of different methods to handle conflict. In each case, a win–win approach is the goal (Fisher, 2008). Win–lose outcomes often create other conflict issues later. From the management view, there are three basic concepts involved in the resolution process:

Problem-solving approach: The first step should be to clearly and concisely identify the issue and its relevant points. It has been estimated that 90% of team conflict is created by a misunderstanding of the situation (Mulcahy, 2005). Once the problem is defined and all agree with the issue defined and the related facts, then try to develop various solutions to consider (without initial judgment as to each one's merit). Even if the parties agree quickly with one of the options developed, it is best to review that option thoroughly to ensure that it does not create another issue elsewhere. Once the issue is agreed on, the solution should be documented along with rationale for the choice.

Use patience and respect: Realize that the existence of a conflict means that the parties have diverse opinions. These will not magically dissolve under the light of inspection and if they do that can also indicate a process problem—probably one of the parties has been intimidated by one of the players and backed out with no change in belief or attitude. So, the first step is to try to achieve an open conversation with as little emotion as possible. Every participant should have a respected opinion even if it is clearly wrong. The problem-solving process is as much a training process as a conflict resolution one. The fact is that team members must learn to discuss controversial issues and seek logical answers without a PM always stepping in to play referee. In many cases the participants know more about the details of the conflict that the PM, so he adds little content in this case. When intelligent participants cannot arrive at some resolution, there likely are some goals or views that are not clearly understood. In this case the PM has to step in and may need to establish some guidelines for the process. A more hands-off approach also helps to build healthy relationships between the parties involved. In addition, better solutions are produced through this type dialogue so long as the players can focus on the task issue and not the individual's role in the opinion.

Construct an agreement that works: On reaching an agreed solution, the identified areas of the agreement should be clearly specified in writing along with the rationale for that option. In some cases, it is best to air the situation to other groups or individuals to see what feedback occurs from these sources before moving forward. The final goal is to not create another conflict with the solution to the current issue. The agreement should be evaluated on the following criteria:

- Enforcement: Does agreement rely on others who were not present in the discussion? Have they commented on the solution?
- Realistic: Do we have resources and expertise to implement the agreement?
- Future oriented: Should we consider other similar issues for common solutions or relationship to this agreement being made?

18.14.1 CONFLICT SOURCES

There are four basic sources of conflict (ALS, 2008):

- *Value conflict:* Values are the beliefs that people use to give meaning to their lives. They explain individuals view as "good" or "bad," or "right" or "wrong," Disputes occur when an issue is viewed by two individuals with different value systems. There is no right or wrong answer to this category, because the individual value system will likely not change regardless of the discussion.
- *Data conflict:* Data conflicts occur when there are different data used to form opinions. This results in a conflict that can be resolved by supplying the same data to all parties (Fisher, 2008).
- *Interest conflict:* Interest conflict occurs when one party believes that, to satisfy their needs, a certain direction should be pursued that is opposed to other views. This class of conflicts can occur due to one of the following reasons:
  - Money, time, and physical resources
  - Method of solving the dispute aids one side

- – Perception of fairness, trust, desire for participation, and respect.
- *Structural conflict:* Structural conflict occurs from forces external to the individuals in conflict. Important factors that can cause structural conflict include the following:
  - – Inadequate physical resources
  - – Geographic constraints
  - – Time to complete task.

Interest conflicts are the most difficult internal issues to deal with because the players are not looking at the team goals, but rather primarily their own. Structural conflicts have an interesting side effect on the team. When external forces create conflicts for the team, the team members often band together to "fight" that external source and in that process team cohesiveness is increased. From a PM viewpoint, the key issue with conflict is to recognize that it will occur and it can either be a destructive or creative force. There are two extremes of team conflict. If team members do not have visible disagreements, they are performing like lemmings following the leader over the cliff, while visible excessive internal personal emotion of other team members' ideas can be disastrous in every sense. The team leader needs to mend his team toward embracing these events as work challenges, not dumb opinions. If individuals can learn to professionally negotiate their ideas to others, their value to the team increases greatly (Fisher, 2008).

18.15   NEGOTIATION SKILLS

The negotiation process involves in dealing with another person or party to settle a matter. "In a successful negotiation, everyone wins. The objective should be agreement, not victory" (Wertheim, 2008). The key goal of team negotiation is to convert the situation to win–win in which both parties feel that they were heard and the solution makes the most sense for the project. Negotiation should result in the settlement of an issue or argument for the benefit of all parties involved in the conflict. In order to achieve this goal, the communication process between the parties needs to be open and honest. Hidden agendas will cloud the result otherwise. Communication is obviously one of the important components required to negotiate an issue.

Before the negotiation process begins, it is important to clearly define what is being negotiated. This process should be done face to face given that a rich flow of information is needed. Relevant data to the issue should be collected and your view of the situation should be clearly formulated. The type of conflict should be evaluated from the list above to judge how strongly you feel about the situation and what a desirable outcome would be. It is also important to recognize that the other party has an opinion that is different from yours; so laying out the two positions is the beginning of the process. Two essential skills required for negotiation are influence and confidence. Influence comes from your ability to be stylistically persuasive in selling your view. Confidence comes from being able to define the issue in terms that the other party understands and show the merits or worth of your view. This approach may be couched in data terms, qualitative goals, technical parameters, timing, cost, customer expectations, or other approaches. In the end the negotiation process should obey the 3Fs—Fair, Fast, and Firm. Some of the successful tactics that are used for negotiation are listed below:

1. Be firm yet polite when making a stand.
2. Emphasize advantages and disadvantages of your approach.
3. Put ego aside and concentrate on the matter at hand.
4. Aim for solutions that are interest based and not based only on what any individual desires.
5. Value time, schedules, and deadlines. Try to not waste time, but be sensitive to the other party's needs to discuss.

18.16 TECHNIQUES FOR HANDLING CONFLICT

In the earlier discussion, we have seen various categories and sources of conflict and a general strategy for negotiating resolutions. However, in real life these come in various forms and the PM needs to be sensitive as to his timing to jump into the fray. One long-term goal is to train team members to resolve their own conflicts and stay out of the way. However, that is only good when a reasonable and timely result occurs and unfortunately that is not always the case. Mulcahy reports that 20% of a PM's time is spent in conflict management, so this activity will be a significant time allocation (Mulcahy, 2005). Specific sources and topics for conflicts come from various aspects of the project and these vary stage by stage. Typical areas are schedule, resources, technical, scope, budget, change requests, and personalities. Thamhain and Wilemon rate these sources over the life cycle (Thamhain and Wilemon, 1975, p. 35). Their research shows that schedules and priorities are the most typical conflict topic, but all items mentioned are also recorded.

Let us see if we can outline how to deal with the mechanics of the conflict process. When a conflict issue emerges the first step is to decide your reaction. Frankly, the natural reaction is to hope that it will go away and do nothing. Unfortunately, experience shows that this is the worst management approach. So, the first point is that conflicts will tend to get worse if left alone and the management mantra for this is to *confront* those events using some measured strategy. There are multiple options to invoke and they may be taken sequentially or iteratively.

1. Withdraw—stay away from the situation.
2. Compromise—get involved with the situation to seek out a solution whereby each of the two parties gets something that they are looking for.
3. Smooth—try to convince each party that some solutions really give each some measure of satisfaction and de-emphasize the negatives.
4. Force—basically, the PM becomes official referee in this case. A decision can be made by either listening to the facts and making a decision, or just making a decision (autocratically).
5. Problem solve—this is the rational mind model that will work assuming that the parties are not mind locked to their position and will react to a set of facts. In this situation the negotiation process unfolds, using the negotiation rules outlined earlier.

The question still remains as to which option to select and when. The preferred choice is to try to minimize team leader involvement in the beginning other than to visibly recognize that you are aware that the issue exists. Also, if necessary, define some resolution constraints such as timing when an answer is needed. If facts are missing, help supply those but then let the team members work on the problem up to the constraint point. When that point occurs the leader has to move to a second strategy and become an active member. The next step involves selecting a negotiation strategy realizing the following likely outcomes for each:

1. *Compromising*—offers some short-term win for both parties, but long term is likely viewed by the participants as lose/lose since neither got what they wanted.
2. *Smoothing*—depending on the skill level of the leader he may be able to placate the parties, but once again this has lose/lose potential for longer term.
3. *Forcing*—in this case the leader uses his formal authority to make the decision in whatever form he wishes. From a management perspective, this is the worst choice of all and if used, time should be spent with the parties on damage control. There may be situations that make this option necessary—timing or executive edict being examples.
4. *Problem solving*—this is the desired option and should be attempted as the most desirable step. Attempts to follow the steps are outlined earlier by first defining the problem and then spending time motivating the individuals to solve the problem. This has the best potential to be long lasting and leaving the parties with a win–win attitude.

Working with the project team on conflict management techniques is a mandatory mentoring activity for the PM. Failing to create this culture in the team leaves the leader with excessive time doing the team's job of getting the work done. Also, recognize that the development of a good plan that properly balances the scope, time, and cost variables will help minimize conflicts later.

18.17 CONFLICT MANAGEMENT SCENARIO CASE

Envision two team members arguing over a technology-based planning issue. Both members are viewed to be technically competent. One person's view is that option A will result in cost savings for the project, while the other view is that option B will result is a technically superior product. It is agreed that option A will in fact potentially produce a higher-performance product and option B will extend the schedule. With these brief facts how would you approach the conflict resolution of this issue under the following scenarios?

1. For step one, what would you say to the two parties and what would be your involvement in the process?
2. As an alternative view, the two parties are very dogmatic in their views and tend to have a narrow focus that is reflected by their individual positions. Neither will change their perspective.
3. All problem-solving efforts have failed up until now. The parties have followed the negotiation process, collected extensive data on their position, and the facts stay as they were first defined with no compromise. Either option A or option B must be selected. An answer to this question is now on the critical path of the planning effort. How would you deal with this situation? This makes a good open discussion scenario.

18.18 LEADER VERSUS MANAGER?

The question often arises as to whether the PM should be a leader or a manager of the team. Warren Bennis once described a leader as one who knows which direction to go (vision), whereas a manager knows how to get there (mechanic). PMs require at various points in the life cycle some of both characteristics. During the early project stages, finding the right vision is a leadership challenge since there is no clear vision as yet. In this situation the manager needs to help supply that direction in concert with user input. Later, the challenge evolves to executing the plan and that involves getting defined work accomplished. This is more of a management activity; however, even in this latter case the role still remains to add a leadership element to the team as they strive to reach the goals. Based on this dual perspective, we are not going to try to segment these two roles here. Instead, the term *style* will be used to reflect how the PM accomplishes the required goal through his team. In one case, he may be behaving as an orchestra leader waving his baton with great music flowing, whereas in other situations he may be more like a military leader saying "follow me over this dangerous hill. This is what needs to be done." In many cases the appropriate manager/leadership style will vary depending on what type of situation and circumstances are present. To be an effective leader you have to know when to cross from one style to another and choosing the correct style at the right time is an important determinant for team success. No one style will work for all situations. As in the case of handling conflicts, style selection is a key skill.

18.19 ATTRIBUTES OF A LEADER

Embedded in the style of the leader are his attributes. These represent qualities and characteristics that collectively make up how he is perceived by the project team. What makes these attributes different from the ordinary is that they stand out in a crowd. These attributes come in different mixes, but include the following: vision, action orientation, attitude, communication skills, motivation, relationships, ethics, responsibility, and confidence. Former Secretary of State, General Collin Powell summarized this idea in a 1996 speech:

> The ripple effect of a leader's enthusiasm and optimism is awesome. So is the impact of cynicism and pessimism. Leaders who whine and blame engender those same behaviors among their colleagues. I am not talking about stoically accepting organizational stupidity and performance incompetence with a "What, me worry?" smile. I am talking about a gung-ho attitude that says "We can change things here, we can achieve awesome goals, and we can be the best." Spare me the grim litany of the "realist;" give me the unrealistic aspirations of the optimist any day (Harari, 2007).

Leaders/managers bring out these attributes in the team. Another attribute of a leader is their persuasiveness that influences groups to follow them, even to the wrong goals. We must recognize that not all leaders lead to the correct goals. Hitler was a great leader and we see other negative international leaders today who would fall into the same category. The PM must also work on finding the right direction to lead. Leaders can move their teams to great heights or low depths. Thus, the job of a PM is to realize that this is his fate.

Good leadership also involves responsibility for the welfare of the team, which means that some will not agree with your actions and decisions. This is an inevitable outcome. Trying to make everyone like you is a sign of mediocrity. To do that you will have to avoid such things as making the tough decisions, confronting the people who need to be confronted, or not offering differential rewards based on differential performance because some would get upset. Ironically, by procrastinating on difficult choices, by trying not to get anyone mad, and by treating everyone equally "nice" regardless of their contributions you will simply ensure that the only people you will wind up angering are the most creative and productive people in the organization (Harari, 2007). Effective leaders know that they cannot make everyone happy with every decision or action that is made. Leaders do not always like to engage in these types of actions but must do so because this sets the tone for other employees in recognizing that the overall team is more valuable than a single individual.

Another aspect of project leadership is taking responsibility for outcomes. When certain things go wrong, something or somebody else usually caused those events. The inclination is to find the source and make sure that everyone knows that it was not you—that is, human nature. Poor leaders want everyone to know that the problem was not caused by them. They do not want to take blame for things that go wrong or take responsibility for the welfare of the group. An effective leader must do both. Effective leaders take responsibility for bad issues and leave credit to others for the good things that happen.

Final point is that a leader may be a good manager but a manager is not always a good leader (this goes both ways). The position of manager may be achieved through a formal job assignment by the organization. Leaders, on the other hand, unite followers with their vision and this can occur outside of formal organizational structures. Regardless, leadership traits as described here are needed in the PMs' skill set and they must use these qualities to their maximum potential to get the team headed in the right direction.

Good management skills actually complement leadership. Management is commonly defined as the process of getting work done through others. Leadership is the process of influencing people to give their energies, potential, determination, and to go beyond their comfort zones to accomplish goals. Management affects work; leadership affects people (Barr and Barr, 1989). These collective traits are the key to moving the HR toward the desired targets.

18.20 TRAINING PROGRAMS

Team skill development can be nurtured through mentoring or various types of training programs. Mentoring is a good practice for emulating skills of more senior team members, but may also perpetuate old habits that you wish to change. Formal training programs can be used to transfer defined information or to build a defined cultural attitude. In any case, the process of skill development needs to be recognized and pursued. For projects that have specified skill requirements that do not exist in the incumbent team, the training needs to be dealt with prior to their arrival on the team, or soon thereafter. One approach to this is to create a skills assessment for the team members and from this decide how to pursue training and in what format.

One helpful approach is to categorize the training into three groups: environment, project, and miscellaneous, then prioritize specific training sessions into time blocks. Examples of these types of training sessions are summarized below.

Environment

- External organization and its goals?
- Basic HR requirements that all employees need to know (benefits, medical, security, IRAs, etc.)
- Pay related
- Employee selection and recruitment, promotion, planning, and management.

Project

- Work processes that the team need to know how to access and use (i.e., project plan, change control, technical documents, time keeping system, status reports, etc.)
- Methodology used for project life cycle management
- Performance of management process
- Skill issues required (this can be both informational and specific).

Miscellaneous other courses: Presentation skills training, business writing, being an effective team member, virtual teams training, decision analysis training, ethics training, various productivity tools used in the team, and so on.

Organizations use different strategies to accomplish training, but one desirable approach is to establish a time allocation and then prioritize the training into that time. Very few organizations overdo the level of training. Once the skill gaps are defined and prioritized, an effective training program can be developed.

18.21   SUMMARY

This chapter has provided a summary overview of the HR process in the project team. Managing HR is a complex undertaking. Every PM needs to go through a self-assessment of his management style and this is a valid internal improvement process that all should undertake. It is important to get to know the project team members and understand their individual goals. Once this is done you can do a much better job on roles and responsibilities, as well as work allocation.

The next chapter deals with communications and this should be looked at as a companion to the HR process. In fact, most HR problems seem to be caused by inadequate communications.

REFERENCES

Academic Leadership Support (ALS), 2008. Conflict Resolution Menu—Step 6. Retrieved October 23, 2008 from http://www.ohrd.wisc.edu/onlinetraining/resolution/step6.htm.

American Management Association (AMA), 2008. Team Building. Retrieved October 23, 2008 from http://www.amanet.org/onsite/team-building/67715.htm.

AST Group, July 12, 2001. The Top 10 Reasons Why Projects Fail. Retrieved October 24, 2008 from http://www.itweb.co.za/office/ast/0107120730.htm [Gijima AST Group Limited (GijimaAst)].

Barr, L. and N. Barr, 1989. *The Leadership Equation: Leadership, Management, and the Myers-Briggs: Balancing Style = Leadership Enhancement.* Austin, TX: Eakin Press.

brij, 2007. *brij Technology Dictionary.* Retrieved November 2008 from brij http://www.brij.net/technology_dictionary.html.

Fisher, R., 2008. Sources of Conflict and Methods of Conflict Resolution. Retrieved on October 24, 2008 from http://www.ohrd.wisc.edu/onlinetraining/resolution/step6.htm.

Gallup, Conflict Management. Retrieved October 24, 2008 from http://gmj.gallup.com/content/532/How-Great-Managers-Define-Talent.aspx.

Hall, K., January/February 2008. The Teamwork Myth. An interview with Kevan Hall published in www.humancapitalmanagement.org.

Herzsberg, F., B. Mausner, and B.B. Snyderman, 1959. *The Motivation to Work*, Second Edition. New York: Wiley.

Harari, O., Quotations from Chairman Powell: A Leadership Primer. GovLeaders.org. 1996. 10–24 April 2007. Retrieved on October 24, 2008 from http://www.govleaders.org/powell.htm.

Maslow, A., 1954. *Motivation and Personality.* New York: Wiley.

McClelland, D., 1965. *Need Acquired Theory.* American Psychologist.

Mulcahy, R., 2005. *PMP Exam Prep*, Fifth Edition. Minnetonka: RMC Publications, Inc.

PMI, 2008. *A Guide to the Project Management Body of Knowledge (PMBOK® Guide)*, Fourth Edition. Newtown Square, PA: Project Management Institute.

Thamhain, H.H. and D.L. Wilemon, 1975. Conflict management in project life cycle. *Sloan Management Review*, 16(3).

Wertheim, E., 2008. Negotiations and Resolving Conflicts: An Overview. Retrieved on December 15, 2008 from http://web.cba.neu.edu/~ewertheim/interper/negot3.htm.

R

19 Project Communica'

The problem with communication … is the illusion that it has been accomplished.

—George Bernard Shaw

19.1 INTRODUCTION

Communication is the fuel that drives project success and the mishandling of this activity is one of the top reasons why projects struggle. Various authors claim that communication is the most important skill for a PM. Clearly, effective communication is required between the PM, the project team, and other stakeholders. An environment that achieves effective communication among the diverse stakeholder group will find an improved outlook for project success. Some of the basic activities involved include interacting with diverse audiences, holding effective meetings, and facilitating communication among external parties involved in the project. This chapter will describe the model processes related to project communication and review basic concepts related to human-to-human relationships.

19.2 ENGAGING EMPLOYEES: A CASE STUDY

Before jumping into the communications theory it would be good to illustrate the impact of "engaging" the employee. The Gallup Management Journal (GMJ) surveyed U.S. employees to discover what effect engagement had on team-level innovation and customer service delivery (Gallup, 2006). The results of this survey indicated that employees who are engaged were more productive and this produced "powerful catalysts for creative thinking in regard to new management and business processes within the company" (Gallup, 2006). Sixty-one percent of those surveyed responded that they were inspired by the creativity of colleagues and an engagement process brought out the best of their creativity, whereas only 3% of those who felt disengaged strongly agreed. Engaged employees were also found to be more communicative with customers and the result of that was a greater sharing of ideas.

Other positive work performance attributes were noted regarding engaged employees having greater self-motivation, confidence to express new ideas, higher productivity, higher levels of customer service, reliability, less employee turnover, and lower absenteeism. The result of this positive impact on employee performance can be translated into finding hidden profits for the organization. On the negative side, the GMJ 2006 survey estimated that disengaged employees costs the U.S. economy about $328 billion in lost productivity (Gallup, 2006). With this knowledge one might ask why organizations are not more visibly concerned with programs to improve these attributes in their employees.

The effect of good communications is hard to measure, but this case illustrates the positive motivational results from an effective program to make employees feel that they are engaged in the organizational processes. The same reaction occurs in project teams, which brings us to the topic of this section. Communications Management is a significant and tangible issue not only for projects, but for entire organizations.

COMMUNICATIONS MANAGEMENT PROCESSES

KAs of Communications Management includes the processes required to ensure timely and appropriate generation, collection, distribution, storage, retrieval, and ultimate disposition of project information. This topic involves processes and methods for transmission of information among the project team members and stakeholders. The *PMBOK® Guide* specifies five basic communications processes (PMI, 2008, p. 243):

1. Identify stakeholders.
2. Plan communications.
3. Distribute information.
4. Manage stakeholders' expectations.
5. Report performance.

As with all of the guide model processes, the first step for each is a plan to define the steps to be followed for the rest of the operational processes. This area follows that basic approach.

19.4   IDENTIFY STAKEHOLDERS

Projects have more stakeholders than first appears. These are groups and individuals who either need to be involved with defining the project, or need to be communicated with in regard to the ongoing status of the project. Stakeholders can reside either inside the host organization or external to it. The key outputs of this process are a *Stakeholder Register* and a *Stakeholder Management Strategy* (PMI, 2008, p. 244). An added item of information that could prove useful later is the attitude of the defined stakeholder in regard to the project (i.e., supporter, resistor, etc.).

19.5   PLAN COMMUNICATIONS

A vital foundation component in the successful management and completion of a project is the creation and implementation of a communications plan at the start of a project. The purpose of this first step is to identify the information and communications needs of the stakeholder population (PMI, 2008, p. 246). As simple as this idea sounds, it is surprising to find that many projects never bother to formally identify their communication targets, timing, content, or receiver's preferred media for the information. One clear message that needs to be discussed in this chapter is that project communications tend to be lacking and this planning step is required to establish the base for improving that situation.

Recognition that communication is a crucial challenge that projects face enforces the motivation to pursue this area with the same diligence as other planning aspects of the project. The primary output of the planning step is a *Communications Management Plan* that attempts to integrate the communications requirements needed to manage project scope definition, technology to be utilized for communications, and the underlying development process within the organizational culture (Richardson and Butler, 2006, p. 293). The *PMBOK® Guide* outlines in some detail the content recommendations for this plan (PMI, 2008, p. 257). For our purposes envision this document as a repository containing information about targets, templates, policies, methods, and individuals who are defined as communications sources.

Effective communications processes are required in order for stakeholders to deal with the human interaction required in achieving the project work. Communications techniques utilized in a project include various formal and informal methods. Examples of formal options include the following items:

- Periodic status reports
- Progress review meetings

- Kickoff meetings (project and stage)
- Executive reports
- Formal presentations to various stakeholder groups
- Project financial status reports
- Government (or external agency) reports
- Issue logs
- Risk logs
- Change request logs
- Role responsibility
- Project organization
- Milestone reports
- Deliverables reports.

Each of these items represents a unique communication strategy for the selected stakeholder groups. One of the basic problems with project communications is that no one method will satisfy all constituents. Timing, media choice, and content requirements vary significantly by audience. Definition of the targets and these issues represent the fundamental planning challenge to resolve.

Once the target stakeholders have been identified and discussions held with them regarding what and how they wish to receive project communications, the documentation step would include the following specifics:

- *Recipient:* The individual or group that will receive the communication.
- *Who:* Identifies the team member who is responsible for the delivery of the communication.
- *What:* Defines the output content of the communication. This includes a defined format in the case of status-type reporting (i.e., Monthly Project Status Report).
- *Location:* Defines where the item will be stored prior to distribution.
- *When:* Defines the calendar time for delivery (i.e., monthly, daily, on demand, etc.).
- *Media:* Identifies the media used for delivery (i.e., e-mail, web, paper, telephone, group briefing, verbal, video conference, etc.).
- *Focus:* An indication of the stakeholder's particular communication focus (i.e., cost, schedule, technical status, subsystem, etc.). This section could also indicate the level of detail desired.

Collectively, this family of formal communication options is intended to keep the project team and its other stakeholders focused on the project and to help maintain proper priorities and guidance information for their respective efforts.

Informal communications tend to be more dynamic and less planned than the formal items outlined above. Many of these will be ad hoc and focused on a specific topic of the moment. The key in these is to attempt to create an open exchange of ideas between the parties and to generate a feeling of trust with the PM.

Beyond the project, internal team key organizational stakeholders are obvious audiences for project communications, but there are other less likely groups that need—or want—project information. So, the search for communication targets needs to include both internal and external stakeholders. The project team is the core of the communication network. Team members work on the project every day and they require active communication. This group is both a heavy user of communication and a significant provider to others.

Management stakeholders are not involved so closely with the project on a daily basis, but they make key decisions about it and for that reason must not be ignored. Typical management groups consist of the project sponsor, project board, change management board, functional managers, future users, and others.

TABLE 19.1
Communication Matrix

Audience	Information Item	Media	Frequency
Project team	WBS status report	E-mail	Weekly
Project team	Project newsletter	E-mail	Weekly
Project team	Risk review	Meeting	Biweekly
Sponsor	Monthly status report	On-line	Monthly
Project board	Change control status	On-line	Daily
Stakeholder A	Project overview status	E-mail	Monthly
Stakeholder B	Technical deliverables	Paper	Monthly
HR	Manpower report	E-mail	Monthly

One example of a supporting group need for status information occurs when the project goal is to create a new commercial product for sale. In this case the marketing department would want to be kept abreast of such things as product features, cost, and availability. The legal department is involved in procurement activities and could benefit from information regarding risk issues and vendor problems. Each supporting organization would have different information interest needs and that issue must be defined in the planning process.

An external communications audience is very diverse in their interests. For example, vendors, suppliers, partners, and their respective project counterparts are often extensions of the internal team in many ways. However, communications with external audiences will be generally in less detail. Investors and regulatory agencies [such as the IRS, Environmental Protection Agency (EPA), or a public utility commission] represent additional examples of external audiences. Governmental organizations may specify the format and schedule required submittals. Once the project communication requirements are defined, it is common to document these in a format called the project *communication matrix*. Table 19.1 is a mock-up of such a format.

The communication matrix helps manage the flow of formal documents out of the project and helps ensure the integrity of the process. In addition to the regular outflow of information, there are various milestones or checkpoints that require additional reporting. These key points are identified in the project plan and are typically management (time and cost) or technically oriented.

Once the communications plan is complete it becomes one of the subsidiary plans compiled into the overall project plan.

19.6   DISTRIBUTE INFORMATION

The physical process of moving needed information to project stakeholders in a timely manner is called *Information Distribution*. This includes the specific items outlined in the communications plan as well as various ad hoc requests for information. The distribution process requires consideration of methods to gather and store the information and related technology to transmit it to the recipients.

Within the project team, the distribution process deals more with intragroup sharing needed information related to the technical aspects of the project. This includes items such as product design architecture, technical drawings, and other information needed by the team to produce the product. Finally, one subtle aspect of the distribution process is lessons learned documentation and the human aspects of communicating with internal and external project contacts (PMI, 2008, p. 261).

19.7   MANAGE STAKEHOLDER EXPECTATIONS

Recall that the term *stakeholder* means anyone involved with the project, or others who have a formal interest in the project. So, when we speak about managing stakeholders this can translate into a wide audience. Anticipation and management of internal team member needs is much more visible to the PM than anticipating what communication a senior manager, future user, or government official requires in regard to the project. There are many examples in project management archives about projects that were completed only to find some significant hurdle waiting at implementation time that cratered the effort. Many of these issues could have been avoided with earlier communications and problem-solving time. There are also many issues that occur during the execution process that require much more complex communication skills than just sending out regular status reports.

Typically, issues that need to be addressed with the external stakeholder population are not those that can be dealt with by simply issuing a formal report, e-mail, phone call, or letter. Many of these encounters will require more face-to-face back and forth verbal negotiation. Sample of these activities follows:

1. Negotiation with the project sponsor or senior management in regard to the time, cost, and scope balance (i.e., insufficient resources approved)
2. Negotiation with a functional manager over allocation of staff (timeliness, quantity, or skills)
3. Critical review of the project by senior management regarding viability of the current effort
4. Disagreement with a customer over whether a requested change was part of the original contractual specifications.

If left alone, many stakeholders would think that the initial Charter or early requirements represented the current project view. As a project evolves, changes occur both within the project and possibly in the outside environment. In order to be judged successful, the eventual project deliverable should match the expectations of the stakeholder group.

If the project drifts away from the desired point, the management process becomes one of negotiating or informing them of the current status. If the external environment drifts, it is up to the project to capture those new expectations and decide how to move forward. Delivering the wrong project with all of the right methodology remains the wrong project. Projects seldom obtain a clear set of initial requirements and move lock step to produce to a happy stakeholder group. This activity should assume just the opposite environment and it would be closer to reality.

19.8   REPORT PERFORMANCE

Much of the Information Distribution process is focused on reporting project status in one form or another. The general theme of reporting involves time, schedule, baselines, quality, risk, and technical aspects of the project sorted into user interest segments. Example for reporting items include the following:

1. Time and schedule variance (ideally with Earned Value)
2. Work performance status
3. Quality related data
4. Change request data
5. Deliverables status
6. Project forecast metrics.

In all reporting activities the PM must maintain an ethical stance of honest reporting to all parties. In many situations, it is tempting to withhold bad information that might trigger an overly negative response with the rationale that this is a temporary problem that will be resolved. The reporting rule says that the facts must be presented along with an action plan to deal with any major issues.

19.9   HUMAN COMMUNICATIONS MODEL

The mechanical communication processes required for project management have been outlined above, but the real communication management issues occur at the human-to-human level. There is an art to effective communication and it is necessary to be sensitive to the general processes that assist in this activity. This certainly involves more than just passing bits through some technical communications channels. All humans communicate with others on a daily basis, but we rarely step back and dissect what is actually happening and how is it received by others. Most technical PMs have never been taught or trained to analyze the communication process, so it is not surprising that it often creates unintended consequences between the parties.

It is relatively easy to learn a simple communications framework that will help to understand the process better and apply it to daily situations. Using this knowledge, we can better deal with many daily work-related issues that result in conflict, errors, or unintended emotions. It has been estimated that 90% of conflicts are created by misunderstandings caused by communications shortcomings (Mulcahy, 2005).

The first step in improving communications skills is to understand the basic communication model and think about its implications in daily information transfer. Figure 19.1 is a schematic view of the model components, and its basic building blocks are a sender, receiver, and transmission media. The sender formulates a message and the receiver attempts to translate that message into understandable meaning. In order to successfully accomplish this, five elements have to be completed:

1. Encode—translating mental images or ideas into an effective format (language) understood by the receiver
2. Message—bits resulting from the encoding step
3. Medium—method used to transmit the message (verbal, visual, tactile, etc.)
4. Noise—anything that interferes with the transmission and understanding of the message
5. Decode—receiver processes the bits transmitted (minus noise) into meaningful thoughts or ideas.

In addition to this, a feedback loop helps to ensure that the message was properly received. This can be accomplished in various ways, but is an important component of the process. If all goes well, the two parties end up after the process with exactly the same mental perceptions for the item being communicated. Unfortunately, there are many places where accuracy can be eroded.

It has been noted from personal experience that speaking plainly to a person who does not speak the same language as you is not very effective no matter how many times you repeat the transmission, or how loudly you speak. In this case the model is working almost exactly as described, but the encoding step results in mental garbage for the receiver. Likewise, noise in the channel can be created by physical noise or other distortions of the message (i.e., speaking too fast, dialects,

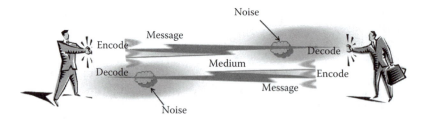

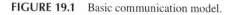

FIGURE 19.1   Basic communication model.

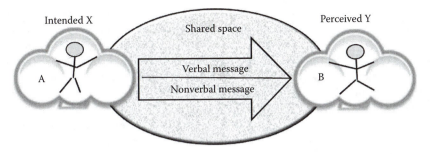

FIGURE 19.2 Actionable communications model.

vocabulary used, and emotion). When any of these distortions occur, the model does not work, and the desired mental perception match fails to achieve its goal. In these and other situations voluminous bits are flowing between the parties but no real communication is occurring. Synchronization between the two has not occurred. Both parties must understand the transmission, whether they agree with the content or not.

In order for real communication to occur, five elements are needed. The process is only partially successful if it reaches the third state (understanding), and is progressively improved at the fourth (agreement) or fifth (action) steps. Figure 19.2 illustrates this process with a slightly expanded communications model.

In order for a successful communications episode to occur the following five steps must occur: transmission, reception, understanding, agreement, and useful action. Let us explore the communication implications of these steps in a little more detail.

1. *Transmission:* In transmitting a message via e-mail or voice mail, the intent is to remotely communicate with the receiver. However, this does not mean the receiver has read or will hear the message. It just means the message has been sent and is assumed to be encoded later. Current technology offers the opportunity to send great volumes of "communication" in this remote forward and store manner, but we often do not consider the other side of the equation. Obviously the link is broken if the message is not received.

2. *Reception:* When someone finally checks their e-mail, voice mail, or signs for an overnight letter, the message is finally received. Once again we still do not know whether the entire message was listened to, understood, or even opened. In each case, the recipient may not have the inclination to interpret or spend any time trying to properly encode the message. While confirmation receipts for the class of transmission indicate physical receipt, they confirm nothing else.

3. *Understanding:* Digesting and interpreting a message's information correctly represents a significant jump in communication compared to simply receiving the message. This step requires cognitive activity to understand the original intent. Depending on the message content, understanding might involve learning something new and it might require other researches on the topic. The message may be ambiguous without the ancillary research and when a question is involved both parties may seek clarification. In this situation, both sides are looking for the completion of a communication receipt and feedback. In some situations the message may contain items that are not understood. For example, if the original transmission was "what do you think we need to do in order to correct problem X?" the receiver may not know there was a problem X. After some research, they can uncover the details about problem X and can then add more content to the next communications cycle. In this example case a feedback communication loop leaves both parties with more knowledge than they initially had. In project communications, it is common for the recipient to

start the process asking for additional information. The understanding step does not occur until some improved level or response action occurs. Also, the research portion of this example can actually spawn other additional questions asked back to the original transmitter or other parties. In this form, the communications cycle is iterative with each cycle adding content to the initial ambiguous message. Moving from a raw bit transmission view of communications to achieving understanding requires more active involvement of the parties than is evident from the schematic model.

4. *Agreement:* Achieving understanding of the message does not mean that a person agrees with what was transmitted. Take for example the statement "you are an idiot!" This message might be quickly understood but would not yield instant agreement. Achieving agreement between two technically oriented, intelligent, and opinionated communicators can be a complex and time-consuming activity, especially if the "facts" related to the issue in question are philosophical or technically judgmental. This class of communications comes closer to what one finds in a complex project environment. When the PM finds himself embroiled in such team discussions the communications channels have already been filled with transmissions, but little agreement. In this situation the challenge is often to first get emotions under control and then work to collect a common set of facts and observations. Until a common understanding of the issues is achieved the conflict and agreement levels will remain.

5. *Useful action:* When we look at the basic reason for project communication, it is to achieve an improved level of understanding and produce some desired action as a result. Team members need to understand what work is required to accomplish the desired goal and take actions as a result of that knowledge. Stakeholders need to use status information to make appropriate decisions regarding the project, even though that decision might result in a cancellation of the project. Failure to achieve an appropriate level of understanding in either case will yield the wrong action. At this level in the model we need to recognize the true role of communications. We all have experienced an aged relative who wanted only to hear himself/herself talk. They were not interested in hearing what you had to say, but the communication process was almost like a brain dump for them. This approach does not work in a project team environment. Failure to successfully accomplish the communication process can contribute to the following undesirable effects:
   a. Project team members will make wrong choices on work efforts
   b. Conflict will be created among the project team by misunderstandings
   c. Stakeholders have erroneous expectations regarding project status
   d. External organizations not receiving the information they require can cause major issues with the project team or the resulting product
   e. Senior management will not have appropriate information to use in regard to the project
   f. Team morale becomes eroded because of low "engagement" in what is going on inside the team.

Good communicators transmit information with the intent of it being understood and then producing appropriate action. Instead of just sending an e-mail, voice mail, or letter and seeing what happens, the effective communicator will carefully analyze each of the five steps described above. They select language, media, and examples that best fit the receiver instead of those most convenient for them. In addition, they follow up to clarify what the likely points of confusion or argument are and identify what actions they want the recipient to take in response. Finally, follow-up occurs until the issue is concluded. Good communication occurs when there is a natural and open sequence of exchanges between the parties. Even when the topic is confrontational, being aware of the communication model framework helps identify where the process is breaking down.

19.10 COMMUNICATION CHANNELS

The communications models outlined earlier have focused on a process occurring simply between a single sender and a receiver. This is a microview of the process, but the more realistic macroview involves multiple communications channels occurring simultaneously across the project. Therefore the communication goal is not just to get a single individual involved, but to accomplish it with all stakeholders. As the size of the project stakeholder population grows, it gets harder to keep everyone "on the same page." Also, the communication flow process is not homogeneous since the various channels are different and each has unique characteristics. There are basically three channel types:

- Upward communication (vertically or diagonally)
- Downward communication (vertically or diagonally)
- Lateral communication (horizontally).

Not only is the approach and content of communication different across these three types but collectively they represent a significant volume of channels to deal with. A mathematical formula for determining how many channels (C) exist within a population of N people is (Schwalbe, 2004, p. 361):

$$C = N\frac{(N-1)}{2}.$$

So, if four people are involved the number of communication channels would be

$$C = 4\frac{(4-1)}{2} = 6.$$

If we increase that number to 10, the number grows to 45 channels. Figure 19.3 shows this schematically.

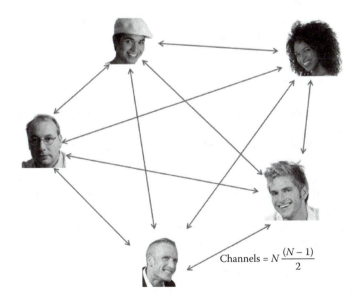

$$\text{Channels} = N\frac{(N-1)}{2}$$

FIGURE 19.3   Channels of communication.

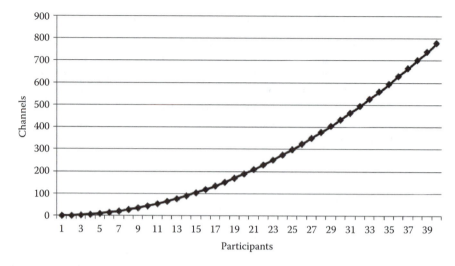

FIGURE 19.4 Communication channel growth versus team size.

Figure 19.4 expands this view to show what the channel count would be for an even larger project group. Given the geometry of this relationship, it should be easy to see why experience shows that optimum project team size is less than approximately seven.

Communications breakdown is considered by many researchers to be the number one reason for project failure (Berkun, 2005). There are five primary causes for project communications breakdown: common language, communication mechanics, personal factors, and workplace factors.

Common language: Global projects often use English as the common language, though this may be a second language for many participants, particularly in an outsourcing situation. Because of this, understanding is decreased as a result of dialect, vocabulary, miscomprehension, or misunderstanding of words. Often, this situation is not resolved because one of the parties does not want to admit they do not understand the information given.

Communication mechanics: The mechanics of effective communication involves not only passively passing messages, but to actively listen to the other person. This involves more of a dynamic interchange between the parties. Clarifying questions and other techniques help to ensure that the process will produce an accurate understanding of the message.

Personal factors: Workplace performance is affected by a wide variety of issues, but external job issues often impact performance on the job. Often times, these issues get into the work communication and become negative noise.

Communication style: Some individuals are easier to talk to than others. The style aspects of the process deal primarily with the setting for the conversation and the way in which the presenters interact with each other.

Workplace factors: To successfully complete a project the deliverables must match the requirements. To produce this outcome the PM is constantly dealing with decisions designed to keep the triple constraints within schedule and budget, while meeting the requirements. Balancing these variables has the potential to send the wrong signals to one or more stakeholder groups.

One strategy to mitigate this problem is to institute a program to sensitize the project team to the issues outlined here. Creating an improved understanding regarding how to produce effective communications is the core requirement. In addition to this review, the source and analysis of conflicts can help with project functioning. Sartor says, "In my 25-year career I have discovered that the most successful projects were the ones where respect for everyone's point of view enabled mature and open dialogue and trust building to work through the project challenges. Quality takes time. That's how great companies are built" (Berkun, 2005).

19.11   COMMUNICATING INFORMATION

Project status information is typically communicated often using tools and jargon-related terms such as WBS, Gantt, and Earned Value (EV), but the primary requirement is mainly about communicating with people. This requires that customers, stakeholders, and team members know what is required to do their jobs in language they understand. Other forms of communications involve negotiating with a broad array of stakeholders, defining team member's assignments, or working with a group to resolve an issue that is impacting the project. All of these scenarios are designed to synchronize various parties communications requirements regarding the project. In order for these processes to work effectively the process requires good communications skills.

19.12   IMPROVING THE EFFECTIVENESS OF COMMUNICATION

Because PMs spend a significant amount of time communicating with individuals and groups, they inevitably carry a broader responsibility for effective communication than other individuals on the team. Success with these activities amplifies the effectiveness of everyone they come into contact with.

Being an effective communicator does not mean that the person necessarily be an extrovert game-show-host personality; nor does it demand a brilliant sense of humor or charismatic powers. Each of these might help, but the first requirement is to understand the important role that this process plays in a well run project. From this initial step there will be more sensitivity to the issue and from that will come an improved understanding of the process.

19.13   EFFECTIVE LISTENING

One of the key attributes of a good communicator is to be an effective listener. The person who can do this successfully will develop a greater degree of mutual respect, rapport, and trust among project participants. Some of the techniques used to improve active communication involve effective listening and some techniques to increase the content of a conversation include the following:

- Asking questions to clarify and gather more focused information
- Paraphrasing what the speaker said to be sure that you understand his point
- Stopping at intervals to review what you have understood up to that point
- Asking the speaker for examples of a point being made
- Ascertaining the speaker's feelings and acknowledging them (e.g., "You seem angry")
- Showing interest by directing the speaker to the most appropriate person to help
- Using nonverbal listening techniques including:
  - Making eye contact
  - Being expressive and alert
  - Moving closer to the speaker
  - Listening for the meanings beyond the words used
  - Using body language to show emotion and agreement.

19.14   BARRIERS TO EFFECTIVE COMMUNICATION

There are many things that a communicator can do to inhibit effective communication. Simply breaking the rules outlined in the communications model is the most obvious. The most important item is to recognize that communications is a two-way process. Input is needed from both sides to improve content. The second major barrier is the physical environment where the communication takes place. Telephone or personal interruptions can negatively impact a conversation. Being in the boss's office versus a nice quiet restaurant can make a world of difference in the process. After the basic environmental issues are resolved, the success of the conversation moves to the topic itself. Two individuals talking about a sports contest can disagree and still enjoy the conversation, while two technicians arguing/discussing their diverse views on a technical issue may become more negative and become a negotiation or problem-solving challenge.

In addition to the mechanical and intellectual aspects of a communication process, the internals in play within the individuals can influence the outcome. For example, consider the following examples (Nokes et al., 2004):

- Judgment—lack of respect for the other's views
- Mind block—already made up your mind on the topic; no more information wanted
- Filtering—picking out some subset of the stream and focusing only on that
- Assumptions—misinterpreting elements of the conversation without confirming
- Side tracking—losing sight of the topic and digressing to others.

Each of these inhibitors will decrease the effectiveness of the conversation. With a little practice each of these can be avoided to a great degree.

19.15   COMMUNICATION TENSION

Beyond the mechanics and attitudinal aspects of the communication process, there are also issues even for the experienced communicator regarding the philosophy behind how the message is created. These difficulties arise mainly from styles pulling in opposite directions, thus the term *tension* (Nokes et al., 2004). Table 19.2 summarizes six examples of this.

In some cases these tension scenarios cannot be controlled or managed because of other environmental constraints. In these situations the communicator must decide how to navigate through the conflicting options. Also, in many cases the receiver has a bias for one of the options and it is up to the communicator to sense this and react accordingly. These tensions occur frequently in the project environment because of time pressure, mixed skill types of the participants, and transient nature of matrix staff.

TABLE 19.2
Communication Tension

The need to communicate the complete story or situation	versus	The need to be brief
The need to tailor the message to the audience, and to simplify	versus	The duty to be open and honest
The need to treat all stakeholders fairly and equal	versus	Competing needs and expectations among stakeholders and the need to release some information over time
The need to listen	versus	Time constraints require a more one-way approach
Demand for the full story now along with all details	versus	The need to release some information over time
The value of being very specific in terms volumes of raw data to back a position	versus	The value of a quicker understanding of the fundamental issues, with details provided later. This requires some faith on the part of the receiver

19.16   COMMUNICATION STYLES

Everyone has a unique communication style that fits their personality and environment. Some will approach the communications process in a friendly manner first and then drift into content, whereas others view it more as a data passing exercise with mostly content. There are basically four model communication styles that can be briefly described as follows:

- *Concrete-sequential: Fix-it.* This style focuses communications on ideas and tasks. They see the project as a set of tasks that need to be completed and the conversation focuses primarily on that.
- *Abstract-sequential: Organizer.* This style uses logical analysis and systematic planning to solve problems. This conversation works to collect data from other sources in order to seek a decision. Organizers are called people and task oriented, which makes them effect team builders.
- *Concrete-random: Explorer/Entrepreneur.* This style relies on people and technology in a search for practical theories and models. Decisions are made after thorough analysis and evaluation. These individuals excel at facilitating planning sessions, discussions, and changes.
- *Abstract-random: Intuitive free thinker.* This style sees situations from different perspectives. They also look at the big picture and the long-term view. Abstract-random communicators make good brainstormers because they can listen actively and enjoy the process of generating new ideas.

All models are abstractions of reality; however, a keypoint is to observe how an individual formulates his conversation, fact gathering, decision making, socialization, and so on. Good communication requires that one party adapts to the other's preferred style. This is called *style flexing*. Once that style is identified, the conversation should be packaged into that form as much as possible.

19.17   COMMUNICATIONS: THE IMPOSSIBLE GOAL?

With all of these variables and contributing factors can effective communications be possible? Hopefully, this discussion has highlighted the reasons why it is not easy and will require careful consideration. If it were possible to write a communications goal statement for the PM and the team, it would contain at least the following six communication attributes:

1. *Project goal:* Significant effort should be made to effectively communicate requirements to the entire project team and confirm that these have been translated into technical work units.
2. *Information distribution:* The project environment is data rich, but can be information poor if an appropriate distribution environment is not created. An effective information environment is important to support project decision processes.
3. *Team engagement:* By having a communications-rich environment, team members will be motivated to achieve project goals. In such an environment, the engaged project team will be capable of doing the right work the right way.
4. *Status:* Various stakeholders have different status needs. Formulate the communication processes accordingly and make sure that the channels are energized. Understandable communications of status will keep expectations in order and help team members focus on the proper targets.
5. *Cohesive team:* There is a selfish reason for the PM wanting a mature project communications environment. When effective communications can be achieved and team members are engaged they do not have to ask questions or guess at the answer. This increases productivity and morale.

6. *Continuous improvement:* Every project needs to have the attitude of making the next project run better than the current one. The communication process does not stop at the end of a project. The lessons learned process is itself a communications process designed to help both the current and future projects run more effectively.

Achieving these six communication attributes should be the guiding focus of the PM. These communications related goals can be achieved if the project team can establish the correct attitudes and processes related to their environment.

19.18 CONCLUSION

This chapter has described the importance of effective communications in the project. Basic models for one-to-one communication and group equivalents were outlined. Research indicates that 90% of the PM's time should be spent in this activity, so failure to effectively accomplish this goal will have adverse implications on project performance. Various surveys grade the lack of effective communications as the number one reason for project failure. In a technical project, accurate results can only occur when there is a clear chain of communications from the initial user requirements through to final delivery. The human communication frailty along this process makes this KA a complex one to deal with.

DISCUSSION QUESTIONS

1. Describe the basic role of project communications as it directly impacts project success.
2. Which method of communications is most effective for change requests?
3. Name some of the communications-oriented attributes a team should have for successful project execution.
4. What do you believe are the most important skills a PM must have to ensure effective team communications?
5. What are some key communications style attributes in dealing with a customer?
6. If you are responsible for coordinating various meetings related to your project during the execution phase, what would be an effective method of keeping track of the outcome of these meetings?
7. What are some of the basic approaches you would use to provide performance feedback to the project team? What communications issues would you predict for this process?
8. One of your subordinates has very good skills in her area of expertise, but for some reason she finds it difficult to perform well on your project. What would you do about this situation?
9. One of your friends is the PM for a large project and is finding it difficult to meet his schedules. What advice would you offer to help in this situation?
10. You are assigned a new responsibility for a project where there is a difficult negotiation going between your organization's management and the external customer. You are now involved in the negotiation process. What can you do to help resolve this situation?
11. You foresee an opportunity with your long-term customer to generate huge additional income through a new market that you are familiar with. As the relationship stands currently your organization will not profit from this suggestion. What will you do?

REFERENCES

Berkun, S., 2005. *The Art of Project Management.* Ann Arbor: O'Reilly Media, Inc.
Gallup Study: Engaged Employees Inspire Company Innovation. *Gallup Management Journal,* http://gmj.gallup.com/content/466/Gallup-Study-Indicates-Actively-Disengaged-Workers-Cost-US-Hundreds.aspx (accessed October 12, 2006).

Mulcahy, R., 2005. *PMP Exam Prep*, Fifth Edition. Minnetonka: RMC Publications, Inc.

Nokes, S., I. Major, A. Greenwood, and M. Goodman, 2004. *The Definitive Guide to Project Management: Every Executives Fast-Track to Delivering on Time and on Budget*. Harlow: FT Press.

PMI, 2008. *A Guide to the Project Management Body of Knowledge* (*PMBOK*® Guide), Fourth Edition. Newtown Square, PA: Project Management Institute.

Richardson, G.L. and C.W. Butler, 2006. *Readings in Information Technology Management*. Newtown Square, PA: Course Technology.

Schwalbe, K., 2004. *Information Technology Project Management*. Newtown Square, PA: Course Technology.

20 Procurement Management

20.1 INTRODUCTION

In many cases, acquiring products and services from an external source is a necessary activity. This can occur for many reasons. One obvious reason is that the buying organization is not technically capable of making the needed item. Also, they may not have the skill to produce the item and they might not wish to invest in the time and effort to create that capability. In other situations the question is one of a trade-off between using in-house capabilities and external sources. Some of the more common reasons for using third-party suppliers are the following:

1. Internal resources are stretched to capacity and an external vendor provides temporary relief from that constraint
2. External vendors have niche skills that are not available in-house
3. External vendors can do it cheaper because this can be a specialty for them
4. An external vendor may have more flexibility to meet the schedule required
5. The buyer organization does not want to hire staff for that class of activity.

Regardless of the reasoning, all organizations deal with third-party suppliers for various goods and services. Some projects are in themselves a procurement activity for another buyer.

20.2 PROCUREMENT MANAGEMENT

Procurement management represents the processes involved in the acquisition of a defined set of goods and/or services from a third party for use in various project activities. Implicit in this process is the existence of a business relationship that will formally define the parameters for both parties. Some of the specific conditions of the transaction regard remuneration, delivery timing, place of delivery, payment details, failure to perform actions, and any other requirements related to the agreement. In this discussion, we observe the procurement process as a rigorous activity and not as one pursued through informal means. This means that a formal contractual document will exist to satisfy the terms of the relationship.

The first question to resolve in this process is an analysis of the types of goods and services that might be beneficial for the project to procure. The listing below contains the major categories that are common for this activity.

Goods: This basically involves tangible raw materials required for execution of the project. Some of these items would be called commodity raw materials (metal, low-level components, chips, etc.) and are sold by numerous vendors with readily definable product descriptions.

Equipment rental/operation: This category represents various supporting items used in the execution of the project. This category is characterized by various tangible items of more or less durable nature, which are needed to ensure the success of the project activities. Examples are proprietary devices, machines, tools, and vehicles.

Finished goods: These items have been passed through a more complex manufacturing or processing method and would be considered more technically complicated to specify and procure. These items may or may not be stocked by the vendor sources.

Services: This group is usually characterized by some form of human service that is performed either in-house or outsourced to a third party external to the physical team. Because of the nature of this activity it can be the most complex of all procurement actions if the requirement cannot be clearly defined. Beyond simple services agreements for building maintenance, temporary administrative help, copying services, and the like, the procurement of technical third-party resources requires thoughtful procurement actions. Examples of these high-end services are as follows:

- *Consultants:* Highly specialized technical resources that have knowledge needed by the project to execute the work requirements. A typical example would be an Enterprise Resource Planning (ERP) project that needed this level of support to ensure that they are following best of class methods.
- *Information/technical/engineering:* Software development services to create, modify, or support application components of the project. Engineering services could involve any technical skill group providing necessary services to the project.
- *Certifications:* These procurement activities involve acquiring external services to independently provide evidence to the acquirer that a product meets contractual or otherwise specified requirements.

20.3 MAKE OR BUY DECISION

Perhaps the first question for the project team to decide is whether to acquire needed goods, services, or personnel from external entities, or to produce them using internal team resources. In order to make this decision, the first order of business is to decide exactly what work activities are needed to be performed and what raw input items are needed as part of that work activities. It is reasonable for a starting place to think of the WBS and its companion dictionary as offering details on this question. At this point the project team would examine these requirements and begin to analyze how best to accomplish that work activity. Each WP has inputs and tools required to complete the task. Likewise, the work defined in the WP could be viewed as best done by an external source. Each of these target issues has its own set of challenges in the procurement planning process. Questions such as:

- Will the raw materials spoil, or need special storage until needed?
- Will internal resources be available to assemble or produce finished products formed from the purchased raw material resources?
- Are semifinished goods available to make the process easier for the project team?
- Do we have appropriate internal resources with the skills or credentials required to assemble the manufactured parts into finished or usable products?
- Would it be more cost effective to receive raw materials and assemble the product using internal resources, or seek out an external vendor to perform the entire process?

Finally, one of the most difficult questions is to resolve whether the buyer organization wants to perform work of this type. It may be low skill and easy to do, but not something that is desired for the internal team. The most general and visible example for this category is to establish a policy that we will procure from external organizations all building cleaning and maintenance since the organization does not wish to staff for that type of work. For project-oriented activities, these decisions may be more subtle, but will have to be resolved as part of the make or buy activity. In some cases this decision is obvious, whereas in others more philosophical. Consider the situation of needing an integrated computer chip as part of the project work. It would not be a long debate to decide that buying one from a competent vendor is the better choice.

Many times the make or buy decision is based on whether the item required is part of the organization's core competency. An organizational core competency is defined as the collective learning and coordination skills required to produce a firm's product lines or services. These competencies represent the difficulty for competitors to imitate strengths and collective know-how and they contribute significantly to the end-product benefits. This does not necessarily mean that the firm owns all the resources that comprise the core competency set, but does represent their collective alliances and licensing agreements. In other words, core competencies represent the true strategic strengths of a company. If the project task falls within one or more of these competencies, is it generally best to seek ways to maintain that knowledge base internally.

The second procurement planning question is to establish whether there are experienced, capable vendors to support the defined need. If the organization has a formal procurement department, this is a good time to involve them with this question. They should either have such vendors on a preferred vendor list, or have capabilities to begin searching for such vendors. It is typically a good strategy to start a new vendor with small transactions and increase the size and scope of those as you gain experience with them. When the buyer has already established a tested relationship, this can be an influential factor when deciding to buy versus make internally. The test of a vendor involves its history of timeliness, quality, service, and flexibility in past relationships (Gray and Larson, 2008).

Through the make or buy process it is good to evaluate if the scope of a WP is adequately defined to either support a make versus buy decision, or whether a vendor would be able to understand the requirement. Project teams often forget that the WBS structure and dictionary should be kept up to date with changing issues and this omission can lead to a bad procurement decision. For this reason, the final test is to confirm that the required task and its output are valid and sufficiently defined. In order to do this, take the reverse view and look at the deliverable and then ask whether the underlying work process is defined.

20.4   PROCUREMENT MANAGEMENT PROCESSES

In keeping with the text theme of a model-based perspective, we need to restate that Procurement Management is one of the nine KAs in the *PMBOK® Guide*. This topic covers all the processes required to purchase or acquire products, services, or results needed from outside the project team to perform the work defined in the project scope. The guide defines four processes for this KA (PMI, 2008, p. 314).

Plan procurements—determining the what, when, and how of purchases and acquisitions.

Conduct procurements—documenting products, services and result requirements, and identifying potential sellers. Obtaining information, quotations, bids, offers, and the like from the sellers. Reviewing bids, choosing among the competing sellers, and negotiating a written contract with the selected seller.

Administer procurements—managing the contract, contract-related changes and the relationship between the buyer and seller, reviewing and documenting the seller's performance, and so on.

Close procurements—completing and settling each contract, including the resolution of any open items and then formally closing each contract applicable to the project or a project phase.

All procurement processes outlined here are discussed from the point of view of the buyer. Secondly, descriptions for the seller may also be referred to as a contractor, vendor, third party, or supplier. Thirdly, the procurement activities have more interaction with the other KAs than any others because of the nature of the transactions.

20.5   PLANNING FOR PROCUREMENT

The results of the make or buy activity target various items for third-party procurement. To proceed beyond this point several assumptions are made:

1. Assuming that a reasonable price can be obtained externally, the defined item is identified as best acquired external to the organization (considering resource constraints, core competency, valid sellers, etc.)
2. A similar vendor exists who is technically and financially viable for delivering the required item
3. The internal organization does not need more control over the item than can be dealt with through a third party. This includes intellectual property (IP), core competency, status, quality, and so on
4. Moving the defined work to a third party does not increase risk beyond the tolerance point
5. A likely third party may have a better skill capability for the defined work than exists internally.

With these assumptions, a base comparison with internal costs will be used for future procurement decisions.

20.5.1   Planning Stage Outputs

The planning stage results in three basic documents that outline the actions to be pursued going forward (PMI, 2008, pp. 317–327):

1. Procurement Management Plan (identifying types of contracts to be used)
2. Procurement Statement of Work (SOW)
3. Risk management issues
4. Make or buy decisions (procurement targets)
5. Procurement documents to be used to solicit proposals
6. Source selection criteria.

A project Procurement Management Plan is produced from information related to the target areas defined in the make or buy analysis and is designed to provide ongoing management guidance and direction to the project team in regard to future procurement actions. In some organizations, a subsidiary document called a Contract Management Plan is used to detail the overall process in regard to vendor relationships.

20.5.2   Procurement SOW

One of the most important planning documents is a work or service specification for the procurement target. This is called a Procurement SOW and one of these should be developed for each target item. This specification is drawn from the WBS Dictionary description of work, but will require more breadth and detail since the target is now to be performed externally and the work involved may be more than a single WP (Heerkens, 2001, p. 68). To be successful the contract SOW should be well defined to the degree that a vendor can reasonably use it to estimate their response. Key categories included in the specification are the following:

1. Scope of work in sufficient technical detail to support a seller response
2. Location where the work is to be performed

3. Performance data—start dates, completion dates, and so on
4. Deliverables schedule
5. Acceptance process—outlines how the deliverable will be accepted
6. Applicable standards—these can be custom defined, company standards, or industry standards
7. Other—there are many unique items that may be relevant and should be included in the document. Examples are special test equipment required, travel, experience required, certification of the seller staff, and so on. It is also good to let the seller know what you are looking for in the way of priorities—cost, schedule, quality, expertise, and the like.

At this point the technical planning process for procurement is complete.

20.6 CONDUCT PROCUREMENTS

The second major stage of the procurement process focuses on obtaining seller responses and awarding a contract (PMI, 2008, p. 313). This activity falls generally into the financial and legal side of the planning process, and key decisions to dealt with at this stage are as follows:

1. Receiving seller responses from SOWs and selecting preferred vendors
2. Negotiating contracts with preferred vendors
3. Communicating status to various stakeholders and processes

Each class of procurement activity has unique characteristics and requires similar treatment in terms of methods to find and deal with the prospective sellers. Over the years, this process has become reasonably well standardized in industry. The following are the typical procurement documents:

- Requests for Information (RFI)
- Requests for Proposals (RFP)
- Requests for Quotation (RFQ)
- Invitations for Bid (IFB)
- Invitation to Negotiation (ITN).

Each of these items is worthy of note to explain their general usage in the process.

20.6.1 REQUESTS FOR INFORMATION

This is a formatted seller response that is intended to allow vendors to uniformly describe how their solutions meet the functional and nonfunctional requirements. The vendor is instructed to provide their answers on the formatted worksheet, which is then combined with a response table received from other vendors to form a master worksheet. This worksheet will then be used to perform a comparative features analysis.

20.6.2 REQUESTS FOR PROPOSALS

Procurement items can be complex, nonstandard, and high in price. This class of items requires additional expertise from the seller, so the response requested is based not only on delivery of the item, but to help define the specification. The RFP method is appropriate when the buyer cannot write clear specifications for the work to be performed and when the seller's expertise is needed. In addition, this method is applied when the buyer wants the opportunity to evaluate vendors' offers that have different approaches, price, and quality. Evaluation criteria are listed in each RFP. Contract award is made to the highest rated proposal that may or may not result in the lowest price.

An RFP typically involves more than a request for the price. Other requested information may include basic corporate information and history, financial information (can the company deliver without risk of financial stability), technical capability (used on major procurements of services, where the item has not previously been made or where the requirement could be met by various technical means), product information such as stock availability and estimated completion period, and customer references that can be checked to determine a company's suitability.

20.6.3   REQUESTS FOR QUOTATION

For those items that are standard, off-the-shelf, and relatively low in price, an RFQ is the most desirable document. For example, when planning the bulk purchase of commodity products such as PCs, printers, modems, and applications software, an RFQ should be generated. The RFQ is used when the most cost-effective solution is the overriding concern. A prepared list of qualified sellers receives the RFQ.

Creative purchasers structure the RFQ to get additional information such as value out of the price. RFQ includes specific and detailed format for basic information, such as price, delivery charges, and other charges appropriate to the quotation. The RFQ should also provide an opening for suppliers to include anything that may affect the price or the value of the goods or services being purchased. A supplier may come back with benefits of which the purchaser was unaware. Detailed RFQs also benefit the suppliers by encouraging them to quote separately. This gives the purchaser better data for determining which services are cost effective and which are not. Additional costs such as those for transportation, packaging, and supplier stocking, and additional benefits, such as extended commitment options, cash discounts, fixed term of price, and reduced lead times can be evaluated on their own merits. The purchaser can then choose which options to take and which to leave.

20.6.4   INVITATIONS FOR BID

Procurement items are standard, but high in price. All items are clearly specified by a SOW. Governments and public agencies tend to advertise their IFBs in newspapers. The IFB notifies the potential vendors about the existence of the project. It is called open competitive selection. In this case, anyone who is interested in and qualified may want to submit a bid. Private businesses rarely advertise bids except for big projects. Some private organizations may have an acceptable bid list of potential vendors who would receive the notifications. This selection process is called closed competitive selection. In this situation, the client will contact several vendors to make them aware of the project. Only those invited to bid in this manner are allowed to bid on the project.

The IFB provides a summary of the project, the bid process, and other brief pertinent procedures for the project. It informs potential bidders of the project, its scope, and ways in which they can obtain further information. The invitation also states whether a security bond is required, how much it will be, and for how long it will be held. The size or length of time for which the bond will be held may discourage some vendors from bidding.

20.6.5   INVITATION TO NEGOTIATION

The purpose of this document is to create a control structure for the technical, legal, and financial negotiation areas. It aids in comparative evaluation of multiple vendor responses. A second major category of the document deal with the evaluation criteria related to each review area (Lamb and Hair, 2006, p. 59).

20.7   BIDDING PROCESS

Once the procurement targets are identified and the class of the activity is chosen, the next step is to seek out sellers. In some cases this will come from existing preferred sellers lists that have been derived by the procurement department, whereas in other cases the requirement will necessitate a more complex vendor search process. In all cases, the goal is to obtain an acceptable response from the seller and from this obtain sufficient information to select a vendor.

20.8 SELECTING SELLERS

Once bids have been received from potential sellers, the process moves to selection. Earlier communication with potential vendors outlined selection criteria that would be used. These criteria are now used to evaluate the bid responses. The basic evaluation categories are technical, financial, and organizational. In some cases the item will be a quick lead time, "in stock" commodity and the issue will be price. In other cases, the criteria will weigh the technical capability of the vendor. Finally, if the procurement cycle is reasonably long, the stability of the vendor becomes a critical evaluation. Obviously, if the vendor organization fails nothing else is relevant. From these evaluations, one or more candidates are selected for further detailed negotiations that are intended to lead to a formal contract.

In most cases, seller selection criteria are directly related to the critical success factors of the project and these include cost, previous business relationship experience with the seller, industry experience, qualification of seller employees, demonstrated understanding of the requirement, financial capacity, technical ability and alliances, industry ranking of its products, and so on. However, in some cases, a seller may be added to the candidate list in order to comply with some corporate or government regulation. For example, a government regulation might require the use of sellers with over 50% of their employees being local or minority vendors. Usually, the selection criteria is created and approved by management before the proposals are sent out. The intent of these criteria is to improve the selection process and strive toward objectivity and fairness. In some bid cases, the selection criteria are defined in the bid package and this helps the seller to know how to respond to the request.

For major procurement items, it is common to receive a preliminary proposal from the sellers and screen the larger group down to a smaller list of qualified sellers based on their data. At that stage, a full RFP is sent to each of these qualified sellers who then prepare a detailed proposal. Based on the evaluation of these proposals, a final short list of sellers is created from which the most qualified seller is selected. Proposals are often separated and evaluated as discrete technical and commercial sections in order to properly evaluate the proposals on the basis of their content.

A weighted selection criteria is one of the most commonly used methods of evaluating sellers. In this method, each selection criterion is assigned a weight (e.g., 1 for poor to 10 for excellent). Then an overall rating score (usually from 1–100) is produced for each seller. The seller with the highest weighted score is judged the most qualified seller. In the event that two or more sellers have the same total score, additional selection criteria may be applied to select the most qualified seller. In some cases, a vendor not rated highest could be selected, but this raises a legitimacy question regarding the evaluation process.

Other techniques that can also be used with the weighted selection criteria include a screening system in which a set of minimum requirements are established for the sellers, and those whose proposals fall short of this requirement are eliminated from further evaluation. For example, the sellers may be required to have in their proposed team some number of PMP-certified team members. The sellers who meet these criteria will form the "qualified sellers list." It is also advisable to use independent estimates to validate the bid cost for each quotation submitted by the sellers. This will help to ensure that the score assigned to the cost criterion for each seller is based on a realistic basis. This is a very important technique to use, because in most situations, the cost criterion carries the most significant weight.

20.9 CONTRACT NEGOTIATION

The role of contract negotiations is to define the specific structure of the relationship from which some form of legal contractual document will be created. Many feel that procurement negotiations are purely technical and price based. Think about this philosophy for a second. If that is all the relationship needed, why write a contract at all? It is a time-consuming process and contains a lot of words that laypeople do not understand. Michael Gold offers the opinion that contracts should be written with the perspective that the relationship will not go well and it will then fall to the contract to resolve the issue (Gold, 2005). Failure to have this protection can place great stress on the relationship, so the negotiation process is the time to deal with these issues as the parties are in a more rational state of mind.

The contract negotiation process involves three functional groups—project team for technical aspects, the procurement specialist for company policy issues, and a legal representative to ensure that the proper legal protections are included. If the item is commodity-like the negotiation is simply price related and involves a standard form. This requires very little involvement of the parties after the decision is made. However, this activity can be tedious and time consuming as the complexity of the work specification grows.

The following are the three stages of contract negotiation:

1. *Prenegotiation:* This includes gathering information about each seller and their bid proposal. Risks to the project plan are analyzed and evaluated to formulate a strategy for cost, schedule, and performance negotiation. The buyer team needs to formulate and understand the strategy and tactics to utilize in the negotiation. It also includes activities to prepare both groups for actual negotiation (meeting invitations and logistics arrangements).
2. *Negotiation meeting:* This ensures the use of proper protocol (introductions and expectations). The process will include probing, tough bargaining, compromising, closure, and final agreements (clarified and documented).
3. *Postnegotiation analysis:* Evaluate your performance in terms of the prenegotiation planning and the actual negotiations.

Contract negotiation is an art and involves significant human interaction in the form of buyer/seller diverse opinion meetings required to reach an agreement between two parties where both are attempting to maximize their value. However, the real art is finding creative solutions where both sides can leave the process as if they got what they needed out of the deal. It should not be a win–lose game as that sets up a bad attitude for the ongoing relationship. Since contract negotiation plays a crucial role in the procurement process, some mature organizations have made a determined effort to train project team members in negotiation practices in order to ensure that their organization achieves maximum value in the process. These practices include

1. Clear objectives: Make a list of goals before meeting the other party.
2. Be prepared: Do your research before going into the negotiation. Know relevant laws, facts, and figures. Also have a first draft of an agreement written before meeting with the other party and use that as your goal template.
3. Have an agenda for each negotiation session: Try to keep the discussion orderly when meeting with the other party. Make a checklist of topics that should be discussed during the session.
4. Make sure that appropriate decision makers are in the meeting: This will help to ensure that final decisions can be made without undue delays.
5. Set your expectations, but be flexible: Consider what you really need to get from the other party and also decide in what areas you are willing to compromise.
6. Build trust with the other party: Trust will aid communication and agreement.
7. Listen to the other party's concerns.
8. Focus on issues, not personalities: Focusing on your goal and treating everyone as an equal will help matters become resolved quicker. By treating all fairly, you will avoid simmering grudges and ill feelings, which can become an obstacle to agreement.

9. When negotiating force majeure (Act of God) clauses, make sure that this clause applies equally to all parties, not just the seller. Also, it is helpful if the clause sets forth some specific examples of acts that will excuse performance under the clause, such as wars, natural disasters, or other major events that are clearly outside a party's control. Inclusion of examples will help to make clear the parties' intent that such clauses are not intended to apply to excuse failures to perform for reasons within the control of the parties.

10. Speak in supportive statements: Attach credibility to your statements by speaking in facts not feelings. Avoid sentences beginning with "I think," "I feel," or "In my opinion." When stating facts, be prepared to quote your sources and elaborate or deflect questions meant to deflate your position.

11. Document meeting minutes: Document all agreements reached at the meeting and obtain buy-in of all concerned by sharing the minutes of the meeting. If possible have the contract ready at the end of the meeting or as soon as possible to get a signature to what has been agreed, so you do not have to retrace old topics.

12. End negotiations on a positive note: Shake hands and smile. Also, take honest notes to yourself on your tactics and see how you can improve for next time.

Complex negotiations involve significant time on the part of the participants and they can drag on for extended periods of time. To reach a successful conclusion, both parties must be willing to compromise. Issues of risk and return will be played out in the negotiation process. Once the various terms are finally agreed to, the last step in the process is to formalize those terms into a contractual document that is legally binding on the two parties and enforceable in a court of law.

The final deliverable of the contact negotiation process is a signed contract. This document represents a mutually binding agreement that obligates the seller to provide the specified product or service and obligates the buyer to pay for it. Depending on the type of item being procured, a contract may also be called an agreement, subcontract, purchase order (PO), or memorandum of understanding (MOU). Beyond the technical and financial aspects, a contract will include legal oriented items like effective date, scope of agreement, quality assurance process, milestones, payment schedules, warranties and guarantees, conflict resolution process, retainage, termination and cancellation provisions, penalties and fees, force majeure, confidentiality, labor rates, IP rights, applicable law and jurisdiction, and so on. Beyond these clauses, Michael Gold recommends defining breach and specific remedies for nonperformance (Gold, 2005, p. 2). Many parties stay away from this topic because of fear that these discussions will set the wrong tone. However, thinking about nonperformance at negotiation time helps both parties to think more clearly about the responsibilities that need to be included. Both parties need to think about the arrangement in terms of their required performance.

20.10 CONTRACTS

In this discussion, the term "contract" is implied to be a formal, singular type document; however, this is not always the case. Think of a contract more as a state of the relationship, not a particular document. In fact, a contract can exist from verbal conversation between two parties so long as that conversation can be verified. In any case, such contracts are dangerous because the specifics would be almost impossible to reproduce. Also, a contract can exist using documents known as agreements, POs, MOUs, and subcontracts. Regardless of the communication mechanism, there are seven elements in a contract:

1. Mutual understanding of the subject area
2. Legal offer
3. Legal acceptance
4. Consideration (something of value)
5. Genuine assent (understanding of the propositions involved; freely entered; no fraud, undue influence)

6. Competent parties (not minor, insane party, or intoxicated)
7. Legal object (not in violation of state, federal, or public policy).

It is not the role of this text to teach contract law. It is however important to understand the importance of this process. Parties that fail to understand the legal consequences of a soured relationship will spend much more than they want either defending themselves or striving to obtain penalty remedies. To counter this problem the contractual documentation should be viewed as the mechanism to structure the relationship and make it clear as to the role of both parties. Every reasonable effort should be made to define the relationship in regard to delivery of the item involved and reciprocal compensation from the buyer. As part of this effort, there are several legal terms that need to be understood and each should be considered for use in your procurement document.

1. Acceptance
2. Agent
3. Arbitration/mediation
4. Assignment
5. Authority
6. Breach/default
7. Change process
8. Force majeure
9. Indemnification
10. IP
11. Material breach
12. Mediation
13. Ownership
14. Payments (details)
15. Principal
16. Reporting
17. Site access
18. Time is of the essence
19. Warranties
20. Privity.

A more detailed definition of each term is included in the appendix to this chapter.

20.11 ADMINISTER PROCUREMENT

After the contract has been formalized the administration period begins. This phase is a monitoring and control-oriented activity and is designed to ensure that the seller's performance meets the contractual obligations. This involves performance monitoring and reporting on cost, schedule, and results. It may also involve the following:

- Quality control to assure conformance to requirements
- Change control related to the project change control process
- Financial control (payments)
- Compliance with special terms for payment and warranties
- Audit activities related to a contract.

Individuals performing the contract administration function should be aware of the legal stipulations in each agreement and know when to elevate deviations for management review. This phase of the procurement relationship can be one of the most troublesome if change in specifications, seller stalls, quality issues, and other variances emerge during the delivery cycle. The main focus of contract administration is to ensure that the buyer acquires everything defined in the contract in regard to quantity, quality, time, and cost constraints. Related to this activity is a stream of formal delivery and payment documentation that needs to be carefully matched.

20.12 PROCUREMENT AUDITS

Procurement audits involve a structured analysis to identify lessons learned and to document the successes and failures of the project (Heldman et al., 2007). In this process, the team gathers information and evaluates the projects' goals versus the outcome of the product or services including all activities and processes that were undertaken during the project. It is important to include any corrective actions taken, outcomes, unforeseen risks, mistakes that could have been avoided, and causes of variances. Whatever the outcome of the project, successful or failure, it should be documented and recorded in the formal archives of the organization. Future team projects can learn from past experiences and make improvements. The PM has to be humble and honest in order to document mistakes that occurred, even if he/she does not want to admit to them. Everybody makes mistakes and the key is to learn from them, not to be afraid of criticism, but gain knowledge and an opportunity to learn and do better next time. Of course, this can only happen when upper management encourages an environment of responsibility linked with trust and a space to make your own decisions.

Even if both parties work to achieve defined schedules, there are several external factors that influence and can affect the project time frame, such as forcing the parties to consider an extension, or even an early termination. They need to evaluate in what part of the project they are to decide if it is more costly to continue with a new end date, or force an early termination and closing the contract and finances at that point. The cost considerations are an increase in budget, penalties, and legal fees.

20.13 CONTRACT REVIEW AND REPORTING

Throughout the contract administration period both parties must evaluate if the overall performance of both parties meets the contractual obligations. Three major control functions must be evaluated: quality control, change control, and financial control. Quality control evaluates the process as producing correct items per the specification and quality assurance procedures must in place handle an overall quality review from the buyer standpoint. Philosophically, the quality process should evaluate whether the vendor has designed quality into their product, and then created that quality into the product during their execution process. One of the major administrative issues concerning quality control is making sure that the proper quality measurements are in place. A poor quality control process can lead to an ineffective product to be dealt with by the buyer. An effective quality control and quality assurance process in the vendor organization will help avoid this problem. For complex procurement, this implies that the buyer must have some access to the internal operations of the seller to confirm this process state.

The change control process must be documented in the original contract and this involves a mini process similar to the original procurement. This means that a captive seller must also agree to this aspect of the contract. Changes often create variances in overall cost and schedule for the deliverables. They also complicate the interpretation of the terms as they stream through time. Changes can relate to anything in the scope of the contract, not just legal requirements, payments, and other small elements. They can relate to the defined work environment, regulations, or business landscape. Through all of this, correct procedures must be followed to ensure that the contract is still valid from both sides. All of the process elements outlined for a legal contract must remain for each of the changes.

The following checklist should be used for each change request:

1. Is the change really necessary?
2. Does the change deliver benefit or value?
3. What is the cost of the change?
4. Has the change been approved by all relevant parties? Have all the contractual issues been reviewed by technical, legal, and procurement staff?
5. Can all the parties still meet the revised obligations with the contractual changes?

The third segment of contract administration is financial control. The purpose of this process is to establish effective cost control and to trigger the vendor payables side of the relationship. The following should be addressed: current authorized budget, expenditures to date, commitments, agreed variations, potential claims, pending change orders yet to be approved, and future changes anticipated.

20.13.1 Record Keeping and Audits

A major component of contract administration is record keeping and audits as the project moves its life cycle. In some situations, an independent auditor will be utilized to handle the review and final processing of procurement contracts. The DoD uses the Defense Contract Audit Agency to perform all of its contract records and auditing process. "These services are provided in connection with negotiation, administration, and settlement of contracts and subcontracts" (DCAA, 2008, Chapter 1). All audits begin with a statement of objectives and those objectives determine the type of review to be performed. The Contract Audit Manual states that "Audits vary in purpose and scope. Some require an opinion of the adequacy of financial representations; others an opinion on compliance with specific laws, contractual provisions, and other requirements; others require evaluations of efficiency and economy of operations; and still others require some of all of these elements." The auditors should review the legality of the contract and verify that the work was performed as specified. If no seller work was performed and only goods were delivered, the auditor should be able to physically find the delivered goods. A full audit of contract status is a good measure to ensure integrity of the procurement process since there is obvious fraud potential in this activity.

20.14 CLOSE PROCUREMENTS

Formally closing the contract is the final step in the procurement process. In this step the buyer verifies that all tasks and requirements defined in the contract have been produced and the contract is completed. Once this process is done, the buyer provides the seller a formal written acceptance stating that the products, services, or other items that the seller was responsible for delivering were completed within the time frame agreed, with the quality desired. The buyer asks the seller to respond to this letter stating that they have received payment as specified in the agreement. Failure to produce the formal acceptance letter and corresponding payment confirmation leaves the door open for future claims, so this activity needs to be pursued to avoid that later when the project team is dispersed and the corporate memory on these events has faded. Outputs of the contract closure process are the formally closed contracts with signed formal acceptance letters from both sides and updates to the organizational asset documentation. All of the contract documents, approval letters, change records, and any other supporting documentation are then moved to an archive contract file usually maintained by the procurement department.

Keeping with the project management model of striving for continuous improvement, the lesson learned report should also be filed outlining vendor performance, process issues, and any other items that would aid future projects.

20.15   PROCUREMENT OF HUMAN SERVICES

Up to this point, we have attempted to describe a classic model-based procurement cycle. The general implication is that some tangible goods were involved. This process is the most standardized and mature KA of project management. This does not say that procurement cannot be mishandled, but standardized practices are defined and used for it. When the item being procured is a human-oriented service, the stability statement is no longer accurate. There are many forms of human services. A typical one is to "farm out" some internal process to an external vendor. This is called outsourcing among other names. Case in point is the trend of outsourcing various information technology services to India, China, and other low labor cost areas of the world. Most of these become complex procurement activities for many subtle reasons. First, dealing in international procurement is complex, but the most typical hidden complexity is involved with the specification issue. First, it is difficult to write the specification for a business process that will likely change over time such as Information Technology. Even more difficult is to describe how to handle the evolution of the process over time? Many firms have been attracted to lure of cheap foreign labor and jumped into outsourcing agreements only to find that they did not understand fully what they were contracting. In spite of some notable failures in this class of procurement there continues to be a migration to global human services outsourcing in many areas—IT, engineering, manufacturing, administrative, and more recently medical. Based on these trends, the PM needs to understand the environment of global human services procurement. This understanding involves more than a comparison of internal versus external labor rates for a defined task. Communications and cultural differences can add to additional cost to the process and sometimes reduce effectiveness.

Procurement of HR leads us to one of the modern challenges of both project and procurement management. Traditionally, these resources were internal project team members. Now, they are 8000 miles and eight time zones away, speaking English as a second language. They are at arm's length in every sense of the word and this adds significant issues to the overall process. The relationship possibilities for this class of procurement are endless, but structuring these agreements and plans in such a way so as to make the outsourced supplier effective and manageable is a great challenge. Interestingly, this text was edited in India at arms length to both the author and the publisher who met only through email.

Putting aside issues of managing outsourced HR, and the costs involved, leaves us with the final issue: the risks in outsourcing. In some cases, outsourcing component parts of a project may effectively reduce risk by supplying needed technical capacity and removing schedule constraints. However, the loss of control and status visibility for these tasks leaves success in the hands of the outsourcer and may well introduce a new set of risks.

One of these risks comes in the form of lost intellectual property (IP). Organizations that indiscriminately outsource projects with patented, sensitive, confidential, or critical IP content are courting balance sheet disaster. Nevertheless, this is happening on a routine basis today in companies around the world. If an outsourcer can learn the organization's marketable IP, what is to be done to keep that organization from becoming a competitor? Today, we can see a great deal of this from China and Japan. Several years ago, the United States sent manufacturing work to these locations because of cheap labor, only to find later that some of these organizations captured the underlying IP. Later, they became not only competitors but also owners of the market. This "bleeding" IP process needs to be considered by the buyer even more when human skills are involved. In general, using external resources to process sensitive information and technologies should be avoided, as this can risk the security of the company itself. Beyond this, quality of the result can be compromised with an improper vendor choice.

Regardless of whether the procurement target is human services or tangible products, the risk analysis for that decision is important. The ultimate goal of procurement planning should be to reduce the uncertainty in the project plan as well as to protect the organization from losing its competitive position as a result of this decision.

20.16   RANKING VENDOR PROPOSALS

Prior to sending out a vendor proposal, you need to have an agreed-upon rating/ranking system that is fair and objective. This will help remove bias or politics during the selection process. Use of a weighting worksheet similar to the sample below can help rate a vendor response. This evaluation approach helps to assess the vendor strengths and weaknesses by category. Keep in mind that the main objective is choosing a vendor that provides the best overall solution based on the criteria that is most important to your organization.

20.16.1   INSTRUCTIONS

The sample worksheet shown in Table 20.1 contains 19 evaluation criteria, their assigned weight scores range from (0–4), and evaluation rankings are also rated 0–4. Note in the calculation column that the weighted score is computed by multiplying the weight times the rank. In the sample calculations, vendor X has the highest weighted score (135) and would be considered the best fit.

As with all evaluation worksheets, the weight values and criteria can be modified for a particular proposal or changed for all proposals depending on the needs.

Use the last two columns to score your sample rating by criteria and then calculate your vendor score.

20.17   SUMMARY

Executing the Procurement Management Plan ensures that the goods and services required to make the project successful are made available, on time, within budget and aligned with the project goal. It involves a series of processes that include requesting seller responses, selecting sellers, contract administration and contact closure. The project team can leverage the expertise of specialist from the functional departments within the organization, for example, procurement, contracts, and legal. These will serve as resources to the project team in developing procurement documents like the RFP, RFQ, IFB, Contract Terms and Conditions, POs, and so on. They can also participate in some of the activities like the contract negotiation, the selection process, contract administration, and closure. This is much easier in organizations that have these departments in place and structured in a manner that enables them to work with various project teams. However, since these resources belong to the functional departments and report to their functional managers, it may pose some challenges with regard to their availability and commitment to the project team; hence some projects add these resources to the project team at this stage of the project.

Outsourcing is an option that can be used to execute the procurement plan. It enables an organization to leverage the expertise of a contractor in procuring goods and services. It also transfers the risk of the procurement process to the contractor based on an agreement. Outsourcing helps to reduce overall cost of production for the organization and directly reducing time for the product delivery to the market. Outsourced contracts needs to be well managed to avoid delays that would impact the project delivery date, which could lead to increased overhead cost. However, some variances to the contract may occur on the part of the seller that will need to be addressed by the project team. For example, the outsourcing contractor for a services contract may not be delivering the right personnel for a project and the project team may have to work with the HR department to obtain some in-house resources to accommodate the gaps. Or the seller is not delivering goods according to the schedule, which may impact the overall project schedule. In this case, the project team will need to adjust their schedule, cost, and communication plans to accommodate the delays. Similarly, the risk that the seller may go bankrupt during a project thereby posing a threat to the supply of resources is something that may need to be put into consideration as part of the Risk Management process.

To be able to successfully execute a procurement plan we recommend following a certain number of best practices to make the plan as effective as possible. When the plan calls for outsourcing

TABLE 20.1
Rating Vendor Proposals

Weighted Proposal Ranking Matrix—Unix Servers

Category–Criteria	Weight 0–4 See Key	Vendor X Ranking 0–4 See Key	Vendor X Weighted Weight × Ranking	Vendor Z Ranking 0–4 See Key	Vendor Z Weighted Weight × Ranking	Your Vendor Ranking 0–4 See Key	Your Vendor Weighted Weight × Ranking
Technical							
Long-term viability	4	2	8	2	8		
Performance	4	2	8	3	12		
Robust administration tools	4	3	12	1	4		
Processor technology	3	3	9	4	12		
Clustering solution	2	3	6	2	4		
Handling unplanned downtime	2	2	4	2	4		
Fault tolerance and disaster recovery	1	2	2	3	3		
Weighted technical total			49		47		
Vendor Support							
Global presence	4	2	8	3	12		
On-site on-line support	3	3	9	2	6		
Ease of doing business	3	3	9	4	12		
Support contract options	3	3	9	3	9		
Self-support programs	2	2	4	1	2		
Flexibility to negotiate terms	2	3	6	3	6		
Record of promised functionality	2	3	6	2	4		
Professional services	1	2	2	2	2		
Service performance reviews	1	1	1	2	2		
Weighted vendor support total			54		55		
Cost							
System	4	2	8	1	4		
Upgrade	4	3	12	2	8		
Support	4	3	12	1	4		
Weighted cost total			32		16		
Weighted proposal total			135		118		

Source: This worksheet authored by Ron Smith and Mike Sowers has been adapted from a similar version originally published in Baseline Magazine, December 2008. The authors approved this version.
Key-weight: 0—intangible; 1—nice to have; 2—desirable; 3—highly desirable; 4—mandatory.
Key-ranking: 0—does not satisfy requirement; 2—satisfies requirement; 4—surpasses all aspects of requirement.

it is best to use a partner approach, provide colocation when needed, and establish long-term outsourcing relationships. These in combination with the above best practices will help your organization make the most of outsourcing. When the RFI/RFP stage comes along it is best to use the RFI to gather information and the RFP to secure the highest quality proposals. It is highly important to give your partners strong and adequate timelines, be firm with your dates. When contracts come around it is most important to do complete due diligence from your side of the contract. Make sure all items are ironed out in the contract no matter how small they seem. Make sure that the contract has clear objectives, is prepared correctly, and is done with trust in your partner. It is also important to focus mostly on the main issues and to set expectations but be flexible. In the administration of the contract make sure that both parties are aware of all of the legal implications and the contract is set up correctly for the current situation. It is extremely important to set up quality control, change control, and financial control to help manage the contract and mitigate any problems throughout the project. Adhering to these best practices will significantly increase the likelihood of executing a successful procurement plan.

DISCUSSION QUESTIONS

1. Core competencies are the strongest skills and assets of a company. Would you think it wise to target improvements in new skills that could broaden those competencies, or should the organization outsource this class of activity assuming that it could be done with no significant risk and at least equal cost and quality?
2. Besides cost, what other sacrifices do you potentially make when you choose to outsource a human-oriented service?
3. What are the advantages and disadvantages of outsourcing?
4. What are the various types of functions and processes that can be outsourced? What distinguishes the better targets?
5. How do you differentiate when to use an RFQ or IFB?
6. What are the critical items that need to be included in an RFP?
7. Which would you utilize if you were trying to give the seller an opportunity to negotiate and work with you—RFQ, RFP, or IFB?
8. Which type of contract provides the highest risk to the buyer?
9. What is a weighted selection criterion and how is it used in the selection process?
10. What is the role of contract negotiation?
11. What are the primary functions of procurement administration?
12. What can happen if contract administration activities are not performed?
13. What is the most important activity involved in formal contract closure?
14. Why do you think it is important to document and archive all activities of the procurement process?
15. What document should be created before procurement planning begins?
16. What are the primary reasons for outsourcing HR?
17. What type of contract minimizes the risk of supplier cost overruns assumed by the purchasing organization?
18. What other costs, besides salary, should be factored into personnel outsourcing decisions?
19. What sources may be used to collect a list of vendors?
20. What are the most important parts of a contract?

GLOSSARY OF PROCUREMENT TERMS

Contract negotiation: A procurement activity designed to produce a formal contract results from a bid that may be changed through bargaining. It involves clarification and mutual agreement on the structure and requirements of the contract prior to signing.

Cost-plus-fixed-fee (CPFF) contract: Cost may vary but fee remains firm.

Cost-plus-incentive-fee (CPIF) contract: Contractor reimbursed 100% of costs. Contractor fee varies between maximum and minimum fee and can have multiple incentives.

Cost-plus-percentage-fee (CPPF) contract: Allows flexibility as the contractor and owner work together on all costs of the project.

Fixed price (FP) contract: Contract with a fixed price or lump sum.

Fixed-price-incentive-fee (FPIF) contract: Target price = target price + target profit. Contractor pays sharing ratio of costs above target cost. Final price never exceeds ceiling price.

Force majeure: Force majeure literally means "greater force." These clauses excuse a party from liability if some unforeseen event beyond the control of that party prevents it from performing its obligations under the contract. Typically, force majeure clauses cover natural disasters or other "Acts of God," war, or the failure of third parties—such as suppliers and subcontractors—to perform their obligations to the contracting party. It is important to remember that force majeure clauses are intended to excuse a party only if the failure to perform could not be avoided by the exercise of due care by that party.

In sourcing: In sourcing is the opposite of outsourcing; that is, in sourcing can be defined as the allocation of work within a project to an internal entity that specializes in that operation. In some cases, this implies that the work is being moved back from a previous outsourcing arrangement.

IP rights: These are a bundle of exclusive rights over creations of the mind, both artistic and commercial. The former is covered by copyright laws and the latter by patents (Ward, 2000).

IFB: The IFB is one of the solicitation documents that companies use in the procurement process when the procurement items are standard, but high in price.

Milestone: An identifiable point in a project or set of activities that represents a reporting requirement or completion of a larger or more important set of activities.

Outsourcing: Outsourcing is subcontracting a process, such as product design or manufacturing, to a third-party company.

Procurement audits: These are structured analyses that identify lessons learned and are used to document the successes and failures of the project (Heldman et al., 2007).

RFI: The RFI is one of the solicitation documents that organizations use to gather information about the potential sellers.

RFP: The RFP is one of the solicitation documents that organizations use in the procurement process when the procurement items are complex, nonstandard, and high price.

RFQ: The RFQ is one of the solicitation documents that organizations use in the procurement process when the procurement items are standard, off the shelf, and relatively low price.

Retainage: In the contracting business, retainage refers to a portion of the payment that is withheld until the completion of a project. The client does not pay the contractor the retainage until all work on the project is complete. Retainage is negotiated up front and is stated as a percentage of the overall cost of the project. A common retainage amount is 10%. Retainage incentivizes the contractor to provide quality work up until the very end of the project.

Selection criteria: This is a set of criteria developed by the project team and approved by management to be used when evaluating the proposals or bids from a qualified list of sellers.

APPENDIX: LEGAL TERMS

The following terms are commonly used in procurement contract language.

1. *Acceptance*—deals with the processes by which the deliverables are accepted by the buyer. An acceptance may be conditional, express, or implied (West's Encyclopedia of American Law).

    Conditional acceptance—acceptance of an offer provided that certain conditions are met (such as specific changes in terms of an agreement).

    Express acceptance—a definite and clear acceptance of an offer without added conditions (as-is).

2. *Implied acceptance*—an acceptance of an offer demonstrated through acts that indicate that the accepting party consented to the terms.

3. *Agent*—any person who is authorized to act on behalf of another party. This representation by the agent brings with it the power to bind that party into contracts, and tie them to liability by the agent's actions (The People's Law Dictionary).

4. *Arbitration*—a small version of a trial held outside of court, conducted by a person or panel who are not judges, in an attempt by the parties in dispute to resolve their issues without going to formal trial (The People's Law Dictionary).

5. *Assignment*—the transfer of rights, interests, or benefits from one benefiting party to another (The People's Law Dictionary).

6. *Authority*—permission to act, or order others to act on behalf of another party (The People's Law Dictionary).

   Apparent authority—authority given to an agent through various signs from the principal to make others believe that the agent has authority.

   Express or limited authority—authority that has certain, defined limitations.

   Implied authority—authority granted from the position the agent holds.

   General authority—"broad power to act for another."

7. *Breach*—a failure to meet agreement obligations, willingly or not (The People's Law Dictionary).

8. *Change process*—this clause defines how contract changes will be processed and names individuals who have authority to approve those changes. The implied authority in this case would be anyone on the buyer team who holds themselves as an agent of the buyer would be accepted by the seller as a legitimate decision maker. This is not a situation that should be left undefined in the contract.

9. *Default*—a failure to respond; to make a payment when it is due; or to file an answer or other response to a summons or complaint (The People's Law Dictionary).

10. *Force majeure*—an event that is caused by forces of nature (West's Encyclopedia of American Law).

11. *Indemnification*—guarantee through compensation for losses or damages (e.g., insurance) (West's Encyclopedia of American Law).

12. *IP*—intangible products of the human mind and creativity such as patents, trade secrets, copyrights, and trademarks (West's Encyclopedia of American Law).

13. *Material breach*—a serious enough breach to destroy the value of the contract, which often leads to legal action (lawsuit) (WordNet 3.0).

14. *Mediation*—the process is designed to use a third party to help resolve disputes of either legal or contractual interpretation. The mediator works to find points of common agreement and helps to find some fair resolutions. Mediation differs from arbitration where the third party (arbitrator) acts much like a judge in an out-of-court less formal setting (The People's Law Dictionary).

15. *Ownership*—legal possession of an entity through title and legal rights (The People's Law Dictionary).

16. *Payments* (details)—deliveries of money or equivalents by indebted parties to parties to whom the deliveries have been promised (West's Encyclopedia of American Law).

17. *Principal*—the grantor of authority to a person or party to act on his or her behalf (West's Encyclopedia of American Law).

18. *Privity*—(access the fourth parties)—"A close, direct, or successive relationship; having a mutual interest or right" (West's Encyclopedia of American Law).

19. *Reporting*—"An official or formal statement of facts or proceedings. To give an account of; to relate; to tell or convey information; the written statement of such an account" (West's Encyclopedia of American Law).

20. *Site access*—permission and freedom to visit a particular location (West's Encyclopedia of American Law).
21. *"Time is of the essence"*—"A phrase in a contract that means that performance by one party at or within the period specified in the contract is necessary to enable that party to require performance by the other party" (West's Encyclopedia of American Law).
22. *Warranties*—statement, assurance, or guaranty that the quality of an item is good, or that a contractual fact is true (The People's Law Dictionary).

REFERENCES

Defense Contract Audit Agency (DCAA), 2008. DCAA Contract Audit Manual.pdf. http://www.dcaa.mil/cam.htm (accessed January 9, 2009).

Gold, M.L., 2005. They signed what? A customer's approach to software development contracting. *Cutter IT Journal*, 1.

Gray, C.F. and E.W. Larson, 2008. *Project Management: The Managerial Process*, Fourth Edition. New York: McGraw-Hill/Irwin.

Heldman K., M.B. Claudia, M. Jansen Patti, 2007. *PMP-Project Management Professional Exam—Study Guide*. Indianapolis: Wiley Publishing, Inc.

Heerkens, G., 2001. *Project Management*. London: McGraw-Hill Professional.

Lamb, Jr., C.W. and J.F. Hair Jr., 2006. *The Distribution of Request for Proposals, Essentials of Marketing*, Fifth Edition. New York: McGraw-Hill.

The People's Law Dictionary. S.v. Agent. http://legal-dictionary.thefreedictionary.com/Agent (accessed December 5, 2008).

PMI, 2008. *A Guide to the Project Management Body of Knowledge (PMBOK® Guide)*, Fourth Edition. Newtown Square, PA: Project Management Institute.

Ward, J.L., 2000. *Project Management Terms: A Working Glossary*, Second Edition. Arlington: ESI International.

West's Encyclopedia of American Law, Edition 2. S.v. Acceptance. http://legal-dictionary.thefreedictionary.com/acceptance (accessed December 5, 2008).

WordNet 3.0, Farlex clipart collection. S.v. Material Breach. http://www.thefreedictionary.com/material%20breach (accessed December 5, 2008).

21 Quality Management

21.1 INTRODUCTION

This chapter emphasizes the importance of and need for project quality management. Major components of this KA are defined as processes for quality planning, quality assurance (QA), and quality control (QC). Explanation of the basic tools and techniques for QC, such as statistical sampling, Pareto analysis, Six Sigma, QC charts, and testing, is provided. The evolution of this topic from its difficult earlier starting period into the international breadth and popularity of Six Sigma today is also traced. All of the various conceptual threads of the quality movement have had a positive impact on the organizations that have been able to successfully implement them; however, the implementation process is often complex and unsuccessful. Quality management should be viewed as a journey rather than a one-time event.

21.2 EVOLUTION OF QUALITY

Quality issues are observed in all aspects of society. We talk about quality in judging art, evaluating the products we make, or describing our experiences. In its broadest sense quality is that which adds value and improves our lives. There is always a vague indefinable immeasurable side to quality and all cultures view quality somewhat differently. We see examples of this in the Australian aborigines, the nomadic tribes of Africa, the Great Wall of China, the Christian cathedrals of Europe, and monuments around the world. Each of these examples is viewed as having a unique form of quality, but each has its own cultural, social, and personal interpretative facets attached to it (i.e., size, beauty, function, etc.).

Achieving high quality requires both art and engineering because we cannot completely separate the aesthetic, indefinable, irreproducible side of quality from the engineering side, which contains the technical, more definable, more measurable, and more reproducible side.

Today, many companies improve their competitive position through an unplanned political process of new idea pursuits. Most organizations have a defined process for proposing an improvement, demonstrating that it has value and then putting it in place through a project-type initiative. Through the evolving stages of quality, we have seen that high quality is achieved by an understanding of certain universal, unchanging principles combined with a practical application of those principles to changing circumstances. Over the past 100 years we have seen these concepts mature and evolve in various ways.

The key to understand quality management is to understand how organizations through time have slowly developed new ways to apply the scientific method to their situations.

Initially, the scientific revolution (roughly 1600–1687) changed our perception of the universe. This was followed by the industrial revolution (roughly 1760–1830, and beyond), which brought a new generation of mechanical technology to our lives. During these periods the relationship between science and industry, particularly the relationship between the scientific method and business management, formed the conceptual foundation for modern quality management principles. This topic

TABLE 21.1
Early Contributions to Quality Management

Year	Person or Organization	Matching Idea/Achievement
1755	John Smeaton	Applied the scientific method to engineering problems
1788	James Watt	First feedback device
Around 1850	U.S. military	Defined the first industry standard
1876	Thomas Alva Edison	R&D laboratory for new inventions and viable commercial versions of products
Early1900s	Assembly lines	Inspection, followed by rework or discard
1911	Frederick W. Taylor	Scientific management
1920s–1950s	Walter Shewhart	SQC
1920s–1980s	Walter Shewhart and Edwards Deming	PDCA model
1940s–1980s	Edwards Deming	Quality principles and management

evolved as engineers, managers, and functional organizational groups responded to the problems of their day. Table 21.1 provides a list of the early quality pioneers who contributed classical perspectives that evolved into quality management as we know it today.

Organizations seek to solve problems related to effectiveness, efficiency, and quality. Through all of this history the theory has remained far ahead of practice. In fact, many of today's quality problems could be solved by proper application of Taylor's 1911 treatise on Scientific Management.

21.3 DEFINITION OF QUALITY

The definition of quality offered here is derived from a combination of four disciplines—Philosophy, Economics, Marketing, and Operations Management. Although such a diverse combination seems unrelated, combining their respective views provides a perspective that includes customer focus, consumer satisfaction, standards, and production efficiency. *Philosophy* offers the broad perspective of human variance. *Economics* looks at quality in terms of value and the fulfillment of needs. *Marketing* looks at customer value and the customer decision-making process to better understand how customers define quality and choose value. *Operations Management* views quality as conformance to specifications with a manufacturing view. Philip Crosby provided one of the classic definitions of quality as

Quality means conformance to requirements (Crosby, 1979).

It does not matter whether or not the requirements are articulated or specified; if a product does not fully satisfy the customer, it lacks quality in some respect.

Other common definitions of quality are the following:

- The degree of excellence of a product or service
- The degree to which a product or service satisfies the needs of a specific customer
- The degree to which a product or service conforms to a given requirement.

Eight lesser used variants of a quality definition are summarized below. These definitions say that quality is

- The ongoing process of building and sustaining relationships by assessing, anticipating, and fulfilling stated and implied needs.

- The customers' perception of the suppliers' work output value.
- Nothing more or less than the perception that the customer has of you, your products, and your services.
- The extent to which products, services, processes, and relationships are free from defects, constraints, and items, which do not add value for customers.
- A perceived degree of excellence as defined by the customer.
- Do what you have to do when you have to do it to satisfy your customer's needs and make your product or service do what it is supposed to do.
- An ever-evolving perception by the customer of the value provided by a product. It is not a static perception that never changes but rather a fluid process that changes as a product matures (innovation) and other alternatives (competition) are made available for comparison.
- In the eyes of the beholder. And in a business environment, the beholder is always the customer or client. In other words, quality is whatever the customer says it is.

From this collection of definitions we can see that the view of quality is quite varied and that makes the theoretical discussion more difficult to structure. The following are the three basic concepts that exist in most views of quality:

- Level of goodness
- Customer satisfaction
- Conformance to requirements.

However, it is important to add to these points that if the original specification is not defined in a proper manner, meeting an inaccurate specification with a quality process will not produce a quality result. One way to develop specifications that will satisfy the customer is to clearly identify the properties that are desired in the final product. In this sense, the term "property" might be thought of as a generic attribute such as strong, durable, or smooth; however, these properties must be translated into some more quantifiable characteristics that can then be engineered and tested to verify conformance. The term "quality characteristic" is used whenever reference is made to a value that is measured for either quality (process) control purposes or to assess product acceptability.

Quality for the customer means that, in selecting and buying the product or service the customer has a hassle-free experience, and in using the product or service it meets or exceeds expectations for as long as they want. If we are providing quality for customers, then, at any moment during or after the process, they would buy more from us or recommend us to others. Key issues include identifying the target market and the needs of that market, establishing effective communication with customers or customer representatives to develop good requirements, providing high-quality sales and customer service so that the customer likes doing business with the company and the interaction, as well as the product or service, and carrying out all these affordably.

Technical project groups can only deliver quality to the customer if the requirements are an accurate, complete, clear representation of the wants, needs, and expectations of the targeted customers and all stakeholders. For the most part, delighting the customer by meeting all needs, wants, and expectations in the product and providing a high-quality customer service experience is strategically beneficial for the company. But, as is the case with many business decisions, success requires balance. It is possible to become too customer focused. All of our quality efforts should focus on delighting the customer and we should be able to demonstrate how each quality improvement will benefit the customer, or not reduce customer value. The other side of this equation is the business requirement to receive commensurate return from the customer. Extreme customer delight can only occur when the quality product is free, thus operational quality requires balance between these two forces.

21.4   PROJECT QUALITY MANAGEMENT

Project quality management is a somewhat difficult area to define. Its primary purpose is to ensure that the project will satisfy the needs for which it was undertaken. To accomplish this, the project team must develop good relationships with key stakeholders and understand what quality means to them. It is through this relationship path that quality will be defined. Many technical projects fail because the project team focuses only on meeting the written requirements for the products originally specified and ignoring other stakeholder needs and expectations. For example, the project team should know what the value will be in customer terms if the item is delivered as specified. Based on the perspectives outlined here, quality must be on an equal level of importance with project scope, time, and cost.

21.5   QUALITY PERSPECTIVE

Any organization that is serious about success must define its quality goals to be consistent with their core customers' needs and the strategic goals of the organization. These goals are decomposed into a set of standard quality requirements that are an integral part, perhaps even the end goal, of the organizational or project quality plan.

For many organizations, minimum quality requirements or merchantability standards are established by a regulating body within the industry and/or government organizations. However, mature organizations in many industries understand that minimum requirements are not sufficient to meet the growing desires of customers in a competitive environment. On the other hand, an organization must be cautious in establishing quality requirements that are impossible to attain or cost prohibitive for the market in which they are competing. The PM and the organization must ask what existing standards are in place for comparison and determine if they are reasonable to achieve. The project team must be familiar with the standard quality requirements that apply to the project and carefully assess how they align with the expectations of the customer for the particular project. Allocating time during the planning phase of the project to negotiate any gaps with the customer and then assessing the impact on the technical direction of the project are important for mutual understanding of expectations and for establishing a realistic scope, budget, and schedule. Some projects, such as those pertaining to the development of a new product or service, may have minimal input from the customer but are highly dependent on winning future business by properly anticipating the needs of these potential customers. The traditional approach of simply matching output to original specifications may not be sufficient or applicable, and therefore the project team must ask the following questions in establishing their project quality requirements:

- What will be controlled?
- How will quality be defined and measured?
- Is sampling acceptable and how will it be carried out?
- What are the standards to be met?
- How are the measures to be compared with the standard?
- How are deviations to be corrected?

Establishing and understanding quality requirements is a key planning step for any PM. It is also incumbent upon the PM to ensure that the customer is also clear on the level of quality that has been agreed upon. Without this foundational agreement, the probability and impact of scope creep due to quality-related misunderstandings can escalate beyond the PM's control. The requirements are baselined and managed with Integrated Change Control as any other project planning output.

21.6 IMPLICATIONS FOR PROJECT PLANNING STAGE

Feedback (lessons learned) from previous projects can add insight on achievable quality. Planning involves identifying the products (deliverables) at the start of the project and deciding the best steps to verify and validate them so that they meet the standards. Applying resources toward output monitoring during each phase will help us identify the problems sooner.

Project quality and deliverable quality are two different facets of focus. Project quality refers to following the correct project management practices and complying with the company objectives (QA), whereas deliverable quality refers to the correct product or deliverables that meets user's needs (QC). A high project quality may have low deliverable quality and high deliverable quality may have low project quality. The PM must manage both aspects of the effort.

The PM and his quality team must be aware of the current quality policy level of the company and evaluate whether this level is appropriate in meeting the quality level specified by the customer. If there is any doubt of meeting customer expectations, then the project management team must be aware of this and take steps toward improving the existing quality approach. This may require creating new procedures or necessitate more resources or tools to improve the quality level. Embedded in this decision process are the associated cost and schedule impacts related to the quality level (cost of quality). An appropriate balance between quality and other performance variables needs to be carefully analyzed and communicated with senior management and the customer. From this, a formal agreement should be documented as trade-offs in some dimension are often required.

Project quality management processes help prevent recurring problems by organizing and managing resources to ensure that project deliverables are completed on time, within budget, and are of high user-perceived quality.

The main tenets of project quality management are as follows:

- Customer satisfaction: Customer satisfaction requires the understanding, evaluation, definition, and management of expectations so that appropriate requirements are established.
- Prevention inspection: Prevention over inspection is the commonsense principle implying that the cost of preventing mistakes is generally much less than the cost of correcting them.
- Management responsibility: Management's responsibility in quality management is to provide the resources needed to sustain success and protect the project team from environmental disruption.
- Continuous improvement: Continuous improvement basically involves following the plan–do–check–act (PDCA) cycle of quality improvement.

The underlying support processes to accomplish this goal are quality planning, QA, and QC that will be discussed in more detail in the sections below.

21.7 QUALITY PLANNING

Quality planning plays a significant role in a variety of business processes. Each organization defines their view of quality in terms that deliver the greatest alignment with their unique business values. For this reason, it is difficult to create a single definition of quality that covers every aspect of every organization. Therefore, it is more appropriate to discuss the framework of quality planning by dividing the topic into two specific components: quality policy and quality objectives.

21.7.1 QUALITY POLICY

The guiding principle behind organizational quality involves creating policies that support specific business values. These policies are produced through the use of quality management processes that seek to deliver substantial benefits by improving the performance of an organization. A number of standard documented approaches for quality policies are available to benefit a broad range of business processes (Microsoft, 2008).

Quality management also refers to any systematic, data-driven approach that organizations use to improve business operations. A few of the more well-known quality management methodologies include Malcolm Baldridge, LEAN, TQM, Balanced Scorecard (BSC), and Six Sigma. Each of these methodologies provides a wealth of distinct and resource-rich information for their quality niche segments.

The majority of these methodologies are comprised of procedures for defining and monitoring key business processes, performing record keeping, checking for output defects, reviewing individual processes, and facilitating continuous improvement. Many businesses combine industry standards and methodologies with their own culture to create local quality polices that align with their own organizational goals. The advantage of using a commercial product as a base for this is that it has been extensively tested and provides quality specifications that are known to be competitive.

21.7.2 Quality Objectives

The first step in establishing quality objectives is to define the organizational objectives for quality. As a prerequisite, the organization has to be prepared to answer the following types of questions:

- What is our objective for quality?
- What do we need to do in terms of QC to achieve the quality goal?
- What quality management methodologies should be adopted to meet the customer needs and support the organization's quality goal?

Each of these questions should be answered during the quality planning process. From this goal definition, specific operational procedures can be defined and implemented.

Customers seek quality as an important attribute in the product and service they receive. As a result high-quality products or services are important competitive differentiators. One way of communicating these requirements in a quality business relationship is to create a "Quality Level Agreement" (QLA). This provides a formal method to match customer requirements with control level deliverables and takes the ambiguity out of that relationship. Table 21.2 provides an example of QLA specifications.

The two foundation components of the quality plan are the QA plan and the QC plan. The former details "quality assurance procedures, quality control activities, and other technical activities that need to be implemented to ensure that the results of the work performed will satisfy the stated performance or acceptance criteria" (EPA, 2001).

The QC plan is "the overall system of technical activities that measures the attributes and performance of a process, item, or service against defined standards to verify that they meet the specifications established by the customer" (EPA, 2001).

TABLE 21.2
QLA Examples

Metric	Standard for Final Acceptance	High Tolerance
Design review to specs gap	≤ 10	>15
Earned value total	$CPI = 1.0$	$CPI \leq 1.2$
	$SPI = 1.0$	$SPI \leq 1.2$
Subsystem A defects	2	<4
Subsystem B defects	2	<4
Final acceptance test defects	7	<10

The quality plan then combines the elements of assurance and control into an organization-specific quality management guideline for the project team. It documents the minimal quality requirements, data collection, measurement and analysis activities, QC procedures, and procedures for communicating and correcting quality defects. It also outlines roles and responsibilities for ensuring proper implementation. Each organization has a standard quality plan that serves as a basic minimal starting point specific to the strategic goals of that organization, their industry, and the products and services that they offer. From this base, it is the project team's responsibility to identify and address any gaps between customer specifications and that of the standard quality plan.

There are a number of specific quality management methodologies that an organization may choose to shape the foundation of its quality plan. Documentation of quality objectives should be seen as an elaboration of quality management principles. These objectives establish the structure required to develop procedures and goals for QC, QA, and quality improvement. It is important to detail the differences between these elements in order to appreciate their relationship and impact on the organization.

21.8   QUALITY MANAGEMENT COMPONENTS

From these somewhat abstract perspectives of quality, we can look at how to define a simpler, more concrete framework for quality management at an enterprise-wide level. This overarching model includes the following five stages—quality definition, quality planning, QC, QA, and quality delivery (customer delight). The *PMBOK® Guide* does not explicitly define the first and last stage, but its model concepts fit that model.

21.9   QUALITY DEFINITION

One of the tenets of quality is that it is customer oriented. In order for this to occur the first step in the process is to produce an understandable specification document. Some excellent practices for obtaining optimum requirements are listed below:

1. Define the goal clearly at the start. The project team must clearly define its deliverable goal in such as manner that meets the customer's product or service needs.
2. Prototype models. Users learn more by being able to try out or play with a sample or prototype rather than verbalizing those requirements.
3. Record everything. If possible, record the specification gathering sessions.
4. Use industry best practice, such as focus groups and requirements definition methods.
5. Learn how to perform good user surveys. Good survey documents are harder to develop than one would think.
6. Study the results. Spend time analyzing the data for trends and variances in views.
7. Test your results. Sample users should be brought in for a review of the results. Running a prototype that emulates the survey is one interesting approach. If that is not feasible a storyboard walk-through can be effective.

Various industry organizations outline the characteristics of quality specifications in regard to documentation characteristics. A well-written quality specification should contain the general attributes summarized in Table 21.3.

21.10   QUALITY PLANNING

Quality planning includes all of the work performed to organize and lay out a plan for all five stages of our quality work. Some of the planning work begins even before requirements definition is finalized. After the requirements specifications are approved by the project sponsor, future

TABLE 21.3
Characteristics of Quality Specifications

Characteristic	Description
Complete	Ensure that all attributes that deal with customer satisfaction are included, defined, and given measurable specifications
Consistent	The specification should contain no internal contradictions across the various categories
Feasible	The specification should deal with the technical and logistical feasibility of its development
Modifiable	The specification needs to be organized in such a manner that it can be kept current over time
Necessary	Each defined requirement should be assessed to ensure that it adds value to the customer
Prioritized	Requirements are ranked based on value. Priority coding is recommended for this purpose
Correct	The specification accurately reflects customers' needs
Testable	Each requirement must be defined in a way that allows for one or more tests of either process or product that will ensure conformance or detect an error
Traceable	Each element is uniquely identified so that its origin and purpose can be traced through the life cycle to ensure that it is properly incorporated
Unambiguous	Each requirement has only one possible interpretation

users, and the project team, it is then possible to complete the definition of quality management actions related to reviews, inspections, and tests. These definitions are focused on more than confirming product validity. They also are concerned with processes for rework, scrap, and process improvement.

The quality planning process does not end with the planning stage. As the product is developed and delivered any information obtained about gaps in quality being delivered is reviewed for inclusion in the plan. By making use of what is learned during the life cycle and feeding that knowledge into ongoing quality planning and then implementing those plans, we energize the continuous improvement cycle and add value for the company. The term "customer delight" used earlier may be a little too much hype, but it certainly has the flavor that a quality organization needs.

21.11   QUALITY ASSURANCE

QA is a set of planned and systematic activities necessary to ensure that a product or a service will satisfy the standards of quality established by the organization. QA covers all activities from design, development, production, installation, servicing, and documentation. This role represents a vital and important part of the overall quality management process. It is designed to ensure that a product or service meets the needs, expectations, and other requirements of the customer. The QA process utilizes various techniques and methods to ensure that the expected quality level will be achieved. Stated another way, it evaluates whether the process design is capable of producing the required quality. QA covers all activities starting from design, planning, development, production, and installation to ensure that the necessary quality standards are in place. It also deals with all levels of the organization, starting from upper management down to the lowest organizational level. QA's goal is to ensure that the organization is doing the right things in the right way. To support this, project quality plans are needed for all related functions and organizational levels to show what is required.

It is important to recognize that quality does not just occur, but requires a technical management activity. The QA function requires technical skills to examine all operational details regarding delivery of products and their services. QA processes use many methods and tools to accomplish its role. One of the most widely used models is the PDCA approach, also known as the Shewhart cycle. This is an iterative four-step problem-solving process first described by Walter Shewhart in the

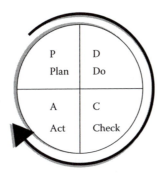

FIGURE 21.1 PDCA model.

1930s and popularized by Edwards Deming in the 1940s and beyond. The meaning of PDCA is explained as follows:

Plan—The phase that establishes all of the methodologies, tools, objectives, and processes that are necessary to deliver the resultant product in accordance with the specifications given by the manufacturer.

Do—The process that implements the plan.

Check—The process that checks, monitors, and evaluates all the processes and their results against the set of objectives and specifications and from these data, reports their status.

Act—The stage in which outputs from the other three stages (Plan–Do–Check) are reviewed and modified in order to improve the process according to the set standards.

The basic concept behind this method is to monitor all processes and continuously improve the quality standards of the organization. Interestingly, this same model was the foundation used to formulate the *PMBOK® Guide* and many other lesser-known project life cycles. Figure 21.1 shows the four steps as a recurring circle, meaning that the process continues forever.

QA may also include activities that investigate certain business processes to ensure that they adequately maintain a desired quality. Lastly, QA focuses on detailed process elements such as documentation and subsequent business processes that define the quality requirements specified by an organization.

21.12 QUALITY CONTROL

In contrast to QA, QC is comprised of activities that measure the quality of a finished product or process. QC involves measurement-oriented activities such as inspections, reviews, and tests that collectively ensure the quality of a product. From this measurement the QC team evaluates expected and unexpected variations of the finished product. The measurement procedures are designed to ensure that the project deliverables adhere to the defined specifications. The foundation of QC is a defined set of specifications and a procedure for evaluating them (see http://www.qualityadvisor.com/sqc/start.htm).

21.13 QA VERSUS QC OPERATIONAL ROLES

QC is a companion process to QA and the two are often labeled together as QA/QC. The basic delineation between these two activities is that QC focuses on the results of the work performed, whereas QA is concerned with the adequacy of the underlying processes, methodology, and standards in place to create the output. QC involves inspections, reviews, and tests to measure status (Mosaicinc, 2001).

Many professional organizations design QA and QC template plans to provide an outline of specific review criteria to ensure that process and product quality are consistent with desired outcomes. For example, the EPA defines a QA plan as a written document that describes the QA procedures, QC specifications, and other technical activities that must be implemented to ensure that the results of the project or task to be performed will meet project specifications (EPA, 2001). Basically, the purpose of these plans is to provide the organization with guidance for measurement and assessment of the quality system performance. In this role, it is important to detail the functional differences between QA and QC and their interrelationships. One key difference is that *QA is process oriented* while *QC is product oriented*. Furthermore, QA addresses the intended actions needed to provide assurance that a product or service will satisfy the known requirements for quality. In contrast, QC involves the efforts needed to uphold the specified integrity of a process (Tudurov, 1996). In both activities, the goal is to look for ways to improve the output of the process or product.

QA is charged with undertaking activities to investigate certain business processes to ensure that they adequately maintain a desired quality. In comparison, QC roles are more limited to activities that evaluate the quality of a finished product. QA works to ensure that the organization is doing the right things in the right way, whereas QC helps to ensure that the results of the work performed meet approved specifications. QA is concerned with creating processes, methodology, and standards that provide the necessary infrastructure to produce the output quality required.

As we have described here, QA focuses on process elements such as documentation and subsequent business processes that define the quality requirements specified by an organization. Basically, the purpose of these plans is to provide the project teams with a formal understanding regarding how to conduct these processes through the project life cycle. Collectively, QA and QC functions are the operational arm of quality management.

21.14 QUALITY GURUS

Modern quality management can be traced back to the AT&T manufacturing organization in the latter 1920s where Walter Shewhart and Edwards Deming worked to develop a statistically based QC model based on sampling theory. This model eventually came to be known as Statistical Quality Control (SQC) and focused on process measurement and evaluation. SQC continues to be used in repetitive processes, but the real contribution from this tandem came from their classical analytical model called PDCA that formed the basis for quality management. Because of Shewhart's conceptual work in these two areas he is often called the Grandfather of Quality, leaving the title of Father of Quality to Edwards Deming who did much to proliferate the quality concepts they formulated.

21.14.1 Edwards Deming

We earlier reviewed Deming's role in the resurrection of Japan's manufacturing infrastructure after World War II. Because of his success in this activity essentially everything he said or wrote became quality gospel. His 14 points of quality became the philosophical foundation for the Japanese movement and continues to be referenced today. A short paraphrasing of the key ideas expressed in these points follows:

- Management leadership required
- Inspection does not create quality
- Training is needed
- Team approach, rather than individual
- Elimination of work standards.

Another concept that Deming offered (but never proved) is that 85% of problems in quality are linked to either management or process. Only 15% is linked to the worker (training or morale type issues). Because of his credibility this statistic has not been refuted, nor has anyone offered another view. This was one more way to say that management must to lead the charge to higher quality.

It is important to recognize that these points are now 60 years old, and a modern comparison does not give credit to the major impact these had when first introduced. The key point to understand is that these statements eventually changed the way organizations thought about quality. It was some period of time before the U.S. management groups embraced these ideas.

In any case, we credit Deming with being the thought leader in the quality movement. He was successful in eventually bringing these ideas to the United States and much of what we see today in the quality arena can be traced back to his work and writings over approximately 50 years.

21.14.2 JOSEPH JURAN

Dr. Joseph Juran was the second major thought leader who contributed to the Japan quality movement. His most visible contribution was the *Quality Handbook*, which is now in the fifth edition. Although this bible of quality is significant, his true lasting contribution was an extension of the Deming view of quality. Juran believed that there were four elements to quality: TQM, quality planning, QC, and quality improvement. Within this structure he saw different levels of the organization having different roles to play. In reviewing his various writings, we see an expansion of the Deming views and one that comes closer to a more contemporary set of notions. For example, the term *customer* is recognized as the driver for quality and he emphasized measurement. Both of these ideas remain in place. A third contribution credited to Juran was the incorporation of the human element into quality equation. A couple of his other focus points were

- Use of the Pareto tool for identifying problem sources
- Quality defined as *fitness for use*.

Many of his ideas evolved gradually over a period of years as he tried to implement the quality concepts in organizations. It is important to note that for both Juran and Deming there was a general lack of evaluation tools. Quality was a somewhat philosophical concept in this beginning period.

21.14.3 PHILIP CROSBY

It seems as though every successful idea or product needs both a conceptual side and a marketing side. Crosby added the marketing side to the quality movement. Up until his point, the quality movement had been considered only understandable by *quality geeks*. Many believed that quality was overhead and you really could not afford to hire any more inspectors. Crosby developed his view of quality from his experience at the Martin Company where he was QC manager for the Pershing Missile Program. In this role he was credited with cutting the product rejection rate by 25% and scrap costs by 30% through an improved quality program. As a result of this experience he published his 1979 book titled *Quality is Free*. This book described and quantified an example showing how producing good quality actually repaid its cost in subsequent savings. At this point the U.S. quality environment was recognized as not being up to international par, particularly with the Japanese who were really beginning to show progress in this area. As a result of this environment, the book was widely read and understandable by the masses. The United States finally grasped the concept that good quality was necessary for competition and not just overhead. This development stimulated the quality movement takeoff in the West. Crosby's mantra for quality was labeled *Zero Defects* and several organizations embraced the idea with some success.

21.14.4 Kaoru Ishikawa

Ishikawa is not as well known as the earlier guru group; however, his contributions are equally valuable. Many would recognize his *fishbone* diagram as a tool to analyze problems (see tool section later). What we see in Ishakawa is a man who wanted to measure quality and evaluate how to improve it. If we have to put a label on him, it would be *toolsmith* although he also coined some additional quality philosophies as well. Nevertheless, when we think of Ishikawa we see technicians trying to identify problems and improve output. He led the effort to identify and teach quality tools and documented seven such tools as basic to the process (see 7QC in the quality tools section).

There are many other contributions that Ishikawa brought to the quality movement, but he stayed below the international radar and does not get the credit that the U.S. trio received.

21.14.5 Armand Feigenbaum

Feigenbaum used statistical techniques at General Electric during World War II to evaluate early jet engine manufacturing. His techniques are titled Total Quality Control and generally fall into the category of an extension of Shewhart's methodologies.

21.14.6 Genichi Taguchi

Taguchi is another guru whom the world generally knows little about, but many prosper from his contributions. In many ways his contributions were similar to Ishikawa but in his manufacturing area the contributions have been very significant. For example, he is generally credited with the conceptual design of the Toyota Production System (TPS) that eventually led Toyota to world dominance in auto-manufacturing. His approach optimizes a customer-perceived value versus the design functionality. Another of his ideas is the Design of Experiments technique to evaluate early performance of processes and designs (see Tools section for more).

21.14.7 Six Sigma

This program does not have a single guru, but does have a similar history in that it started based on the needs of an organization. The Six Sigma program was launched by Motorola in 1987. The engineer typically credited with the creation of this program is Bill Smith, but he died in 1993 before ever knowing the broad impact that this program would have. So, the credit for popularity of the program goes to the host organization and two high-level managers who publicly touted it. The roots of the idea came from a Motorola team of engineers who were seeking a way to achieve higher quality levels than were defined in the traditional three sigma models.

So, in the Six Sigma case we have something more akin to an organizational guru than a single individual. Beyond the complex underlying theory and mechanics of this program, the popularity is often credited to organizational leaders Larry Bossidy of Allied Signal and Jack Welch of General Electric (Isixsigma, 2008). As a result of this high-level sponsorship and advertisement of successes at Motorola, some leading electronic companies such as IBM, DEC, and Texas Instruments launched Six Sigma initiatives in the early 1990s. However, it was not until 1995 when GE and Allied Signal launched Six Sigma as strategic initiatives that a rapid dissemination took place in nonelectronic industries all over the world. In early 1997, the Samsung and LG Groups in Korea began to introduce Six Sigma within their companies. We will see a more detailed discussion of the Six Sigma methodology later in the chapter.

21.14.8 Other Gurus

Each of the individuals mentioned here contributed to moving the concepts of quality along through its 60-year evolution. There were other pioneers in the quality era that are not credited here.

Individuals such as James Harrington and Shigeo Shingo are examples, but our goal is to understand the evolution and the major ideas have been captured in the concepts developed by Shewhart, Deming, Juran, Ishakawa, Taguchi, and Six Sigma.

21.15   QUALITY MANAGEMENT PROGRAMS

There are a number of well-known quality management methodologies that an organization may choose from in shaping the foundation of its internal quality initiatives. The most commonly recognized include ISO 9000, Zero Defects, TQM, Six Sigma, BSC, and Lean methodology. A brief overview for each of these is shown below.

21.15.1   ISO 9000

This is an internationally known quality program that defines the organizational requirements for development and documentation of a quality plan "consisting of a set of specification, procedures, and tests for checking the production processes." This quality plan is intended to be applied every time to ensure consistent quality of products or services produced (Sharma, 2007).

21.15.2   ZERO DEFECTS

This concept was the first quality program known to most American organizations. It emerged from Philip Crosby's view that quality defects were not acceptable and everyone should "do things right the first time." Interest in this quality initiative emerged primarily during the period of 1970s.

Philip Crosby coined this phrase in his 1979 book titled *"Quality is Free"* (Crosby, 1979). There were two major contributions to the quality movement that resulted from this event. First, the book made American managers sensitive to the quality topic in recognizing that there was also a significant cost to bad quality. Crosby's view of quality also focused on the idea that errors were not inevitable and could be sorted out and improved. Second, quality output requires conscious work on the part of the organization, even though the idea of "zero defects" is not an achievable goal. We now look at this program as more of a philosophy than a methodology or even a program, but nevertheless it is an important step along the quality trail.

21.15.3   TOTAL QUALITY MANAGEMENT

This program emphasizes a systematic and integrated organization-wide perspective involving everyone and everything, not just the end products. The heart of the TQM philosophy is the prevention of problems and an emphasis on quality in the design and development of products and processes. A formal quality planning process is integral to the TQM philosophy.

A TQM program strives for one basic aim: providing the best possible services and products to the customers. It also stresses both current quality and the continuous improvement of products and services directed at increasing business and reducing losses due to wasteful practices. The methodology aims at management and employees working together to achieve these goals. "TQM is a management approach for an organization, centered on quality, based on the participation of all its members and aiming at long-term success through customer satisfaction, and benefits to all members of the organization and to society" (ISO 8402:2000). The design focus of TQM is a customer-first orientation rather than focusing on internal activities and constraints (Sharma, 2007). In order to achieve these goals, there is a strong focus on process measurement and controls to identify means of achieving current quality and continuous improvement. The four integrated process component subsets of TQM are derived from Deming's original 14 quality points. Many feel that TQM was the first formal quality management methodology and had a significant international impact on quality perceptions and teaching. To balance this view we need to recognize that the implementation history of TQM is not all bright. One survey reports that only 36% of organizations that undertook TQM were at least partially successful (John Stark Associates, 2008). This statistic is another way of saying that significantly more than half did NOT successfully implement the program. However, it is also significant to note that organizations that did successfully implement TQM were more successful in the marketplace than their competitors. One final lesson learned in the TQM experience is that these programs are not quick fixes and must be conceived as long-term organizational strategies.

21.15.4 Six Sigma

This program is currently the most popular of the quality programs and encompasses much of what has been learned over the previous 60 years. Based on this, it receives the bulk of discussion. The two key methodologies embedded in Six Sigma are DMAIC and DMADV (described below). Both of these are inspired by Deming's PDCA model.

Based on the extremely low defect rate defined for Six Sigma it is a program aimed at the near elimination of defects from every product, process, and transaction. One commonly asked question is "where does this term come from?" Sigma (σ) is a letter in the Greek alphabet that has become the standard statistical symbol to signify variation in data (measure of dispersion). In quality terms sigma describes the variability of a quality measure as defects per unit, parts per million (PPM) defectives, and probability of a failure. Classic QC from the original Shewhart era used a three-sigma variation to imply a reasonable measure of variation in a process. This meant that 99.7% of the process variations are within this range, which equates to three outputs out of 1000 being outside of this range. When Six Sigma was defined, this allowed level of variation was shrunk to produce no more than 3.4 defective PPM. This compares to 66,000 defectives that would have been allowed in the traditional model. Tightening the operational performance to this degree has drastic process quality implications. For this reason Six Sigma is viewed as more than an incremental change in quality perspective.

Pursuit of the Six Sigma model implies three things: statistical measurement, management strategy, and a quality culture. In practice, it also requires a focused target established at a high level in the organization. In this mode, it is a new strategic process under the leadership of top-level management to create quality innovation and total customer satisfaction. It provides a means of doing things right the first time and working smarter by using data information.

21.15.4.1 Statistical Definition of Six Sigma

In many organizations the Six Sigma methodology is simply defined as a measure of quality that strives for near perfection. But the statistical implications of a Six Sigma program go well beyond the qualitative eradication of customer-perceptible defects. It is a methodology that is well rooted in mathematics and statistics. If the design goal is to produce no more than 3.4 defective PPM the process has a very narrow band between the upper and lower control points. Looking at variation in this way means that we are concerned with six standard deviations of variance, hence the name Six Sigma; however, the bandwidth is now very small compared to the classical three-sigma view. This represents a great leap forward in terms of process management and the underlying quality culture of an organization.

As the process sigma value increases, we reach a point that essentially equates to *zero defects*. Certainly, the attitude of the organization culture must view it that way. Figure 21.2 shows the effect on allowable process variation changes from three sigma to six sigma (see Figure 21.2).

21.15.4.2 Six Sigma Fascination

There are several reasons for the international popularity of Six Sigma. First, Six Sigma is regarded as a fresh quality management strategy that incorporates the best of earlier programs such as TQC, TQM, and others. In a sense, we can view the development process of Six Sigma as shown in Figure 21.3. The second driver for Six Sigma popularity comes from its view as having a more

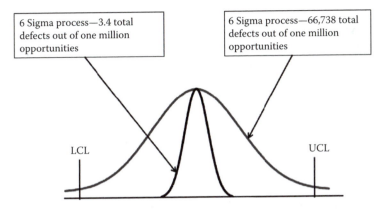

FIGURE 21.2   Statistical interpretation of Six Sigma.

robust systematic, scientific, statistical, and smarter (4S) approach for management innovation that is better aligned for use in a knowledge-based information society.

Six Sigma methodology represents an integration of four elements—customer, process, manpower, and strategy—to provide management innovation. The tools and methodology of Six Sigma provide a scientific and statistical basis for quality assessment. The defined methods allow comparisons among all processes, and describe the state of a process. Through this information, management can better discern what path to follow to achieve process innovation and customer satisfaction.

Third, Six Sigma provides a method to efficiently allocate and cultivate HR toward formalized high-level goals. It employs a belt certification system in which the levels of mastery are classified as green belt, black belt, master black belt, and champion. As a person obtains certain training they will be recognized with a belt level. It is common to require at least a black belt level to lead a Six Sigma project team and several green belts would be mentored as part of that team.

Fourth, the many published Six Sigma success stories have been a major stimulus for organizations to learn more about the technique. Nothing spreads in organizations more than the claim of success with a new idea.

Lastly, Six Sigma is capable of embracing common problem elements facing organizations today. These are the 3Cs of change, customer focus, and competition. The pace of change during the last decade has been unprecedented, and the speed of change in the new millennium is perceived to be equally fast. Most notably, economic power is now in the hands of an increasingly sophisticated customer who has access to more quality-related information than ever before. The producer-oriented monopolistic industrial society is over, and the customer-oriented information society is the new model. At this point the customer has all the power to select competitive sources for order, evaluation, and purchase of their goods and services. This is especially obvious in the e-business domain. Ready availability of comparative information related to quality and price is ever increasing through international Internet access. Second-rate quality goods and service cannot survive in such an environment.

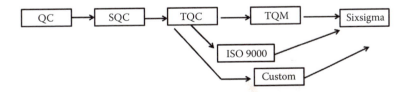

FIGURE 21.3   Six Sigma evolution.

With all of these trends Six Sigma is recognized as a popular tool for organizations to use in identifying and pursuing quality improvements. The embedded 4S processes—systematic, scientific, statistical, and smarter—provide the required methodology to make required changes in customer-focused output to remain competitive.

21.15.4.3   Basic Methodologies

Six Sigma includes two key methodologies called DMAIC and DMADV. DMAIC is used to improve existing business processes, and DMADV is used to create new product or process designs for predictable, defect-free performance. DMAIC methodology consists of the following five steps:

- Define the process improvement goals consistent with customer demands and enterprise strategy
- Measure the current process and collect relevant data for future comparison
- Analyze to verify relationship and causality of factors, determine the key relationships, and ensure that all factors have been considered
- Improve or optimize the process using techniques like Design of Experiments
- Control to ensure that any variances are corrected before they result in defects. Set up pilot runs to establish process capability, transition to production, and thereafter continuously measure the process and institute control mechanisms.

Figure 21.4 expands this basic view by showing a sample of some of the tools and techniques that are employed in each of these steps.

The basic DMADV methodology has a set of steps similar to DMAIC.

- Define the goals of the design activity consistent with customer demands and enterprise strategy
- Measure and identify critical to quality (CTQ), product capabilities, production process capability, and risk assessments
- Analyze to develop and design alternatives, create high-level design, and evaluate design capability to select the best design
- Design details, optimize the design, and plan for design verification
- Verify the design, set up pilot runs, implement production process, and handover to process owners.

As far as a corporate program framework is concerned, Six Sigma embodies five elements that have been recognized in other quality initiatives. These are top-level management commitment, training schemes, project team activities, measurement systems, and stakeholder involvement.

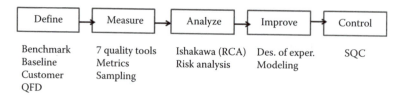

FIGURE 21.4   DMAIC framework.

At the core of the Six Sigma framework is a formalized improvement strategy that involves one of the two process groups outlined above. For illustration purposes, the DMAIC process will be briefly summarized in order to obtain a better sense of what Six Sigma involves.

21.15.4.4 DMAIC Process Overview

DMAIC is often called the process improvement methodology. It was designed to deal with a problematic process related to a product and/or service offering in order to reshape its cost, quality, and/or cycle time. Such fixes often include defects or failure targets, excess process steps, and deterioration. The method was designed to detect a problem within an existing steady-state process. The five defined steps are Define, Measure, Analyze, Improve, and Control. This method is primarily based on the application of statistical process control, quality tools, and process capability analysis. It is not intended to be a product development methodology, although it can be used to help redesign any process. Implementation of the method requires four components:

- A measurement system (a gauge)
- Statistical analysis tools (to assess samples of data)
- An ability to define an adjustment factor (to put the response on target)
- A control scheme (to audit the response performance against statistical control limits).

Let us briefly review each of the five steps, their requirements, and typical combinations of tool–task–deliverables.

21.15.4.4.1 Define

The objectives of this step are twofold: to confirm the existence of a problem or opportunity and to define the improvement project boundaries and goals. Formal goals are identified (in measurable terms), and a high-level project plan is created. This is very similar to the initiation process outlined for a typical project. Just as in all projects, scope definition for this effort is critical to control the project. This step produces the following deliverables:

1. Identify and quantify the problem or opportunity
2. Develop a high-level process map
3. Gather business requirements
4. Develop a communication plan
5. Finalize the project charter
6. Select a project sponsor and members
7. Identify stakeholders
8. Gain approval and necessary funding to conduct the DMAIC project.

21.15.4.4.2 Measure

This step gathers the necessary data to understand and measure the current state. Collecting or establishing the baseline of current state performance and process metrics is the crux of this step.

21.15.4.4.3 Analyze

This step identifies the root cause of the problem. From this definition the project team delves into the operational details in order to enhance its understanding of the process (or product/service offering) and the problem. Analytical tools are used to dissect the root cause of process variability and separate the vital few inputs from the trivial many (Pareto distribution).

21.15.4.4.4   *Improve*

This fourth step involves developing solutions targeted at confirmed causes. The two primary objectives are to verify that the confirmed causes (or critical inputs) are statistically significant and to optimize the process or product/service with the defined improvements. Prioritization and validation are important components in selecting a recommended fix to the problem. The project team must quantify key variables effects on the process and develop an improvement plan that modifies the key factors to achieve the desired process improvement. The short list of tasks within this step includes the following:

1. Develop potential improvements or solutions for root causes
2. Develop evaluation criteria
3. Select and implement the improved process and metrics
4. Measure results
5. Evaluate whether improvements meet targets
6. Evaluate risk.

21.15.4.4.5   *Control*

The fifth and final step is to complete the project work and transition the improved process to the process owner with procedures for maintaining the improvements as an ongoing operation. In preparation for the transition, the project team and operations work together to verify the ability to sustain the improvement's long-term capability and plan for continuous process improvement. The summary list of tasks to be completed within this step follows:

1. Document the new or improved process and measurements
2. Validate collection systems along with repeatability and reproducibility of metrics in the new operational environment
3. Define the control plan and its supporting plans
4. Communications plan outlining the improvements and operational changes to the customers and stakeholders
5. Implementation plan
6. Risk management (and response) plan
7. Cost/benefit plan
8. Change management plan
9. Train the operational stakeholders (process owner and players)
10. Establish the tracking procedure in an operational environment
11. Monitor implementation
12. Validate and stabilize performance gains
13. Jointly audit the results and confirm financials.

This section has been a brief overview of Six Sigma to show some of its general characteristics. Six Sigma has matured the use of various analytical tools as well as the encapsulation of many classical quality concepts. In many ways this new program is an extension of previous quality programs. Park offers the following summarization of Six Sigma attributes (Park, 2003, p. 126):

- Bottom-line business results delivered
- Senior management leadership role defined in the process
- Disciplined approaches (DMAIC and DMADV)
- Rapid completion of target actions
- Clearly defined measures of success
- Defined roles for participants

- Clear focus on customers and processes
- Use of sound statistical tools.

21.16   *PMBOK® GUIDE* QUALITY PROCESS MODEL

The PMI model guide for quality is philosophically aligned with the models discussed here. Three basic quality processes are outlined: quality planning, QA, and QC. Process specifications for each are consistent with the definitions described in the standard models, except that these processes are focused on the project level only. So, quality planning would address how the project would deal with specific quality standards for that specific project.

The guide's QA and QC processes require the same type of project customization as the high-level quality planning model. The only uniqueness of the guide's approach is in its level of specificity for the project. For example, each of the three-quality processes has defined inputs, tools and techniques, and outputs. Embedded in these are tangible management items such as

- Quality management plan
- Quality metrics
- Quality checklists
- Quality baseline
- Change management activities.

The *PMBOK® Guide*'s QA and QC functions are basically an operational level implementation of the quality program and the techniques described are very similar to the general quality program model discussions outlined here.

21.17   OTHER PROGRAMS

Two other popular programs impinge on the quality management domain. These are BSC and Lean (manufacturing).

BSC: Popularity of this program at high organization levels impacts quality management by integrating quality metrics and other measurements into the overall organizational planning and reporting system. However, BSC does not offer a defined process for structuring a quality organization, so it is viewed more as a reporting medium (see Chapter 29 for more specifics on this program).

Lean: "Businesses and other organizations use lean principles, practices, and tools to create precise customer value—goods and services with higher quality and fewer defects—with less human effort, less space, less capital, and less time than the traditional system of mass production" (see www.lean.org for more details on this). The TPS is an operational example of this model today and it is described later in another chapter.

Interest in various quality programs evolved from the 1950s through the 1990s, with Six Sigma being the current most popular one. Each of these ushered in a new phase of QC and management. Also, each increasingly sensitized organizations to the concept that quality was a strategic organizational issue for competitiveness and even survival.

21.18   EVALUATING QUALITY

There are several techniques used to measure quality. Each of the three methods/techniques summarized below have a role in aiding organizations in their quest for quality. The following three items selected for this discussion:

- Benchmarking
- Continuous improvement
- Failure modes and effects analysis.

21.18.1 Benchmarking

21.18.1.1 What Is It?

… benchmarking … [is] … "the process of identifying, understanding, and adapting outstanding practices and processes from organizations anywhere in the world to help your organization improve its performance."

—American Productivity & Quality Center

The process of benchmarking represents a strategic management process in which an organization evaluates its processes with respect to other competitors or industry standards. By doing this type of comparison, the organization can evaluate areas in which they need to seek improvement for competitive purposes. Benchmarking techniques use both quantitative and qualitative standard measurements for comparison with other organizations in order to gain a perspective on organizational performance. Benchmarking can be performed by independent consumer-oriented organizations or within the organization. Obviously, obtaining valid data is the major issue with this type of analysis (Sixsigma, 2008).

Benchmarking can provide qualitative and quantitative data to answer various process and product questions. Perhaps more important, benchmarking can offer guidance on how to achieve improved results. When performed across major competitors it provides an external reference and best practices on which to base evaluations and standards to design work processes.

Benchmarking is the search for best practices, the ones that will lead to superior performance. Establishing operating targets based on the best possible industry practices is a critical component in the success of every organization. (Camp, 1995, p. 4)

Benchmarking is a managerial tool used to measure and compare the organization's internal processes in terms of cost, time, or quality against those factors from another organization considered to be best-in-class. These results lead to the identification, understanding, and adaptation of exceptional practices. Although benchmarking within the same industry is important since the enterprise should have a good understanding of its own business, it is even more essential to compare customer and external perceptions of products and services.

21.18.1.2 Types of Benchmarking

There are essentially three types of benchmarking: strategic, data-based, and process-based benchmarking (ASQ, 2008). They differ depending on the type of information you are trying to gather. Strategic benchmarking looks at the strategies that companies use to compete. Benchmarking to improve business process performance generally focuses on uncovering how well other companies perform in comparison with you and others, and how they achieve this performance. This is the focus of data-based and process-based benchmarking (ASQ, 2008). Many business processes are common throughout industries. For example; NASA has the same basic HR requirements for hiring and developing employees as does American Express. Likewise, British Telecom has the same customer satisfaction survey process as Brooklyn Union Gas. These processes, albeit from different industries, are all essentially similar and can therefore produce comparable benchmarking metrics. This multiindustry view is called "getting out of the box."

21.18.1.3 J. D. Power's Quality Benchmarking

J. D. Power and Associates is a well-known commercial benchmarking organization. Each year they publish the Initial Quality Study (IQS) report that serves as an automobile industry benchmark outlining the comparative quality of new vehicles (J. D. Power, 2007). This report documents defects, issues, and malfunctions experienced by owners during the first 90 days of vehicle ownership.

The annual IQS report has existed as a quality icon for the past 20 years and has promoted continuous improvement throughout the automobile industry. In fact, the study shows that the automobile industry quality trend has been to improve quality metrics at a rate of about 6% per year on average and automobile problem counts by 50% every 7–8 years (J. D. Power, 2007). An extraction from the IQS annual report shows some of the product metrics captured.

> J. D. Power and Associates Report: 2007 IQS Ranking Highlights Mercedes-Benz and Toyota models capture three segment awards each. Toyota models that receive awards are the 4Runner, Sequoia and Tacoma. Mercedes-Benz models that earn awards are the E-Class, SL-Class, and the S-Class. The S-Class ties with the Audi A8 (total) for having the fewest quality problems in the industry, with just 72 problems per 100 vehicles. "Mercedes-Benz shows dramatic improvement, particularly with its newly-redesigned S-Class, which improves 63 PP100," said Oddes (sic). "Overall, Mercedes-Benz improves its nameplate rank by 20 positions—the greatest rank increase of any nameplate in the study. All Mercedes-Benz models in the study improved substantially, and the breadth and speed of these improvements demonstrates the Mercedes-Benz commitment to quality." For the second consecutive year, Porsche tops the overall nameplate rankings, averaging 91 PP100. Following in the rankings are Lexus, Lincoln, Honda and Mercedes-Benz, respectively. Honda, with the fewest problems per 100 among non-premium brands, improves in the ranking to fourth from sixth since the 2006 study and earns awards with the Civic and CR-V. Among non-premium brands, Kia posts the largest improvement in ranking, moving from 24th in 2006 to 12th in 2007 and earning an award for the Kia Rio/Rio5 for the second year in a row. The most improved nameplates in the study are Land Rover (increasing in initial quality by 34 PP100), Saab (improving by 30 PP100) and Mercedes-Benz (increasing by 28 PP100). Other nameplates receiving model awards in 2007 include Chevrolet (Express and Silverado Classic HD), Lexus (RX350/RX400 h) Pontiac (Grand Prix) and Porsche (Boxster). *Source:* http://www.jdpower.com/corporate/news/releases/pressrelease.aspx?ID=2007088 (accessed June 1, 2008)

21.19   CONTINUOUS IMPROVEMENT

This term represents more of a philosophy than a set of tools or techniques. The key idea is that today's quality is not good enough for the future, so the focus must be on continually and incrementally moving to new quality levels that "excite" the customer. This can be reflected in cost or performance efficiency, or some new functionality in the product. In order to accomplish this goal, processes and methods of production or service delivery are tested continuously and shortcomings removed, thus improving the process or product in a continuous manner.

21.20   FAILURE MODE AND EFFECTS ANALYSIS

This procedure is used with physical products to analyze failure characteristics in the design. The results of these tests classify the impact of the failure and rationalize strategies to improve taking into account the quality goals versus the cost. Failure causes can result from any error or defect in the process, design, or execution (manufacture) of a product. Effect analysis refers to studying the consequences of those failures in regard to the customer experience with the product. This type of activity goes beyond the process and inspection-type components in that it uncovers the limits of the product design. Modern examples of this would be impact to passengers in a car crash test, or tests to evaluate mean time to failure for an item. The performance of a product is checked and tested under increasing stress until it fails to work. This process exposes the weakest points of the product as it is vibrated, dropped, heated, or otherwise abused. From these tests the quality management team can decide whether it is viable to improve the design or process to improve the resulting product. Sometimes small changes in a product or process can have significant impact on the resulting overall quality of the product.

21.21   QUALITY TOOLS

As the quality movement evolved from a more philosophical bent in the latter 1940s to a more analytical one, the need for measurement and analysis of support tools emerged. Around 1960 a group of Japanese quality professionals defined a core set of tools for general use. Over the years since then many such visual, statistical, and descriptive support items have been developed and used. In more recent times maturation of tool usage grew significantly in the various life cycle processes such as those defined by Six Sigma's DMAIC and DMADV structures (more details presented later). Within each of these stages various tools and techniques are commonly used. A reasonable cross section of these will be described in this section. The first set of tools described here is classified as the core quality tool kit, which is represented by seven tools (a.k.a. 7QC). There are actually eight tools shown in this list because some industries replace stratification (7A) with flowcharting (7B). Collectively, these tools are used to review various quality aspects; however, their primary role is in quality measurement and analysis activities. The core tool set is as follows:

1. Cause and effect
2. Check sheet
3. Control chart
4. Histogram
5. Pareto
6. Scatter
7a. Stratification
7b. Process flowchart (process mapping).

Sung Park offers an excellent overview of this area in *Six Sigma for Quality and Productivity Promotion* (Park, 2003).

1. *Cause-and-effect diagram*   The cause-and-effect diagram is an effective tool for use in the problem-solving process. It is also known as the Ishikawa or fishbone diagram. This problem-solving technique is useful to trigger ideas and promote a balanced approach in group brainstorming sessions where individuals list the perceived sources (causes) with respect to outcomes (effect). Figure 21.5 illustrates how this tool is used. In this example, the quality variance is shown in the box on the right-hand side, and the potential causal areas are listed in boxes on the lines. Essentially, the box is the effect and the notes on the various lines are meant to portray possible causal factors from that area (i.e., man, machine, etc.). When constructing a manufacturing-oriented diagram, it is often appropriate to consider six main causal areas that can contribute to an outcome response (effect). These are called the 5Ms (man, machine, material, method, and measurement), plus the environment.

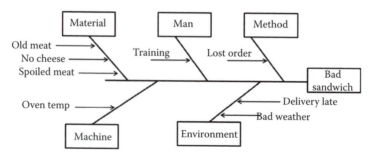

FIGURE 21.5   Example fishbone diagram.

In Figure 21.5, the sample diagram models a hypothetical problem of producing a bad sandwich. In this case, potential causal factors are collected for each of the items and attached to the figure. Obviously, in this case the root cause would not be difficult to identify but the example does show the process in understandable form.

2. *Check sheet*   A check sheet is a template used for data collection of any desired characteristics of a process or product. It is frequently used in the measurement phase of the DMAIC process.

3. *Control chart*   The control chart is a very important tool in the "AIC" phases of DMAIC. In the analysis phase, control charts are often used to judge whether the process output is in the predictable range. In the improve phase, it aids in providing evidence of special causes of variation so that they can be acted upon. In the control phase, it is used to verify that the performance of the process is under control. Recall that the original control chart was first proposed by Walter A. Shewhart in 1924 as a core component of his SQC model. Since that time the tool has been used extensively in industries to evaluate repetitive processes. These control charts offer an excellent means to study process variation. Observations failing outside of the control bands often give early identification of an unexplained or abnormal variation so that there can be timely corrective action before the process goes out of control and unusable products produced. Shewhart's control charts track processes by plotting data over time as shown in Figure 21.6.

4. *Histogram*   It is meaningful to present data in a form that visually illustrates the frequency of a value's occurrence. During the analysis phase, histograms are commonly applied to help understand the distribution of the data by value. The classic example of a histogram is a document of the values thrown by two dice over 200 observations. Each value would be shown on the x-axis (i.e., 1–12) and the corresponding number of times each value was thrown would be represented by a vertical bar (y-axis). If these dice were "honest" we would expect to see a bell curve for this experiment. The typical role of a histogram is to simply document the results with no conclusion.

5. *Pareto chart*   An Italian economist originally derived the Pareto chart in the later 1800s, who used it to describe the maldistribution of incomes across the population. Juran reintroduced it to the quality world in the 1940s as a way to distinguish the "vital few from the trivial many." It is now better known as the 80/20 rule—80% of the problems stem from 20% of the causes. The Pareto chart has many implications both in the quality arena and elsewhere. One of its quality applications is for selecting appropriate improvement targets during the define phase. Figure 21.7 shows a sample format for this chart.

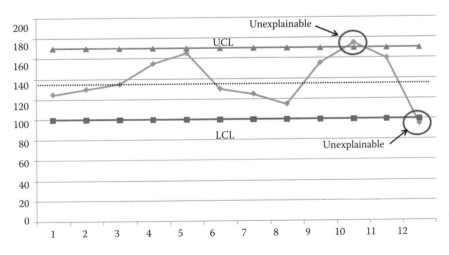

FIGURE 21.6   Shewhart control chart format.

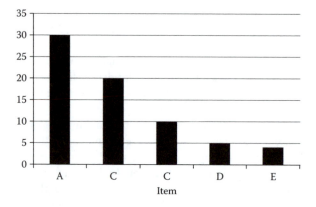

FIGURE 21.7   Sample Pareto chart.

6. *Scatter diagram*   A scatter diagram is a useful way to show a display between two factors plotted on an *xy*-axis. An important feature of the scatter plot is its visualization of a relationship pattern. As part of the improve phase one often searches the collected data for *X*s that have a special influence on *y*s. By examining the status of such relationships, it is possible to identify input variables that appear to have causal relationships.

7a. *Stratification*   Stratification is a tool used to split collected data into subgroups in order to determine if any contain special cause variation. Using this method, data from different sources in a process can be separated and analyzed individually. Stratification is mainly used in the analysis phase to organize data in the search for special cause variation.

7b. *Process flowchart and process mapping*   Some types of processes and structures can be better understood through visualization of flow patterns. A flowchart is a method used to provide a picture of the steps and relationships in a process. This technique would be considered one of the classical diagnostic tools used in areas such as information systems, quality management systems, and employee procedural handbooks. The flowchart can also be used in the measurement phase as a procedural aid, as well as in the analysis phase for identifying improvement potential compared to similar processes. It can be useful in the control phase to aid in describing new processes and would help guide those executing that process.

21.22   OTHER QUALITY ANALYSIS TECHNIQUES

There are a variety of measurement and analysis tools and techniques that have been developed to support various aspects of the quality process. Three quality techniques examples are notable in their use of analysis tools and for that reason they need to be included in our discussion. These are design of experiments, quality function deployment (QFD), and quality auditing.

21.22.1   DESIGN OF EXPERIMENTS

This is a quality planning technique that helps identify which variables have the most influence on the overall outcome of a process. Understanding which variables affect an outcome is a very important part of quality planning. Quality planning also involves communicating the correct actions for ensuring quality in a format that is understandable and complete. In quality planning for projects, it is important to describe important factors that directly contribute to meeting customer's requirements. Organizational policies related to quality, the particular project's scope statement and product descriptions, and related standards and regulations are all important input to the quality planning process. The main outputs of quality planning are a quality management plan and checklists for ensuring quality throughout the project life cycle.

21.22.2   QUALITY FUNCTION DEPLOYMENT

QFD is a structured technique to ensure that customer requirements are included in the design of products and processes. In the Six Sigma methodology, QFD is mainly applied in improvement projects related to the redesign of products and processes (Park, 2003). Shigeru Mizuno and Yoji Akao originally developed this technique in Japan during the late 1960s. It was first applied in ship-building in 1972 and then migrated into the Japanese auto industry some years later. It eventually was used in U.S. manufacturing in the mid-1980s.

The QFD process is an important technique for use in product design. Its process relies heavily on a formal translation of customer requirements into product and process characteristics including specified target values. Basically, QFD can be divided into four phases:

Phase 1: A market analysis to establish knowledge about current customer requirements and the related translation into product characteristics.

Phase 2: Translation of critical product characteristics into component characteristics, that is, the product's parts.

Phase 3: Translation of critical component characteristics into process characteristics.

Phase 4: Translation of critical process characteristics into production characteristics, that is, instructions and measurements.

These four phases embody four standard units of analysis that evolve in the following order: customer requirements, product characteristics, component characteristics, and production.

Quality audits. This review technique is used to confirm that the system is working and producing output as required. The two forms of audit are classified as internal and external.

Internal audits. The goal of the internal audit program is to review the entire organizational quality system specific assessments and shall include at least one management review of the overall effectiveness of the quality system. These are typically conducted at scheduled intervals. The following areas should be reviewed as part of the project schedule.

Documentation audit. This will include a review of the quality manual, all control documents, and records.

Management responsibility audit. This includes a planning and management review to ensure that management is committed to the company's policies and values.

Resource management audit. This focuses on items such as employee turnover and competency of personnel. Outsourcing will also be included for areas that are subcontracted.

Product realization audit. This covers a wide array of business processes and customer-related processes such as customer communication and other related activities.

Design and development audit. This includes areas such as planning, inputs, outputs, design review, verification, and validation. This audit is one of the most important to review owing to the design changes that typically occur. Scope creep and budget issues usually impact these areas if the design changes are not conducted according to the outlined processes.

Purchasing audit. This focuses on the purchasing process and the verification of purchased products as outlined in the quality system. The company must be able to verify that the products purchased do not turn into scrap. Product and service provisions as well as the control of measuring devices are also included in this area.

General audit. This includes all remaining areas such as customer satisfaction and the monitoring and measuring of processes. The audit process will also identify the need for continuous improvement and will include both corrective actions and preventative actions to eliminate further discrepancies in the future.

External audits. This will be conducted on all external suppliers and vendors in order to evaluate whether the products delivered will meet company specifications. This is normally a graded level audit. For example, the more critical suppliers require a detailed audit classified by letter "A" and the less critical suppliers will be identified as a "B" supplier. This method may also be used for service companies. All vendors graded of high importance will be required to have a quality system in place and must be able to demonstrate their effectiveness during the audit and should be able to meet many of the requirements of the internal audit criteria as mentioned above.

21.23   ORGANIZATIONAL ROLES AND RESPONSIBILITIES

The process of creating a quality organization would appear to be somewhat mechanical from the discussion thus far; however, experience shows that this is not the case. Installing a quality culture in an organization is difficult for many reasons. Probably the most obvious being that it takes time and the results may not show for some period. If management is not convinced that it is worthwhile they will lose interest and pull resources to other activities. If we assume that the goal is to produce such an organization there are clear steps in that process.

The first step in creating a quality organization is to convince senior management that it is a worthy goal. They must be the formal initiator and driver for the subsequent activities required. As with all major projects, a formal definition of goals is necessary. In this case the goal definition involves approving a policy document that includes a statement or definition of the organization's quality objectives. This document should also address what will be done to comply with the standard after implementation (Besterfield et al., 1995). The intent of the quality program should be formally documented with a goal of making the quality objectives clear, precise, and easy to understand. In this regard, the formal quality policy is considered to be the foundation for a strategic roadmap of continuing quality improvement and business success. Included in the document are key processes needed for the quality management system and explanations showing how the overall system will be applied across the organization.

According to the ISO process, creation of the operational system first involves gathering all existing policies, procedures, work instructions, and forms that are presently in use. Every document should be reviewed and approved for use by the management and made to fit into one of the elements. As the new documents are produced, the implementation team becomes the review committee. As changes are necessary, suggestions are made and reviewed with the team (QSAE, 2008).

From an organizational structure point of view, some management representative should be placed formally in charge of shepherding the quality implementation process through the entire structure. Implementation of the quality system should involve everyone in the organization. Once the formal process definition has been communicated to the functional management groups, the next step is to develop a quality awareness program. This process will affect the day-to-day operations and the potential benefits (QSAE, 2008). After everyone has been informed of the organization's intentions to develop the quality system, an implementation team should be assembled, with members drawn from all levels and areas of the organization.

Successful implementation of a quality system requires full support of the organization from the top to the very lowest levels. It must be driven by a formally documented and approved system. Quality emphasis is placed on problem prevention, rather than detection (inspection), in all organizational areas from customer sales through installation and service after delivery. Appropriate responsibility and authority must be defined for all personnel affecting quality, and they must be given the freedom and authority to take the actions necessary to prevent proliferation of nonconforming products or services. This activity includes root cause analysis and problem correction. Management review is required to ensure that the system remains effective and this process makes use of information from customers and internal audits, as well as process and product performance. From the results of these reviews, management can determine if a change is required in the organizational structure or the operations of the organization to improve the quality system.

21.24 ISSUES IN QUALITY MANAGEMENT

Throughout history, quality has been associated with two important aspects: measurement and inspection activity. However, the concept of quality management now is expanding into a broader perception. Today, it is associated with a wide variety of indicators to describe perfection, consistency, waste elimination, delivery time, policies, procedure compliance, and customer satisfaction, to name a few. All of these factors relate to the development of a product or service designed to be *fit for use* and to ensure that the execution process is meeting customer needs.

The purpose of quality management is to ensure that the product and services exceed, or at least fit, customer's expectations. This concept is finally consolidated during the execution phase of the process. Quality management is an important element of the planning phase since it provides the framework and direction for the execution of quality. However, it is during the execution phase of the project where the conceptual philosophy of project management is matched with the quality view.

The perception of quality can be quite variable among stakeholders. Therefore, it is necessary to formalize a general understanding between these perceptions and some measurable criteria for control purposes. Quality management is implicitly associated with the parameters used to compare the product quality with these predetermined standards. There are several QC and QA techniques that are utilized in performing this activity. One of the most visible is quality testing.

Quality testing effectiveness is based on compliance with a standard and therefore effective quality measurement requires that the standard is accurate. When the measurement of the goods or service does not fit the standard, it creates a variance, which needs to be corrected or modified in order to meet customer expectations. There are several areas where these discrepancies can occur:

Misalignment of customer expectations and translated specifications. The project team might believe that they understand the customer's perception of compliance, but in reality this might not be the case. As a result all the subsequent deliverables would be flawed.

Mismatch between quality standards and defined specifications. The resulting design does not properly translate the customer requirements.

Inconsistency between designed and actual specifications. This implies that manufacturing and engineering designs must execute quality in a coalescent effort.

False expectations by customers. Customers should only be promised what can realistically be produced based on the actual capacity of the quality of the deliverables.

All identified discrepancies should be exhaustively evaluated throughout the life cycle of the project. The goal in performing these reviews supports the organizational goals of delivering promised project performance that meets or exceeds customer's expectations. There is no known better example of the merits of building a quality organization than the culture created at Toyota.

21.25 TOYOTA QUALITY PERSPECTIVE

The internationally recognized flagship quality organization is Toyota. This organization now reaps the economic benefits of a long-term quality culture and today stands at the top of the automobile manufacturing world as a result. Jeffrey Liker in his book *The Toyota Way* outlines their 14 quality principles (is this an accident that Deming had the same number?). In this review, he summarizes the main ideas that drive this culture as follows:

- Base management decisions on a "philosophical sense of purpose."
- Use long-term planning
- Have a formal problem solving process
- Develop employees
- Recognize the value of continuous improvement.

Toyota has taken the classic ideas of quality and developed its own quality management program. The assembly line segment is known as the TPS (Toyota, 2008). TPS represents a custom set of principles, philosophies, and business processes designed to enable the manufacture of quality products while utilizing the leanest manufacturing techniques (i.e., quality and value combined). Additionally, Toyota exercises business process management (BPM) to help identify opportunities for improvement throughout its business activities and then ensuring that appropriate actions are taken in order to benefit from those opportunities.

The origins of quality improvement in the Toyota Group date back to 1902 when founder Sakichi Toyoda invented a textile loom that monitored defective production by halting operation if snapped threads were detected. This helped prevent the creation of defective products and developed into a process known as *Jidoka*. In the 1930s, Toyota developed what is now known as the just-in-time (JIT) supply chain system. This system allowed Toyota to draw its resources efficiently by only acquiring necessary inventory from suppliers at appropriate intervals rather than having unneeded supplies pushed onto them by other manufactures. Provided here are some examples of how Toyota applies these balanced quality and value principles today in their assembly line (Liker, 2004):

JIT. After new cars are painted, a computer-controlled system sends a production request to its seat manufacturers with details regarding color, materials, quantity, and other configuration options. The seat manufacturer then processes the requests thereby minimizing costs due to wasted materials.

Jidoka. Both humans and automated workers have the capability to stop a vehicle's progress at any stage of the assembly line if defects are detected. Action is taken to correct the issues and then production is continued. This prevents costly defects from occurring later in the life of the product and improves long-term customer satisfaction.

Kaizen. Kaizen is a tool originally used by Toyota to foster continued improvement within its TPS. It began as "Quality Circles," a means of factory shop floor employees to solve quality issues within a structured team framework, using specific new tools. It is now used around the world by many companies and has been adapted to suit their individual needs and customs.

This system is directly related to the promotion of continuous improvement by eliminating waste from processes. For example, the layout of an assembly area may need to be reorganized so that a worker does not have to waste time and energy by excess motion around the assembly floor in order to finish working on their task. This promotes greater efficiency throughout the assembly line.

Muda. This deals with the concept of avoiding waste. According to the concept, the 10 forms are as follows:

1. Waste from overproducing (making more just because the material is available, a machine or HR is available)
2. Waste of time (excessive lines—batch and queue mentality)
3. Waste from transporting (excessive movement of materials)
4. Waste from overprocessing (process is too complex process or ineffective)
5. Waste caused by excessive work in process inventory
6. Waste from excess motion of operators and workers (lack of good job design)
7. Waste from scrap and rework
8. Human underutilization (poor resource allocation processes)
9. Improper use of technology
10. Working to the wrong metrics.

Andons. This process uses visual signals and controls to display the status of the assembly line. As a result operators and supervisors are aware in real time the current status of all manufacturing processes. The andons can also display critical information about a faulty machine and the action required to keep the assembly running smoothly.

PokaYokes. This process utilizes a number of devices that help to prevent defects. Electronic sensors scan for predetermined movements and send out warning signals when appropriate. For example, PokaYokes are used to remind the assembly workers to use all the components needed in the final assembly of a larger part.

Genchi Genbutsu. This process requires employees to investigate a problem directly, thereby promoting teamwork and collective action toward the resolution of an issue.

By promoting continuous quality improvement throughout its business processes, Toyota has learned to identify and manage the factors that are advantageous to its business processes and to take out the factors that impact the company negatively. Utilization of the TPS and BPM processes has yielded direct benefits such as increased customer satisfaction, lowered costs, and the production of reliable products. These benefits have moved the company into the number one brand position in their industry with internationally recognized quality and sales volumes. It is important to recognize that this has not been a short-term process, but one pursued consistently over many years. This is the nature of building a quality organization.

21.26 FUTURE OF QUALITY MANAGEMENT

This chapter has traced the evolution of quality management and concepts from the 1930s with Shewhart's process model to the currently in vogue Six Sigma version. If we examine the basic focus of the various quality program targets through this period, a high-level perspective emerges. The key quality concepts discussed thus far can be generally summarized as follows:

- Defect filtering by inspection models—Shewhart
- Creating a quality organization management culture—Deming
- Process improvement tools—Ishikawa (analyzing status)
- Customer focus—various
- Organizational improvement targets—Six Sigma.

These summaries are oversimplified in that the human view is not specifically shown, nor is training of the worker specifically shown. One might conclude that we are still in the maturation process for the concept and more evolution will follow. The breadth of this topic defies simple definitions or projections to clearly predict what that evolution would look like. Up to this stage we have seen quality management start with a philosophy (Deming) and work toward a broader integrated view (Six Sigma). It seems now that the latest stage has essentially combined the fragments of the past into a single view and matured the analytical processes. More subtly put, there is a broadening of the concept of quality, so projection of what the next iteration might entail becomes risky. One view would suggest that it would have to minimally incorporate customers, processes, training, measurement, analysis, products, and goals. But what might be added beyond that is vague. Given the lack of a crystal ball, projecting focus areas of the next quality wave is conjecture, so the discussion that follows should be taken simply as thought provoking. Only time will answer this question in reality.

One reasonable projection is that the future quality domain should include a formalized decision process whereby organizations can improve selection of product targets that better fit consumer desires (the market view). This has the characteristic of PPM on a broader scale. In relation to this selection process the organization would profit by having efficient work planning and management processes to produce those products (including efficient process design and implementation). Accomplishment of this goal requires an effective resource allocation process that is much more robust than what exists in most organizations today. Also, a second potential target is developing techniques to improve worker productivity through some form of formal skill improvement process similar in design to the Six Sigma "belt" certification methods. Third, the growing popularity of international virtual organization structures has highlighted weaknesses in overall management across the physically dispersed groups. A new organizational structure option would focus on building project teams from multiple organizations into a homogeneous view of processes and management. In this new model the actual division of work may be more of a peer relationship among the players than a hierarchical buyer to contractor relationship. Fourth, the Six Sigma model has introduced formally trained layers of quality professionals to guide the organizational efforts. This formal process-mentoring approach would seem desirable in future projects or quality models.

So which of these ideas will emerge in the next era beyond just the maturation of the current ideas outlined above? Whatever evolves we can expect it to be a complex undertaking and thus involve broad changes across the organization.

In the project arena there are two areas that have interesting promise. First, the current waterfall life cycle management methods by which project objectives are produced would seem the most suspect. Related to this would be better Portfolio Management techniques and the PMO. Each of these concepts offers potential to help the organization identify and execute appropriate targets more efficiently. Embedded in this change is the need to re-engineer the classic network management model along the lines of the critical chain (CC) model, as described in Chapter 18 of the book. Specifically, finding more effective methods for task management clearly helps to deal with the timely completion of projects. Dealing with any of these niche area changes will impact how the concept of project quality will be viewed in the future. The most logical prediction is that it will involve finding more effective methods for managing workers in the project—better processes for the allocation of resources to accomplish defined goals.

The final and wildest conjecture of this quality future is the most abstract, but worthy of thought. This view deals with an analytic-based methodological approach to be used for allocating organizational resources to selected targets. As an example, IBM is currently exploring methods of quantifying human metrics of their workforce in regard to their skills and other variables. If successful, these metrics will support their ability to more effectively allocate team members based on cost, skill, and availability factors. The evolving technique called *numerate* is an advanced form of metrics capture and exotic analytical systems related to the work force analysis. Further discussion of this concept is beyond our scope here, but the essence of the idea is that a lot more employee information will be used in future decision making. As a result of these metrics, a more effective job of allocating proper and cost-effective resources to specific task assignments can be achieved. Obviously, this level of change in human profiling will not happen in a short time horizon; however, pragmatic approaches for more effective resource allocation and management procedures can be developed in reasonable time frames. A full analytic view will take significantly longer.

Can we categorize these new ideas into the realm of future quality programs? Maybe the future title for these will change to something more expansive. Note that the Six Sigma program does not have the term "quality" in it, so we might have already begun to drift away from the narrow view of the term. The more important consideration is what management implications need to be considered to make the organization more competitive and successful. This is the real meaning of quality. Many of the terms outlined throughout this book have this same flavor. So, regardless of the next installment, it is likely that all of the components listed here will be somehow involved in that future definition.

21.27 WORKSHEET EXERCISE: ARE YOU MEETING QUALITY GOALS?

Table 21.4 shows a demonstration worksheet to illustrate project QA reviews. There are three types of QA reviews: deliverable, compliance, and health. The role of each one is as follows:

1. *Deliverable:* This activity can range from an early client approval of the project plan to a later approval of a defined outcome product.
2. *Compliance:* Covers areas such as meeting the requirements of a project and complying with established standards and processes.
3. *Health:* Evaluates how well a project is doing; if it has run off the rails. The team can use this review to decide which actions to take to get the project back on track.

TABLE 21.4
Quality Analysis Worksheet

Item	Metric	Example/Comments	Problem Weight: 0–5 0 = Minor 5 = Major	Your Project
		Deliverable		
A	Road map	The solution, or technical outline, is one week behind schedule, which means implementation will be delayed	3	
B	Milestones	There are two late milestones: 1. Approve project plan 2. Approve success criteria	5	
C	Resource changes	There has been some minor staff turnover; no additional staff changes are anticipated	1	
		Compliance		
D	Workflow/reasonability	The project plan has a good WBS, which defines the organized elements needed to meet the project's scope and deliverables	0	
E	Critical path	There are a number of tasks on the critical path (i.e., a group of related tasks that have no slack time between them) that are delayed, on average, by one week	5	
F	Client processes	Client is slow in getting its field people to validate its requirements. PM will follow up	2	
		Health		
G	Hours in project plan versus financial plan	For both plans, the actual hours are over by 150 h; expected remaining hours are on track	3	
H	Earned value	The Earned value or overall project performance, is below target, see above item	3	
I	Risks and issue	The risk plan is up to date. Two issues—procurement and network response—remain open and needed to be resolved	3	
J	Impact of changes	Two project change request are awaiting the client sign-off. If approved, the project plan (schedule and/or cost) will need to be updated	2	

Instructions: Fill in your own numbers under Your Project, compute your Total and compare results with the ratings at the bottom. If your score is in the mid-20s or higher—which is the case below—the project could be headed for trouble; the table uses potential trouble spots to generate the QA grade.

Interpretation: If the rating values are in the range of 1–19, the project is on track; between 20–39 it needs attention and values higher than 39 indicates that the project is headed for failure.

Source: This worksheet is adapted from a similar version published in *Baseline Magazine*, October 2006. The version shown here is approved for use by the author Ron Smith.

Besides serving as an external check-and-balance on the project's performance, the QA staff will ensure that a project and its related artifacts are being developed according to an acceptable process.

Every project should have periodic or "mini," reviews to rate its progress against certain defined metrics such as staying within budget. The worksheet below combines these three types of project reviews into one minireview that can be used to determine whether the project is proceeding smoothly.

DISCUSSION QUESTIONS

1. How would you describe a quality organization or project team?
2. What are some of the key components of quality management?
3. How does Juran's Quality Trilogy compared with Deming's 14 points?
4. Briefly describe the key methodologies of Six Sigma. How does Six Sigma differ from traditional quality ideas?
5. What is the ultimate goal of a quality program?
6. Why is it important for the PM to understand the differences between project-specific requirements and the quality policies of the organization?
7. What can be gained by involving personnel from all functions of the organization in the quality planning stage of a project?
8. When the organization does not consider the quality function to be important what would be the best way to emphasize why you were attempting to sell the idea to senior management?
9. What is the role of data collection and analysis in quality management?
10. How can you justify the overhead cost of a quality management program? What are the cost issues in the quality arena?

REFERENCES

American Society for Quality (ASQ), 2008. Benchmarking. www.asq.org/learn-about-quality/benchmarking/overview/overview.html (accessed January 9, 2009).

Besterfield, D.H., C. Besterfield-Michna, G.H. Besterfield, and M. Besterfield-Sacre, 1995. *Total Quality Management*. Englewood Cliffs, NJ: Prentice Hall.

Camp, R., 1995. *Business Process Benchmarking: The ASQC Total Quality Management*. Milwaukee: ASQC Quality Press.

Crosby, P.B., 1979. *Quality is Free*. New York: McGraw-Hill.

Environmental Protection Agency (EPA), 2001. EPA Requirements for Quality Assurance Project Plans (EPA QA/R-5). http://www.epa.gov/quality/qs-docs/r5-final.pdf (accessed December 15, 2008).

Sixsigma, 2008. Benchmarking. www.isixsigma.com/library/content/c020815a.asp (accessed November 8, 2008).

ISO 8402. Quality Management and Quality Assurance (ISO 9000:2000). http://www.iso.org/iso/iso_catalogue/catalogue_ics/catalogue_detail_ics.htm?csnumber=29280 (accessed December 15, 2008).

J. D. Power, 2007. IQS Ranking Highlights. http://www.jdpower.com/corporate/news/releases/pressrelease.aspx?ID=2007088 (accessed June 1, 2008).

John Stark Associates, 2008. The Ten Step Approach to PLM. www.johnstark.com/fwtqm.html (accessed December 15, 2008).

Liker, J.K., 2004. *The Toyota Way*. New York: McGraw-Hill.

Microsoft, 2008. Adopting Quality Management for Business Success. Projects at Work. http://www.projectsatwork.com/article.cfm?ID=242007 (accessed November 30, 2008).

Mosaicinc, 2001. What is the Difference Between Quality Assurance, Quality Control, and Testing? Selected Risk Management Tip. http://www.mosaicinc.com/mosaicinc/rmThisMonth.asp (accessed November 30, 2008).

Park, S.H., 2003. *Six Sigma for Quality and Productivity Promotion*. Tokyo: Lean Sigma Institute, Asian Productivity Organization.

Quality and Standards Authority of Ethiopia (QSAE), 2008. Implementing ISO 9000 Quality Management System. http://www.qsae.org/web_en/pdf/ISO9000ImpSteps.pdf (accessed January 9, 2009).

Sharma, S., 2007. Quality Management for Business Success. www.gantthead.com/article.cfm?ID=236228 (accessed December 15, 2008).

SixSigma, 2008. www.isixsigma.com/library/content/c020815a.asp (accessed November 15, 2008).

Toyota, 2008. Toyota Production System. http://www.toyota.co.jp/en/vision/production_system (accessed December 15, 2008).

Tudurov, B., 1996. *ISO 9000 required: Your Worldwide Passport to Customer Confidence*. Portland: Productivity Press.

22 Risk Management

22.1 INTRODUCTION

One view of risk is that it involves the full spectrum of uncertainties in the project. These uncertainties can be related to schedule, cost, and quality variability of the end deliverable. Risk management is the means by which this uncertainty is systematically identified and managed in order to increase the likelihood of meeting project objectives. One might argue that project risk is represented by all of the things that make the project plan inaccurate. Based on this view project management becomes risk management. If nothing ever went wrong with the plan variables, we would have to call the process "project checker" because that would be what the person would be doing with their time. A more technical view is that the term "risk" can be interpreted to represent an uncertain opportunity (positive) or threat (negative). When organizations consider whether to introduce a new product to the market, the decision to pursue the idea contains both opportunity and threats. Decision makers have formally struggled with the opportunity side of this subject for many years as they pursue organizational goals; however, it has only been in the past few years that techniques to formally assess the threat side have been structured into some usable and teachable forms. This is not to imply that the art of risk assessment has reached a scientific state of accuracy, but major progress has been made to reveal techniques that aid in minimizing project risks. Risk profiles are now being defined for various project types and these will help in at least better tracking of the event even if there are not so many techniques to identify them.

One of the complex issues related to risk is that it is a *potential* event that may or may not occur. This means that it cannot be recognized in the project work plan if it does not exist. Beyond this dilemma there is the related issue that we do not know when it will occur or the impact—unfortunately, those are the two basic elements to accurately deal with the problem. It is little wonder that a problem with this characteristic has been less formally considered and thus avoided in the project management sphere for so long. One lesson that we have observed in previous situations is that problems that are not looked at carefully will remain unsolved. This is the case with risk. Organizations are finding that the more they look at this area of the project the better they become in identifying the general situations that can be monitored. Once the identification process improves other supplementary techniques are derived.

After this stirring description of the risk problem you may be asking "If I learn this material will I be able to avoid project risk in the future?" The answer to this question is clearly "no." But this is a valid model to evolve in maturing toward a better capability later—continuous improvement. Being sensitive to the risk issue and monitoring risk situations is a critical skill for the project manager. In order for the risk management process to be truly effective it needs to be proactive throughout the project life cycle. This implies not only trying to anticipate and plan for the event, but to monitor activities later in order to minimize them when they do occur.

The ultimate management vision of risk management is to understand the realm of potential problems that can occur in the project and how each class of them can affect project success. Project risk management is the art and science of identifying, analyzing, and responding to risk events throughout the life of a project. Its objective is to increase the probability and impact of positive events, and decrease the probability and impact of events adverse to the project. All risks can never be fully avoided or mitigated because of financial and practical limitations. *Risk management* is the systematic process of planning for, identifying, analyzing, responding to, and monitoring project risks. It involves processes, tools, and techniques that will help the project manager maximize the probability of a predictable outcome.

The term *risk* in the basic dictionary refers to the possibility of loss or injury. This definition highlights the negativity often associated with risk and proves that uncertainty is involved. However, risk is an uncertainty that can have negative or positive effect on meeting the objectives of the project. The uncertainties can be regarding the schedule, the costs, or the quality of the end project. Risk management is the means by which this uncertainty is systematically managed in order to increase the likelihood of meeting project objectives. It helps in meeting the project objectives effectively. Negative risk management is like a form of insurance whereas positive risk management is like investing in opportunities. The main objective of project risk management can be viewed as minimizing negative risks and maximizing positive risks.

A technical project consists of a myriad of activities required to achieve its required goals. Due to the complex nature of these projects they often encounter problems that were not planned, which results in missed schedules, budgets, and substandard outputs. There is no known management technique that can cure all of these ills; however, the risk management process described in this chapter is one that must be considered as a mandatory activity if one desires to avoid many of the potential negatives waiting in the project life cycle. The risk management process represents one of the newer and less practiced management processes in the toolkit, but one that is being increasingly recognized as being an important element in the overall management process.

The problem with managing this aspect of the project is that we do not know when or if any of these will occur. We just know that something like this might occur. These events are called *known/ unknowns*. Most of the literature on this topic focuses on the negative (threat) side of risk, but we must keep in mind that there is a positive (opportunity) side as well. The same theoretical decision processes work in either direction.

The risk management process model provides a means by which these known/unknowns can be systematically identified and managed throughout the project life cycle.

- Risk management analyzes the project deliverables, environment, and stakeholders from a critical perspective to find situations that could have an adverse effect on the project. Examining these weaknesses allows refinements to various aspects of the project plan in order to minimize the impact of the critical events.
- Risk event planning items are translated into the project plan as updates to various work unit statements of work, responsibility roles, communication plans, or new tasks on the project schedule.

This process is a structured set of steps progressing from the initial risk event assessment, developing impact assessments, defining mitigation strategies to deal with the selected events, and then tracking status of the event. On the threat side of the equation, risk management represents an organizational discipline for dealing with the possibility that some future event will cause harm to the enterprise or one of its projects. The ultimate concept behind risk management is to help better manage potential problems that may occur in the project, and from this develop techniques to minimize their impact. In its present state, it is both an art and a science of identifying events, analyzing their impact, and then responding to those events throughout the life of a project. Identifying potential events prior to their occurrence provides time to either minimize their impact or avoid them completely. This technique is called mitigation.

There are many real-world examples where some catastrophic event destroys the success of an organization or a project. For example, approximately 50% of the businesses that occupied the World Trade Center did not survive the terrorist attack, whereas other similar organizations were able to resume operations with minimal impact. How can this be? At the foundation level all organizations should have contingency plans to cover loss of their facility and infrastructure— buildings, technology, equipment, and people. Each of these categories represents classes of events that would be catastrophic to the project if they occurred. One does not anticipate this event, but do you want to bet your project on the fact that it might occur? In project situations the risk events tend to be less dramatic and generally less severe, but in the cases where the entire organization was impacted the project got carried along. For this reason it is hard to separate enterprise level risk from project risk.

The domain of risk is large and uncertain. It is impossible to deal with everything that can go wrong in the project life cycle. For that reason the domain is divided into two categories: known risks and unknown risks. A brief definition of each is as follows:

- *Known risks* are those that fall into a class that could be logically expected to occur and for which some general probabilities and impacts can be estimated. These risks can be reasonably dealt with by effective risk management techniques and can be minimized by following those techniques.
- *Unknown risks* are not in the domain of predictable events and are not generally anticipated. These are not generally identifiable or predictable. These occur at quite random intervals.

Classification of risks into these two categories is difficult in reality. Basically, the only difference is in the handling of the event after the fact. There is a somewhat amusing example of this classification problem (i.e., if you are not a big fan of owls). One of the popular certification exam questions for project management talks about two owls that have taken up roosting in one of the exhaust towers of a nuclear power just as it is ready to go into operation. Environmental group pressure keeps the plant shut down for an extended period of time until the owls decide to leave. The question is what type of risk is this, a known or an unknown risk? The author's Texas students say that this is not a risk at all; it is dinner (minor workaround item to be handled when it occurs). An environmentalist would say that the project team should have known that mating owls were common in this area and accordingly made provisions for their roosting by setting up a suitable alternative place and then monitoring them coming into the plant where their first choice would be discouraged (monitoring issue). Others might say that knowledge of owls is not reasonable to have anticipated the issue and they would therefore classify the issue as an unknown risk. Although the example is unusual it makes a good point on type classification issues and their corresponding event results (i.e., impact on the plant varies depending on recognition and treatment of the event). As this example illustrates, how a risk is classified lies in the level of ability to identify known events. In order for projects to be successful, its associated risk items need to be successfully identified and then managed proactively and consistently throughout the life cycle. Even if the events are not fully defined an active monitoring process is still required. Not all risks can be taken out of the equation, but an active management process is mandatory.

Organizations perceive risks as they relate to project success or to opportunities to enhance chances of project success. An opportunity risk may be acceptable if it is in balance with the reward that can be gained by taking up that risk. Risks that are opportunities have the potential to benefit the project or organizational objectives.

As we have indicated, the biggest problem in dealing with risk is guessing which set of events is most likely to occur and at what probability level. If we could do that accurately the rest of the process is not difficult. It is important to recognize that doing a great job of risk management on the wrong set of events has marginal to negative value. The only way an organization is going to get better at this process is to start doing it and learn from its experience and that of others in similar efforts. In the case of enterprise level risk, is there a potential for a facility fire? What about extended power outage? Certainly there exists a common set of general environmental risk events common to all activities. At the project level, the risk identification becomes more focused to activity internal to the project. When new technology is involved, that immediately becomes a risk target to assess. Why then do we observe the type of failures outlined above? In many cases the victim organizations did not have a risk culture. That is, they simply did not think about the issue or were naïve to its existence. Why do older employees typically believe in risk more than younger ones? The answer probably lies in the fact that they have seen these negative events occur and are more sensitive to them.

22.2 RISK MANAGEMENT PROCESS

The *PMBOK® Guide* defines six major processes in the risk management process (PMI, 2008, p. 273):

1. *Plan risk management:* This involves how to approach, plan, and execute the risk management activities for a project. The main output of this process is risk management plan.
2. *Identify risk:* It involves determining the risks that are likely to affect a project and documenting the characteristics of each. The main output of this process is risk register.
3. *Perform qualitative risk analysis:* This involves prioritizing risks for subsequent further analysis or action by assessing and combining their probability of occurrence and impact. The main update to this process is updates to the risk register.
4. *Perform quantitative risk analysis:* This involves numerically estimating the effects of risks on project objectives. The main output of this process also involves updating the risk register.
5. *Plan risk responses:* This involves developing options and actions to enhance opportunities and to reduce threats to project objectives. Using the outputs of the above processes various risk response strategies are made, which result in updating the risk register and project management plan as well as risk-related contractual agreements.
6. *Monitoring and control risks:* This process involves tracking identified risks, monitoring residual risks, identifying new risks, executing risk response plans, and evaluating their effectiveness throughout the project cycle. The main outputs of this process are recommended corrective and preventive actions, requested changes, updates to the risk register, and revised project management plan.

This process structure is illustrated in Figure 22.1.

Completion of the six processes summarized above involves the collective efforts and skills of multiple persons or groups based on the nature and requirements of the project. These analytical processes may occur once in the project or be formally repeated for each project phase, but it is important to recognize that risks can be identified at any point in the life cycle and for that reason one should think of this as an ongoing requirement. The sections below will further operationalize these steps and show their interrelationships.

The Environmental Risk Management Authority (ERMA) of New Zealand is chartered by their government to foster improved risk management and they offer a real-world operational view of this process to show how the basic processes are grouped and interrelated (AS/NZS 4360-2004, 2004). Figure 22.2 melds their view of the process into a model similar to *PMBOK® Guide* view.

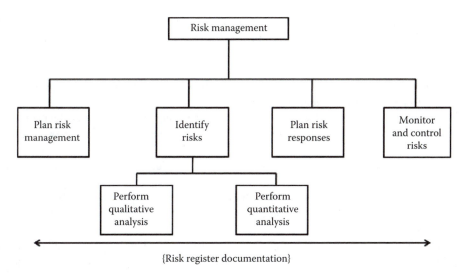

FIGURE 22.1 *PMBOK® Guide* risk management structure. (PMI, 2008. *PMBOK® Guide*, 4th Edition. Newtown Square, PA: Project Management Institute.)

22.3 RISK MANAGEMENT PLANNING

The risk management plan describes how this aspect of the project will be structured and performed. It is included as a subsidiary plan of the project management plan. This is the portion of the planning process where decisions are made regarding how to formulate, plan, and execute the risk management activities for a project. The process involves a systematic approach to planning the risk management activity based on the premise that careful planning enhances the possibility of project success. The main output of this process is a risk management plan that documents the procedures to execute and manage the risk-related activities throughout the duration of the project. Careful and explicit risk planning helps guide the project team's reactions when a particular event occurs later. Developing a plan to outline how the risk management steps will be executed is essential to ensuring the proper level, type, and visibility. The risk planning process is embedded in the basic project planning process activities and interrelates with those activities as decisions are made regarding

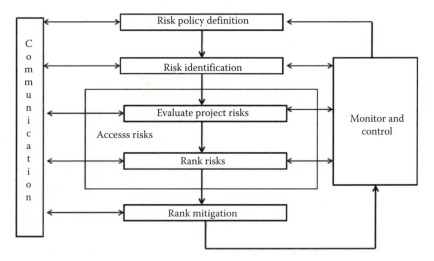

FIGURE 22.2 ERMA risk management process.

handling of specific work activities. Various knowledge area plans can be impacted and modified as a result of these actions. A summary of topics addressed in the risk management planning model is as follows:

- *Methodology:* This involves the various approaches, tools, and data sources that are to be used to perform risk management on the project.
- *Roles and responsibilities:* Defines the event manager, support, and risk management team membership for each type of activity in the risk management plan, assigns people to these roles, and clarifies their responsibilities.
- *Budgeting:* Assigns resources and cost estimates for inclusion in the project contingency plans.
- *Timing:* Defines when and how often the risk management process will be performed throughout the project life cycle, and establishes risk management activities to be included in the project schedule.
- *Risk categories:* Provides a structure that ensures a comprehensive process of systematically identifying risk to a consistent level of detail and contributes to the effectiveness and quality of Risk Identification. An organization can use a previously prepared categorization of typical risks. An RBS is one approach to providing such a structure, but it can also be addressed by simply listing the various aspects of the project.
- *Definitions of risk probability and impact:* The quality and credibility of the Qualitative Risk Analysis process requires that different levels of the risks' probabilities and impacts be defined. General definitions of probability levels and impact levels are tailored to the individual project during the Risk Management Planning process for use in the Qualitative Risk Analysis process.
- *Risk documentation:* This segment of the plan defines the formats and processes that are going to be used to document the risk items identified.

There can be various types of risks. Generally, the risk events fall into broad categories such as market risk, financial risk, technology risk, people risk, and structure/process risk.

The output of the risk planning stage is an RBS, which is a useful tool to determine potential risk categories for the project. This structure guides the project team in considering the risk categories for their respective project. In addition to this, checklists and templates are often used to review areas and stimulate thinking for risk identification.

Apart from identifying risks based on the nature of the projects produced, it is also essential to identify potential risks according to project management knowledge areas such as scope, time, cost, and quality. Examples of various risk conditions associated with different project KAs are as follows:

- *Integration:* Inefficient planning, improper resource allocation, poor integration management, and lack of post project preview.
- *Scope:* Maximized scope, poor definition of scope, and poor definition of work packages.
- *Time:* Errors in critical path calculations, early release of competitive products, errors in time estimation, errors in calculating resource availability, poor allocation, and management of float.
- *Cost:* Errors in estimating cost, inadequate productivity, change, or contingency.
- *Quality:* Inadequate quality assurance program, substandard design, substandard materials, and substandard workmanship.
- *Human resources:* Poor conflict management, inadequate leadership qualities, and poor organization of responsibilities.
- *Communication:* Insufficient communication with the key stakeholders and improper planning.

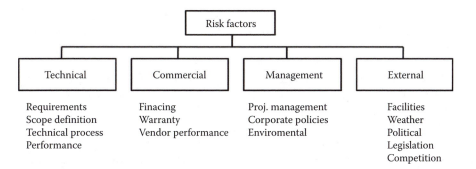

FIGURE 22.3 Sample RBS.

- *Risk:* Less interest in risk management, less concern toward insurance management, and improper analysis of risk.
- *Procurement:* Unenforceable conditions and contract clauses, and hostile relationships.

The output of the Risk Management Planning process is the Risk Management Plan that identifies and establishes various procedures for managing risk throughout the project. The responsibility of Risk Management Planning is to determine the scope of the risk management activity (e.g., determine at what level the potential risks are considered to be relevant), map out the other risk management activities, develop the risk management plan, and communicate these plans to the appropriate stakeholders. The document outlines the *what*, *who*, and *why* questions and establishes an operational context for the remaining steps.

22.3.1 DEVELOPING AN RBS

One major aspect of the risk plan is categorization of risk elements and one approach to outlining a macroview is to use an RBS as illustrated in Figure 22.3.

The use of an RBS aids in categorizing potential risk groups for a project. Each project will have to modify this structure, but it does offer general guidance for a high-level view. The specific example areas listed underneath the major risk groups represent subareas that can spawn this type of risk. The general RBS is a simple chart to help ensure that the project team considers important risk areas.

22.4 RISK IDENTIFICATION

The second and most complex step is *risk event* identification that consists of identifying and documenting potential risk events to the project. This process is performed by the project manager, project team members, risk management experts, business process experts, customers, end users, other technical resources, stakeholders, and outside experts. It is an ongoing, iterative process because new risks may become known as the project progresses through its life cycle. The frequency of iteration and who participates in each cycle will vary from case to case. The project team should be involved in the process so that they can develop and maintain a sense of ownership and responsibility for the risks and associated response actions. In this phase, the context of the risk is an important consideration. Issues such as organization, political, economic, time frame, and other such trigger mechanisms will help to understand why the item is perceived to be an opportunity or threat. It is crucial to identify potential risk events as early as possible in the life cycle of the project. A risk cannot be properly managed if it is not identified—surprise events lead to a chaotic environment later.

Through experience, a localized checklist possibly structured into an RBS format can be developed to guide the project team even more specifically to items relevant for the local environment.

Related to the RBS view of risk identification is a checklist format. This is illustrated by a sample checklist of risk-related questions. This list is adapted from the software development risk methodology used at Hill Air Force Base (GSAM, 2003):

1. Is the project manager dedicated to this project, and not dividing his or her time among other efforts?
2. Are you using a proven development methodology?
3. Are requirements well defined, understandable, and stable?
4. Do you have an effective requirements change process in place and do you use it?
5. Does your project plan call for tracking/tracing requirements through all phases of the project?
6. Are you implementing proven technology?
7. Are suppliers stable, and do you have multiple sources for hardware and equipment?
8. Are all procurement items needed for your development effort short-lead time items (no long-lead items?)
9. Are all external and internal interfaces for the system well defined?
10. Are all project positions appropriately staffed with qualified, motivated personnel?
11. Are the developers trained and experienced in their respective development disciplines (i.e., systems engineering, software engineering, language, platform, tools, etc.)?
12. Are developers experienced or familiar with the technology and the development environment?
13. Are key personnel stable and likely to remain in their positions throughout the project?
14. Is project funding stable and secure?
15. Are all costs associated with the project known?
16. Are development tools and equipment used for the project state-of-the-art, dependable, and available in sufficient quantity, and are the developers familiar with the development tools?
17. Are the schedule estimates free of unknowns?
18. Is the schedule realistic to support an acceptable level of risk?
19. Is the project free of special environmental constraints or requirements?
20. Is your testing approach feasible and appropriate for the components and the system?
21. Have acceptance criteria been established for all requirements and agreed to by all stakeholders?
22. Will there be sufficient equipment to do adequate integration and testing?
23. Has sufficient time been scheduled for system integration and testing?
24. Can software be tested without complex testing or special test equipment?
25. Is the system being developed by a single group in one location?
26. Are subcontractors reliable and proven?
27. Is all project work being performed by groups over which you have control?
28. Are development and support teams all collocated at one site?
29. Is the project team accustomed to working on an effort of this size (neither bigger nor smaller)?

The intent of such a checklist is to trigger a defined risk response based on that area. A review of the checklist against each WBS element would aid the risk review process regarding specific event discussions about that area. From this type of review one or more specific risk events might be identified. As stated earlier, no predefined checklist will capture all such possibilities, but they do represent a way to get started with the process. As items are identified, they are documented in the *risk register*, which becomes the master repository source for the rest of the process. Figure 22.4 illustrates the steps that follow the identification process.

The first operational step in the process is to identify and document potential events that could negatively affect or enhance a particular project's ability to achieve its objectives. This activity

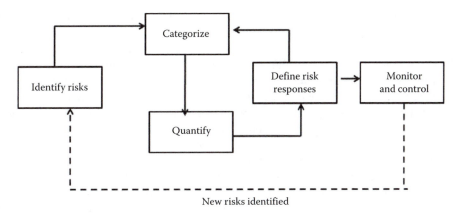

FIGURE 22.4 Risk response planning steps.

should be ongoing during the planning stage and continue throughout the project life cycle. One of the key themes of risk management is to identify potential risks that can affect the project outcome and manage these events in an improved fashion over what would have been the case if that process had not been undertaken.

Potential risks can be identified based on a general understanding of common risk sources and by reviewing project scope documents, the WBS details, environmental factors, and other organizational issues. Participants in risk identification activities include the project manager, project team members, appropriate subject matter experts, customers, end users, other stakeholders, and risk management technical experts. The Risk Identification team should consist of people with appropriate project and domain experience (Glazewski, 2005). Several tools and techniques are useful in the identification process. A sample of these is included below (DAU, n.d.):

- *Personal interviews sessions:* Structured interviews sessions with experts help identify target areas for further analysis. This class of activity will develop high-level ideas and potential solutions. The goal of this activity is to gather ideas spontaneously and without judgment. Individual interviews are an effective way to work on more detailed issues and group interviews can help gain consensus on various aspects of the process.
- *Work breakdown structure:* The WBS serves as a checklist of work activities to review for risk. Each box in the structure represents a work effort with associated technical, resource, and other risk aspects. Think of it this way; if each of these work units were completed on schedule and within budget, the project would achieve its goal. Conversely, if that were not the case a risk event probably entered the scene. We reiterate that the WBS is the single most important management artifact and this is just one more of its roles in the life cycle.
- *Documentation review:* Artifacts created by previous projects can provide a wealth of information regarding risk. Specifically, lessons learned documentation from an earlier project team could provide specific insights into issues that might translate to the new initiative.
- *Root cause analysis:* Often times a visible event is the result of some hidden relationship. For this reason it is not adequate to just say "this bad thing might happen." What is more relevant to this process is the root cause behind those potential "bad things." This level of understanding enhances the manageability of the event and facilitates grouping risks by source. From this, effective risk responses can be developed.
- *Means-ends analysis:* A chain of underlying lesser items can trigger many significant risk events. As an example, the threat of a catastrophic fire could be lessened by several

underlying mitigation processes (fire alarms, sensors, trained staff, extinguishers, etc.). In the analysis process this means-ends view becomes part of the required activity. By understanding the means-ends relationships we can better decide how to manage the event.

- *Checklist analysis:* As organizations deal with a common project type they develop a clearer view of risk events. From this understanding, checklists can be developed to guide the analysis process and help ensure that potential items are not overlooked. A pilot's checklist before takeoff is designed to ensure that some critical activity is not overlooked. A checklist can form the redundant core of analysis, but they do not deal with the unique characteristics of a project. Only human intelligence will deal with that.
- *Brainstorming:* This is a free-form idea collection process in which project team members, subject matter experts, stakeholders, and anyone else who might have information or knowledge about this project meet to identify potential risks for consideration.
- *Delphi technique:* This technique involves experts in the area being reviewed to identify an event, or the potential for an event. Participants rank their answers and the results are reviewed by the group. After multiple iterations a consensus tends to occur and this is used as expert input on the topic.
- *Strengths, weaknesses, opportunities, and threats (SWOT) analysis:* This is an orderly review for assessing the strengths, weaknesses, and opportunities of a particular risk event. The SWOT technique ensures examination of the project from each of the four perspectives, which increases the breadth of considered risks.

Details related to the risk events identified in the identification phase are added to the original event data originally captured in the project *risk register.* Conceptually, this is an evolving database of all identified risks and their associated status information. These data items represent information sources for the ongoing planning and control activities related to risk management. Data elements captured for each risk event would include

1. Reference number
2. WBS impact area(s)
3. Description of risk—possibly a short and long version
4. Statement of consequence—a code reference could be used here
5. Likelihood of occurrence
6. Impact of occurrence
7. Frequency—one time, monthly, and so on
8. Other items, as the process evolves.

The outputs of the Risk Identification phase are as follows:

- *Triggers:* Triggers are early warning signs that a risk has occurred or is about to occur. Any triggers that are identified during this process are documented and the list is updated and used as a monitoring and control aid.
- *Risk register:* This repository is a formal source for capturing the project knowledge regarding risks and their status. It is a document that contains results of various risk management processes and is a tool for documenting potential risk events and related information.

Each risk event identified in the register should contain an identification number, a severity ranking, a description of the risk event (probably both a short and a long one for different purposes), the category under which the risk event falls, triggers for each risk, potential responses to each risk, the risk owner or person who will own responsibilities for the risk, the probability of the risk occurring,

TABLE 22.1
Sample Risk Register Format

No.	Rank	Risk	Desc.	Cat.	Root Cause	Triggers	Potential Responses	Risk Owner	Probability	Impact	Status
R6	1										
R43	2										
R21	3										

and the impact to the project if the risk occurs. Sample risk register data elements are summarized in Table 22.1.

22.5   QUALITATIVE AND QUANTITATIVE RISK ANALYSIS

These two steps are designed to provide a measure of quantification to the identified risk events. Risk assessment involves the determination of qualitative and quantitative rating values for the identified threat or opportunity. The first pass through the risk event list will qualitatively label each item as to their likelihood of occurrence and impact. These are somewhat subjective grades shown typically as H, M, L, or numeric grades from 1 to 5. The goal at this stage is to select the highest probability and impact grade items on the list, realizing that some sizing partitioning is going to be required in order to focus on the most likely and troublesome ones.

At this point we hit a critical operational problem with the methodology. If it were possible to assign numeric values to each event in terms of probability and dollar impact, the mathematics for risk would be quite reasonable. Before moving further with the model methodology let us look at what the pure math side of this activity would look like. If a risk event was estimated to occur one time in the project with a probability of 10% and if it occurred the impact would be $100,000, we could model the expected impact of that event at $10,000 (0.10 × 100,000). Using the same method that an insurance company would use to calculate customer premiums, we would set aside in a contingency fund $10,000 to cover the expected impact of this known/unknown event. It is always confusing to those exposed to this idea for the first time as to how $10,000 can cover a $100,000 event. A short answer to this is that we have a lot of such probabilistic events and all of them will not happen. In mathematical terms, if all of the estimates have been just right and we have a sufficient number of them, the contingency fund will be exhausted at the end of the project with zero remaining. Obviously, given our rough method of defining probability and impact, our math is not that accurate in this case, but this is the game we are playing in defining the contingency amount to be set aside for these events.

22.6   RISK ASSESSMENT

Once the potential risk event list is developed, some additional judgment is required. The assessment progresses in two stages: qualitative review, then quantitative. Given that the number of risks identified will be too numerous to review in detail, the goal of the preliminary phase is to categorize their population into more manageable groups. Basically, this often results in a high, medium, low grouping or some numerical grouping (e.g., 1–5). From this level the more detailed quantitative analysis will follow.

Qualitative Risk Assessment: This process essentially involves a grading of the defined events by grading the likelihood and consequences with the "pseudo-mathematical" scoring scheme as outlined above. This process should also assess other factors such as time frame of occurrence and risk tolerance based on project constraints of cost, schedule, scope, and quality. The techniques involved here are as follows:

Risk potential and impact assessment: This assessment explores the likelihood of each risk event occurring, whereas the impact assessment explores how such an event could affect a project objective such as time, cost, scope, or quality. This includes both negative effects for threats and positive effects for opportunities. Risk potential and impact are rated according to the definitions given in the risk management plan. In some cases an event with obviously low ratings will not be analyzed further, but will be kept on a watch list for future monitoring.

Qualitative risk analysis matrix: This evaluation matrix is defined in the risk management plan and is used accordingly at this point to categorize the risk events on a relatively crude basis. This matrix provides the scoring logic for combining scores for the likelihood and consequences of a risk. A sample five-level qualitative analysis matrix is shown in Table 22.2. A risk class score for a given event is defined by the matrix intersection using the assigned likelihood and consequence descriptors.

In order to populate the risk matrix, the following three steps should be considered for organizing the risk view:

Risk categorization: Project risks can be developed by categorizing potential sources of risk using the RBS. This is basically a view of the WBS sorted into risk groupings. This type of organization can be useful for identifying risk events or developing effective risk responses.

Risk data quality assessment: Analysis of the quality of risk data is a procedure to assess the degree to which the data about risks are useful for the risk management process. It involves examining the degree to which the risk is understood and the accuracy, quality, reliability, and integrity of the data about the risk.

Risk urgency assessment: Risks requiring imminent responses may be considered more urgent to tackle. Risks that should be addressed immediately are indicated with priorities such as time to affect a risk response, symptoms and warning signs, and the risk rating.

TABLE 22.2
Qualitative Risk Analysis Matrix

	Impact				
	1	2	3	4	5
Probability	Insignificant	Minor	Moderate	Major	Catastrophic
Highly likely	H	H	E	E	E
Likely	M	H	H	E	E
Possible	L	M	H	E	E
Unlikely	L	L	M	H	E
Rare	L	L	M	H	H

Note: E = Extreme; formal risk assessment must be performed.
H = High; formal risk assessment should be performed.
M = Moderate; for possible and above ratings, a formal risk assessment should be performed.
L = Low; formal risk assessment at discretion of project manager.

Upon conclusion of this process risk events are identified, graded using the matrix codes, and recorded in the *risk register*.

Quantitative risk assessment: Once the risk events are categorized according to the qualitative process, quantification of these is needed to help decide how to deal with the items. Some items may be sufficiently severe to necessitate some amount of mitigation effort to remove or minimize the risk. Alternatively, some will be so small that further mitigation work is not justified. In all cases, the quantification results are captured in the risk register. Risk quantification requires two analytical components: the impact of the potential loss and the probability that the loss will occur.

The Risk Matrix is a simple tool to help prioritize risks and is used to translate the qualitative risks into impact/probability groups that can be further analyzed. The normal approach would be to deal with all of the high impact events (dark grey), many of the middle tier, and probably few of the less significant ones (light grey) (see Table 22.3).

One potential method of quantifying risk is to assign a numeric value to the risk matrix groups and to multiply these values times the impact estimate for the risk event to form an expected value for the event. There are many operational problems with this approach as actual probabilities are seldom known, but as one gains experience with the process it may be possible to develop forecasting metrics of this type.

The Risk Matrix serves two basic purposes. First, it guides the risk assessment process by the color-coding and helps establish policies as to what level of risk needs specific treatments. In the ideal case, it is conceptually a method to develop risk probability values that would be useful for computing expected loss estimates. In such a case, the value in the cell would be multiplied by the estimated dollar impact of the event to determine an amount to be set aside as a risk contingency fund.

To assist in the risk event grading process, there are several supporting analytical techniques used in quantitative risk assessment including data gathering, quantitative risk analysis, and simulation modeling techniques. The commonly used tools and techniques are as follows (Elyse, 2007):

1. *Decision tree analysis:* Probabilistic evaluation of an event can be modeled using the decision tree diagramming technique. This method traces the various outcomes of their projected results and from this one can judge the appropriate course of action. It encompasses the cost of each available choice, the probabilities of each possible scenario, and the rewards of each alternative logical path. Solving the decision tree provides the expected monetary value (EMV) for each alternative, when all the rewards and subsequent decisions are quantified. A sample decision tree is shown in Figure 22.5.

TABLE 22.3
Qualitative Risk Analysis Matrix

	Impact				
	1	2	3	4	5
Probability	**Insignificant**	**Minor**	**Moderate**	**Major**	**Catastrophic**
Highly likely	0.15	0.15	0.20	0.30	0.35
Likely	0.10	0.15	0.15	0.20	0.30
Possible	0.07	0.10	0.15	0.20	0.20
Unlikely	0.07	0.07	0.10	0.15	0.20
Rare	0.05	0.07	0.10	0.15	0.15

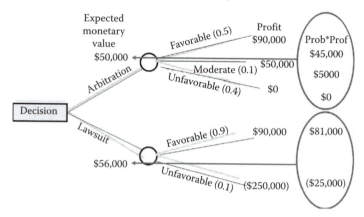

FIGURE 22.5 Decision tree example.

    This decision tree example compares a strategy of pursuing a lawsuit or using arbitration to settle a dispute. Each of the branches of the tree outlines the options that are considered and the corresponding probability of that individual outcome. At the end of each branch is the monetary result. By multiplying the probability of the event times the outcome and summing the branch we see that the decision to pursue a lawsuit receives the greatest expected value. Also, note that it has the highest potential for loss at $250,000. In some decision cases the avoidance of a major loss is more desirable than achieving maximum return.

2. Modeling and simulation: Creation of simulation models is a sophisticated quantitative risk analysis technique. The typical method used to simulate project outcomes is to assign variable time and/or cost probabilities to project activities. Then using these probability distributions the model is executed many times (say 1000 cycles) to produce a probabilistic outcome for the project. The net result of this is a histogram of defined results and from that data appropriate risk-related decisions can be made.

3. *Sensitivity analysis:* Sensitivity analysis is used in determining that risks have the most potential impact on the project. One typical display of sensitivity analysis is the *tornado diagram* which ranks the project activities in order of significance for, say, critical path. From this view one can decide where to focus attention.

4. *EMV analysis:* EMV is a common calculation used in probabilistic applications to measure decision outcomes. This variable is calculated by multiplying the value of each possible outcome by its corresponding probability of occurrence and then adding the results. The EMVs for opportunities are expressed as positive values, whereas those of threats are shown as a negative.

5. *Data availability:* A common risk analysis mistake is to base risk ratings on incomplete specification data. It is always necessary to review the data used in the assessment as part of the process.

6. *Data quality:* Use of lessons learned data for comparable projects is a valid approach so long as that data are timely and relevant. Comparing a current project using data from a project 20 years ago is not normally a viable practice.

7. *Data integrity and reliability:* Qualitative Risk Analysis is imprecise and its ratings reflect subjective opinions and judgment. However, with this fact in mind, it is important to obtain the most accurate and unbiased information available. For example, if the previous project fought a war of politics among key stakeholders, it may not be appropriate to use that environment for comparison purposes.

TABLE 22.4
Contingency Fund Calculation

Risk	P (Risk Probability)	I (Cost Impact) ($)	Risk Contingency ($)
A	0.80	10,000	8000
B	0.30	30,000	9000
C	0.50	8000	4000
D	0.10	40,000	4000
E	0.30	20,000	6000
F	0.25	10,000	2500
Total		118,000	33,500

Source: TenStep, 2008. Management Tip of the Week, www.Tenstep.com, (accessed January 21, 2009.)
With permission.

22.7 RISK CONTINGENCY BUDGET

After all risks have been dealt with, the final planning challenge is to set aside sufficient reserves to support these future events. The simple example shown in Table 22.4 illustrates some sample mechanics of this process.

Of course the challenge is in developing the probability and impact values for this calculation. Experience will help with this, but recognize that there is an error built into the process. Also, recognize that other risk events will occur that have not been identified. Mathematically, the risk contingency fund would be sized at $33,500; however, given the calculation estimate error and other unknown events, the fund should be sized higher. Tom Mocal, CEO of TenStep says "I believe the project manager should request an additional 5% of your total budget for risk contingency to cover the undefined risks that you will encounter later. This is in addition to the risk contingency of the known risks that have already been identified" (Mocal, 2008).

The project manager will need to justify the size of resources set aside to cover risk events and management will likely challenge all reserve funds.

22.8 RISK RESPONSE PLANNING

Envision at this stage of the process that the full slate of potential risk events has been defined, along with some measure of their probability and impact. The key question now is what to do with this list. While it is true that the risk environment across industries, organizations, and projects may have common elements, it is also true that different organizations will likely react to these similar events in varying ways. Terms such as *risk tolerance* or *risk aversion* are used to describe how one organization will go to great lengths to avoid a potential event, whereas another will judge the same type event to be tolerable and not spend funds to minimize the event. In this situation, both organizations may be making perfectly valid decisions for their particular environment. Such is the nature of this activity.

As risks are identified they are entered into the *risk register*, which is essentially a database containing all information about the status of the known/unknowns. At this point, the known data are the event itself, but the detail level grows as more data are added through the process. More specifically it will contain data related to the following four activities:

1. Attached to the initial list of risks will be a scoring of their characteristics including their likelihood, mitigation potential, and person in charge of dealing with the risks
2. Rank the risks identified in order of priority using the Risk Matrix coding
3. Calculate the impact on the project if the identified event does occur
4. Response to each risk will be identified and in some cases a contingency reserve will be defined for a particular event

Risk Response planning involves developing responses to named risk events. The goal of these responses is to enhance opportunities and reduce threats to the project's objectives. The result of this activity is to define the action taken for specific risks and the associated steps that are required to carry them out. It should not be difficult to see that these actions have a high potential to add scope, time, or cost to the original project. Each mitigation strategy selected is designed to help ensure project success at minimal cost, but this is a probabilistic statement. These strategies thus represent resources spent to avoid significant bad events. Collectively, these actions represent the risk mitigation plans for the project and these should be documented in the risk register for future access.

It is up to the project team to identify which response strategy is best for each risk, and then design specific actions to implement that strategy. As threats and opportunities are identified, the goal is to minimize threats and maximize opportunities for the project (Elyse, 2007). The decision options for the threat events are as follows:

- *Avoid:* This involves changing the project plan to avoid the impact of a threat on project objectives such as time, cost, scope, and quality. This can be done by changing some element of the work requirement or reducing the scope of the project. Also, this can be accomplished by removing the risky item and replacing it with another more tested version. Replacing new technology with a more stable, older version is typical in this case.
- *Transfer:* Risk transfer requires transferring or shifting the negative impact, along with responsibility for its management to a third party. The process reduces the risk only if the third party is more capable of taking steps to reduce the risk. It typically involves payment of a risk premium to the party taking on the risk.
- *Mitigate:* Risk mitigation is a process designed to reduce the probability or impact of a potential risk to a more acceptable level. This may also include reducing the consequences of the risk.
- *Accept:* Some risks are so small and easily dealt with that it is not economical to spend time developing a response mitigation plan. In these cases the event is simply put on a watch list for monitoring, but nothing else is done to minimize the potential occurrence. On the other hand, there are some risks that are significant, but cannot be mitigated or avoided. They are embedded in the fabric of the project itself. Obviously, in this case the event monitoring process is much more aggressive and should be kept visible to all concerned.

In similar fashion, the three strategies to deal with positive risks (opportunities) are as follows:

- *Exploit:* This strategy is selected for positive risk impacts where the organization wishes to ensure that the opportunity is pursued. This strategy seeks to eliminate the uncertainty associated with a particular upside risk by making sure that the opportunity will have a higher potential to be successful.
- *Share:* Sharing a positive risk involves allocating partial ownership to a third party who is best able to use the opportunity for benefit of the project. The examples of sharing include forming risk-sharing partnerships, mutual teams, working with elected officials, joint ventures, and joint ownership companies.
- *Enhance:* Enhancing a positive risk involves changing or modifying the size of the opportunity by improving its probability and/or impacts, and by identifying and maximizing key drivers to positively influence these items.

Decisions made at this point in the process represent the culmination of the planning stage. The answers to the response questions are used to populate the remainder of the Risk Response Plan elements. Specifically, the wrap-up activities involve the following:

- *Risk register (updates):* Risk Register updates of the appropriate risk response.

- *Project management plan (updates):* Project Management Plan updates occur as response actions are added after being processed through integrated change control.
- *Risk-related contractual agreements:* Risk-related contractual agreements for insurance, partnerships, and services will generate language specifying each party's responsibilities.

22.9 RISK MONITORING AND CONTROL

The risk monitoring and control process represents the ongoing management activities involved for this aspect of the project. During the course of the project some of the identified risks will likely occur (trigger) and the risk plan details will aid in working through the management of that event. In addition, new risks may arise during the life cycle and they will need to be processed as previously described for the planning phase steps.

Activities related to the risk monitoring and control process often lead to plan changes, updates, and revisions. If carried out properly, this process improves the overall effectiveness of the project outcome by providing workable reactions to negative events.

22.10 RISK EVENTS VERSUS ISSUES

There is a general confusion in the project management lingo regarding the difference in a risk event and an Issue. The *PMBOK® Guide* defines an Issue as follows:

> A point or matter in question or in dispute, or a point or matter that is not settled and is under discussion or over which there are opposing views or disagreement (PMI, 2008, p. 429).

Said another way, an Issue is an event that is causing some disruption to the project, but is anticipated to be resolved in a manageable time frame. The sticky definitional area here is that Issues look a lot like risk events. Some would argue that an Issue could evolve into a risk event if the anticipated resolution goes beyond the manageable threshold.

Let us try to illustrate these definitions through the example of a piece of equipment needed for some project work task. During the planning phase we might consider the availability of this item a potential risk event and in fact allocate a contingency reserve for a major outage of this item. Later, during the execution phase the equipment in fact does break down but the initial prognosis is that the failure is minimal and the local technician believes that he can have it working in 4 hours. This outage does not disrupt the project. At this point we define this situation as an Issue and record it for tracking purposes in the Issue Log; however, 4 hours later it is determined that the problem is much worse than forecast and will take much longer to repair. At this stage the event is classified as a risk event and we trigger the risk contingency plan. In this example, an Issue became a risk as it evolved into a more significant outage.

Projects have a seemingly endless list of Issue items that need to be resolved daily. These do not represent regular work items as defined in the WBS, but are more often supporting activities—a drawing needs to be fixed, a vendor is slightly late in delivery, or a key team member is ill, and a replacement is needed for 1 month (felt to be available).

Regardless of how one views these two terms, there is likely to be confusion over how to handle those that begin to look like a risk event. This cataloging would be even worse if there was no risk event defined and it became major. In either case, the event (Issue or Risk) needs to be resolved so that the project can move on without major disruption.

22.11 PROJECT RISK ASSESSMENT WORKSHEET

During the early stages of a project it is good to get a general assessment of the overall risk of the effort. One way to do this is through a high-level weighted criteria worksheet. This method provides a quantitative assessment of the aggregate risk as well as a checklist for the project team as they deal with the internal detailed risk management process. Later, during project execution these same parameters can be computed to see if there is an increased view of overall risk compared to the initial values. Another use of the worksheet data is to provide a mandatory monthly risk assessment and make that part of the standard reporting data.

The use of this type of assessment can provide initial insights into overall project risk level and help assess areas where the risk profile is changing during the execution phase. Table 22.5 shows an example risk analysis worksheet. As with all worksheets of this type, the parameters and measuring scales can be altered over time to fit the local project profiles and experiences. Relevance of a particular score will be learned over time as these numbers are compared to actual projects.

22.12   RISK CASE STUDY

22.12.1   Mishap Foils Latest Attempt at a 25-Mile Skydive

Note: This case excerpted from a *New York Times* May 28, 2008 article written by Matt Higgins.

Michel Fournier, a retired French Army officer who had hoped to fly a helium balloon about 25 miles above the Earth and parachute down, has failed again. As spectators watched, his 650-ft-high balloon inflated and then suddenly floated away, leaving the gondola with Fournier inside on the ground. The damaged balloon was recovered 25 miles away.

The launch team is investigating whether static electricity might have led to the setting off one of charges at the release point between the capsule and the balloon, the agency said.

Fournier had planned to climb into the pressurized gondola of the balloon and make a 2-hour journey to 130,000 ft. Then he planned to step out of the capsule, wearing only a special spacesuit and a parachute, and plunge to Earth in a 15-min free fall. If successful, Fournier would fall longer, farther, and faster than anyone in history.

Fournier has spent 20 years and nearly $20 million in pursuit of the milestone. He sold his house and most of his belongings and solicited funds from sponsors to finance his project.

He has attempted the feat twice already, but technical and weather-related problems foiled the efforts before he left the ground. The most recent attempt, in 2003, failed when his balloon ruptured before takeoff.

If money was no object, outline what you would have done to bring this project in on time (note that the schedule had already been delayed by a previous problem with the balloon).

22.13   CONCLUSION

Before we exit the theoretical risk discussion let us leave you with this final project management point. All projects should go through a risk management assessment process as outlined here. The less maturity an organization has with this process, the less accurate the project results will likely be. However, experience with the methods outlined here can help develop that maturity level, and simply thinking about risk in a formal manner will help sensitize the organizational culture to its existence and the associated fact that mitigation strategies exist to minimize their impact. The last step in the maturation process will be developing techniques to produce a quantification process closer to a true expected value mathematical form. Nevertheless, value is gained even through just the risk assessment process.

Formalized risk management is now becoming widely recognized in both the public and private sectors as an integral facet of effective business practice as it provides management with a deeper insight and wider perspective regarding effective management of the organization activities within its dynamic environment. Also, risk management at the project level is an essential contributor to success as it focuses attention on issues that potentially affect achievement of its objectives.

TABLE 22.5
Project Risk Analysis Worksheet

Tool: Tracking Your Project Risks How Likely will this Adversely Affect Your Project?	A Project's Risk (Filled in Example)				Your Project			
	Low (1)	Medium (3)	High (5)	Risk Level	Low (1)	Medium (3)	High (5)	Risk Level
1. Application complexity			5	5				
2. Baselines	1		1	1				
3. Contract or statement of work	1			1				
4. Customer expectation			5	5				
5. Customer involvement		3		3				
6. Customer acceptance	1			1				
7. Design level of detail		3		3				
8. External dependencies like deliveries	1			1				
9. Hardware, that is, switching to new technology	1			1				
10. Software, that is, switching to new technology	1			1				
11. Interfaces with other systems		3		3				
12. IT experience with system applications		3		3				
13. Productivity rates of IT team members		3		3				
14. Project management		3		3				
15. Project planning/scheduling	1			1				
16. Project resources		3		3				
17. Requirements change		3		3				
18. Requirements definition	1			1				
19. Subcontractor involvement		3		3				
20. System performance requirements			5	5				
21. Telecommunications/network		3		3				
22. Workload estimate of IT team members		3		3				
Monthly totals								
Current month	8	33	15	56				
Last month	8	30	30	68				
First month	6	24	50	80				

Source: This worksheet is adapted from a similar version published in *Baseline Magazine*, July 2007. The version shown here is approved for use by the author Ron Smith. With permission.

Instructions: If an item is not relevant to your project delete it and add new ones as needed. Otherwise, grade each item for risk using a 1 for low, 3 for medium, and 5 for high. For each item, enter your assessment grade and also enter the same number in the appropriate risk level column. When done, compute the column totals.

Analyze the results looking at the total risk and the distribution of the risk levels. Pay particular attention to the items graded as high risk (5s). During the execution phase high-level grade sources should be carefully monitored and negative trends may indicate potential trouble spots. Review your risk mitigation plans that deal with the high-risk areas and take corrective action wherever possible.

The key to effective risk management is to be able to identify, measure, and minimize probabilistic events affecting project execution. A structured risk management approach involves event policy planning, risk identification, qualitative risk analysis, quantitative risk analysis, risk response planning, and risk and monitoring and control. The benefits of adopting a structured approach for managing risks are significant. Sample advantages are more successful projects, fewer surprises, less waste, improved team motivation, enhanced professionalism and reputation, increased efficiency and effectiveness, and more.

Project risks are a potential collection of uncertain events or conditions that, if they occur, create a positive or negative effect on at least one objective of the project. Experience suggests that the risk management process does not have to be complicated or time consuming to be effective. By following a simple, tested, and proven approach, the project team can prepare itself for the events that may occur. There is no tool that can avoid or fix all potential risk events; however, risk management offers a process to produce higher success in navigating through this minefield.

DISCUSSION QUESTIONS

1. What could a project team have done to mitigate the known/unknown issues causing the latest baloon failure from the case study outlined above
2. If a project is dealing with a new technology, what risk mitigation strategies should be considered?
3. When must a risk event be accepted?

REFERENCES

AS/NZS 4360-2004, 2004. The Australian and New Zealand Standard on Risk Management (Tutorial Notes). Retrieved on January 19, 2009 from http://www.broadleaf.com.au/pdfs/trng_tuts/tut.standard.pdf

Defense Acquisition University (DAU), n.d. Risk Management. Retrieved on January 19, 2009 from https://acc.dau.mil/CommunityBrowser.aspx?id=17607&lang=en-US

Elyse, 2007. Risk Response Planning. Retrieved on April 15, 2008 from http://www.anticlue.net/archives/000820.html

Glazewski, S., February 23, 2005. Risk Management (Is Not) for Dummies, *Crosstalk, The Journal of Defense Software Engineering*, www.stsc.hill.af.mil (accessed January 19, 2009).

GSAM, 2003. Guidelines for Successful Acquisition and Management of Software-Intensive Systems. Chapter 5, http://www.stsc.hill.af.mil/resources/tech_docs/gsam4.html (accessed January 19, 2009).

Mocal, T., 2008. Personal correspondence, January 23, 2009.

PMI, 2008. *A Guide to the Project Management Body of Knowledge (PMBOK® Guide)*, 4th Edition. Newtown Square, PA: Project Management Institute.

TenStep, 2008. Management Tip of the Week, www.Tenstep.com (accessed January 21, 2009).

23 Plan Review and Approval

23.1 REVIEWING PLAN COMPONENTS

The final project plan derived from the planning stage activities is the result of a complex interaction of the nine KAs. This chapter focuses on the five KAs that primarily support the creation process. At this point you should be comfortable with the basic role that each of the KAs play in the overall scheme of planning. To review this logic the construction of a raw project work plan consisting of scope, time, and cost was described in Part III of the book. In Chapters 18 through 22, each of the supporting KA roles is described in regard to its impact on the final project plan. Each of these chapters focuses on a single KA and looks outward to its impact on the other KAs. For instance Chapter 18, HR Planning, looks at the planning issues related to personnel (availability, skills, etc.). Likewise, Chapter 20 does the same for procurement. Across these two views we learned how a shortage of HR resources could be covered by procurement actions to acquire third party resources to fill the gap. This is an interaction example across KAs and represents the balancing required during the planning process.

At this point, all of the various KA decisions have been made and we have approached the end of the planning cycle in regard to a technical description of the project. All of the KA perspectives have been integrated into the plan and also integrated with each other. The assumption of the planning at this point is that the plan produced will support a successful outcome if the component parts are executed accordingly. The team has collectively negotiated and resolved all required internal issues dealing with the KAs. In order for this stage to have produced a successful plan, it is necessary that all KA perspectives to be integrated together and any known disagreements be resolved—not necessarily agreed with, but resolved sufficiently to support the version as defined. The final draft plan resulting from this activity undoubtedly involves compromise in various ways. So, the last step in the planning stage is to obtain formal approval of the plan and formally close out the stage.

Once the plan is approved, it serves as a road map for execution in the same way that a highway road map serves to guide a planned trip. The map may not be 100% correct, but hopefully the detours will be minor and we will always know the near vicinity of where we are. To accomplish this goal the plan must have accurately captured the stated requirements, mapped a work pathway to achieve those requirements, incorporated various necessary technical items into the plan to support the output, and balanced all of these together such that all necessary components are in synchronization. To the best of the team's knowledge, this plan, if followed, will achieve the goal. Just as in any marketing effort, the plan now needs to be packaged in such a format to make it salable.

The role of a project plan lasts beyond the approval stage. In the execution stage the management philosophy will be to "work the plan," so we have to be sure that all of the needed elements for doing this are included in the plan. In order to match this need, many of these final steps relate to setting up the project for next stage management and control activities. In the ideal case, the final plan would be so accurate that all that will be required of the execution team is to simply allocate defined resources to defined work units. Since no surprises would occur there would be nothing else to do. Coming out of this dream state we know that this is not the upcoming environment, but still realize that the more accurate the planning effort the less confusion later.

23.2   PLAN APPROVAL PROCESS

Five major groups and 21 activities are defined below as being required to move the plan through the final planning process stage.

Review major planning artifacts

1. Review final WBS structure and related dictionary items; match WBS work units to requirements statement
2. Review underlying planning assumptions and constraints
3. Review schedule, budget, and critical path issues
4. Review integrated change control process; obtain formal approval
5. Review HR staffing and training plan with resource owners
6. Review risk management plan
7. Review procurement management plan
8. Review quality management plan

Financial and control structures

9. Define CAPs and variance tracking process
10. Define contingency plans and related management processes

Documentation plan packaging

11. Complete subsidiary plan documentation and obtain approval from SMEs
12. Sponsor signoff of plan
13. Key stakeholder signoff of plan
14. Draft executive summary overview

External communication

15. Project board briefing—plan and board responsibilities
16. Formal senior management approval process
17. External plan presentation to key users and external stakeholders

Planning stage close

18. Set plan baseline—multiple performance control items possible
19. Document planning lessons learned
20. Archive planning documents in project repository
21. Activate the execution-phase activities

The sections below will elaborate further on the activities in these five major areas.

23.3   REVIEW MAJOR PLANNING ARTIFACTS

During the document review process, each of the KA subsidiary plans is validated for completeness and clarity. In the final version, each of these will be attached to the summary overview to supply details for those that need more explanation. From a technical viewpoint, the one issue that needs to be confirmed in this review is that all of the documented requirements are included in the WBS. Tagging the WBS to show where each requirement is produced is a good confirmation of this process and may have been done initially; however, it needs to be validated for insurance. Later, a review of these artifacts with the new project team will be performed and that activity is a good introductory technical activity. Since the project team may not yet be formed, this might have to be delayed until later. Recognize that this document set will be used throughout the project life cycle, so its value should not be underestimated.

Every project plan will not include a full complement of these artifacts because of small size or other factors, but every project planning effort should have considered all eight KAs and their implication discussed whether a full subsidiary document is included or not. For example, some projects would have no procurement requirement, so not much is necessary in that case. Many projects do not perform a formal risk assessment, but to leave out the discussion all together seems inappropriate.

23.4   FINANCIAL AND CONTROL STRUCTURES

Even though there are only two items on the summary list for this category the substance of those items is significant. First, identification of formal CAPs determines the level of granularity of future monitoring. The higher up in the WBS these are chosen the less visibility to lower-level performance details. If you need to refresh this subject, refer Chapter 12 where the role of CAP is outlined. The basic trade-off decision lies in the need for status visibility versus administrative work. The lower in the structure, the more data collection is required to capture and analyze the results. There is no right or wrong answer regarding location selection of CAPs, but they should be located low enough in the structure to aid in reasonable status tracking and variance analysis.

The enterprise accounting system will most likely be the official system of record for capturing actual resource costs, and the operational issue will be to establish whatever formal accounts are required in this system to supply needed actual. The enterprise system will identify the project within its chart of accounts codes. Within this structure the project CAPs will need to be set up. As resources are charged against this account, it will be possible to link from the accounting system back to the project WBS structure. This will supply needed actual performance information. Beyond this linkage, there are also issues regarding the granularity of data related to fund types (i.e., expense versus capital), material, hours, and so on. Hopefully, the organization has established a useful system to perform this service so that the only requirement will be dealing with CAP codes. We have used the metaphor multiple times before that the organization is like a flowerbed to the project. The better the flowerbed, the easier it is for the project. This is a perfect example of that abstraction. If the project has to invent its own tracking system, it will be time consuming and crude.

As documents emerge in the planning process, they should be placed in a formal repository and changes should be monitored. Appendix B describes a model project document repository architecture, but recognize that every organization will be unique in their systems and approaches to store and retrieve project data (another flowerbed metaphor example). In any case the PM needs to ensure the flow of plan and actual data into an appropriate repository to support various control activities at the CAP level. Large projects need to have automated linkages for this purpose to minimize the manual handling of data.

It may be a surprise to learn that the complete project budget has not been completed as yet. Final structure of this document is both financial and political. The financial components are deterministic, but the method of showing these becomes more of a political nature. The final budget logically consists of incorporating various resource groupings, some of which are not part of the basic work units. Figure 23.1 illustrates these logical components schematically.

These five budgetary components exist in every project to some degree. Also, if the project is being performed under contract, then profit is also a part of the budget structure. Based on these elements the budget package needs to address and incorporate the following additional budget items as dictated by the organization policy and format:

1. Scope reserve
2. Contingency reserve for risk events
3. Level of effort allocations for various support charges
4. Overhead allocations
5. Management reserve
6. Profit—if the project is done under contract.

FIGURE 23.1   Budget components.

Another way of looking at an aggregate budget is as a layer of components as illustrated by Figure 23.2.

The implication of this view is at that the base project costs are derived from low-level WPs, which are then aggregated upward into control accounts and various level of effort items. This aggregation collectively represents the base budget. Above this level the reserve layers are identified to account for the dynamics described in Figure 23.1. Finally, a profit margin is shown at the top. It is also important to recognize that overruns in the lower levels invade upward to consume reserves and profit.

The general definition and theory behind these budget terms are introduced in Chapter 14, Cost Management, so at this point we just need to be sure that appropriate amounts are included in the budget structure to cover the defined categories. The budget model proposed here is that the project budget lives inside of a planned envelope within which the cost structure is controlled. Total measured resources cannot exceed the envelope value without a formal management review (with good explanation to go with it). Any time the total budget exceeds this envelope value it is vulnerable to shut down. This means that all anticipated costs for the project need to be recognized in the budget and the project management system will keep this value visible in future reporting.

Some of the most difficult items to justify in the budget are the various reserve items surrounding change, internal buffers, risk, and management reserves (in that order). These items are often viewed as padding. For this reason, budgets seldom show such reserves as outlined here and they are typically frowned on by management reviewers thinking that they do not represent real events. It is true that scope change, triggered risk events, and management reserves are not real work yet but some form of them will be. Each of these items cannot be defined as clearly as the defined work units at this point in time. However, to reject these as valid categories is ignoring the real world. The alternative is to hide these categories in various lower-level work units and that clouds the picture from a control viewpoint.

Recognize that the reserve items outlined do not represent padding, rather a new form of planning and control. In previous chapters we discussed the topics of scope creep, risk, and estimating

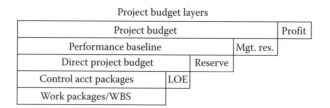

FIGURE 23.2   Project budget layers.

variability, so there is a basis for understanding how these reserves would be needed during the execution stage. Mature organizations try to understand the phenomenon they are managing and work to better deal with them.

Let us examine what happens when the reserve class of activity is hidden in various WBS units. As subsequent scope, risk, and estimating errors began to surface, their impact would be absorbed with this padding, even though the placement of the padding would not fit the event. When the project is completed, the actual charges will not reflect where and how they actually occurred and the cost traceability is lost in the process. Not only has the organization not learned from the experience, it has confused itself. A more logical approach would be to estimate the future impact of these items and then develop a mechanism to compare what actually happens in each category compared to the plan. Some new components in the budget structure are needed to help with this process. Basically, this translates to defining a method of showing how the reserve pools are used. The discussion below offers a brief comment on each of the six nondirect budget components that need to be included.

23.4.1 Scope Reserve

Some level of scope creep is normal for a project and needs to be planned for. The integrated change control process is established explicitly to manage and control the flow of such activity. However, scope work units are not explicitly part of the budget, hence cannot be shown in the WBS view, and for that reason should not be embedded in the project direct work budget. However, later when a scope change is approved by the project board that amount of resource will be needed to execute the work, so some budget mechanism is needed to support the activity. One straightforward method to handle this would be to extract the approved change amount from the scope reserve and move that into the WBS structure as a normal work activity. It would now become part of the base budget. More importantly, the overall budget for the project would not have changed and the controlled flow of funds fits the management requirement. A related issue for this is the impact on the project baseline. Conceptually, all changes in scope represent changes in the approved project baseline, but in most cases this is not recognized unless the change is significant. The normal method is to keep the original baseline for comparison purposes. Further discussion of this is beyond our scope for now. Movement of funds from reserve pools to the base budget would be under the control of the project board and the PM. In the case of scope, the project board approves the change and this could be authorization for the PM to move the funds.

23.4.2 Risk Reserve

The technical name for this fund is *contingency reserve* and it represents a level of funds estimated to cover future risk events that are evaluated during the planning risk assessment process. These events are formally defined in Chapter 22 as *known/unknowns*. Even if no formal risk assessment process is performed during planning some budget reserve should be set up for this class of event based on history, gut feel, or whatever technique the organization wishes. Movement of funds from this reserve pool should also be under the control of the project board or the PM and could actually be mechanically controlled as part of the change control process.

23.4.3 Level of Effort

Project support resources often come from various parts of the organization and their charges will either be part of general overhead or they will be charged in some form. These charges are difficult to map to WBS work units so would be best kept in one or more general budget categories. The level of the charge could be simply allocated to those accounts based on time spent in the activity, or the project could negotiate a fixed cost arrangement to supply these skills when and where needed. For the sake of management control it does seem reasonable to show this class of charge under a line item with a defined name. Control of this class of account would simply be comparing actual with plan because the concept of work completed does not fit the activity. These items can be shown as one or more defined lines in the budget (i.e., help desk, desktop support, office space, copy center, documentation department, etc.).

23.4.4 OVERHEAD

Projects will almost always show organizational overhead as part of their budget. These cost elements come from various sources and represent the overall cost of support. Organizations own buildings, have executives and managers, various support organizations, employee benefits, basic infrastructure, and so on. The normal accounting approach is to attempt to show costs in proportion to their impact on the organization, so a full budget view would recognize these costs. Some overhead items fit into the WBS structure and some do not. For example, team personnel have benefits beyond their salary so rather than simply showing direct salary we would likely reflect the full cost of the employee. This overhead amount would be embedded in the work structure. Other items are not so easy to link to work efforts. In those cases the budget might simply apply a factor called a burden rate to some portion of the budget to cover these expenses. These items need to be categorized on the budget structure outside of the working portion of the project because they are not controlled internal to the project.

23.4.5 MANAGEMENT RESERVE

This reserve category is defined to handle *unknown/unknowns*. This group is different from the risk contingency because these items were not anticipated as part of the planning exercise. There is a certain amount of mystique surrounding this reserve type. Some organizations do not admit that such a thing exists and basically just ignore budget overruns so long as an explanation is supplied. As in previous examples we believe that the budget should be able to show variances on individual work units, but still have mechanism in place to keep this from being a total budget overrun. There are several ways in which this class of project resource can be budgeted and controlled. One way is for management external to the project to hold the funds and require explanations of all variances. This is not a practical solution and would invoke many poor behavior traits within the project to minimize this. This approach is also a poor use of management's time. A more practical approach is to allow some small percentage of the budget for undefined overruns. These variances would be managed by the PM without external oversight up to the limit of the reserve fund amount. A second component of this could be kept at some higher level and either be known to the PM or not as the organizations sees fit. This fund could be included in the external category but still in the overall control envelope for the project.

A managed approach for the management reserve category with explicit allocation to the PM is recommended here. The first tier would then serve to handle normal WP cost variances. When these occur, the overrun would be left with the WP showing its variance, but the overall project work budget would be protected by the reserve amount. Operation of how to mechanically handle such overruns requires some thought and planning. In theory, when a WP overruns that amount should be extracted from the reserve and tracked accordingly. This process would serve to balance the budget books appropriately, but does require financial mechanics not often found in organizations. Regardless of the method used, strong consideration should be given to an internal reserve in the neighborhood of 10% to protect the work budget. At the same time, there should be an increased emphasis on developing work unit cost estimates that do not include padding. If that could be achieved the role for this reserve takes on new meaning. In both of the described cases, this reserve would protect the budget from normally occurring variations without padding those events during the planning cycle.

Although not common, a second management reserve allocation should be recognized above the amount described for internal project buffer. This segment would be focused on unknown/unknown events external to the project work itself. Because of the nature of these events, allocation of these funds would be more tightly controlled by management and might not be visible in the published budget, or even known to the PM. As described here, it is designed to reflect some additional aggregate amount of resource that management is willing to let this project consume based on its environment. This could represent some high-level risk assessment external to the project or could reflect that the requirements for a particular project are vague and management is not willing to release funds into the scope reserve just yet. For projects operating in a high-risk environment, management should ponder what overall value to attach to the project. This second tier reserve could represent that assessment.

As indicated earlier, most management reserves are not visible even though the concept of unknown/unknowns is real. Lack of formal recognition of this budget category places the PM in a tenuous position in that when the project budget begins to overrun and because of these events the budget overruns. Often this will force asking for more budget resources from the sponsor when the actual variances are small and expected in any project environment. In a professional trust relationship both parties should understand these dynamics and provide the proper operational environment for the PM to do his job.

Based on this interpretation the management reserve concept becomes a split responsibility between the PM and his sponsor who has funded the effort. Both parties understand that the items in this category are not anticipated, but some of them will occur. The management philosophy is that it is better to encourage honest work unit estimates and manage the resulting potential overruns rather than to bury resource padding in all such units and ignore the problem. If we look at this issue politically rather than technically the real problem is how to show it on the visible budget, not the concept itself. For that reason it may be necessary to keep management of the reserve group off of the visible view, but make them known in some format as generally described here. It is primarily up to the project manager and his project board to ensure that these funds are properly managed.

From an academic viewpoint the project budget should be considered a contract between the PM and the sponsor. It also represents an initiative to achieve some higher-level organizational goal, so in that perspective the contractual view spans upward even further. The sponsor and the organizational management support units have shown willingness to commit some amount of their total resource pool to the project effort and in return anticipate receiving the defined benefits. In this view, the PM must be looked at as the change agent who will get that job done as specified by the plan and a budget design should support that requirement.

23.4.6 PROFIT

Inclusion of this budget item is only applicable when the project is being pursued for a third party under contract. Similar to the management reserve issue, showing profit on a budget is often avoided or visible only to certain management levels. Profit targets are often established by senior management, so this number would be derived from previous conversations with them on this topic.

23.4.7 BUDGET STRUCTURE AND FORMAT

Organization of the formal budget line items may be standardized within the enterprise in order to fit their accounting structure. However, from a project viewpoint, the format that would make the most sense is one based on the WBS organization. During the planning process, the majority of cost estimates were made on this basis at a WP, planning package, or summary-level activity level; however when presenting the budget to senior management it may be necessary to catalog this view into the enterprise format. This could also include dealing with types of funds (capital, expense, etc.). During the planning process these were likely not the paramount concerns, but they must be resolved before the formal approval is given.

23.4.8 Control Structure

After all of the budget components and machinations described above are properly formatted into the plan, the final item in this category is to define how the various budget categories will be controlled. For example, the PM needs to understand what types of financial flexibility he has. The budget components outlined above provide a structure for resource plans and the accounting system will record expenditure of resources versus these categories. Each of the budget groups represents potentially different control models.

As described here, the base work budget would use its associated WBS structure to compare planned versus actual resource status. This comparative data would drive the various formal project monitoring and control processes that were described in depth in Part VI of the text. The other more nondirect budget categories discussed here fall primarily into a sizing status view. In other words, control actions for these would be more in tracking actual status compared to planned values. For level of effort and risk categories, there is a time phase view for this status, whereas in the case of scope change it is more of rate monitoring. The direct project work segment of the budget will be controlled by comparing actual resource consumption in comparison with baseline planned values.

Another aspect of budget management involves controlling the movement of funds across categories and the authority to make decisions that create expenditures (i.e., buy material, contract resources, hire, etc.). The PM cannot view the total amount of these approved resources as his open fund to spend or manipulate in any form desired. For example, the scope reserve fund has been approved to handle anticipated scope changes. Resources from this fund should only be moved into the base budget with an approved scope change indicating the amount associated with the change. This is a formal reallocation process. Likewise, a risk event occurring would be analyzed along with the associated impact amount. From this, the project board would authorize a resource allocation out of the risk reserve into the base budget. This is also a formal process. Level of effort and overhead categories would be controlled depending on how these resources are shown in the budget structure. If overhead is embedded into discrete project budgets sections such as personnel, then the overhead would be automatically taken care of in the resource accounting process as time is charged to work units. Alternatively, if benefits and other items are budgeted separately, then these overhead amounts would be shown in that manner. One tracking method for the level of effort budget groups would be to report actual charges into their assigned budget line items. An alternate method would be to charge these items at some summary level. From an analysis viewpoint, it would be best to keep this work group separate from the basic WBS WPs, but that might not be feasible. Finally, the management reserve control options have been discussed earlier. Regardless of the way these funds are to be handled during execution, the process must be formally defined. Defining how much overrun is allowable before external management review is an important process question.

The final budget structure including all of the categories described here is a formal document that will live with the project through its life cycle and it basically represents the scorecard performance template for the project. Specific methods to deal with the dynamics related to scope, risk, and unanticipated overruns have been covered in this model view (not always so in the traditional budget). At this point, the PM and the project board are committing to the organization that they will produce what has been shown in the plan and will report their performance in a manner described here. They are now asking senior management for the resources to pursue this venture.

A final comment on budgets: Simply reporting a total budget plan value versus actual resource consumption does not satisfy the planning or control requirements of modern project management. A budget and control structure similar to what is described here needs to be considered as the appropriate planning and control view.

23.5 DOCUMENTATION PLAN PACKAGING

During the planning stage, numerous notes and spreadsheets are created to support the final plan. At this point, the documentation set needs to be segmented into two basic groups—work papers and final plan elements. Work papers should be archived and stored until later. Some of these documents have no future value, but for now they should be kept. Other items become a part of source data to be included in the formal project documentation. Items to consider for inclusion in the final plan are summarized in the following 23-item checklist:

1. Project Charter
2. High-level assumptions and constraints
3. Project Objectives
4. Scope statement
5. WBS
6. Labor and Time Estimates with background notes
7. Work units defined and their dependencies
8. Resource Allocation
9. Network Diagram
10. Schedule
11. Staffing Plan
12. Communications Management Plan
13. Risk Management Plan
14. Procurement Management Plan
15. Quality Management Plan (quality objectives and approach)
16. Budget/Spending Plan
17. Documentation Plan
18. HR Management Plan (staffing, skills, and training issues)
19. Testing Plan (component testing, system testing, stress testing, etc.)
20. Configuration Management Plan
21. Baseline plan for all control attributes (time, cost, performance, quality, etc.)
22. Performance Analysis Plan (control process and metrics to be used)
23. Define project document archival structure.

Local needs will modify the format of the final plan. In the pure model view, the project plan consists of a subsidiary plan for each KA and a summary overview. Depending on the underlying technology related to project goal there may need to be supplemental information collected for this aspect. Not only does the plan prescribe how the work will be performed, but a performance measurement plan should be included to outline how the project will be monitored and controlled. Appendix B outlines a conceptual document architecture for a project storage and retrieval system. Something of this form would satisfy the goal of item 23 above. Regardless of the technique used it is important to formally define a project artifact storage strategy.

The final project plan package consists of a summary overview outlining the key attributes of the plan and all eight KA subsidiary management plans describing management details about each area. A summary overview lays out the issues of greatest interest and concern to the various review groups. It is an important document in that many readers will not wade through the full set of subsidiary documents, so this becomes the face of the project for most.

23.6 EXTERNAL COMMUNICATION PROCESS

Packaging of the project plan documentation for presentation must be sensitive to the communication value of this activity. For that reason the presentation material will have multiple versions in

326 Project Management Theory and Practice

order to focus on the needs and interest of a particular audience. Senior management typically wants a summary version of project objectives with planned milestone dates. Cost is usually their main focus, although showing how the project will contribute to some aspect of organizational goal alignment will be of interest to management. Second, user presentations focus more on the projected deliverables functionality and related schedules. This review segment will want to see how the original requirements or vision statement was translated, particularly if some original requirement was deleted through a planning rationalization process resulting from resource, cost or time constraints. Third, if there is a PMO in the organization they may well want to review many of the plan details (see Chapters 32 and 33 for a detailed overview of this functional group roles). A final review group will be the new project team. Once the project team is formed an upcoming kickoff session provides the opportunity to go through the work plans in significant detail.

The key requirement in presentation content packaging is to communicate to the respective audiences how the planning process translated the original project vision. Significant changes could have been approved during this phase and these need to be explained to the appropriate user community. This communications cycle is the time to clear up planned deliverables and level set expectations. If requirements changes did occur, be prepared to discuss why they occurred and in some cases be prepared to go back to management and report stakeholder concerns. The user presentation process also should outline where they are needed for support to make the project successful. Activities such as design reviews, product prototype reviews, test acceptance, and training are typical user interface points requiring some defined action on their part. The presentation process has a defined audience sequence as summarized below:

1. Step one: Review the final plan package details with a broader group of formal stakeholders that either have a major involvement in providing resources to the project or will be major users of the output from the project. The goal of this step is to obtain formal approval of the final plan version and to gain commitment for their support of that version. If issues occur in this review cycle they must be resolved before going forward to the next step.
2. Step two: Package the planning document into a format suitable for management review (i.e., higher-level details with key points). The primary external audience for this formal plan approval would consist of the project sponsor and senior management who have authority to formally approve the project or who need to understand what the plan entails. In some high visibility projects this can include a Board of Directors level overview. Once appropriate management has approved the plan, the last step of the review process is designed to communicate the results to other key stakeholders.
3. Step three: The final review step communicates the plan details and should strive to obtain a positive buy-in from key parties. Most of these groups will have some active involvement in the execution or implementation phases and it is important to obtain buy-in. Accomplishment of the buy-in goal will help the effort move successfully through its remaining life cycle. Remember that failure to obtain user support is one of the top two reasons projects fail.

23.7 PLANNING STAGE CLOSE

At this point, in the close activity the plan has now been formally approved and it is time to move the project into the execution phase. These last steps fall into the category of stage closing and preliminary execution activation. First, up to this point, changes in the plan have been relatively easy as various planning decisions were added to the set, but from this point the plan details are frozen and we set the formal *project baseline*. Assuming that automated software is being used, this is a simple process involving setting a logic flag in the planning utility. Second, moving into the execution phase usually expands the size of the project team and may activate other geographical teams and vendors. Prior to this point each of these groups should have been kept in the communications loop as to timing of a probable start date. If internal resources are used to staff the team they need to be housed. If the team is being collocated, this will involve physical space for which the requirement should already have been anticipated. All of these related transition actions are on the fence between planning and execution. These also illustrate how the project phases overlap. A good PM must always be thinking ahead and trying to anticipate what will be needed before the project has to stop and wait. To not deal with these transition issues until the end of this phase would simply delay the schedule.

Before leaving the planning stage the lessons learned should be documented and stored with the project archives.

Finally, various individuals have been involved with the planning process. The last step involves being a good HR manager. These individuals need to know how much you appreciated their contribution and this should be transmitted both personally and formally to their manager. Lastly, find a suitable nonwork activity to celebrate the successful completion. A nice lunch or dinner (whichever is more appropriate) is often used for this purpose. If the planning team has had to work long hours to finish the effort it is also nice to include the spouse in this celebration. And in some cases a gift, bonus, or other memento is appropriate. All of these gestures are part of building a team culture now and into the future.

DISCUSSION QUESTIONS

1. What project issues do you see resulting from the planning process that need to be communicated and to whom?
2. Do you agree with the level of planning detail outlined in this part of the book? What issues do you see with this model theory versus your view of reality?
3. Assuming that your management instructed you to move into execution before you were comfortable with the plan what would you do?
4. How would you react to receiving significant negative response to the plan after you have received formal approval from management as outlined in this chapter? Discuss your strategy and actions.
5. What is the purpose of a project baseline? Do you think that all scope changes represent new baselines?
6. Given that most real-world projects do not create the level of planning or budget documentation outlined in this chapter, what rationale do you attribute this to?
7. Most PMs say that they have never seen reserve pools used as outlined here. Do you agree that these reserves should be managed this way or do you see another option that would work better? Does the real-world culture impact this question?
8. How would you sell the idea of building a common project document archive for the organization?

Part VI

Project Execution—Managing the Plan

LEARNING OBJECTIVES

The execution and control learning objectives for this section are taken verbatim from the *PMI Handbook of Accreditation of Degree Programs in Project Management*. They are as follows:

EXECUTING THE PROJECT

The following management processes are utilized in this phase of the project:

1. Commit project resources by utilizing the work authorization/release system and procedures to initiate and monitor the performance of work in accordance with the project plan. Assess work in process to ensure that all activities and only those activities required to produce the project deliverables are performed
2. Implement the project plan by authorizing the execution of project activities and tasks required to produce project deliverables
3. Manage project progress by applying performance reporting, analysis, and progress measurement techniques in order to ensure that activities are executed as planned so that project objectives are achieved
4. Communicate project progress by producing project reports to provide timely and accurate project status and decision support information to stakeholders
5. Implement/carry out quality assurance procedures by performing project control activities to meet project objectives.

CONTROLLING THE PROJECT

The following management processes are utilized in this phase of the project:

1. Measure project performance continually by comparing results to the baseline in order to identify project trends and variances
2. Refine control limits on performance measures by applying established policy in order to identify needs for corrective action
3. Perform timely corrective action by addressing the root causes in the problem areas in order to eliminate or minimize negative impact
4. Evaluate the effectiveness of the corrective actions by measuring subsequent performance in order to determine the need for further actions
5. Ensure compliance with the change management plan by monitoring response to change initiatives in order to manage scope
6. Reassess project control plans and practices by scheduling periodic project and change control system reviews with stakeholders in order to ensure their effectiveness and currency. Update the plans and practices as required
7. Recognize and respond to risk event triggers in accordance with the risk management plan in order to properly manage project outcomes
8. Monitor project activity by performing periodic inspections to ensure that authorized approaches and processes are followed or to identify the need for corrective action.

Source: This list of educational monitoring and control objectives is from PMI, *Handbook of Accreditation of Degree Programs in Project Management*, Project Management Institute, Newtown Square, PA, 2007, pp. 17–20. Permission granted for use in this context.

Project execution involves not only following the details created during the planning stage, but also monitoring and controlling the actual results as they occur. Execution and control processes consume a major portion of the total project life cycle. The execution activities involve work activities to produce the desired deliverables and the control processes involved deal with tracking status of work products and work processes of the established baselines. Collectively, these two sets of processes deal with the real-world dynamics and attempt to help correct variances from plan and influence desired outcomes for the project. Major topics discussed under this banner are as follows:

- Team acquisition
- Team development
- Managing the team.

Each of these processes and techniques provide vital management components to support the overall project management monitoring and control process.

24 Project Execution and Control

24.1 INTRODUCTION

Other than the impact of the real world there is nothing new to be added in this chapter. Unfortunately, Murphy, Parkinson (i.e., the mythical laws of these and others), and all of the other villains of the real world can make this story seem like the project environment is filled with vampires and we are living through the plot of a cheap movie. All through the other chapters of this book, various model ideas of project management have been defined with more than subtle hints that each of these processes can go awry. We also know that the success rate of projects is not good and there must be a good reason for that (a lot of these have been described). If approximately 50% of the projects are judged to be unsuccessful what is going on? Our theme here is to highlight a reasonable cross section of ways that project performance can drift away from the defined path. As a project manager, our goal is to produce the desired product in spite of these undesirable events that invade our model world.

24.1.1 MAGIC TWELVE SUCCESS INDICATORS

Step 1 in this discussion is a summary of the *magic twelve* visible indications that the project is going well. We get a positive check when each of the following is observed:

1. Planned milestones events are being met.
2. Budget is under control.
3. Quality control results are within specifications.
4. Change control process indicates minimal requests for change.
5. Project resources are being supplied per schedule and skill levels are adequate.
6. Project team appears to be cohesive and reasonably happy.
7. Users seem satisfied with the progress of the work.
8. Top management remains visibly supportive of the project goals.
9. Third-party vendors are delivering quality items on schedule.
10. Risk events are under control and nothing unusual is appearing.
11. Project training program is progressing according to plan.
12. Relationships with support groups appear to have no identifiable issues.

Let us describe this list as a wish list because it likely will not be the real-world case. However, it does provide a visible overview of desirable attributes representing project execution status. If we observe one or more of these not going so well, it would indicate some underlying management issue that needs to be understood. A second point that emerges from this list is that all of the nine knowledge areas are in play during the execution phase and the list above contains visible components from those knowledge areas.

The *PMBOK® Guide* has defined several model operational execution and control-oriented processes for the project. Previous knowledge area-oriented discussions of these have been discussed within a single knowledge area in various chapters of the text, but these now need to be combined into one working set. They basically contain the communication and control levers that the PM has at his disposal. If there is a problem with the project, one or more of these processes represents the control knobs to be used to correct the deviation. In many ways one can look at this array of knobs (processes) as helping an airplane pilot navigating through rough weather. There is a visual display of data providing flight status information and an ground radio providing less structured information. From these two sources the skilled pilot (manager) manipulates the knobs with the goal of successfully landing the plane. This metaphor provides the right initial abstraction for the execution and control process.

Let us first look at the PMI summary process list for the execution and control phase. Table 24.1 summarizes the 17 *PMBOK® Guide* processes for these activities.

Prior to the project going into execution, the work of project management has been more "soft" in the sense that plans are derived, communications are visionary, and nothing visible has gone wrong. One could describe this as a whirlwind of activity, but that does not help the reader. Let us see if we can segment some of the key issues into more of a directory approach, but do keep the information display and control knob analogy because that is the management paradigm at play here.

24.2 STATUS VIEW

There are many formal and informal methods for receiving status. Some project managers walk around and observe, while other methods are more quantitative and formal. For example, we have

TABLE 24.1
Execution and Control Model Processes

ID	Knowledge	KA Process
4.3	Integration	Direct and manage project execution—work the plan
4.4	Integration	Monitor and control project work
4.5	Integration	Perform ICC
5.4	Scope	Verify scope
5.5	Scope	Control scope
6.6	Time	Control schedule
7.3	Cost	Control cost
8.2	Quality	Perform quality assurance
8.3	Quality	Perform quality control
9.2	Human relations	Acquire project team
9.3	Human relations	Develop project team
9.4	Human relations	Manage project team
10.3	Communications	Distribute information
10.4	Communications	Manage stakeholder expectations
10.5	Communications	Report performance
11.6	Risk	Monitoring and control risks
12.2	Procurement	Conduct procurements
12.3	Procurement	Administer procurements

Source: PMI, 2008. *PMBOK® Guide*, 4th ed., Project Management Institute, Newtown
            Square, PA. With permission.

Note: The ID reference number is the guide number for description of the process. Note that all
            of the nine knowledge areas are referenced in this activity.

seen the role of various measurement techniques to highlight schedule, cost, and quality status. Earned Value parameters evaluate cost and schedule performance versus the plan and these are a key part of the status measure, but other metrics also play a key role. Observations related to the volume of change requests, status of risk contingency funds, technical performance measures, test results, and variance analysis all help paint the overall status picture of the project. Beyond these more quantitative and mechanical values unstructured conversations with team members, sponsors, users, and senior management add an additional level of understanding to the status view (in some cases, a more timely and better one).

24.2.1   STATUS-TRACKING PROCESSES

Status tracking basically attempts to map how the project is moving in comparison with the approved plan. There are two important execution and control concepts that have to be in place for the tracking process to work. First, a well-defined plan with quantifiable objectives is necessary to establish comparison targets. Without this base there are no measurable targets that can be used to identify the project goals. Second, the control process involves measuring actual performance against these planning targets. These two concepts are called the *Siamese twins of management*. Another important control concept is that of a baseline. Baselines can be established for any parameter of the project (i.e., cost, time, speed, weight, etc.), although they typically focus on scope, time, cost, and quality metrics. One of the management problems with comparing ongoing status to the original baseline parameters is that approved changes by the project board can logically change the original baseline values and by that action create a variance that it not produced by the actions of the project team. This means that the Integrated Change Control (ICC) process becomes a vital component of management and control.

If we could assume that there were no changes through the life cycle, the control process would be mostly one of comparing the ongoing values to the originally approved baseline values. The point to be made here is that any changes that are approved to the baseline values need to be incorporated into the control process, but that is often not done. Let us illustrate the potential magnitude of this problem. Scope creep in technical projects can be in the range of 2% per month. For a 1-year project this amounts to a 24% increase in project scope, which in turn would likely have a significant impact on schedule and budget. Does this mean the project has done a poor job? Maybe the requirements definition process was done poorly, or maybe the environment changed. At any rate, simple comparisons with the original baseline may not be measuring the right variables. Keeping the original baseline and adjusting the baseline for approved changes would seem to give a more balanced set of data for evaluation. The original baseline comparison would show the project variance compared to the approved version and the current baseline would help to show performance compared to the formally approved scope of work. As a PM one of the most important aspects of control is to get formal sign-off of the initial requirements. Given this step, the changes that occur after that are primarily related to external factors from the project team standpoint. This adds more credibility to the notion that the current baseline should be the team measurement metric.

As the project gets underway, work units are completed in various WBS segments. In order to evaluate how the project is progressing, several formal measurement, control, analytical, and informational processes are used to evaluate status and decide what management actions are needed. Summarized below are 14 formal project control-oriented activities that will collectively produce status data and analytics. These collectively are the project manager's information dashboard from which he will use a major input source to decide on any needed corrective action. They are as follows:

1. *Scope control:* This involves various factors that influence project scope changes and corrective actions associated with that activity. The change control system is part of this process (PMI, 2008, p. 119).

2. *Scope verification:* This deals with formal customer acceptance of the completed project deliverables (PMI, 2008, p. 118).

3. *Schedule control:* This involves the use of various tools and techniques to evaluate and influence the status of the project schedule and related variances. This activity is a portion of the ICC system (PMI, 2008, p. 93).

4. *Cost control:* This involves the use of various tools and techniques to evaluate and influence the status of the project baseline budget and related variances. It is also a portion of the ICC system. The goal of this activity is to keep the project within budget limits (PMI, 2008, p. 170).

5. *Quality control:* This "involves the monitoring project results to determine whether they comply with relevant quality standards and identifying ways to eliminate causes of unsatisfactory results" (PMI, 2008, p. 190). Another major activity is to influence process quality improvements.

6. *Project team performance:* This involves a number of quantitative and qualitative processes to track team and individual performance and provide feedback.

7. *Status meetings:* Status meetings provide an unstructured communication forum for team members to share their experiences and inform other team members of their plans.

8. *Risk monitoring and control:* This involves monitoring the risk register and other sources for the emergence of both identified and unidentified risks. Once previously identified or new risk items emerge, there is a management activity required to properly handle each category (PMI, 2008, p. 264).

9. *Third-party deliverable status:* Monitoring of third-party activities is essentially done through the contract administration process. Key metrics here are deliverables, funds flow, and status reports.

10. *Risk audits:* This examines and documents the effectiveness of risk responses in dealing with identified risks and their root causes, as well as the effectiveness of the overall risk management process.

11. *Reserve analysis:* This compares the status of contingency and management reserves for the project in order to ascertain whether the remaining reserves are sufficient.

12. *Variance and trend analysis:* This process is used to translate the raw data into meaningful decision format. Both the degree of the variance and the trends are important metrics for the project manager. These also lead to corrective actions and completion estimates for various project parameters.

13. *Technical performance measurement (TPM):* This represents a set of techniques to identify deficiencies in meeting system requirements, and provides early warning of technical problems and related technical risks. This process attempts to compare the ongoing project with specified performance goals to estimate the degree of success in achieving the project's scope.

14. *Performance data:* This involves collecting data from various sources regarding project performance and then comparing these values to established baselines. This data summary regarding scope, schedule, budget, quality, risk, procurement, and team resources is used to evaluate overall status. Collectively, these parameters aids the project manager and team and is used to distribute key components to appropriate stakeholders.

The status data and analytic-oriented processes outlined above are internally focused on project results versus the baseline plan. In addition to these sources, additional insight into the project status is obtained from a less formal set of external sources. These relate to the project sponsor, senior management, future users, and possibly outside regulatory-oriented sources. Each of these perspectives can have a bearing on the project future and need to be assessed by the project manager, even though that data may be difficult to obtain and less formal in format. One of the most likely external issues that can affect the project relates to the original vision. In some cases, the organization objectives change before the project is completed and/or the market for the product changes as a result of some economic or technology event. In any case, this means that the existing specifications may be wrong and it is up to the PM to air these issues with appropriate management. A PM must be willing to shut the project down or change directions if that is in the best interests of the company. The paramount consideration is that the project must further organizational goals and it must stay aligned with that notion. One of the dangers of a project is that it will finish successfully according to the plan only to find that the reason it was chartered is no longer valid. This situation is difficult for the PM to deal with, but he needs to be sensitive to the point (i.e., landing his plane successfully at the wrong airport). With this dual focus recognition the execution management goal portion is to complete the project according to an approved plan that aligns with *current* organizational goals.

24.2.2   Turning the Management Control Knobs

Through all of the defined communications paths we have defined a collective status delivery perspective. As some aspects of the project begin to show a troublesome variance, some measured management actions are needed. We liken this to turning a specific control knob with the intention of influencing the status back to the planned value. Just to complete this thought, the control panel will light up with such events and they do not come one at a time for handling. They often come in waves, so a second management issue is to decide which items to deal with first and will fixing one create an even worse situation elsewhere. In other words, the knobs are not independent and changing one variable in the process can cause other situations to change as well. The level of variance chaos to expect is correlated with how accurate the original plan was. If the plan was well thought out and the support organization is mature, then one might expect the work to flow somewhat according to that plan with a reasonable level of issues to deal with. Otherwise, a poorly created plan can leave the PM fighting all 14 status issues at one time.

Since our management control knobs are metaphorical we have to translate this into the physical management reality of the knob-turning process. When some execution event needs to be corrected, the management process most often deals with a decision involving HR either internal to the team or external. This can involve a physical resource allocation decision or a communications action. We have described in previous examples how a work unit or deliverable does not occur on schedule. When that happens, the project schedule and possibly budget or other items are in jeopardy. How do we react to this? If that work unit is on the critical path we might decide to add resources to other activities to the critical path in order to bring the project back in line. However, if budget is more critical than time we might judge that it is better to let the project schedule slide. As a second example, when conflicts emerge, we have indicated that they need to be confronted. The style of resolution can have wide ranging positive or adverse impacts on the project situation (review the 14 events above to evaluate how conflicts impact each area). The management decision in the conflict case likely arose not from the formal status system, but from informal sources. Once again the management issue is which control knob to turn. The option in this case is the management knob that has the project manager's authority and influence attached (there are some knobs that are not available to the project manager). Our two examples have shown that control mechanics can be a tangible decision (money or human reallocation) or a communications decision that is designed to influence behavior. In all such control scenarios, the goal is to influence the project variables to a more desirable state—time, cost, scope, quality, morale, and so on.

One of the control concepts that is often difficult to understand initially is the interaction across the impact areas. As an example, the late delivery of an item by a third-party contractor can impact several different internal areas (i.e., schedule, budget, resources, activities, etc.). Having a team member do a customer a "quick favor" by making an unauthorized and unapproved change can result in schedule, budget, quality, and risk issues. So, the last step before turning a control (change) knob is to attempt to evaluate the consequences of making a particular resource allocation or communication move.

Hopefully, the use of a physical airplane analogy for the information dashboard and control knobs has not been too abstract. A much more idealistic metaphor is the orchestra maestro who waves his baton and the orchestra members follow the "plan." If the musical "plan" is well designed the outcome can be beautiful based on the skills of the members. This metaphor is correct up to a point, but falls apart when we have to admit that some of the players are not in synch. In that case, waving the baton will not correct the problem. This view does not show the control aspect of the orchestra leader very well. However, both examples are still interesting enough to keep in mind as a management perspective. We would like to develop the team to the point where we can be more of a maestro and less of a control knob turning oriented manager.

24.2.2.1 ICC Process

One of the major reasons why projects get into stress during execution is the negative impact that changes have on the plan and the project team. As the volume of these grows, there is increased coordination resulting from the related resource allocation changes required and the interaction of the change across the overall project becomes unwieldy to manage. Ideally, the PM would like to edict that no changes will be made in the project but that is seldom a viable option. The best compromise is to establish a formal and rigorous change control process. The formal name for such a process is ICC and it deals with the following activities (PMI, 2008, p. 93):

- Define the process for analyzing, reviewing, executing, and implementing a change
- Review and decide how to handle all change requests
- Manage the overall project baselines and management constraints
- Ensure that appropriate documentation and configuration control are followed.

The ICC process should be under the management control of a project board that represents the organizational management and project sponsor. It is their responsibility to protect the project from excessive changes and to help keep the project on track from an overall perspective.

The ICC process represents more than just the scope change control activity. This process also is the focal point for recommended internal project process changes that can improve operational productivity and quality. A third area of activity for the ICC is document configuration management control. During the course of a project's life cycle, a great deal of documentation is produced. Lack of version control for these artifacts can be just as damaging as lack of scope control.

The PM deals with the ICC process primarily by managing the change request impact analysis, providing technical assistance to the decision process of the project board, replanning the project based on approved changes, and then executing the work approved. A second thread of activity would occur in the improvement side of the ICC role. These actions could originate from lessons learned sources or other inputs from the team, which are then converted into change requests. Handled properly, the ICC process brings a measure of project operational stability that would not otherwise be possible.

24.3 HUMAN RELATIONS AND COMMUNICATIONS ISSUES

Beyond the general control mechanics described earlier, a great deal of the project manager's time is spent with the softer decision and influencing type of activities involving various humans internal or external to the project. Winters says that poor communication, vaguely understood project goals and objectives, and poor leadership move projects on their path to failure (Winters, 2003). From these and other studies we can conclude that communications failures are at the heart of this. We have previously stated that 90% of a project manager's time should be spent in various forms of communication. For this discussion, we will focus on the communications issues around the major management processes related to team acquisition, development, and management.

24.3.1 TEAM ACQUISITION

The basics of the team acquisition process are discussed in Chapter 18. The major item to add here is that this process can be ongoing through the execution phase if team members leave the organization or the functional support departments fail to live up to their staffing commitments. Also, failure to properly staff the work activities is the most direct way to lose control of the project objectives. If we assume that the original estimates were accurate and we fail to staff that work unit for 1 month, the expectation is that it will make the project 1 month late, if the activity is on the critical path. This same result can occur if the functional managers do not live up to their skill level commitment for staffing. Given the direct impact that staffing has on the project this adds one more tracking variable needed by the project manager. That is, tracking and managing the allocation of skilled team members to work units. Of all the mechanical management tasks, this one is the most vital.

24.3.1.1 Developing the Project Team

Project team development relates to the processes involved in improving productivity of the team and the individuals. This is both a short-term and long-term management activity. Short-term actions can improve the process for the current project, but long-term actions have the potential to transform the resource skills along the path of continuous career improvement that impact future project performance.

As project teams are initially formed, there is typically internal confusion as to their roles and responsibilities. Later, with proper management, the team begins to form into a more cohesive unit and productivity grows. This process has been labeled as a four-step evolutionary change from forming, norming, storming, and performing (Wong, 2007). In the forming stage, each member of the team focuses on the leader, accepting his leader's guidance and authority and maintaining a polite but distant relationship with the other team members (Wilson, 2008). In the storming stage, team members are often more concerned with the impression they are making than the project in hand. They want to be respected and often battle feelings of inadequacy, wondering who will support or undermine them and, above all, proving to the leader their value to the team (Wilson, 2008). The third stage moves the group into a defined work group (norming) in which in-group feelings and cohesiveness develops, new standards evolve, and new roles are adopted. At this stage, personal opinions are being expressed more freely (Tuckman, 2001). In the fourth stage, known as performing, a less structured environment is needed because the team members understand their roles and group energy is channeled into the task. At this point, structural and relationship issues have been resolved, and the resulting structure now generates the required performance (Smith, 2005). Not every project completely goes through this evolution, but in any case, management of the project team requires involvement as the team members find their individual identities and roles. These are not necessarily mechanical management roles for the project manager, but often more of a behavioral relationship approach to dealing with the team. Several management actions are embedded in the team building stage. Examples of these are as follows:

1. Training programs to improve team and individual skills
2. Interpersonal activities to build morale and improve performance
3. Using recognition and reward to motivate positive performance
4. Developing an effective team communication environment
5. Creating a climate conducive to high productivity
6. Providing appropraite leadership
7. Resolving conflicts that occur
8. Providing timely team and individual performance feedback.

Note that most of the items outlined above are more behavioral relationship in nature and they occur throughout the full life cycle. Also, many of these can occur in concert with some other management action, such as dealing with an unrealistic change request or some controversial technology-oriented decision. Collectively, the method by which the PM creates the work environment for the project team will dictate many other aspects of the project outcome both for the project deliverables and the HRs.

24.3.1.2 Team Development

Evaluation of overall team performance can lead to various strategies for improving outcomes. This evaluation can determine such things as tasks are being frequently delivered late, not delivering what has been requested, making poor use of the tools and resources, or not integrating work efforts well. Each of these represents visible signs of a need to improve some aspect of the team. The Father of Quality, Edwards Deming, stated that the cause for poor performance was most generally related to poor processes, inadequate training, or bad management. So, based on this observation, much of the team development potential comes from improving project processes and team skills. This means that the PM must focus attention on the work processes being used and an individual assessment of each worker's skills. If this evaluation is not viewed by the individual as punitive, it can be a positive motivational event. The fact is that every person has some developmental need and they should be introspective enough to know what that is. The same can be said of every process, so constant attention is needed to both skill and process areas where changes can make the most improvement. Skill changes can be dealt with through formal training or mentoring activities, while process issues would fall into the ICC process domain for consideration.

Once the basic team analysis is done, a training program should be defined in concert with the team members. This activity needs to be viewed as a positive exercise for career development and not punishment by being sent to school. Training can be provided in many forms, and in some cases, on-the-job training is even more effective than going to a class (Munby, 2008, p. 37). The operational reality of training is that it takes the individual away from "productive" project work and consumes resources that are not actively focused on the defined deliverables. Organizations are fortunate if they have adequate budgetary funds to execute these types of improvement programs as they contribute to motivation levels and help build a stronger work force over the longer period. Companies that shun training tend to lock their methods and skill base with the long-term result that they become less competitive. Good project managers contribute to their organizations future by improving their team member skills so that when the individual moves to the next project they are better skilled than before. In this view, the "team development" goal is both designed to improve the current project and the future career of the individual.

Team development is not just mechanical skill based. The technically focused project managers often ignore improvement of soft skills. Other valuable skills involve topics such as effective listening, conversational skills, communication dynamics, presentation skills, and working with diverse cultures, among others. In order for team members to grow in the organization capabilities in other disciplines, mentoring and coaching are valuable techniques to consider. These are not only good training techniques for the junior team member, they build the confidence and knowledge levels for the senior member as well (Ivancevich et al., 2008, pp. 411–433). Each of these formal development strategies contributes to the overall skill development of the project team and represents a decision item for the project manager. If done properly, this class of initiative helps build a productive culture for the team and improves employee skills for future work efforts.

There is one final point on team development before we leave this topic. The younger generation does not have the same view of organizations, management, training, and a host of other attitudes that reside in middle age project managers. In order to be effective with skill development, it will be necessary to find out how different the employee base is from the traditional. The new generation of worker is very bright, technology savvy, and they may be more comfortable approaching the learning process in a different manner. Recent reports indicate that dealing with younger employees has forced some employers to rethink how they evaluate employee performance (Hite, 2008).

24.3.1.3 Team Skills and Capabilities

When evaluating the solutions and action plans to develop the team, a similar approach to the one used for individual development can be used. If done properly, the development and contributions of individual employees should align with the team goals. Identifying the right set of skills for the team is one key for success and improvement in overall team performance. However, the one additional required feature is the interaction processes between team members versus simply dealing with an individual's current skill level. Despite the implementation of programs such as formal training, mentoring, and coaching, a team's development plan is heavily constrained by time and cost. All team leaders should strive to develop the group of individuals they lead into a productive collection who are able to blend their complementary skills and talents to achieve far more than they could ever aspire to individually (McCabe, 2006, pp. 116–121). This is basically the goal we are after in the team development activity.

24.3.2 Manage Project Team

The execution process deals with a wide variety of processes, including task integration, quality assurance, risk, communications, budgets, schedules, and procurement. Improper management of the team's HRs can lead to adverse results in all of these associated knowledge areas since it is humans who create the output.

24.3.2.1 Motivation Level

Motivation is certainly an issue that most managers deal with throughout the execution phase of any project. Classical motivation theory attempts to explain how the work environment affects motivation and productivity (Bjørnebekk, 2008, pp. 153–170). Myers research shows that there are first- and second-level factors affecting motivation (Myers, 1964). Only when employees have their own intrinsic generator can real motivation exist (Herzberg, 2003, pp. 87–96). One way to determine an employee's job motivation level is by monitoring his or her performance. The opportunity, capacity, and willingness to perform collectively predict an employees' actual performance. Team performance is linked to the individual's tasks-related skills, abilities, knowledge, and experiences (Ivancevich et al., 2008). Consequently, if the employee does not know what needs to be done, or how and when something needs to be done, high levels of performance are not possible. Once again, we see management role in the productivity equation beyond just building member level skills.

Another aspect of an individual's productivity is related to their personality profile. Some are motivation seekers, whereas some others are maintenance seekers. For motivation seekers, "the greatest satisfaction and strongest motivation is derived from achievement, responsibility, growth, advancement, work itself, and earned recognition" (Myers, 1964, pp. 73–88). Maintenance seekers are motivated by the nature of their environment and tend to avoid motivation opportunities. "They are chronically preoccupied and dissatisfied with maintenance factors such as pay, supplemental benefits, supervision, working conditions, status, job security, company policy and administration, and fellow employees" (Myers, 1964, pp. 73–88). These attributes go back to the Herzberg research results described in Chapter 18. The PM needs to understand both aspects of the positive and negative side of these factors, while trying to maximize the motivation items and minimize the dissatisfiers.

There are three components that make up individual motivation: direction, intensity, and persistence (Kidman and Hanrahan, 2004). Direction indicates if the person is motivated toward (i.e., trying to create) or away from (i.e., trying to avoid) an object or experience (Di Rodio, 2002). For example, if the employee is asked to work on a project with another group or individual, and there are issues between them, the employee may opt to do it, or to simply work alone. If the employee does agree, their primary motivation comes from the desire to comply with management directions. Conversely, if the employee chooses the second alternative of doing the job alone, this is still motivation but in this case it does not show a desire to comply with management directions. The intensity of motivation may be thought of the total amount of effort a person will make to satisfy a motive over time (Brehm and Self, 1989, pp. 109–131). Each employee's motivational response is different and it may be slow or fast. As an example, what happens to motivation when a team member working identifies an issue, communicates it to the project manager, and the manager fails to act quickly to resolve it? In this case the employee concludes that not meeting a specific requirement is not as important as they thought due to the lack of the managers concern. Thus, their intensity of the motivation will decrease over time. The last component of motivation, persistence, is defined as the ability to maintain action, regardless of the employee's feelings. Employees press on even when they feel like quitting (Pavlina, 2005). Managers who are capable of identifying and understanding the presence of these individual motivational differences will be more effective at determining proper ways to correct such issues. For example, a reward and recognition program that links performance and behaviors can increase both the level of intensity and persistence.

24.3.3 MEETINGS AS INFORMATION SOURCES

Traditional meetings involve bringing all or part of the team together for a face-to-face information sharing and this is one common strategy for team communication. There are several forms of group meetings and these sessions can be also used for both technical and social purposes activities (birthdays, promotions, or special occasions). Technically oriented meetings are used to disseminate information to the group that requires common understanding, technical problem solving, or status presentations. It should be recognized that group socialization and is a required process and a valuable part of the team building process. The role of these sessions is to improve the interpersonal relationships of the group and rigorious information focus is not the main agenda.

The second form of meeting is group dissemination of key information. In this case, the topic is judged to be significant and complex enough to warrant holding the session. This can involve some new company policy, major initiative, reorganization, or other such event that has interest to the group.

A third form of meeting may be required to work out details related to the mechanics regarding how to execute the planned work. These tend to be more work sessions than meetings but can involve a reasonable number of team members.

The fourth form of meeting is the one most often misused. That is, meetings in which groups of individuals get together to discuss some status-oriented view. These meetings obviously can be of different forms and goals. If we assume that there is a critical issue that requires multiple people to discourse and offer suggestions in a problem-solving mode, then such a meeting can be productive. In such cases the meeting must have an agenda and should have a limited time block unless it is considered to be a workshop. Meeting participants begin to become nonproductive after the first hour and few can handle a 2-hour session. In any case, at the conclusion of the meeting the chair of the session should assign follow-up to appropriate individuals, be sure that the meeting minutes are recorded and distributed, and then follow up with the delegated action individuals. If a follow-up is required, the action item status reports would be sent to the members for review prior to the next session.

As a modification to this problem-solving form, it is common to bring groups together to discuss status of their individual components and this becomes the *dangerous meeting*. The reason why it is dangerous is because it wastes valuable time with no return value. What often happens is for these to turn into the opportunity for various players to talk about past status events that have no relevance to the future. This will take the form of each player describing their past activities, and immediately the nonspeakers will begin to worry about what they are going to say when their turn arrives. Along this path some members will pontificate about issues that no one cares about, bragging about results, and the like. Others will simply execute their tasks with minimal content. The issue that the PM needs to think about in this and all meetings in general is whether the session is worth the cumulative cost and time of the individuals and is the subject matter worthy of a meeting. In many cases, the answer to this question would be a clear no. Most professionals indicate that they get more out of the free conversation that occurs before and after the meeting than in the meeting itself.

24.3.3.1 Meeting Structure

What else is wrong with this form of communication besides lost cost and resource time? The issues of agenda and time block remain important for all meetings. After that, we get to three more fundamental issues. These are

- Punctuality
- Rules of order
- Subject matter.

Punctuality: The first two items are procedural in nature. In some organizations the number of meetings is so vast that they are scheduled hourly. This means that those leaving the previous meeting will be late to the next one, while others await their arrival. Regardless of the organizational culture, the meeting schedule needs to run on the allotted times. It is impolite to the group to hold up the session. If the meeting was worth having, it was worth the participants coming on time. When this does not happen, an already suspect communication mechanism goes in the ditch in terms of value.

Rules of order: Some organizations use formal rules of order for voting and other aspects of the meeting. Normally, such rigor is not required; however, multiple concurrent speakers are a mandatory no-no. It is up to the meeting chair to limit side conversations. If the goal is to share information, trying to listen to multiple conversations at one time is impossible. Establishing a team culture that all meetings will start and end on schedule is a good management precedent, as well as polite behavior among the participants.

Subject matter: As suggested above, having a circle of individuals verbalize their status for last month is seldom a worthwhile exercise for the team. So what is worthwhile? One rule of thumb is that if it has already happened, we cannot change it. Doing a lessons learned from past actions to discuss what might be done to improve that situation may be worthwhile. But the personal favorite is to discuss with key team members the issues that are most likely to affect the project performance in the coming period (next 2 weeks to a month ahead). An example of one of these forms would be the major risk events that might be in play during this time period. Also, other troublesome operational issues that are now known could be viable topics.

The key often-ignored issue with meetings is that they are expensive and seldom produce the value compared to the cost. When they do occur, they should be focused and short. One of the popular software development methodologies requires a daily meeting to discuss key items for that day. It lasts for less than 30 min and everyone stands up for the duration. Normally, this frequency is not justified, but the standing up rule might be interesting to try.

24.3.3.2 Management Involvement

Another factor affecting execution is the level and style of management involvement. While management may be highly involved during the planning phase, this often declines as the project moves into the more technical aspects of the work. In many cases, the lack of management visibility begins to be interpreted as a lack of commitment toward the project, whereas the manager might be thinking that the project was moving along so well that he is not needed and would be in the way. The visibility and style of management involvement is important for various reasons. They do have a coordinative role in the process, but also have a motivation and support role that may well be even more important. Remember, the Hawthorne experiment taught us that workers like to feel important.

Having management involved helps with that process, if the manager understands their role. Simply stated, that role should be to help the project team overcome roadblocks to success.

In regard to this coordinative role there are multiple issues, ranging from simple to complex, that team members alone do not have the breadth of authoritative view to resolve. For example, two engineers may differ regarding the way a particular design decision should be made. One engineer may be thinking from the design point of view, whereas the other may think from a manufacturing point of view. For the most part, managers should serve as mediators and help provide proper and timely support to ensure that the facts are properly interpreted. Conflict management is clearly one of these roles and occupies approximately 20% of the typical project manager's time. Proper involvement in the conflict resolution process is necessary, but can be a team demotivator if not handled properly. Being a trusted and fair participant in the problem solution can stimulate and motivate the team. It can also serve as a mentoring opportunity in the hope that team members will get better at internal problem resolution. In general, active management involvement will improve the overall image of the project in the team member's eyes and will serve as a style example to follow.

Failure to be an active participant can result in resentment toward management and this can be reflected in several ways. Based on the Tuckman's four-stage development model described earlier, the needs and roles of management change through the stages (McCabe, 2006). In the early project stages, management involvement is needed to handle the role and relationship gaps that have yet to form. As the fourth stage emerges, the role of manager becomes more of helper than manager. Obviously, it is important to be able to recognize what the team needs and wants from you. So many times, the manager gets accustomed to being the knowledge base for the team, when at some point the team has built their knowledge base beyond the project manager. Many senior executives believe that because they have been promoted to a high level in the organization, they now have God-given knowledge on all subjects, when in fact they have become so isolated that they know very little about their internal organization. To a lesser degree, this is what happens as the project reaches the fourth, performing, stage of development.

One interesting characteristic of fourth stage teams is their high internal cohesiveness and tendency to operate within their own established guidelines and resist cooperation with management. This situation can be very disturbing to some project managers who are not treated as royalty any longer, but in fact may be a sign of a very productive team. It takes a very secure management professional to deal with a team at this stage. The team is now aware that they know more about how to get the job done than the PM does, so his suggestions will not be greeted with open arms. The problem-solving process becomes more internal to the team and less involved with the manager. Project managers who cannot change their styles from the more hierarchical model will be frustrated by this event. However, the same behavior can result in a team that is not highly productive, and when that occurs a more aggressive management style must take over to change the result. So, the PM must understand the team and recognize which form of management is needed. If the negative situation occurs, textbook behavioral theory is not so relevant and more drastic authoritative action is required. Once other positive motivational avenues have been explored, it may be necessary to transfer one or more team members out and in extreme cases dismiss them from the organization (Mind Tools, 2008). Regardless of the strategy needed, a good manager must be able to handle a wide range of situations (Mind Tools, 2008).

24.4 PROJECT TEAM MEMBER DISSATISFIERS

Herzberg (Chapter 18) provided an extensive set of potential job dissatisfiers. The five largest impact items are (in order of severity) as follows:

1. Company policy and administration
2. Supervision—technical
3. Interpersonal relations with superiors

4. Salary
5. Working conditions.

Previous discussions have described various aspects of managerial style in regard to the project team. Company policy and administration is above the level of the project and will be difficult for the PM to change, although he should use every opportunity to minimize whatever the irritant is.

Salary and working conditions are two items more in the domain of the project, although both still have higher-level organizational constraints as well. The most controversial item among the dissatisfier list is the role that salary plays as both a motivator and a demotivator/dissatisfier. The discussion offered here will likely not change any personal views on this topic, but may offer some insights into the subject.

It is common for many companies to have a central HR-derived compensation program that is to be applied consistently to all employees. Experience suggests that the amount offered by these programs is never enough to satisfy the individual. Even if they get the maximum amount allowed that only will to be a positive motivator for a few days. So, how is salary viewed in organizations? HR managers often say that they cannot pay enough to really be a motivator and still keep the company competitive, so their compensation plan is designed around two targets: the low target is to try to pay competitive enough salary to keep the individual from being motivated to leave the company. At the same time, raise amounts would be set to provide some small measure of positive motivation. The net effect of this is that in most cases salary will not be part of the overall motivation package that the PM can use to any great degree.

Ideally, it would be nice to have a performance bonus for project participants based on some defined set of completion metrics. But this does not appear to be a common strategy, or at least is not advertised as one. Salary issues at best will be a neutral situation in most cases and other motivational strategies will be required.

Working conditions: This term can have a wide variety of interpretations. Issues such as an office with a window would be considered a working condition perk in many organizations. Ditto an office with a wall versus an open cubicle. Employees often say that their building is old and dingy as though this is a dissatisfier, while at the same time a refinery engineering group sitting in the middle of a potentially dangerous environment with nothing but gray paint seems perfectly ok with their environment. The author once worked on a project that got moved to the back lot in an abandoned Quonset hut with scrap furniture. This facility had zero class or status, but that team was one of the most happy and productive teams ever experienced. We had our own world. The social environment was great and we got a lot done. Review the attitudes described for the fourth stage team, and you will see some of these characteristics. So, working conditions are in the eyes of the beholder. It is up to the PM to evaluate this and see what can be done to mitigate the negatives. Even free pizza for lunch every Friday can go a long way to make working conditions better.

24.5   PROJECT TEAM MEMBER MOTIVATORS

The following are the top seven Herzberg motivators in general order of severity:

1. Achievement
2. Recognition
3. Work
4. Responsibility
5. Advancement
6. Salary
7. Possibility of growth.

At least the top four items on list are viable options that the PM has at his disposal. The goal of allocating rewards should be predicated on performance. We have described the control knob model earlier for the project control strategy. A similar data collection process is needed for the reward system for team members. If you do not know what the individuals are doing in some measurable form, you simply have no equitable way of allocating whatever rewards are available.

24.6 CONCLUSION

The execution phase brings to bear all of the knowledge area issues at one time. Failure to have in place a viable plan can spur all of this into a variance condition, thereby creating chaos for the project manager. However, if we assume a reasonable project plan, then management of the execution process becomes more coherent. Some of the management actions are monitoring and analysis oriented, whereas others deal with the softer side of human interactions.

As a project team matures the appropriate management style must evolve to match the team requirements. During the initial team acquisition phase, managers should be aware of the candidates' individual skills and how they may contribute to the overall achievement of the team goals. In addition, the PM requires similar evaluation and consideration. Depending on the situation and project stage, different manager/leader styles are required to ensure successful performance.

During the execution phase, teams go through different development phases, referred as forming, storming, norming, and performing. As these stages unfold, the development strategies for the team also evolve. Teams will be developed through various types of training and process development. In addition to this, the PM needs to be aware of various motivation and dissatisfier variables that can impact the team's performance in either a positive or a negative manner. It is important to provide the team members with the required training and tools, but those alone are not sufficient. Capacity, lack of trust, and compensation are other issues that may also affect the team behavior and consequently their performance. If not addressed properly and in a timely manner, those issues may result in lack of motivation, development of negative group cohesiveness (groupthink), and resentment toward management within others. Understanding differences in opinions and behaviors within team members is a key factor when dealing with such situations. Each individual has a unique way of thinking and processing the information. Consequently what will work for some individuals, will not work for all. Similarly managing all teams the same way will not provide the same results.

Several questions surface in regard to execution management issues. For example, will a manager be able to identify all the different circumstances that would determine a specific individual's attitude? Would a manager be able to apply all these concepts when working with a team? What should the HR department do to facilitate this process? Do project managers require special training?

Execution issues require a root cause analysis to identify the problem, as well as the reasons behind the situation or behavior. Only then will a manager be able to act and provide a feasible solution to the problem. When working on those decisions, managers should consider that one minor mistake might cause major issues with the team; hence the importance of developing team-building exercises, having defined and consistent communications guidelines, and ensuring management involvement. If the PM can activate these recommendations, it should increase the opportunity to have a highly successful and performing team.

DISCUSSION QUESTIONS

1. How do you assess the general status of a project?
2. Describe two of the *control knobs* that a PM might use to correct a project's direction.
3. Can scope creep be controlled? Define some measures for this.
4. What is the role of ICC?
5. Is salary a motivator? Explain your position.
6. Are meetings effective collaboration processes?

REFERENCES

Bjørnebekk, G., 2008. Positive affect and negative affect as modulators of cognition and motivation: The rediscovery of affect in achievement goal theory. *Scandinavian Journal of Educational Research*.

Brehm, J.W. and E.A. Self, 1989. The intensity of motivation. *Annual Review of Psychology*.

Di Rodio, W., 2002. An exploration of the concept 'motivation' as a tool for psychotherapeutic assessment. Practical philosophy. *The Journal of the Society for Philosophy in Practice*.

Herzberg, F., 2003. One more time: How do you motivate employees? *Harvard Business Review*. (January), 87–99.

Hite, B., 2008. Employers rethink how they give feedback. *The Wall Street Journal,* (October): http://online.wsj.com/article/SB122385967800027549.html?mod=googlenews_wsj (accessed June 10, 2008).

Ivancevich, J., R. Konopaske, and M. Matteson, 2008. *Leadership in Organizational Behavior and Management*, 8th Edition. New York: McGraw-Hill/Irwin.

Kidman, L. and S. Hanrahan, 2004. *Creating a Positive Environment, The Coaching Process: A Practical Guide to Improving Your Effectiveness*. Wellington: Dunmore Press.

McCabe, M., 2006. Accelerating teamwork: A personal reflection. *Musculoskeletal Care*.

Mind Tools Ltd, 2008. Coaching for team performance: Improving productivity by improving relationships. *Mind Tools Ltd.*, http://www.mindtools.com/pages/article/newTMM_66.htm. (accessed June 10, 2008).

Munby, S., 2008. Learning on the job makes good leaders. *Times Educational Supplement*.

Myers, M.S., 1964. Who are your motivated workers? *Harvard Business Review*. 42, 73–88.

Pavlina, S., 2005. Self-discipline: Persistence. http://www.stevepavlina.com/blog/2005/06/self-discipline-persistence (accessed June 10, 2008).

PMI, 2008. *A Guide to the Project Management Body of Knowledge (PMBOK® Guide)*, 4th Ed., Project Management Institute, Newtown Square, PA.

Smith, M.K., 2005. Bruce W. Tuckman—forming, storming, norming and performing in groups, the encyclopaedia of informal education. http://www.infed.org/thinkers/tuckman.htm. (accessed June 10, 2008).

Tuckman, B.W., 2001. Developmental sequence in small groups. *Group Facilitation: A Research and Applications Journal*.

Wilson, C., 2008. Training Journal. BNET Business Network (February). http://findarticles.com/p/articles/mi_6778/is_2008_Feb. (accessed June 10, 2008).

Winters, F., 2003. The Top Ten Reasons Projects Fail (Part 7) (August). http://www.gantthead.com/article.cfm?ID=187449 (accessed June 10, 2008).

Wong, Z., 2007. *Key Stages of Technical Development, Human Factors in Project Management: Concepts, Tools, and Techniques for Inspiring Teamwork and Motivation*. San Francisco: Wiley.

Part VII

Monitoring and Controlling Techniques

LEARNING OBJECTIVES

This part will focus on techniques for accomplishing the following eight monitoring and control process objectives:

1. Measure project performance continually by comparing results to the baseline in order to identify project trends and variances
2. Refine control limits on performance measures by applying established policy in order to identify needs for corrective action
3. Perform timely corrective action by addressing the root causes in the problem areas in order to eliminate or minimize negative impact
4. Evaluate the effectiveness of the corrective actions by measuring subsequent performance in order to determine the need for further actions
5. Ensure compliance with the change management plan by monitoring response to change initiatives in order to manage scope
6. Reassess project control plans and practices by scheduling periodic project and change control system reviews with stakeholders in order to ensure their effectiveness and currency. Update the plans and practices as required
7. Recognize and respond to risk event triggers in accordance with the risk management plan in order to properly manage project outcomes
8. Monitor project activity by performing periodic inspections to ensure that authorized approaches and processes are followed, or to identify the need for corrective action.

Source: This list of monitoring and control objectives is adapted from the PMI, *Handbook of Accreditation of Degree Programs in Project Management*, Project Management Institute, Newtown Square, PA, 2007, pp. 17–20. Permission granted by PMI for use in this context.

Monitoring and control processes cover the entire life cycle of the project, with major focus on the execution phase. The monitoring process involves measuring results and tracking status of defined items versus the established baseline, while the control activities are more corrective in their focus.

General control approaches have been described in previous discussions, primarily in Parts IV and V. This part, Part VI, will deal with those control aspects that are not specifically KA based and deal with evaluating project execution. Each of these techniques provides insight into some aspect of project control or status reporting. Following are the major control-related topics discussed under this banner:

- Measurement and tracking of actual performance
- Establishment of performance metrics
- Change control process
- Earned Value Management (EVM)
- Enterprise reporting.

Each of these topics introduces some important aspect of monitoring or control. If we decompose the basic underlying logic of the five major chapter titles in this part, we would find a general logic process similar to the following:

- Define the items that identify performance status (metrics).
- Measure the defined items throughout the life cycle (tracking).
- Rigorous control change process focusing on the approved plan (stabilize).
- Measure actual performance as compared to plan (performance analysis).

Each of these topics represents a critical control concept, but surprisingly many organizations do not accomplish these goals very well. For example, failure to identify meaningful metrics to show the status simply means that the wrong performance units are reported and the receiving stakeholders are left with an ineffective understanding of project status. The classic example of misuse is a status presentation showing planned versus actual resource consumption (by itself) as though that is a measure of good or poor performance. The Earned Value discussion will show why such comparisons by themselves are wrong. The Earned Value technique emphasizes that in order to be effective the control process must be couched with a performance measure to compare against resource consumption.

The third control item described in this part is that of change control. Too often, scope creep occurs because excessive or subtle unapproved changes creep into the work. Each such event adds incremental time and resource effort to the project with the result being an out of compliance conclusion. Taking a hard position on which one of the control processes is most important would be controversial. However, the management of scope change has to be considered a mandatory control requirement if the goal is to stay within planned boundaries.

Personal bias says that the PM should not be the one deciding on which change request to approve. This should be done by a higher-level management board who has authority to approve or reject additional work and who has a clear understanding of the business goals linked to the project. Decisions to approve a change request involve many factors, both technical and organizational, so the project board needs to have good understanding of the project and its future operational environment. In addition, the project board needs representation from both the business and technical side in order to properly consider the spectrum of future proposed change requests.

However, there is much more going on in this domain than passive monitoring. It is clear that the PMI model management philosophy also focuses on "influencing" factors that either can affect or are affecting project performance. This implies a much more dynamic control process than simply measuring the output. In addition to this, the monitoring and control (M&C) activity list specifies that stakeholders are to be managed and performance reported, so there is an information sharing component as well.

The various operational systems described in this part can all be important tools in achieving project success. The role of the five techniques discussed in this part is briefly described below.

Integrated Change Control

One of the most important control subsystems is ICC. This process formalizes the flow and approval process for all change requests. The absence of a comparable system tends to leave the project in chaos.

Metrics

A metric is a quantitative measure used to compare and report project status. This chapter summarizes the use of these in the monitoring and control process. Management and other stakeholders use these measures to evaluate various aspects of the project. Traditional metrics were output oriented and dealt basically with schedule and cost-related issues, but the more contemporary metrics approach is to use a broader set to focus organizational attention on more strategic objectives. The adage "What gets measured gets done" summarizes how the use of metrics can shape the day-to-day activities of various organizational units.

Earned Value Management

EVM is a project management concept that is now gaining broad industry acceptance. Interest in this concept has been static for much of the past 40 or so years and continues to be a difficult one to implement in many organizations; however, with increased management maturity there seems to be growing acceptance of the concept as it has now become an ANSI standard tool. For the modern PM, this tool must be understood in its potential role for monitoring and control.

Tracking Project Status

Regardless of the techniques involved, one of the basic management requirements for every project is to communicate status. The internal project team needs to understand areas of deviation from plan and similar elements of this are also important to external management and stakeholders. This chapter will outline the theoretical constructs to be considered in project tracking.

Enterprise Reporting Using the BSC

Since every project should be created to help the enterprise achieve some defined goal, it is important to align the project with the higher-level goals of the organization. For this reason the concept of enterprise reporting should link with project monitoring. It is one thing to say that project goals should be consistent with organizational goals, but it is quite another to actually find the mechanism to accomplish this. The BSC model represents just such an approach and it is described in this chapter.

Final Thought

A final philosophical thought on control—you cannot control what has not been planned. Each of the control items mentioned in this part and throughout the book has this fundamental requirement built into its internal design mechanisms. The basic premise of project management is that the execution phase is designed to produce and deliver what has been planned. In concert with this, the control phase oversees that progress and attempts to influence corrections required based on observed variations from the plan. We must be sensitized to the reality that every project will experience some degree of variation, but the essence of control is to minimize the impact of these variations. The collective tools discussed in this part represent KAs that the PM must have in his pursuit of this goal.

25 Change Management

25.1 INTRODUCTION

Effective change management integration is a vital and key ingredient to the success of projects and project management. Successful integration of change management into all aspects of a project life cycle is one of the most critical processes related to a successful outcome. In highly technical projects, where cutting-edge technology and requirements changes are a constant, PMs must be highly regimented in their approach to dealing with the related dynamics. As part of this process it is important to track, understand, and quantify the impact of each change to the approved project plan baseline. This suggests that the defined change management process must be formal and rigorously followed by all participants. Leadership, team focus, and effective communication are very important aspects associated with an effective change control process. Improper change procedures will result in confused expectations from various stakeholders as the initial project goals change. A key ingredient of the change process is to not only control the changes being proposed to the project, but to communicate the status of those proposed to appropriate organizational elements.

25.2 INTEGRATED CHANGE CONTROL

The ICC process involves identifying, evaluating, and managing changes throughout the project life cycle (Schwalbe, 2006, p. 151). One major operational objective of ICC is to ensure that the changes that are made after the plan is approved are supportive of the current organizational needs. A key question is to determine whether the value of the change is greater than the additional resources required and the disruption that it brings. "The original defined project scope and the integrated performance baseline must be maintained by continuously managing changes to the baseline, either by rejecting new changes or by approving changes and incorporating them into a revised configuration baseline" (Love, 2004).

The PM must be vigilant to the various change activities that occur on a daily basis. Each change represents a miniproject in its own right in that each has a life cycle from proposal to implementation. No change should be made to the original requirements unless it is deemed worthwhile, taking into consideration the downside of interrupting the existing plan. Scope creep is a term used to describe the expansion of project requirements over time. Without an effective change control system this can crater the projects chance for success. Even the use of an effective change control system can suffer if the initial requirements were poorly done. Also, there are situations where the project target environment undergoes a radical change and this can cause levels of scope creep that challenge the validity of the project.

Uncontrolled changes have subtle effects on the underlying project variables, which over time negatively reflect the customer's view of the project. Generally speaking, a change has high potential to create additional work for the project team, which then impacts schedule, project costs, resource plans, and additional risks. The following is a sample list of possible implications from a single change:

- An expansion or reduction of project scope
- An expansion or reduction of product features
- An expansion or reduction in performance requirements
- An expansion or reduction in quality requirements
- A significant change in the target milestone or completion dates
- A shift in the implementation or deployment strategy
- An increase in resource costs
- An expansion or reduction in the project budget
- A change in any of the project objectives
- A change in any of the final acceptance criteria, including ROI forecasts
- A change in any of the project assumptions, constraints, or dependencies, especially regarding resources and work effort estimates
- A shift in project roles or responsibilities, especially on projects with contractual arrangements
- A decision to reset the performance baselines due to an unrecoverable performance variance.

One metaphor for the change process is that it is like dropping a large rock in a pond. The waves reach across the entire surface. Even small changes can have a similar impact.

25.3 PROJECT MANAGEMENT PLAN

The project management plan is a document used to coordinate all project components and to aid in directing activities. Because of this role the plan is an important input to the ICC process. As changes are approved the project plan should be adjusted to reflect their impact. The project management plan is the blueprint for all project activities; it offers insight when considering changes and deciding whether to proceed with the change. As the plan evolves, it maintains the view of both approved baseline and current plan.

The performance measurement baseline (PMB) is the approved plan baseline parameters for a project, activity, or deliverable. Baselines can be established for any project attribute and these simply provide for a comparison value through the life cycle. Typical baselines relate to cost and schedule, but additionally some technical performance goals could also be established in the plan and status comparisons would be based on these through the execution phase. Metrics of this type help the PM to communicate product performance status to various stakeholders.

Performance reports are an effective method of conveying project status information. They are also useful for determining whether there is a need for change. Typical formats for this class of reporting include histograms, Gantt charts, S-curves, schematics, colored graphical icons, and tables.

Histograms are useful to show distribution of data by category. If the data are maldistributed the best method of showing results is the Pareto diagram (Figure 25.1).

A Gantt chart is the most used task and schedule communication media format in organizations today. It is successful because it is graphical, simple in structure, and provides needed information without a requirement for training. The example shown in Figure 25.2 is a typical project Gantt bar plan format.

Figure 25.3 shows a sample S-curve. This type of data display is often used to show time-oriented rate information such as coding rate, change requests received, defect correction, completion of WPs, and so on.

The rapid and constant evolution of cutting-edge technology has made information technology projects some of the most difficult to accomplish successfully. In this environment PMs are dealing with a constant state of change since stakeholders, users, and engineers are frequently at odds over

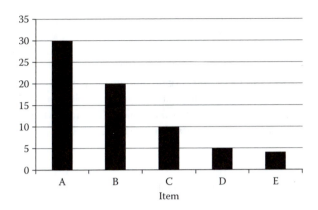

FIGURE 25.1 Pareto example.

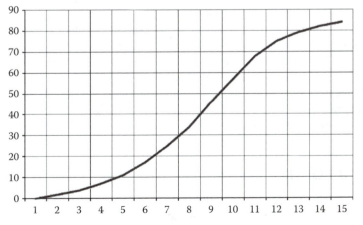

FIGURE 25.2 Example of a commercial Gantt chart. (Output from Kidasa Software, Inc., www.kidasa.com.).

FIGURE 25.3 Example of S-curve.

requirements and technology usage to achieve those objectives. Recognition of this dilemma was one of the major drivers in the life cycle management movement that is reflected in this text. The result of this created a stronger push to better define requirements prior to execution and a corresponding focus on change management through the project.

Based on research carried out by various organizations, there is a consistency in factors leading to project failure. Common among these are unclear goals and objectives, estimation mistakes, and high levels of scope changes requested during the project. The following list reviews some other frequently observed factors previously discussed in the book:

- Poor planning
- Unrealistic time or resource estimates
- Inadequate executive support and user involvement
- Poor project team management
- Inappropriate team skills.

The FBI's Virtual Case File project makes a graphic example of what poor change control can do to the deliverables. In hindsight, FBI management admitted that the underlying design technology had failed to meet the operational requirements and as a result five years of development and $170 million was lost (Neimat, 2005). In a 2004 postaudit, the National Research Council saw no evidence that a technology contingency plan had been formalized for this project. The development vendor claimed that they delivered the first phase of the project ahead of schedule and under budget, but the requirements for the software had changed significantly after the September 11, 2001 terrorist attacks. Subsequent research revealed that major requirements changes had also been ongoing since the start of the project. In addition, there was a high turnover of top management during this period. The final audit conclusion revealed that the requirements were never satisfactorily defined or stabile and this created many of the downstream issues. The factors outlined in this situation are not uncommon, but their results are often more clouded than in this example. When a project either does not know what the goal should be or does not properly define what the user needs, the result will be similar to the one outlined here. This example also shows that much of these difficulties are often related more to the people process than the technology itself, even though technology may also contribute to poor results (Neimat, 2005).

PMs often have the feeling that their project will never settle down and stabilize. Change requests seem endless. When this happens, it is likely a result of some dynamics either in the original specifications, or in some changing environmental variable (management, technology, marketplace, economics, organization, etc.). To sit back and simply try to incorporate the changes may not be the best strategy. In some cases, it may be necessary to halt the project and do an assessment of its initial scope compared to the current view. If the organization is not sensitive to these dynamics, they could well be assisting in creating a failed effort. The Virtual Case File system obviously followed this path with both uncontrolled and unmanaged scope and feature creep.

Scope creep refers to previously unplanned and unexpected changes in user expectations and requirements as a project progress, whereas feature creep refers to similar requested additions of features to a system. In both cases changes of this type may appear minor, but often bring with them unseen consequences resulting from underlying technology failure or inability to manage the changing work requirements. PMs who are not sensitive to these subtle interactions inside the system will approve what appears to be isolated changes, only to find out later that there were other related implications that further increased project work (i.e., interactions with other areas, risk, quality, etc.) (Neimat, 2005).

25.4 CHANGE CONTROL SYSTEM

Implementing an effective change control system is vital and necessary for a PM who understands the negative impact of this aspect of his project. Failure to define an effective change control process could well be the single biggest management mistake one can make. This process should be identified as part of the Scope Management Plan and agreed with by all management participants. This process needs to be formal, documented, and inviolate. It specifies how changes will be requested and the associated processes for evaluation, approval, and implementation during the execution phase. A *Change Control Board* will be specified as the approval authority and this group should be external to the project team. Also, the process will include some formal repository to maintain status for all requests.

The Change Control Board Charter should include the authority to approve changes of some specified level and to manage the allocated funds set aside to handle this class of activity. Within the board structure, roles and responsibilities will be defined including chair and permanent members. Typical members include the PM, sponsor, senior users, and support manager. As new changes are requested, the project team reviews the request in regard to resources and work impact, as well as other related impacts regarding risks, quality, and rewards related to the proposed change. They also analyze how the change affects the overall schedule, costs, and feature set (Miller, 1992).

In all cases, the change control policy should provide guiding review principles that provide the basic approach for ensuring that appropriate issues have been fully coordinated prior to being submitted to the board for review. In addition to the basic decision processes outlined above there should also be a higher-level appeals process for rejected changes (Love, 2004).

Below are some sample guidelines for the change approval process and the list details how to best apply change management:

- *OK to say "No":* If the project team has done a good job of collecting initial requirements and managing the life cycle, many features have already been reviewed and prioritized before execution begins. If a newly proposed requirement is not worth the time to analyze it, then it is not worth the time to implement it. Therefore, it should be rejected immediately. Normally, a good Change Control Board will say "No" more times that it says "Yes" in order to promote a more stable work environment and ensure that only critical changes are implemented.

- *Changes should be bundled:* A large number of small changes, when done independently, can greatly affect the project timeline because each one affects many areas of the system such as testing, user documentation, support, and so on. To gain economies of scale and effective use of personnel, it is normally more efficient to bundle a set of changes into one change package and include them in a single approved change. When handled in this manner, a master umbrella change number would be created to control implementation. In this manner, it may be possible to include some minor changes at near zero cost once a major change is approved. In any case, this needs to be done to help the project team better manage their time with this activity.

- *Eliminate bureaucracy:* Some Change Control Boards are made ineffective by individuals who just like to say "No" because they do not want to be bothered by the change. Even though change is a negative event to the team, it may well be a required event for the success of the project. Excessive rejection creates ill will to the project when it seems as if the board is not making decisions that are in the best interest of the user community. To eliminate this problem, it is important to educate all members of the Change Control Board regarding their roles and constraints. Ensure they understand that there will be frivolous changes that will be submitted as well as legitimate ones necessary for the product to be marketable and to meet the needs of the business. So as each new change is suggested, it is important to produce a risk/rewards discussion that analyzes the impact to the project and documents the business reasons for the change. An important part of the change process is to document and publish the findings and rationale for each change request, whether it is approved or rejected. This postreview is important to create buy-in from stakeholders that requested the change. Rejection of a change request is particularly sensitive since it by definition represents a conflict of opinions as to its value.

- *Change process documentation:* Each change request should be identified by a unique tracking number. This scheme is important to ensure that all requests are dealt with and it helps to monitor the flow of the request through the process. Mechanics for managing the flow of these documents are important as well. One approach is to have the PM or his delegated team member initially screen the requests. In some cases, a change represents a misunderstanding and the requirement is already being dealt with. In this manner, it may be possible to delete those without further handling simply by explaining the situation to the requestor. There are other possibilities that also may be able to be screened without detailed analysis. For example, a large change request might be routed to the board to see if they would defer it outright as a change in Charter (assuming that they have the authority for such actions). All of the requests that are not filtered out by this preliminary review would need some level of detail to produce the analysis of risk/rewards for the change. One approach for the second review would be to do a quick assessment to see if the request is viable technically or within constraints. If not, a discussion could be held with the person submitting the request to show them why this needs to be deferred. If successful, that request would be simply left in the system as a deferred item that would be reviewed after the project is completed. This could be a discussion topic for a follow-up project phase. The third review level would be a full analysis of the request as described. The results of this step would be documented and reviewed by the PM prior to formal submittal to the project board. Ideally, all of these review options would be supported by an on-line system to allow oversight into status.

25.5 CONFIGURATION MANAGEMENT

The configuration management process is designed to address the control of project-related documentation throughout its life cycle (Appendix C outlines the basic requirements for a project document repository). Collectively, the goal of a configuration management process is to ensure that project artifacts are properly defined and tracked as changes occur. There are four general activities involved in configuration management:

1. Identifying and documenting an item or system's present functional characteristics, physical characteristics, or a combination of both, which represents the configuration baseline
2. Controlling changes to the functional and physical characteristics of an item
3. Monitoring and reporting status of changes to the system
4. Supporting an audit function to verify conformance to requirements. This involves a physical inspection of the product to ensure that it meets the required standards, and to ensure consistency of the released product.

Communication of change requests is an important factor in change control because it affects the work definition for the internal team and sets expectations for the users. It also can have significant impact on project deliverables. Also, at the upper extreme, a major change can impact the planned project schedule and budget. As we have seen in this discussion, it is necessary to use both written and oral communication to identify and manage the project change process; however, status tracking of this must be done formally. Within the communication framework, the goal is to manage the overall change process and communicate this to appropriate parties. As we have emphasized, the underlying goal of the change management process is to keep the project as stable as possible in regard to teamwork assignments and to protect the project's baseline parameters as much as possible.

One of the basic management process questions involves the media and methods to communicate changes throughout the project environment. A process based on moving paper through the organization is fraught with error and using traditional email has operational limitations as well. Mature organizations develop some form of on-line system to manage change configuration-related issues, including not only the initial change request submittal process but also the overall tracking of the status as it moves through the system. Typical status would indicate state of the request, who has it now, and any comments that are relevant. The goal should be to not let these lay dormant and a service level response time should be stated in the system design. The three major service level issues would be initial response, analysis cycle time, and time for the board to act. Hopefully, the discussion of actions needed to manage the change process has highlighted the level of work that goes with this activity. A medium-sized project can consume a full-time person in this role if the initial requirements are not reasonably well done and that level of overhead is noticeable in the overall budget level. Appendix C will offer some additional insights into a theoretical approach to structuring a project repository.

We have stated previously that creation of an effective change control process is fundamental to project success. In many cases this system is defined as a paper form with loosely followed procedures. There are many ways in which this type of system can be structured, but the key is to be sure that all requests are properly handled and all approved requests are properly tracked and implemented. Any failure in these areas invites miscommunication, misdirection, and frustration (Pritchard, 2005).

25.6 CHANGE MANAGEMENT WORKFLOW

A change requirement can be stimulated by many events and various organizational sources, including the internal project team. The one cardinal rule of change management is that no ad hoc change is made to the planned deliverable. All changes will be approved at some defined management level, which could be the PM, project management board, or some higher-level management entity, if the change is of significant magnitude to require a new project Charter. If anyone asked for "one little change" the stock answer is to submit a formal request.

The change management process needs to be approved at project inception. Failure to do this leaves the project team vulnerable to an uncontrolled environment. If you believe that this is one of the major sources of failure, does it not make sense to take a hard stand on keeping it under control? Remember, the original plan was approved by management. It is important to not let future changes violate that approval.

As indicated above, the change requirement could come from almost any source, so the first step in the process is to collect the details on the change. This would typically be done using a paper from or an automated system. If the project stakeholders are geographically distributed, the latter option is preferred and all mature organizations should be supporting a robust change system with a formal system. The arrival of a change request requires a quick response (say 24 hours). These would be reviewed by the PM or his delegate. At this filter point the request will be assessed as to size, criticality, timing, and so on. The submitter will be informed that the team is looking at the request. If it is clear that this will not be approved, a political negotiation with the submitter should be undertaken. It may be that the requirement is already in the work plan or it may be that the cost of doing what is proposed is clearly out of balance.

Step two in the process involves a technical and economic assessment of the impact. This should include all aspects of the change and not just time and cost, but risk, quality, skill requirement, and the like. The ideal technical system to handle this would be an automated workflow from the initial filter person to the individuals assigned to analyze the request. This may be handled by one or more individuals. As the complexity grows it may be necessary to hold a technical review meeting to resolve the response. This result is passed back to the PM or his delegate for communication back to the submitter.

Step three involves a discussion of the findings with the submitter. This is very similar to taking your car to the garage and some technician examine it and come back with essentially a SOW. You may or may not want to get that part of your car fixed at that price and the same can occur with a change. If the price or impact is too high the submitter may agree to withdraw the change (forever or until a later time). If it desired to continue, the process moves on.

Step four involves a management assessment. If the PM has access to funds for scope he may be delegated some level of flexibility to approve small changes. In other cases, the project board would handle the approval. Technically, an approved change should bring with it more resources to handle any work involved, but in many cases this is not done (a mistake in the author's opinion). A rejected request would be communicated back to the requestor with the logic for rejection. Otherwise, we now have an approved change to execute.

Step five requires insertion of the new work into the project plan and WBS. Details of the change will need to be communicated to appropriate team members. Conceptually, the change is now embedded into the work process just as though it had been planned that way from the beginning.

In looking back over the flow of work required to handle a single request it should be very obvious how this class of work could sabotage productivity of the team. It essentially bleeds the resources away from the planned to plan the new component. Also, band aiding new work into the planned work is often complex and in some case portions of the original work will have to be undone.

25.7   EXTERNAL COMMUNICATION ISSUES

Given the analytical complexity of change requests it will often be necessary to hold face-to-face meetings to iron out issues and negotiate a settlement strategy. In some cases, larger group meetings are also used for this activity, but they are often misused and are ineffective as problem-solving strategies. Regardless of the negotiation method, it is important to recognize that the flow of information through the change cycle stresses typical communication channels. This is caused by the wide organizational scope of this activity and the underlying complexity of the issues.

Poor communications between departments and the project team are a major source of project frustrations and failure according to a report released by Unilog, an independent pan-European consultancy and service company (CRM Today June 11, 2003). This study also documented that only 10% of PMs had a formal business sponsor identified at the outset of a project and 100% had experience with a project that had failed to meet all its objectives. Although PMs work hard at avoiding many common implementation pitfalls related to failure, they are often made scapegoats for failed projects that have had inadequate input from the business side or excessive changes from that source. In order to mitigate this result the wise PM would be on the lookout for the following *Seven Deadly Sins* (CRM Today June 11, 2003):

1. Poor project scoping and undefined project objectives, roles, and responsibilities—leading to the setting of unrealistic expectations
2. Lack of communication between the project team and the business—resulting in a mismatch of requirements and expectations
3. No senior business sponsor
4. Technology put before people—no or minimal involvement of key users during the scoping phase and lack of regular communication with them throughout the project
5. No project success metrics defined
6. No risk assessment or contingency plan created
7. Lack of regular checks to ensure that the project is on track per the user views.

This list is essentially identical to other surveys discussed previously in the text, but the logic of repeating it is to emphasize that these same symptoms appear in various survey forms as causal effects leading to change requests and potentially failures. A proactive PM should focus on these aspects of the project early to establish a more stable execution environment later.

More than one quarter (28%) of PMs polled cited the lack of communication between the project and the business as the primary reason for the failure of their most recent project, a figure that rose to almost a third (32%) for companies with a revenue of $350 million or more. Expectations not properly set (or communicated) and inadequate project scoping were given as joint secondary reasons for project failure, with each being identified by 20% of the respondents (CRM Today June 11, 2003). Once again, each of these factors points to poor planning at the outset of the project.

The traditional view of change management is essentially one of a reactive process. That is, managing the stream of change requests that come floating through the system and they are dealt with in some fashion. This reactive mode is not conducive to either efficiency or success. The earlier rationale for going back through some statistics on project failure factors is to restate one of the cardinal philosophies of project management. That is, the PM should take proactive positions to influence desired outcomes, not sit back and reactively try to manage the flow of documents through the project. Proper requirements definition in the beginning and effective "influencing" the external user community during execution can help stabilize the project work. If users are aware that a change can elongate a time critical project, they are often willing to wait and consider the change later. By avoiding the underlying factors that contribute to project failure the change management process becomes less controversial. This means that more verification of requirements needs to be undertaken early in the life cycle and more user communication related to those requirements exercised.

The balancing act that is involved in the change process is one of attempting to complete the project as defined, while at the same time trying to hit a reasonable portion of a moving target. Successfully completing a project that no longer reflects business value is just as bad as not completing it at all. This is a fundamental mindset that goes with this topic. In order to wrestle with this two-headed equation, business managers must be involved from the outset and must be in partnership with the decision process. Project failure rates and high change volume will remain issues so long as an ineffective communications gap exists between the two (CRM Today June 11, 2003).

25.8   PROJECT OPERATIONAL INTEGRATION

Up to this point the discussion has viewed the change process primarily from the perspective of initial plan through execution. There is yet one more project layer in which the topic of change is critical. That is, once the project deliverable is completed, it needs to be moved into an operational status in some user environment—this is the planned process state change that was discussed in Part II of the text and represents the reason that the project was initiated in the first place. This deliverable can be a new software system or some new product widget. Regardless, in most cases the process of moving the project deliverable into the user environment upsets the status quo of that organization unit. Humans are asked to change the way they currently perform tasks and therein lies the management change challenge.

Technical project teams often become convinced that the item they are delivering will be loved by all, only to find out later that this is not the case. These teams are also often not sensitive to the fact that the new user will not intuitively know how the new item works, even though the project team feels that it is easy because they created it. In this environment the change process is not oriented toward analysis and approval; rather it is oriented more toward motivating and teaching the new user group to be successful in using the item.

Several years ago the author was involved in implementing a word processing system into a government organization typing pool. Everyone today knows what such a system can do and it would probably not be resisted, but back in the 1970s period computers were relatively new tools for end users and the new word processing system seemed pretty complicated to the traditional typist. Their current system was essentially a standalone set of typewriters and there was near zero computer literacy in the group that was led by a senior person with almost 30 years of typing experience. As a relevant side note, this individual was not a particularly good typist and her documents were often filled with correction fluid (used to blot out typos). When the initial system was introduced to a small group there was a lot of resistance to its "complexity." In particular, the typing pool supervisor said that it was too complicated to use and her attitude impacted the acceptance of system by others. Management pressure continued on the group to work with the system until one day the supervisor found that she could make corrections to her document simply by backspacing. No more correction fluid was needed and her document looked just as good as everyone else's. Magically, from that point onward the rest of the project moved smoothly into production. The "complexity" did not disappear, but suddenly the value of the new system became greater than the pain of learning how to use it. If value is not obvious, there will be limited acceptance of new systems or products. Experiences like this sensitizes PMs to understand that the methods used to introduce new technology into the organization need to be motivation based and finding a way to make the human user believe that they will benefit from the new process is a key element. Once trained and productive, the user will once again become hesitant to move to another environment where they become a rookie user again. This same reluctance to change phenomenon has been observed in the introduction of technical systems related to email, document repository systems, and a host of other applications where new screen formats and key stroking are part of the requirement.

Our original conceptual view of a project was that it was designed to move an organization from State A to improved State B. The project team works to produce the defined goal, but that goal is not achieved until the new State B function is in operation. This last implementation step is part of the project and should be represented in the project plan and managed as a work activity much like the regular core tasks. Project teams cannot throw the results over the wall and walk away as they often want to do.

25.9 SUMMARY

The change management process has been described as covering the entire life cycle from requirements definition through implementation. Successful change management integration depends on identifying, evaluating, and managing change events in a project and eventually in the user environment. Classic change management deals with understanding and quantifying the requested change impact on the approved project plan. This involves complex impact analysis, management oversight, and effective communication to be successful.

High technology and complex projects are particularly prone to high change request rates since requirements definition is more difficult to do well in the planning phase. The incidence of project failure is particularly notable when new or cutting-edge technology is involved. Since the requirements definition process is never perfect some degree of change is inevitable; however, the specifics of acceptable change must be managed carefully with a formally defined process that weighs the benefits, risks, and costs. The basic change management system components are controlled by a project board decision, supported by a formal workflow process and appropriate communication technology.

It is essential that the project board and the PM maintain focus on this aspect of the project as the change dynamics emerge. Changes create instability ripples that can lead to overruns or failure if not properly controlled.

25.10 CHANGE REQUEST CHECKLIST

The following is a detailed checklist of tasks required to process a requested change through the project organization. A review of these steps shows why changes are disruptive to the overall work flow of the project and should be avoided wherever possible. It is understood that some change requests must be dealt with, but the total impact on the team must also be considered. Table 25.1 defines the steps and responsibilities that should be followed for each change request. Project

TABLE 25.1
Change Management Process Checklist

No.	Action	Owner
	Immediate Response	
1	Clarify change details with clients	PM
2	Validate business value of change with appropriate management	PM
3	Involve appropriate team with initial request review	PM Core team
4	Document preliminary impact assessment	PM
	Formal Assessment	
5	Identify deliverables and acceptance criteria, either added or deleted	Core team
6	Identify remaining deliverables impacted	Core team
7	Identify risks related to change	Core team
8	Document risk contingency issues	Core team
9	Identify team work-related issues	PM
10	Assess financial impact on project	PM
11	Assess possible reuse of existing work	Core team
12	Ensure solution's technical feasibility	Core team
13	If not feasible, assess redesign impact	Core team
14	Assess resource reallocation	PM Team leader(s)
15	Assess impact on project schedule	PM
16	Initiate draft SOW for proposed change	PM
	Project Board Approval	
17	Present change request to project board	PM
18	If approved, move forward	PM
	If rejected, archive	
	Internal Change Management Activities	
19	Communicate results to customer	PM
20	Obtain appropriate signoffs	PM
21	If necessary, reallocate project resources	PM
	If necessary, manage additional resource allocation	Team leader(s)
22	Update project plan	PM
23	Update project schedule: tasks, resource allocation, optimize, redefine critical path, rebaseline schedule.	PM
24	Communicate results to core and extended stakeholders	PM

team members must also be restricted from handling any change other than through the formal process. This process should be formally approved with the Scope Management Plan during the planning phase.

REFERENCES

Love, B., 2004. Operational Change Control Best Practices (July), *The Project Perfect White Paper Collection*. http://www.projectperfect.com.au/downloads/Info/info_occ.pdf (accessed April 1, 2007).

Miller, S., 1992. Topic: Change Management: Extending Your Reach, *Pragmatic Software*. http://www.pragmaticsw.com/pragmatic/SE_SP_change_management.asp (accessed April 1, 2007).

Neimat, T., 2005, Why IT Projects Fail (November), *Project Perfect*. http://www.projectperfect.com.au/info_it_projects_fail.php (accessed April 7, 2007).

Pritchard, C., 2005. Project Communications: A Lack of Tools or a Lack of Proper Use? (March), *ESI International*. http://www.esi-intl.com/Public/publications/html/20050301HorizonsArticle1.htm (accessed April 1, 2007).

Schwalbe, K., 2006. *Information Technology Project Management*. Canada: Thomson Learning Inc.

26 Project and Enterprise Metrics

26.1 INTRODUCTION

Metrics help guide project results toward organizational alignment by establishing common goals, defining project health, enabling predictive analysis, shaping user perception, and supporting decision making. The concepts, mechanics, and best practices presented here are intended to provide a framework for designing a custom metrics program. Realized benefits of such programs will vary depending on the validity of selected Key Performance Indicators (KPIs), the organization's maturity level, and the management of the program over time.

An enterprise typically measures and compares its overall project performance by quantification through the use of defined metrics. Much of this activity is carried out under the auspices of the overall monitoring and control processes. In some cases the collection of performance variables is used to compare one organization or project entity versus another. This activity is called *benchmarking*. In both situations, a defined metric's role is used to identify gaps in the project or organization versus planned values for that variable. Development of gaps between actual and planned (or desired) performance is the focus element that management uses to establish future corrective actions. In the case of project monitoring, this process is very dynamic and lies at the heart of the control process.

26.2 FUNDAMENTALS

A metric is a quantitative property of a process or product whose possible values are numbers or grades; a measure is a specific value of a metric (Parth and Gumz, 2003). Metrics translate vision and strategy into tangible targets and focus work activity on formal critical success factors. Given the potential influence metrics may have on project activity, they should be relevant, straightforward, and quantifiable. Failure to define and analyze measured results and adjust project direction can lead to less than desired results.

26.2.1 ALIGNMENT WITH ORGANIZATION GOALS

Every project has some outputs that are measurable. Albert Einstein summed up this by saying "Not everything that can be counted counts …" Focus should be given to those factors that are key to the organization's success, which is in turn defined as the realization of a specific goal. To this end, Reh (2008) maintains that project KPIs should be defined and limited to factors essential to the organization reaching its long-term goals. He reasons that defined KPIs serve as the basis for a metrics program and tend to stay somewhat constant over time. However, specific KPI targets may change as the organization gets closer to achieving a particular goal, or as the organization's goals change.

Reh advises that a limited number of standard KPIs should be selected so that everyone is focused on achieving the same primary goals. To accomplish this, a metrics program in total should consist of only 3 or 4 well-thought-out KPIs. Each project can support this view by defining 3 to 5 specific

KPIs that support the organization's overall KPIs. Projects that impact multiple business units will have to incorporate appropriate consolidation metrics. As an example, the organizational KPI of "Increase Sales" would translate downward to lower-level units into one or more KPIs that contribute to that higher-level goal. For instance, a customer relationship management (CRM) program could contribute to increased sales by improving the linkage between the sales force and the customer. A CRM development project metric would also fit a departmental goal. An example of an organizational goal that might be translated across multiple business units is as follows:

- Product Department KPI—"New Product Time to Market"
- Sales Department KPI—"Number of New Customers"
- Marketing Department KPI—"Number of Telemarketing Calls Completed"
- IT Department KPI—"Online Website Up-Time"

It is generally recognized that properly translated departmental KPIs help these organizations focus on providing an overall organizational scorecard KPI.

It is beneficial to study the best practices of successful organizations and the Canadian Transportation Agency (CTA) has a well-developed process for ensuring that the activities they undertake are linked to strategic outcomes (CTA, 2008). In a recent request for government appropriations, the CTA developed a "basic results chain" to demonstrate how resources, activities, and outputs all link to strategic outcomes and ultimate results (Table 26.1). The CTA utilizes this results chain throughout a multiyear campaign to show progress toward achieving higher-level goals. Additionally, they develop results chains for each key initiative with the overall goal of providing a map that outlines the linkage between resources and outcomes across various initiatives (CTA, 2008). To implement this concept, the CTA uses a standard template for iterative development of a basic results chain. The template helps develop the basic results chain by linking an ultimate result goal downward to defined lower-level outputs.

In a similar fashion the Department of Energy (DOE) identifies the types and categories of metrics as outlined in Tables 26.2 and 26.3.

TABLE 26.1
Constructing a Basic Results Chain

Ultimate results	Why?	Why do we carry out this program or initiative?
		What is it we ultimately expect to achieve, recognizing that it may take years, even decades to achieve this ultimate result(s)?
Strategic outcomes	What?	What do we expect to see or hear as a result of our outputs and activities?
	Who and where?	Who do we need to engage and reach and where?
		The strategic outcomes are often referred to as the behavioral changes that arise as a result of our work.
Outputs	How?	Outputs, activities, and inputs are effectively the operational elements required in order to achieve the strategic outcomes.
		What outputs (i.e., decisions and orders, codes of practice, etc.) in order to achieve the expected strategic outcomes?
Activities		What key activities do we need to undertake in order to effectively contribute to the strategic outcomes?
Inputs		What inputs (financial and human) do we have to carry out key activities?

Source: Canadian Transportation Agency (CTA), 2008. Performance measurement framework for the Canadian Transportation Agency. http://www.cta-otc.gc.ca/about-nous/excellence/performance/index_e.html (accessed March 20, 2008). With permission.

TABLE 26.2
Types of Performance Measures

Process metrics	Increase capability level (i.e., SEI maturity levels)
	Do more with less (shorter schedule, less resources)
	Improve quality (less defects, less rework)
Project metrics	Track project progress
	Assess project status
	Award contract fees
Product metrics	Determine product quality
	Identify defect rates
	Ensure product performance

Source: Department of Energy (DOE), 2002. Basic Performance Measures for Information Technology Projects (January 15), pp. 4, 7.

TABLE 26.3
Typical Metric Categories

Schedule	Actual versus planned
	– Schedule and progress
Budget	Actual versus planned
	– Resources and cost
Functionality	Delivered versus planned
	– Product characteristics
	– Technology effectiveness
	– Process performance
	– Customer satisfaction

Source: Department of Energy (DOE), 2002. Basic Performance Measures for Information Technology Projects (January 15), pp. 4, 7.

26.3 ALIGNMENT WITH ORGANIZATIONAL MATURITY

KPIs and other metrics should also be aligned with organizational maturity objectives. Metrics designed to track relatively new or immature processes and procedures should start with basic metrics. For example, if the organization is in the early stages of implementing a project management program, it makes sense to collect basic project management metrics versus advanced project management ones. Parth and Gumz (2003) comment that the following are examples of basic project management measurements:

- Completions of milestones for project activities compared to plan
- Planned activities effort and duration compared to the plan, work completed, effort expended, and funds expended in the project.

Conversely, the use of basic metrics within a mature process may cloud some of the advantages of the more mature process and restrict the organization from further improvements. Improper use or definition of metrics can drive the organization in either good or bad directions; hence it is important to understand how to formulate the metric to focus on the desired outcome. So, in considering appropriate metrics for an advanced environment, it might be more appropriate to include the following areas (Parth and Gumz, 2003):

- Project management administrative effort compared to plan
- Frequency and magnitude of replanning efforts, given requirement changes
- Realized risks compared to estimated loss
- Frequency, number, and size of unanticipated impacts to the project.

The point of using a basic versus advanced view is that the latter is focusing on more than the former and the basic issues may not reflect that. In the list above, one of the metrics is looking at contingency reserve status. A lower maturity process would not have dealt with the topic.

26.4   DRIVERS OF PERFORMANCE AND CHANGE

The adage "What gets measured, gets done" summarizes how the use of metrics can shape the day-to-day activities of various organizational units. Defined correctly, a KPI succinctly communicates what management deems most critical. Based on their understanding of the metric, employees will likely take action and adjust their activities according to the highlighted metrics and established targets. Expanding on this principle, Eckerson (2006) describes metrics as levers that executives can pull to move the organization in new and different directions. He states "as powerful agents of change, metrics can drive unparalleled improvements or plunge the organization into chaos and confusion" (Eckerson, 2006). This underscores the importance of establishing KPIs that are directly tied to what the organization is trying to accomplish on a strategic level and linking metrics to specific KPIs (Parth and Gumz, 2003).

26.5   SUMMARY CATEGORIES

Leading or lagging indicators: KPIs are generally characterized as leading or lagging indicators. Leading indicators are those represented by activities that have a significant effect on future performance, while lagging indicators are those that reflect output of past activity (i.e., most financial metrics). From a management point of view, leading indicators help the most with decision making and should be a measure of future business trends. They may also measure activity in its current state (i.e., current sales revenue growth) or future state (i.e., the number of customer contacts scheduled for next 2 weeks). Metrics based on future state are valuable and powerful as they provide an opportunity to have an insight into influence outcomes (Eckerson, 2006).

Simple or composite indicators: Metrics can be expressed as counts, percentages, ratings, numbers, or trends. Simple metrics commonly denote items completed, such as lines of code, number of widgets produced, or tests performed. Composite metrics combine one or more factors and are usually expressed as ratios. Examples are the number of change requests received per work unit or the number of defects found per work unit.

Quantitative or qualitative indicators: Whether simple or composite, metrics are generally graded using quantitative or qualitative valuations. The various numeric and statistical metric examples cited above fall into the quantitative category. The major concern with any metric is whether it means anything or not. In some cases they can be misleading. The classic example of this is the metric of planned versus actual resource consumption. At some future status meeting someone will show a chart or metric that says that the actual budget for the project is below plan, therefore this project is doing quite well. Taken by itself that metric may send the wrong message. This project could be hopeless behind and doing nothing, therefore no expenditure of resources. Do not use and do not be misled by seemingly straightforward quantitative metrics. On the positive side, a measurement showing that a milestone date has been met has an obvious value and recording hours for work performed on a particular WBS ID may have a value, but falls into the gray area described above. As a general rule, a simple quantitative metric by itself will only tell part of the story.

Qualitative metrics are judgment or perception measures and they may also be translated into some form of quantification a less granular scale. Samples of such measures could be a measure of customer satisfaction or team morale. Upon completion of the requirements definition process the room could be polled for a measure of satisfaction. Effectiveness could be presented as the number or percentage of respondents who felt the process was successful or not. A typical method of translating perception-based measures is the classic five-stage Likert scale (i.e., very good, good, neutral, not good, and poor). In most cases, quantitative metrics based on objective measurements are inherently more objective and useful; however, in some situations use of qualitative metrics is necessary to provide a general status in an area difficult to quantify (i.e., team morale). So, the total of metrics suite will likely contain some of each type.

Application to project management: In a more specific categorization, White (2001) categorizes project management metrics. The process of correlating metrics across project phases is performed by establishing a metric baseline value and comparing progress against those baselines as a means of monitoring and predicting the final outcome.

Phase level metrics: Tables 26.4 through 26.8 illustrate examples of metrics by project phase. In each of these examples it will be necessary to decide how to use such parameters and even if they are worth collecting. These are offered here only to help illustrate the types of variables that make sense to consider.

During the planning phase, the major issue with data collection is that the primary variables are simply showing consumption of time and resources with no tangible deliverable to compare it with.

TABLE 26.4
Planning Phase Metrics

Planning Phase

Schedule estimate

Meaning	Estimated amount of elapsed time required to complete the project
Measure	Number of planned work days based on work effort and required resources
Benefit	Establishes a baseline to support comparison during later project phases

Cost/hours estimate

Meaning	Amount of resources (dollars, people, equipment) it will take to produce the project
Measure	Number of planned work hours and estimated costs based on work effort and required resources
Benefit	Establishes a baseline to support comparison during later project phases

Defect rate

Meaning	Anticipated amount of rework (numbers, average time to repair)
Measure	Based on past experience, number of defects to be incurred based on size of product
Benefit	Establishes a baseline to support comparison during later project phases

Component size

Meaning	Anticipated size of products to be delivered
Measure	Based on past experience, number of functional modules, and documentation Pages to be developed
Benefit	Establishes a baseline to support comparison during later project phases.

Quality

Meaning	The metric that will be used to determine acceptability of the end products
Measure	Varies from project to project—could be expressed as a factor of defects, defined Number of defects measured at various test points
Benefit	Removes ambiguity about product acceptance

Source: White, K.R.J., 2001. Measuring and Managing Success, *PM Solutions*, pp. 4–8. http://www.jamesheiresconsulting. com/IT_Project_Metrics.pdf (accessed February 22, 2008). With permission.

TABLE 26.5
Execution Phase Metrics

Execution Phase

Actual hours

Meaning	Actual labor hours spent to date on project activities
Measure	All labor hours, including those of support personnel and contractors
Benefit	Provides comparison to budget and business case and supports schedule analysis

Actual schedule

Meaning	Schedule performance to date
Measure	Number of days behind or ahead of schedule
Benefit	Supports early determination of potential late delivery

Actual cost

Meaning	Actual costs associated spent to date on project activities
Measure	True total costs spent to date, including all labor, software, and hardware costs
Benefit	Provides comparison to budget and business case

Defects per peer review

Meaning	Quality of work produced to date, prior to testing phase
Measure	Number of defects per peer review
Benefit	Early measure of quality of product; indication of a training or specification problem

Staff productivity

Meaning	Average staff productivity
Measure	Number of function points per staff hour
Benefit	Determine rate of work to be anticipated in remaining software build activities

Source: White, K.R.J., 2001. Measuring and Managing Success, *PM Solutions*, pp. 4–8. http://www.jamesheiresconsulting.com/IT_Project_Metrics.pdf (accessed February 22, 2008). With permission.

Beyond the planning stage, most metrics defined would be formulated based on requirements or desired future levels. The basic role of the metrics defined would be to help track progress through the subsequent project phases. Execution phase metrics enable the project team to determine how the project is proceeding compared to plan, or to aid in evaluating that some corrective action is needed (White, 2001). The metrics White emphasizes here are primarily leading indicators. In the same spirit of monitoring leading indicators, White identifies testing phase metrics for a software development activity that are linked to its planning activity targets. Implementation metrics indicate the readiness of the product for production release and may represent good predictors of future customer satisfaction. Table 26.8 summarizes the approach defined by the DOE methodology for creation of performance measures that link from focus and project purpose into a defined measurement variable. Also, note that a measurable target level is specified for each metric.

26.6 METRICS EVALUATION CRITERIA

Although there is no right or wrong method for selecting appropriate metrics for status tracking, the following criteria should be reviewed when selecting a candidate:

1. *Validity:* Is the granularity of the value adequate for interpreting results?
2. *Relevance:* Does the metric actually relate to the area of concern?
3. *Reliability:* Is it a consistent measure over time?
4. *Simplicity:* Is information available to capture a value?
5. *Affordability:* Is it cost effective to collect and analyze the data?

TABLE 26.6
Testing Phase Metrics

Testing Phase

Schedule estimate

Meaning	Schedule performance to date
Measure	Number of days ahead of or behind schedule, and amount of float
Benefit	Ability to predict actual completion date, and approximate risk

Cost/hours estimate

Meaning	Amount of resources (dollars, people, equipment) it will take to produce the project
Measure	Resources/cost spent to date
Benefit	Predict total cost of the project

Defect rate

Meaning	Determines the quality of the work produced to date
Measure	Number of defects per some predetermined value (functional modules)
Benefit	Determine rate of future rework

Response time

Meaning	Ability of the application to handle volume in a timely manner
Measure	Response time in seconds per hundreds of users
Benefit	Advance notice of performance problems

Average time to repair

Meaning	Amount of duplicate work due to errors
Measure	Number of hours and dollars spent for correcting the problem
Benefit	Reduce unnecessary costs and increase work efficiency

Source: White, K.R.J., 2001. Measuring and Managing Success, *PM Solutions*, pp. 4–8. http://www.jamesheiresconsulting. com/IT_Project_Metrics.pdf (accessed February 22, 2008). With permission.

TABLE 26.7
Deployment Phase Metrics

Deployment Phase

Defect rate

Meaning	Determines the quality of the work produced to date
Measure	Number of defects per some predetermined value (functional modules)
Benefit	Determine rate of future rework

Response time

Meaning	Ability of the application to handle volume in a timely manner
Measure	Response time in seconds per hundreds of users
Benefit	Advance notice of performance problems

Quality

Meaning	The metric that determines acceptability of the end products
Measure	Varies from project to project—could be expressed as a factor of defects Response time; number of users supported
Benefit	Removes ambiguity about product acceptance

Average time to repair

Meaning	Amount of duplicate work due to errors
Measure	Number of hours and dollars spent for correcting the problem
Benefit	Reduce unnecessary costs and increase work efficiency

Source: White, K.R.J., 2001. Measuring and Managing Success, *PM Solutions*, pp. 4–8. http://www.jamesheiresconsulting.com/IT_Project_Metrics.pdf (accessed February 22, 2008). With permission.

TABLE 26.8
Sample Summary of Project Metric Groupings

Category	Focus	Purpose	Measure of Success
Schedule performance	Tasks completed versus tasks planned at a point in time	Assess project progress	100% completion of tasks on critical path; 90% all others
		Apply project resources	
	Major milestones met versus planned	Measure time efficiency	90% of major milestones met
	Revisions to approved plan	Understand and control project "churn"	All revisions reviewed and approved
	Changes to customer requirements	Understand and manage scope and schedule	All changes managed through approved change process
	Project completion date	Award/penalize (depending on contract type)	Project completed on schedule (per approved plan)

Source: Department of Energy (DOE), 2002. Basic Performance Measures for Information Technology Projects (January 15), pp. 4, 7.

Selecting project metrics should not be a once-and-done activity. Procedures should be put in place to continually assess whether the current metrics are sufficient or excessive, prove useful in managing the business, and drive the organization toward strategic goals (National Public Review, 1997). Beyond this assessment, modifications may be necessary to respond to market conditions or regulatory requirements. From the review process some metrics may be discarded, or replaced over time. If a metric is not focusing the organization on meaningful targets, it should be considered for retirement.

26.7 ESTABLISHING A BASELINE AND SETTING TARGETS

A comparative baseline should be established for selected metrics. This may be accomplished internally, through information gleaned from past experiences, or externally in accordance with industry standards or a benchmarking partnership. Many organizations utilize this opportunity to set stretch goals to entice performance improvements. Comparing internal metrics with industry equivalents will not only add motivation to improve, but will validate that those goal levels are attainable. For example, a 100% customer satisfaction goal sounds admirable, but when the industry standard is 80%, there may be factors in play that are not easy to overcome. Logic would suggest a more realistic target since setting unrealistic and possibly unachievable levels can be demotivating.

Figure 26.1 illustrates a hypothetical example to show how a company might use various industry benchmark performance metrics to compare their internal field performance. Comparisons of this type are often used to evaluate needed areas of improvement. Also, this same presentation format could be decomposed into geographical units for a lower-level comparison. The automobile industry uses similar metrics for analyzing their performance regarding time to produce a unit, comparative labor costs, defect measures, and so on.

26.8 BEWARE OF THE METRICS PITFALLS

6th Sense Analytics (n.d.) identifies the following five reasons why metrics fail to promote strategic results (6th Sense Analytics Document Library, 2006):

1. *Use of inappropriate metrics or measures:* Measuring the wrong things and creating an artificial incentive system (i.e., measuring lines of code produced that leads to a greater

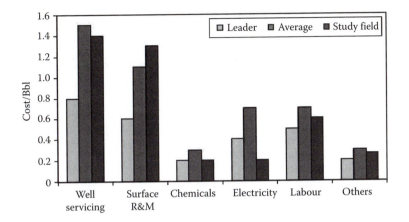

FIGURE 26.1   Oil field productivity benchmarking.

number of short lines of code or verbose code that is optimized for size rather than function or quality).

2. *Garbage in, garbage out:* The metric that has no meaning.

3. *Poor support processes:* The process of tracking activities and collecting information is inefficient and distracts from productivity and progress (i.e., excessive overhead in the collection process).

4. *Information is not timely:* Metrics are not produced or evaluated in a timely manner and out-of-date results lead to bad decisions (i.e., accounting data that might be six weeks old).

5. *Inadequate follow through:* No actionable results or specific decisions are identified based on analyzing results.

Overriding all of the combined mechanics of metrics creation, collection, and use is the human element surrounding this process. According to a National Public Review survey, employees in best-of-class organizations focus on achieving organizational goals by using performance measures to gauge goal achievement, but they do not focus on the measures *per se* (National Public Review, 1997). Performance measurement overall is seen as a means, not an end. Establishing metrics that motivate employees in a particular direction and establishing a sense of camaraderie around meeting performance targets may be the most important tactics related to a successful metrics program.

26.9   MECHANICS

Table 26.9 illustrates how various metrics can be translated into a scorecard format. This scorecard contains quantitative and qualitative evaluations of schedule, cost, technical performance, and quality measures for selected WBS units. From this basic set of metric, it is also possible to produce a WBS-weighted assessment or an overall project equivalent. Computation of an overall composite score would require some experimentation with different approaches.

In a scorecard model such as this the terms used will have to be understood by the receiving audience. In this case the two Earned Value (EV) columns would need some interpretation as to how that data are collected and what they mean. We will discuss that in Chapter 27. In this case, the use of qualitative scores is easier to understand, but are these value auditable? Who defined the value and what was it based on? These are two classic interpretation issues for metrics usage. Use of this type-reporting scorecard by one project would likely create confusion given its cryptic coding. This typically means that common metrics need to be defined and used by all projects both for comparability and for understanding.

TABLE 26.9
Example Project Scorecard

WBS	SPI	CPI	Tech. Perf.	Quality
A	1.10	1.10	5	4
B	1.25	0.79	3	3
C	1.30	0.65	4	3
D	0.90	1.10	5	4
E	0.75	0.65	2	5
F	0.75	1.10	5	4
G	0.75	0.65	2	5
H	0.75	1.10	4	4
I	0.75	0.60	4	4

Note: WBS—A work breakdown structure unit.

SPI—a quantitative measure of schedule performance computed by EV (see Chapter 27).

CPI—a quantitative measure of cost performance computed by EV (see Chapter 27).

Tech. Perf.—a qualitative measure of technical performance based on a scale of 1 = poor to 5 = exceeds requirement.

Quality—a qualitative measure of estimated current status quality based on a scale of 1 = poor to 5 = exceeds requirement.

26.9.1 Miscellaneous Issues

Metrics should be reported in such a way as to help the receiver comprehend his project status needs. If these data are to be used internal to the project team, a low level of detail might be required; however, giving the same data to a senior manager would not be appropriate. So, a fundamental design issue is level of granularity. For instance, should the detail shown be the first level of a WBS, or something deeper into the structure?

In addition to this consideration, there are subtleties related to phase tracking. For example, tracking of quality and technical performance is difficult if not meaningless during early design; however, cost and schedule can be better tracked. Should quality, technical performance, and weighting metrics be omitted during the early phase and then picked at some later point in the life cycle when the values can be more accurately determined? The answers to these issues are couched in the method by which the project life cycle is defined. In the case of software development some of these variables would be hard to capture before the execution phase and even then would probably be based on subsystem level results. These design concerns are more complex to resolve than appear, given the goal is to produce a system that can be used by all projects and not custom built for each.

One final point before we leave this example. As metric reporting moves out of the project level and is used by other organizational entities, the focus moves to become higher levels and in some cases more narrow perspectives. One example of this is to translate a quantitative scorecard table into colored balls for each of the major parameters. The color will be based on some threshold definition. For instance, if the overall project score for schedule were less than 0.90 the ball would be red; between 0.90 and 1.0 yellow; otherwise green. Each parameter could be specified in this manner so that the senior manager looks at status by looking at color rather than numbers. There might be a capability in the reporting system to drill into any red area for more detail. In this fashion status is converted to color rather than numbers or graphs.

When all of the specification process is dealt with, the issue of stakeholder reporting format remains. Should this metric table be sent to all stakeholders or is there a graphical format that would be best for the masses? Dealing with issues of this type highlights the complex nature of a metrics program.

The following three chapters (27, 28, and 29) outline essentially parallel project monitoring and control approaches that are commonly used—EV, basic tracking using KPIs, and the Balanced Scorecard. In addition to these techniques, traditional financial measures are often used to show either planned status or current projections of the project effort.

Business value measurement methodologies: Technology in various formats is now embedded in almost every business process and projects are the common mechanism to transform a business process. One common project management tactic is to measure the incremental value of technology-enabled business processes (Symons et al., 2006, p. 4). In regard to IT projects, Symons et al. (2006) state that standard methodologies that help organizations more accurately predict returns from their investments and overcome many of the weaknesses in using simple financial metrics. Other industries follow similar patterns with common WBS structures, life cycle standardization, common metrics, and the like. Two weakness areas related to these strategies are lack of ability to handle risk and assessing tangential opportunities. To see how various organizations handled more complex evaluations Forrester compared the following four different custom methodologies.

- Business Value Index (BVI)—Intel
- Total Economic Impact (TEI)—Forrester
- Val IT—IT Governance Institute
- Applied Information Economics (AIE)—U.S. Government methodology.

A brief summary of these approaches is included below.

Intel's business value index (BVI) methodology was developed in 2001 with the goal of incorporating common business value measures, efficiency, and financial attractiveness of the initiative. Business value is measured by factors such as customer need, business and technical risks, strategic fit, revenue potential, level of required investment, innovation, and learning generated. Efficiency is established by how well the project complies with standards, architecture, and core competencies. Also, projects are scored using weighted criteria from each of the three vectors and depicted on a three-dimensional Business Value Chart to enable all projects to be compared with one another. The final evaluation grid maps business value versus IT efficiency. This technique was considered the most traditional and simplest of the four compared. A white paper outlining details of this method is available on the Intel IT web site and is a good background source for this general topic area.

The second methodology called Total Economic Impact (TEI) was developed by Forrester. It is similar to BVI in that it includes use of a business case, valuing intangibles, and calculating financial returns (Symons et al., 2006, p. 7). More specifically, TEI utilizes three economic areas: flexibility, business value, and technology cost. This is then matched against risk factors to produce a total organizational impact. Cost impacts are quantified by the incremental change in costs associated with the effort versus maintaining the status quo. Business benefits are quantified by their impact on various business units and this reflects the impact of the initiative on the overall user organization. This metric indicates the level of effort required to absorb the new system across the organization. Flexibility is based on the value of the option to take subsequent future actions. The specific future option does not have to be defined; rather a value is associated with the ability to take action in the future. Risk analysis provides "risk-adjusted" costs and benefits that are then used to compute a risk-adjusted ROI. Projects are scored across each of the categories by weighted criteria. The weighted scores are summed to represent a single quantitative number, which is the TEI for that proposal.

The IT Governance Institute (ITGI) as a complement to their COBIT governance framework created a third methodology called Val IT. The goal of this process is to provide a means to optimize the realization of business value from the organization's portfolio investments in information technology. This approach consists of practices related to the project life cycle, project selection, and investment management. There are 11 key management practices, 15 portfolio management practices, and 15 investment management practices. Through these combined 41 practices, ITGI claims that the value of IT investments, overall portfolio, and individual IT investment programs can be measured and optimized (Symons et al., 2006, pp. 10–11). This type of methodology is an example of layered metrics and would require a very mature organization to manage the approach. Forrester judges this as a work in process; however, given the broadly recognized sponsor level of ITGI as a result of their role in Sarbanes-Oxley and COBIT it could well be accepted as it matures.

The fourth and final methodology reviewed is called Applied Information Economics (AIE). This methodology appears to be mostly used in governmental organizations and contains the most quantification of the review group. Its design goal is to "clarify, measure, and optimize" investment alternatives even when there are "intangibles" (Symons et al., 2006, p. 12). AIE removes definitional ambiguity by focusing on variables that can be expressed as units of measure. One key differentiator of this method is use of a risk/return metric to compare a proposal across all investment categories. One of the discrimiating characteristics of project selection methodologies is their ability to evaluate proposals in this manner. Many technology-based strategies and their related metrics are very difficult to forecast as compared to evaluating a new building, purchased equipment, and the like. The techniques embedded into AIE are similar to those used by financial and insurance organizations for portfolio selection. EPA used this method to calculate a desktop replacement strategy, but not a lot of real-world results were available to assess the technique further. It is the most complex of the four examined and requires analysis expertise.

To summarize, the four custom methodologies reviewed differ most notably in their emphasis on breadth of topic coverage and use of quantitative or qualitative assessments. The general assessment is as follows:

- BVI is the simplest to implement
- TEI values flexibility as well as traditional metrics
- Val IT takes a layered governance approach
- AIE offers the greatest rigor and deals with risk and cross-functional-type initiatives.

The point made in looking at a cross section of valuation methodologies is that essentially all organizations are looking for techniques to accomplish this goal. None have proven to have found the optimum method and all seem to be focusing on slightly different goals in their selection process. Once again, we see that metrics are designed to guide direction and organizations have different views on that direction.

26.10 INDUSTRY STANDARD METRICS FOR MONITORING AND CONTROL

The Earned Value (EV) methodology is defined as an ANSI standard and in that role enables the project manager to establish a sophisticated project status analysis in forecasting completion costs and schedule, based on actual progress relatively early in the project (see Chapter 27 for more details). Interpretation of these results can be compared across all project boundaries. EV parameters do not cover all required aspects of project performance, but certainly should form the core of a metrics program in regard to cost and schedule performance.

EV standard parameters enable early warning if trends indicate that the project will be over budget or extend beyond the targeted deadline. Likewise, EV metrics can assist in justifying spending ahead of plan given that the project is progressing ahead of schedule, or can help the project team evaluate areas within the WBS where progress is lacking.

26.11 CONCLUSION

Defining key project performance indicators, collecting data, and producing status metrics throughout the project life cycle are only valuable to the extent that the resulting information is communicated to the organization in a manner which drives the project toward appropriate objectives. Use of a formal metrics program influences behavior in those defined directions. For this reason, the selection of reported values needs to be considered carefully. Examples shown in this chapter have demonstrated some of the key issues related to definition and use of this strategy.

Best-of-class organizations include metrics as part of individual and departmental performance review and reward systems. In order to successfully implement a formal metrics program careful time and attention must be paid to the formulation of the metric, data collection techniques, and training of the organization to utilize the results.

REFERENCES

6th Sense Analytics Document Library, 2006. The 5 essentials of effective software development metrics. *6th Sense Analytics White paper.* http://www.6thsenseanalytics.com/wp-content/assets/pdfs/whitepapers/6SA-5-essentials-for-software-dev-metrics.pdf (accessed March 23, 2008).

Canadian Transportation Agency (CTA), 2008. Performance measurement framework for the Canadian Transportation Agency. http://www.cta-otc.gc.ca/about-nous/excellence/performance/index_e.html (accessed March 20, 2008).

Department of Energy (DOE), 2002. Basic Performance Measures for Information Technology Projects (January 15), pp. 4, 7.

Eckerson, W., 2006. How-To: Making sense of metrics. http://searchcio.techtarget.com/columnItem/0,294698, sid19_gcill57781,00.html?track=NL-48&ad=539699 (accessed February 15, 2008).

National Public Review, 1997. Benchmarking study report. Retrieved March 22, 2008 from http://govinfo.library.unt.edu/npr/library/papers/benchmrk/nprbook.html

Parth, F. and J. Guma, 2003. How project metrics can keep you from flying blind. http://www.projectauditors.com/Papers/Whiteprs/ProjectMetrics.pdf (accessed March 22, 2008).

Reh, J., 2008. Key performance indicators. http://management.about.com/cs/generalmanagement/a/keyperfindic.htm (accessed February 15, 2008).

Symons, C., Orlov, L.M. and Sessions, L., 2006. Measuring the business value of IT, *SearchCIO.* http://searchcio.bitpipe.com/detail/RES/1201804423_89.html?psrc=TPP (accessed December 30, 2008).

White, K.R.J., 2001. Measuring and Managing Success, *PM Solutions*, pp. 4–8. http://www.jamesheiresconsulting.com/IT_Project_Metrics.pdf (accessed February 22, 2008).

27 Earned Value Management

EV is a contemporary project management technique that has gained broad industry acceptance over the past few years. This acceptance process took over 40 years of scrutiny and prodding, mostly by DoD to its contractors. Evidence now shows that EV provides one of the most effective and meaningful tools available today to measure and report project cost, schedule, and performance. EV has the unique ability to combine cost, time, and scope completion measurements within a single integrated methodology.

EVM was originally developed within DoD to support their control processes with large defense acquisition programs. In 1965, the U.S. Air Force defined 35 management criteria for their acquisition projects. Two years later, DoD adopted these same criteria as their Cost/Schedule Controls Systems Criteria (C/SCSC) as part of Department of Defense Instruction (DODI) 5000.1 and these were required in various contracts over the next three decades. In 1996, private industry accepted the basic concepts behind EVM and rewrote and published the revised C/SCSC regulation under the title of EVMS. In 1998, private industry, through the sponsorship of the National Defense Industrial Association (NDIA), obtained formal acceptance of EVMS by the ANSI through their publication of the ANSI/EIA-748 Standard (Fleming and Koppleman, 2006).

Application of the EV model cannot be accomplished without a reasonable degree of management operational maturity. Initially, EV was viewed as excessive project overhead and some contractors were able to negotiate it out of their requirement set through this claim. Like many government regulations EVMS was initially overdefined without proof of value. Lack of industry ownership, inadequate training, and awkward technical jargon further contributed to its early implementation problems. Eventually, even though EVMS was intended to serve both cost and schedule needs, it came to be seen as a purely financial exercise due to the fact that financial personnel managed the process (Christensen, 1998, p. 5). During the 1990s governmental procurement groups streamlined their acquisition regulations and EVM not only survived the reform movement but it became strongly associated with the reform implementation itself (Haugan, 2003, p. 70).

The concepts of EV were eventually adopted by the National Aeronautics and Space Administration (NASA), the U.S. DOE, the U.S. Office of Management and Budget (OMB), the construction industry, and several foreign countries. In 1987, the PMI included an overview of EVM in their *PMBOK® Guide* and expanded that discussion in subsequent editions. Efforts to simplify and generalize EV gained further momentum in the early 2000s, and to date, experts in the field continue to validate it as an effective tool for management of projects of any size and risk (Christensen, 1998, p. 13). In the hands of a trained technician it is currently the most robust project management tool available for evaluating project status and forecasting completion cost and schedule.

EV computation requires a cost and schedule baseline, ideally defined by WBS WPs, and accompanying baseline schedule. From this base level definition it is possible to compare project plans versus actual status in a more meaningful way than the traditional planned versus actual resource presentation. Variances between the actual cost (AC) of the work and the current progress provide a timely warning of performance problems at both the project level and lower levels of granularity.

27.1 BASIC PRINCIPLES

Based on its underlying conceptual model, the EV principles incorporate many project management areas including scope definition, resource allocation, scheduling and budgeting, accounting, analysis, reporting, and change control. EV's specific mechanics include the use of the WBS, PMB curves ("S" curves), and a defined set of work unit metrics. Before delving further into the computational methodology, it is important to review one underlying concept critical to EVM implementation—that is, the WBS.

As we have described in earlier chapters, the WBS is a fundamental technique for defining and organizing the total project scope into a hierarchical tree structure. The WBS defines a set of project deliverables and related work units that collectively represent 100% of the project scope by definition. At each subsequent level, the children of a parent node represent 100% of the scope of their parent. The lowest level of decomposition for each parent node is called the WP and represents the lowest level of control in the structure. Figure 27.1 shows a skeleton representation of the WBS. In this type of structure, each of the WPs is cost estimated, resources allocated, time scheduled, and linked into the next higher-level WBS. In this fashion, the project schedule and budget are defined and once approved by the stakeholders this will become the baseline against which the project performance will be measured. In EV terminology the WBS boxes represent the Planned Value (PV) for that work unit and this value will be used for comparative purposes.

In the baseline schedule WBS work units become discrete measures for assessing periodic progress. As the project progresses, work unit planned performance through time is measured against actual work accomplished to yield EV. AC for the work unit is then used to provide both cost and schedule measures for the project at that status point.

To illustrate the concepts of EVM, it is useful to review a more traditional method for tracking project performance. If the project has been planned in detail, a graph as shown in Figure 27.2 can be used to represent the planned cash flow for the defined work units. The graph would also typically show the cumulative actual recorded cost for the project. In this example, the planned versus actual curves indicate that ACs are below plan, so we should conclude from this that the project must be doing just fine, right? Well, maybe not! The missing ingredient in this view is a coherent measure of work accomplished.

If the level of planned work was actually completed by the status date (July), then the project would in fact be underbudget and ahead of schedule. Conversely, if the project was only 70% complete by the sixth month and then only 25% complete in activities planned for the seventh month, it would be significantly behind schedule. If we assign planned dollar credit for each work unit that has been completed and compare that value to the planned work units up to that point and actual resources consumed values, the earlier status chart might look as shown in Figure 27.3.

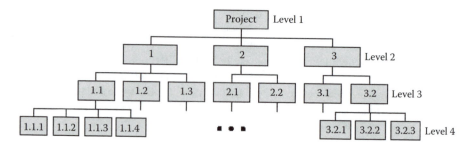

FIGURE 27.1 Work breakdown structure.

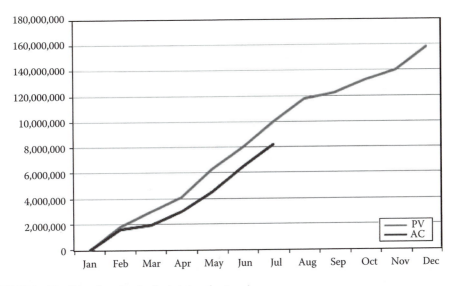

FIGURE 27.2   Traditional project budget status chart.

Inclusion of an EV curve in the status view supports an improved evaluation of the current project cost and schedule in a quantitative and objective fashion. As an example of an improved status analysis for July let us say we have the following (see Figure 27.3):

PV = $10 million
AC = $8.2 million
EV = $6.2 million

This data better quantify that the project is not doing well. It has basically accomplished about 75% (6.2/8.2) of its budget. Said another way, it is running about 25% higher than budget for the work produced. From the schedule view, it is running at about 62% (6.2/10) of planned "speed." This says that the project is approximately 38% behind schedule. We will see more on calculation mechanics for these metrics and their interpretation in the next section.

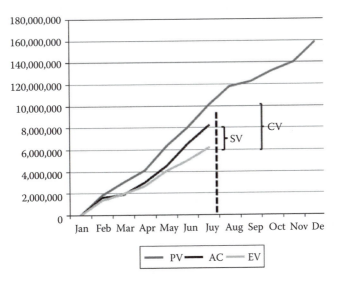

FIGURE 27.3   EV progress tracking.

27.2 CALCULATING EV PARAMETERS

Three status variables form the basis of EV metrics. They are the AC, EV, and PV. As a memory device remember these in alphabetical order (A, E, and P). We will show how this leads to the calculation formulae shortly.

At the project status point (July) it is necessary to calculate the three parameters as defined below:

PV: The sum of planned baseline cost for all WPs scheduled to be accomplished up to the status point.

EV: A measure of completed WPs and partially completed portions of WPs up to the status point. For example, a 50% complete WP would receive 50% of the PV as earned.

AC: The AC incurred for the planned WPs up to the status point.

Using these three parameters we can calculate four basic historic cost and schedule performance metrics: Schedule Variance (SV), Schedule Performance Index (SPI), Cost Variance (CV), and Cost Performance Index (CPI). As a second memory device, remember the performance variables in alphabetical order (C and S). So, we now have three forecasting parameters (A, E, and P) used to derive two status variables (C and S). This structure represents the EV calculation model. A simple way to structure this is to use a "memory table." This table consists of three columns and two rows laid out as shown below (Table 27.1).

The first challenge in EV analysis is to calculate the basic cost and schedule performance parameters. Once this is done the rest becomes relatively easy. The memory table can generate EV Schedule and Cost formulae by starting from the middle of the table (with EV) and move outward toward either the Cost (left) or Schedule (right). So, CV would be EV − AC. The operator is a minus for variance formulae and a divisor for an index formula. All formulae either have EV as the first variable or the numerator. Examine the index and variance formulae above and compare them to the memory table and the associated memory tricks described. Experience suggests that having a method to organize the raw data will save having to memorize formulae. Once you have mastered these basic mechanics the next challenge is to learn how to interpret the quantitative results. This is the challenge for the section below.

27.3 INTERPRETING EV PARAMETERS

Working definitions for the basic EV performance parameters are defined below:

SV: Is the difference in dollar value between the amount of work that should have been completed in a given time period and the work actually completed.

TABLE 27.1
EV Memory Table

	AC	OPER	EV	OPER	PV	
	←		•		→	
CV	?	—	?	—	?	SV
CPI	?	/	?	/	?	SPI

EV Formulae: CV = EV − AC
SV = EV − PV
CPI = EV/AC
SPI = EV/PV

CV: Is the dollar value by which the project is either overrunning or underrunning its estimated cost.

CPI: The ratio of cost of work performed (EV) to AC. CPI of 1.0 implies that the AC matches to the estimated cost. *A CPI >1.0 indicates that work is accomplished for less cost than what was planned or budgeted. CPI <1.0 indicates that the project is running over the cost planned.*

SPI: The ratio of work accomplished (EV) versus work planned (PV), for a specific time period. SPI indicates the time rate at which the project is progressing. Index values have the same interpretation as those for cost.

Budget At Completion (BAC): The total WP budget for the project. Be aware that a project budget may contain other components such as reserves, overhead, profit, and the like. Those components will have to be analyzed as part of the total project cost, but they should be removed from the status analysis part since they are not "productive" in the sense of the WPs. If material costs are a significant part of the project budget, these should also be removed for the same reason.

Estimate At Completion (EAC): Represents a forecast of the total WP portion of the project costs based on values of CPI and SPI parameters, plus other assessments of past and future conditions. Four alternatives for forecasting EAC are described below:

$$\text{Standard forecast: EAC} = \frac{\text{BAC}}{\text{CPI}}. \tag{27.1}$$

This approach is most often used when it is predicted that the future project will progress the same as previous history. The estimation formula for this would be

$$\text{Flawed estimating: EAC} = \text{AC plus a new estimate for all remaining work.} \tag{27.2}$$

This approach is used when past performance has shown little relevance to current conditions, or some change in the project environment would indicate that new estimates are needed.

$$\text{Poor start-up: EAC} = \text{AC} + \text{plus remaining budget.} \tag{27.3}$$

This approach is most often used when current actual cost variances are seen as atypical and the project management team expectations are that similar variances will not occur in the future.

$$\text{EAC pessimistic estimate: EAC} = \frac{\text{BAC}}{(\text{CPI*SPI})}. \tag{27.4}$$

When CPI and SPI values vary significantly from 1.0, this indicates an anomaly from the plan. Even if one variable is below 1.0 and the other is above 1.0, there is reason to question the plan structure. More typically, both indices are less than 1.0, which means that both cost and schedule are below the plan at this point. If there is no identifiable reason for these trends to change, a pessimistic estimate can be generated from this formula.

The proper EAC formula to use is based on a forecast of the future environment. If there is no reason to believe a change will occur, the first formula 27.1 or the pessimistic formula 27.4 should be used.

Estimate To Complete (ETC): This parameter is an estimate for the project remaining cost from the status point. It is calculated as the difference between EAC and AC. Senior management and financial groups are the ones most concerned about this number.

Variance At Completion (VAC): This is the difference between BAC and EAC. VAC represents the amount forecast for the project to be over- or underbudget.

PMB: The sum of all PV WPs aggregated by time period for the total duration of the program. The PMB forms the time-phased budget plan against which project performance is measured. The PV curve in Figure 27.2 is a PMB display.

For large projects Christensen found that the CPI values tended to stabilize by the 20th percentile point into the project to within ±10% accuracy (Christensen, 1998, p. 10). This means that once the project reaches this point, CPI is a reasonable predictive parameter.

Use of EV values provides great insight into cost and schedule performance for the project. For instance, if the CPI was reported at 0.8 this would imply that the project was using resources at approximately 20% above plan (i.e., running at 80% of plan efficiency). In similar fashion, the SPI metric describes schedule performance. So, an SPI value of 1.2 would suggest that the project was running ahead of schedule by about 20%. General interpretation of the index values is as follows:

Index values >1.0 is good (better than plan performance)
Index values <1.0 is bad (below plan performance)
Negative variance values are bad (below plan efficiency)
Positive variance values are good (better than plan efficiency)

27.4 EVM CRITERIA

The original DoD definition of EVM was very complex for their contractor community to deal with in the late 1960s. Looking back, we would say that these organizations had a fairly low operational maturity; however, we also need to recognize that computer technology was still in an early growth phase and its use was restricted to formal operational systems that did not extend to project management. Plans could be stored and the accounting system could deliver some measure of ACs, but in general neither the operational infrastructure nor the associated project management processes were capable of collecting the required data. In addition, the original specification contained 35 management criteria (eventually reduced to 32) and very few contractors could satisfy the full set of these requirements. The reader can explore in ANSI/EIA-748 for the full details of this set of specifications (see www.webstore.ansi.org).

As a result of this complexity overkill there was strong resistance to many of the processes that we see in operation today, but did not fit the management mindset of that early period. Over time, there was an acceptance of a subset of this requirement and eventually two lesser levels of implementation were loosely defined. The first one is characterized as "simplified" for small projects and a second one appropriate for the typical commercial project of moderate size. These two modified options are summarized below.

27.5 EVM SIMPLIFIED

Since there are many more small and simple projects than there are large and complex ones, a simplified version of EVM is desirable for the masses, while still offering the basic EV metrics benefits. The simple criteria are contained in three steps as outlined below (Christensen, 1995, p. 156).

1. Scope definition. Advocates of formal project management techniques would typically do this using a WBS for requirements definition; however, the work may also be defined just through a simple list of tasks. In either case, the work definition needs to be comprehensive and decomposed into reasonably sized WPs that are mutually exclusive.
2. The second step is to assign a PV or budget for each WP. This budget could be in terms of units of currency (e.g., dollars), or in abor hours, or both. For very simple projects, each WP may simply be assigned a weighed "point value" instead of a budget number, so progress is measured in terms of points rather than planned cost values.
3. The third step is to define the earning rules for each WP. One simple method is to apply just one earning rule, such as the 0/100 rule, in which no credit is earned until the WP is complete. Other rule variations may be more applicable to WPs that have greater time

duration, such as the 50/50 rule, where 50% credit is earned as soon as the WP is started and the remaining 50% is earned upon completion. These can be further modified to 25/75, 20/80, or any other work credit rules desired. Typically, these modified earning rules serve to allocate performance credit to WP completion, yet still gives motivation for the project team to get a WP started. These combination rules help focus the team's attention on work completion. Nonlinear earning rules tend to work well in situations where WPs are short in duration (e.g., an average of 2 to 4 weeks duration). One final rule option is to estimate percent completion for each WP. The challenge here is to get honest status evaluations. From a pure mathematical view, this would provide the most accurate status measure if one can assume an accurate completion assessment. Regardless of the method selected, the process requires a calculation of an "EV" for each work unit. Everything else in the process fits the calculation mechanics described earlier.

Using the steps above, each WP would yield a cost and schedule performance EV metric. Using even crudely calculated metrics, such as those described here, have benefit because they provide both a measure of current status and a projection of forward progress. Overall, this process helps to provide a scorecard for the project team.

27.6 EVM FOR COMMERCIAL APPLICATIONS

Fleming and Koppelman (2006) studied control requirements contained in the ANSI/EIA-748 Standard, and from this review distilled ten basic management steps necessary to implement a reasonable form of EVM that will satisfy control requirements for most moderate to small projects in any industry. These management process steps are summarized below:

Step 1: Define the scope of the project. One of the most useful tools available to the PM is the WBS and it is critical to the EV method.

Step 2: Work definition. Determine who will perform the defined work, and identify all critical procurement. The PM must evaluate and assign internal or external HR by balancing constraints such as experience, cost, and whether or not the desired expertise resides in-house. These choices are called the make or buy decisions, and selecting those items that must be bought for the project is an essential extension of the scope definition process. If the work is outsourced, proper care must be given to legal arrangements to adequately protect the project. Whether the HR are internal or external, the measurement and reporting mechanisms must be defined. The project must be able to continuously measure the EV of a work unit versus the AC of the work being performed.

Step 3: Plan and schedule defined work. EV methods represent little more than a good scheduling system, with authorized resources (budgets) embedded into the schedule. The schedule must reflect the authorized scope and time frame. From this base, earnings are calculated as work is accomplished. A formal work scheduling system is thus necessary because it is the vehicle that translates the project scope into a time sequence that is necessary for the status calculations.

Step 4: Estimate the required resources and formally authorize the budget. Once the project scope has been fully defined and subsequently planned and scheduled, the next step is to estimate the resource requirements (budgets) for all the defined work units. Each defined WBS element must have a resource value estimated for the specified work. Management will then assess the requested resources and approve this in the form of an authorized project budget. Individual WBS budgets will not contain contingencies or management reserves in the EV analysis portion. They are held separately.

Step 5: Determine the metrics to convert PV into EV. The challenge for this step is to identify viable methods to quantify the authorized work units and then measure the completion of the authorized work. There are various methods of measuring project performance; the most respected ones use some type of discrete measurement. Specific completion milestones representing points in time can be assigned values, which are earned when fully completed. Also, tasks are assigned values, which can be measured as they are partially completed and assigned some EV for the reporting period.

Step 6: Define a PMB and determine the points of management control. EV requires the use of an integrated project baseline containing the defined work, baseline schedule, and authorized budget. Parameter calculation takes place within each of the defined WBS elements. Management control then occurs at focal points placed at selected WBS elements and are referred to as Control Account Packages (CAPs). See Chapter 12 for a review of the CAP concept. A CAP can be best described as an arbitrary point in the WBS where AC data are collected. On some projects, the total baseline cost may sometimes include such things as indirect costs and even profits or fees to match the total authorized project commitment. The cost baseline must include whatever executive management has authorized for the project, but realize that EV analysis only applies to the direct work portion and not so much for items such as level of effort tasks and material.

Step 7: Record all direct costs consistent with the authorized baseline and in accordance with the organization's general accounting structure. This criterion requires the PMs to have access to the current level of expenditures at the level of detail required. It is essential that direct costs be tracked to a work unit or CAP. In order to employ EV metrics on any project, the ACs must be aligned with the baseline budget. For instance, EV must be relatable to AC by work units in order to determine the CPI, which is the single most important EV performance metric.

Step 8: Continuously monitor EV parameters to determine cost and schedule departures from the base plan (i.e., determine EV variances). Projects employing EV must monitor their cost and schedule results against the authorized baseline throughout the duration of the project. Management should focus their primary attention on exceptions to the baseline plan, particularly those work units that exceed previously defined acceptable tolerances. In this role, EV is a management by exception concept. As previously indicated, the single most important aspect of employing EV is its ability to monitor ongoing cost and schedule status of the project at WP levels of detail.

Step 9: Forecast the final project status. One of the more beneficial aspects of EVM is that it provides the capability to forecast cost and schedule of the project based on current performance. The use of CPI values is particularly valuable in the cost analysis and forecast aspect. However, there is a subtlety in using the SPI metric and one must be more careful using it. Recall that the SPI formula is EV/PV and the rules of EV dictate that the EV of a work unit is its PV at completion. As a project moves passed the 60th percentile point, this formula begins to lose its predictive value and by definition starts to approach 1.0 at the end of the project. This means that SPI is not a good schedule predictor past this point. We will see another option to deal with this a little later.

Step 10: Manage authorized scope by approving or rejecting all changes, and incorporating the approved changes into the project baseline in a timely fashion. The project PMB set at the project start is only valid as a benchmark value so long as no changes occur. Once a change is approved, the PMB becomes only a historical comparison point. Since changes complicate the status interpretation process they must be handled carefully and effectively. An ICC process is a mandatory operational process. Even with this process in place, the EV calculation process becomes more complex as new work units are added to the WBS during the course of the project.

Utilizing the 10 process steps outlined above will provide the typical PM with a reasonable mechanism to satisfy most basic monitoring and control needs. If the scope and complexity of the project dictates, these steps can be expanded to satisfy the more rigorous criteria contained in the ANSI/EIA-748 Standard.

27.7   EMERGING APPLICATIONS OF EVM

This section will describe two variations of the EV system that address peculiarities or shortcomings inherent in the traditional EV metrics. These are presented as an extension of the theory and are considered complementary to the basic analysis provided by EVM.

27.7.1 EARNED SCHEDULE

Historically, this EV measure for cost performance has been based on work unit completion compared to the resource budget plan. This approach works well for cost performance, but as mentioned above poses limitations in the analysis of the project schedule for the reasons outlined below.

EV formulae for time performance will approach 1.0 by definition regardless of the actual schedule situation. So, calculations of SV and SPI after the 60th percentile begin to generate values that have lost analytical value. So, SV and SPI are not a time-based metrics as the names implies. In this sense, SV is really a measure of the volume of work accomplished, versus the volume of work planned. With the benefit of hindsight, these measures might better be called an "Accomplishment Variance." In order to correct this computational issue Lipke conceptualized an alternative approach to evaluating schedule performance (Lipke, 2003). Following this period Lipke and Henderson have been key proponents for maturing the concept under the banner of *Earned Schedule (ES)* (Lipke and Henderson, 2006). ES is an extension of EVMS and is designed as a time-based measure. ES is analogous to EV, except that it is formulated from a duration basis instead of a cost-based measure.

The ES concept is simple and is illustrated graphically in Figure 27.4. Basically, the SV(*t*) represents the time variance from the baseline plan (5–7 in the example). This is a measure of how far ahead or behind the plan the project is at the status point. The advantage of this geometry is that it holds valid through the entire life cycle rather than collapsing as the traditional formula does.

The current time point is indicated by the status point (7) and the ES value is found by moving horizontal from that EV point to the PV curve. This shows that the actual accomplishment is equal to only time period 5, therefore the project is two time periods behind the plan. This description may seem a little confusing at first, but becomes very simple by tracing the geometry on Figure 27.4. Based on this view, the time-based metric values translate to a similar interpretation as the traditional EV metrics would be (Lipke and Henderson, 2006):

$$SV(t) = ES - AT,$$

where ES is the calculated ES value and AT is the current time; for example, SV(*t*) = 5 – 7 in the example figure.

$$SPI(t) = \frac{ES}{AT}(SPI),$$

$$EAC(time) = actual\ time + (planned\ duration - ES)*CF,$$

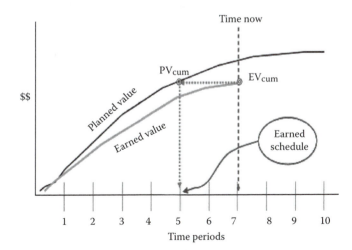

FIGURE 27.4 Graphical view of ES. (From Lipke, W., 2006. Earned Schedule—An Extension to Earned Value Management, www.earnedschedule.com, [accessed February 5, 2009.] With permission.)

where CF is a confidence factor used to forecast how future performance will progress compared to plan. If no change in project efficiency is anticipated the value of SPI(t) could be used for this value.

Values generated from these formulae are quite similar in interpretation to the original EV formulae, except that this view is created more from a time perspective that does not suffer from formula degradation as the project goes beyond the 60% point. Translating this view from the graphical view into a more formal quantitative notation is beyond our scope. The interested reader can find more details on this method in the references from Lipke, Henderson, and Fleming.

27.8   ES MATHEMATICAL FORMULATION*

Let us begin the mathematical explanation of the ES concept by reviewing the words in the upper left of Figure 27.5: *"The basic idea behind ES is to determine the time at which the EV accrued should have occurred."*

Now, we will use the graphic details to explain what these words are attempting to say. At time period 7, Time Now, we have a cost value for EV. The question is, "When should this amount of EV have been earned?" The answer is when an equal amount of PV appears on the PMB. Graphically this is determined by projecting EV onto the PV curve (arrow A), and then dropping a vertical line to the time axis (arrow B). The duration from the project start to the red vertical line is the cumulative value of ES. For this example, ES is equal to five time periods. This is not so difficult. From this view we visually see what ES represents.

The example shown in Figure 27.5 illustrates that ES is a whole number; however, this is normally not the case. More typically, the dotted line B would fall between two time periods. When this occurs, ES is determined by first counting the number of PV periods completed and then adding the earned fraction of the incomplete period. The amount earned in the partially completed period is determined by linearly interpolating from the known costs (EV and PV) as shown in Figure 27.6.

Subscript C identifies the PV period at which $EV_{cum} \geq PV_{icum}$. Only the incomplete period is shown in the graphic. Also shown are the three budget values, EV, PV_C, and PV_{C+1}. For this discussion, months are used as the time period. But, the schedule increment could be weeks or any other time unit desired. From the schematic in Figure 27.6 and some trigonometry relationships, we know

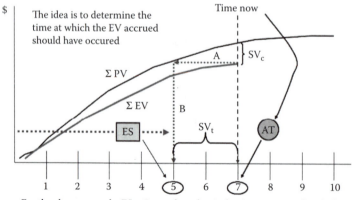

For the above example, ES = 5 months a that is the time associated with the PMB at which PV equals the EV accrued at month 7.

FIGURE 27.5   Earned schedule concept ES (From Lipke, W., 2005 [June]. *Crosstalk*, www.stsc.hill.af.mil/crosstalk/2005/06/0506Lipke.html. With permission.)

---

* This section is written by Mr. Walter Lipke who is generally recognized as the inventor of the ES concept and continues to be a strong proponent. More of his material on this topic can be obtained through http://www.earnedschedule.com/Papers.shtml.

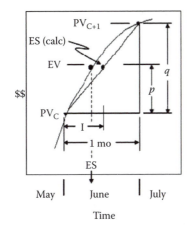

FIGURE 27.6 ES interpolation. (From Lipke, W., 2005 [June]. *Crosstalk*, www.stsc.hill.af.mil/crosstalk/2005/06/0506Lipke.html, [accessed October 17, 2008]. With permission.)

that *I*, the value we are trying to determine, is for one period, or one month in the example, as *p* is to *q*. So, *p* and *q* are easily determined since

$$p = EV - PV_C, \text{ whereas}$$

$$q = PV_{C+1} - PV_C.$$

After algebraic substitution and solving for *I* the resultant equation provides a method for calculating the fractional portion of the ES measure. These steps are summarized below:

$$I/1 \text{ month} = p/q.$$

$$I = (p/q) \times 1 \text{ month}.$$

$$p = EV - PV_C.$$

$$q = PV_{C+1} - PV_C.$$

$$I = (EV - PV_C)(PV_{C+1} - PV_C) \times 1 \text{ month}.$$

ES and ES(calc) are shown in the diagram to illustrate the error inherent in the linear interpolation. This calculation substitutes a straight line for one time period of the PV baseline curve. The error from this assumption is small, and is negligible.

27.8.1 ES Formulae

ES_{cum} is the number of completed PV time increments EV equals or exceeds, plus the fraction of the incomplete PV increment. So

$$ES_{cum} = C + I,$$

where *C* is the number of time increments for EV ≥ PV,

$$I = (EV - PV_C)/(PV_{C+1} - PV_C),$$

$$\text{ES}_{\text{period}}(n) = \text{ES}_{\text{cum}}(n) - \text{ES}_{\text{cum}}(n-1), \quad \text{and}$$

$$\text{ES}_{\text{period}}(n) = \Delta\text{ES}_{\text{cum}}.$$

27.8.2   ES Indicators

From the two measures, ES and AT, the SV and the SPI are formulated as shown below.
SV(time)

$$\text{SV}(t) = \text{ES}_{\text{cum}} - \text{AT}_{\text{cum}}, \quad \text{where AT is the actual time.}$$

$$\text{SV}(t)_{\text{period}} = \Delta(\text{ES}_{\text{cum}} - \Delta\text{AT}_{\text{cum}} \quad (\text{normally } \Delta\text{AT}_{\text{cum}} = 1).$$

SPI(time)

$$\text{SPI}(t) = \text{ES}_{\text{cum}}/\text{AT}_{\text{cum}}.$$

$$\text{SPI}(t)_{\text{period}} = \Delta\text{ES}_{\text{cum}}/\Delta\text{AT}_{\text{cum}}.$$

Note that these indicators are set apart from the traditional EVM schedule indicators by appending a (t) to each. When the (t) is seen, you know that the indicator comes from ES mechanics rather than the traditional approach.

This section has illustrated how the ES parameters can be produced both graphically and mathematically. The interested reader should review further writings from Lipke and Henderson at www.earnedschedule.com.

27.9   EVM PROS AND CONS

While there is general consensus that EVM concepts are useful and meaningful tools for project monitoring and control, it is also acknowledged that their implementation, as defined by the original C/SCSC criteria, or the ANSI/EIA-748 Standard, are likely too complex and costly for most commercial projects. DoD project studies also seem somewhat polarized on this topic. One supportive study contends that the full implementation of the C/SCSC criteria is significant and represents the requirement ranked third among the top 10 cost improvement drivers in projects (Fleming, 2006, p. 16). Another study supports this view in finding that stronger implementation of the EVM principles results in greater levels of project success (Goodpasture, 2004, p. 178). A third study concludes that, ultimately, the measure of whether the benefits provided by EVM systems exceed their cost is not quantifiable but rather subjective in nature. The same study also states that the most compelling evidence that the benefits of EVMS exceed the cost is the major increase of EV usage outside of DoD by other agencies, commercial companies and other countries (Henderson, 2003).

Fleming and Koppelman outline a list of EVM benefits derived from three decades of government contracts and assert that the application of the criteria is essential for large, cost-reimbursable projects by providing a single management control system providing reliable data. This research showed that EVM had benefit in the following nine areas (Corovic, 2006):

1. The integration of work, schedule, and cost using the WBS
2. A database of completed projects useful for comparative analysis
3. The cumulative CPI as an early warning signal
4. The SPI as an early warning signal
5. The CPI as a predictor for the final cost of the project
6. An index-based method to forecast the final cost of the project

7. The To-Complete performance index to evaluate the forecasted final cost
8. The periodic (e.g., weekly or monthly) CPI as a benchmark
9. The management by exception principle to reduce information overload.

The list above does not separate the benefits of EV metrics from the general benefits derived from improved overall supporting management processes. One might infer that the application of mature management practices also contributes to these benefits.

27.10 CONCLUSIONS

EV methodology offers the PMs an integrated system for tracking project cost, schedule, and scope and is effective in influencing project performance. There is consensus that the EVM criteria defined by ANSI/EIA-748 Standard, while suitable for large, cost-reimbursable projects, are considered overkill for small- to moderate-sized initiatives. Since the majority of the projects fall into the small- to medium-sized category, the key to broad-scale successful integration of EVM in commercial practice is to define a subset of management support processes that are simplified, yet effective in providing accurate and reliable performance data necessary for the proper utilization of EV metrics. In this sense, the "lighter" implementations of EVM described here may offer a more feasible solution for most projects. Fleming and Koppelman's 10-step simplified approach outlines the rudiments of a fundamental and sound project management process, which should be practicable for most organizations. Use of this would move the basic theory into an operational mode and help derive improved monitoring and control metrics.

APPENDIX

This section contains a summary of definitional terms and mechanical details that are useful in implementing EV at the introductory level.

A. SUMMARY REVIEW OF EV METRICS AND PERFORMANCE PARAMETERS

Three quantities form the basis of EV metrics: the PV, the EV, and the AC. These parameters were originally called the BCWS, BCWP, and ACWP respectively, but have now been shortened to the more easy to remember two letter titles (PV, EV, and AC, respectively).
   Definitions for the basic parameters are summarized below:

PV: The sum of budgets for all WPs scheduled to be accomplished within a given time period.
EV: The sum of budgets for completed WPs and completed portions of open WPs.
AC: The AC incurred in accomplishing the work performed within a given time period. For equitable comparison, AC is only recorded for the work performed to date against tasks for which an EV is also reported.

The performance and forecasting terms defined below often found in the EV literature are summarized here.

PMB: The sum of all PV WPs for each time period calculated for the total duration of the program. The *PMB forms the time-phased budget plan* against which project performance is measured.
BAC: The sum of all the budgets allocated to a program. In addition to the PMB normally there will be additional funds allocated for reserves, profit, and so on. The BAC consists of the PMB direct project costs plus all management reserves.

TABLE 27.2
Summary of EV Formulae

Name	Formula	Interpretation
CV	EV – AC	NEGATIVE is overbudget, POSITIVE is underbudget.
SV	EV – PV	NEGATIVE is behind schedule, POSITIVE is ahead of schedule.
CPI	EV/AC	I am (only) getting _____ cents out of every $1.
SPI	EV/PV	I am (only) progressing at ___% of the rate originally planned.
EAC	1. BAC/CPI	As of now how much do we expect the total project to cost $_____.
Note: There are multiple ways to calculate EAC	2. AC + ETC	Used if no variances from the BAC have occurred.
	3. AC + BAC – EV	Actual plus a new estimate for remaining work.
	4. AC + (BAC – EV)/ CPI	Used when original estimate was fundamentally flawed. Actual to date plus remaining budget.
		Used when current variances are atypical.
	5. AC + budget	Actual to date plus remaining budget modified by performance.
		Use when current variances are typical.
ETC	EAC – AC	How much more will the project cost?
VAC	BAC – EAC	How much overbudget will we be?

SV: The difference between the work actually performed (EV) and the work scheduled (PV). The SV is calculated in terms of the difference in dollar value between the amount of work that should have been completed in a given time period and the work actually completed.

CV: The difference between the planned cost of work performed (EV) and AC incurred for the work. This is the actual dollar value by which a project is either overrunning or under running its estimated cost.

CPI: The ratio of cost of work performed (EV) to AC. CPI of 1.0 implies that the AC matches to the estimated cost. CPI >1.0 indicates that work is accomplished for less cost than what was planned or budgeted. CPI <1.0 indicates that the project is facing cost overrun.

SPI: The ratio of work accomplished (EV) versus work planned (PV), for a specific time period. SPI indicates the rate at which the project is progressing.

B. EV Formulae and Interpretation (Table 27.2)

C. Using a Summary Project Plan Spreadsheet to Calculate EV Parameters

The spreadsheet shown below in Table 27.3 illustrates how a summary level project plan can be used to produce aggregate EV metrics. Reviewing the sample calculations would be a good way to test your knowledge of the concepts.

D. EV Earning Rules

One of the basic tracking issues for work units is to define what portion of the work has actually been completed. The EV methodology requires that some formal consideration be given to this issue and it should be applied consistently across all projects. Summarized below are a few operational rules to consider when making this selection.

WPs: The general rule of thumb for a WP size is 2/80, meaning that it will be sized in the general range of 2-week duration and 80 h of work effort. Other values can be used as guidelines, but effort should be made to keep these as short as possible since large WPs can hide overruns.

Earned Value Management

Earned Value Calculation Spreadsheet

Activity	Jan	Feb	Mar	Apr	May	Jun	Jul	Aug	Sep	Oct	Nov	Dec	PV	% Complete
Soil Work	24,123												24,123	100
Engineering		954,544	1,189,225										2,143,769	100
Equipment and Fabrication				1,851,278	2,328,825	1,708,280							5,888,384	70
Construction							1,730,460	1,766,129	900,042				4,396,631	25
Instrument and Electrical										797,011	786,970		1,583,981	0
Completion Work												1,580,061	1,580,061	0
Planned Value (PV)	24,123	954,544	1,189,225	1,851,278	2,328,825	1,708,280	1,730,460	1,766,129	900,042	797,011	786,970	1,580,061	12,452,907	
**Cumulative Planned Value (PV)	24,123	978,667	2,167,892	4,019,170	6,347,995	8,056,276	9,786,736	11,552,865	12,452,907	13,249,918	14,036,888	15,616,949		
Monthly Actual Cost (AC)	45,000	1,500,000	1,250,000	1,500,000	2,000,000	1,100,000	1,200,000							
Actual Cost (AC)	45,000	1,545,000	2,795,000	4,295,000	6,295,000	7,395,000	8,595,000	8,595,000	8,595,000	8,595,000	8,595,000	8,595,000		
Monthly Earned Value (EV)	24,123	954,544	1,189,225	1,295,895	1,630,178	1,195,796	432,615							
Cumulative Earned Value (EV)	24,123	978,667	2,167,892	3,463,787	5,093,964	6,289,761	6,722,376							

Project EV as of July 31	6,722,376
Project PV as of July 31	9,786,736
Project AV as of July 31	$ 8,595,000
CV=EV-AC	$ (1,872,624)
SV=EV-PV	$ (3,064,360)
CPI=EV/AC	78%
SPI=EV/PV	69%
EAC cost = BAC/CPI	$ 19,967,298 (original plan divided by CPI)
EAC time = BAC/SPI	17.47 (original plan divided by SPI)

Level of effort: This class of WP can cloud the project status analysis since these charges are independent of accomplishment. Productivity studies should be done without these in the mix.

Material costs: As with level of effort, material costs variances are caused by factors other than productivity. If material costs represent a significant percentage of the total budget, these should be extracted from the analysis.

Work accomplishment rules: There are several options for recording WP accomplishment. The list below summarizes the implications of the most used:

    a. 50/50: This rule gives a 50% credit once the WP is started and the last 50% is withheld until the unit is completed.

    b. 0/100: This rule penalizes WIP since no credit is given until the unit is completed.

    c. Percentage: This is the most accurate mathematical calculation of accomplishment if one can make the assumption that this can be done and will be done honestly.

    d. Level of effort: This class of resource charge is normally allocated to the WP as defined in the support agreements or as billed to the project based on actual charges.

PROBLEMS

  1. *Basic Gantt Plan*

    The simple Gantt chart below contains three activities. PVs for each activity are shown in bold font on each bar. Total completion estimates are shown as a percentage. Time period 2 is the status point. The accounting system has collected ACs for the three activities as 6, 12, and 6, respectively. Compute the EV parameters for this model. What is your prediction for cost and time completion of this project?

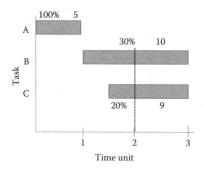

  2. *EV Dog Pen Exercise*

    Assume that you are the PM for the Dog Pen project. The goal is to build a four-sided dog pen. Each side is budgeted to cost $100 to complete and require 1 day of your time. As the project progresses the following results are recorded:

        Day 1: Side A completed; spent $100
        Day 2: Side B completed; spent $120
        Day 3: Side C 50% completed; spent $60
        Day 4: Side D not started; spent $0

    a. By inspection and without resorting to EV formulae what is the general status of this project?

        Cost at completion?

        Schedule completion?

    b. Calculate the following EV parameters:

        PV =

        EV =

        AC =

        BAC =

        CV =

$$SV =$$
$$EAC =$$
$$ETC =$$

c. Using your knowledge of EV interpret the status of this project.

3. *Microsoft Project Presentation*

Microsoft Project is a popular project planning utility. The table below shows a sample output from that utility. The columns labeled Baseline Duration and Baseline Cost are equivalent to PVs for time and cost in the EV notation. The dashed line appearing after Activity/Task B Gantt bar reflects the status point. Use this vocabulary translation and compute the EV parameters for this slightly different presentation format. *Note: The Microsoft Project software can generate these parameters, but training is needed to ensure that the program is using AC in an appropriate way, so these internal metrics are not demonstrated here.*

0	Task Name	% Complete	Baseline Duration	Duration	Cost	Baseline Cost
1	⊟ Total Project	50%	30 days	39 days	$39,000.00	$30,000.00
2	✓ A	100%	10 days	13 days	$13,000.00	$10,000.00
3	B	50%	10 days	13 days	$13,000.00	$10,000.00
4	C	0%	10 days	13 days	$13,000.00	$10,000.00

Questions:

a. What is the planned project cost and duration?

b. What is CPI and SPI for this model?

c. What are the EV calcuations for project cost and schedule at completion?

4. *EV Spreadsheet Template*

One of the more common methods to compute EV parameters is through the use of a spreadsheet. The sample below contains summary level activity planned costs by month for a selected level of the WBS. Percentage completion estimates for each activity are recorded in column "O." The status point for EV evaluation is May as indicated by the vertical bar. AC data are recorded in rows 15 and 16. Using your knowledge of EV answer the following questions.

a. Translate these data to produce the project EV status.

b. What is the condition of the project based on these metrics.

c. What is the estimated cost and schedule at completion?

	Activity	Jan	Feb	Mar	Apr	May	Jun	Jul	Aug	Sep	Oct	Nov	Dec	PV	% Comp	EV
1																
2	Plan and staff project	4,000	4,000												100	
3	Analyze requirements		6,000	6,000											100	
4	Develop ERDs			4,000	4,000										100	
5	Design database tables				6,000	4,000									100	
6	Design forms, reports, and queries					8,000	4,000								50	
7	Construct working prototype						10,000									
8	Test/evaluate prototype						2,000	6,000								
9	Incorporate user feedback							4,000	6,000	4,000						
10	Test system									4,000	4,000	2,000				
11	Document system											3,000	1,000			
12	Train users												4,000			
13	Monthly Planned Value (PV)															
14	Cumulative Planned Value (PV)															
15	Monthly Actual Cost (AC)	4,000	11,000	11,000	12,000	15,000										
16	Cumulative Actual Cost (AC)	4,000	15,000	26,000	38,000	53,000										
17	Monthly Earned Value (EV)															
18	Cumulative Earned Value (EV)															
19	Project EV as of May 31															
20	Project PV as of May 31															
21	Project AC as of May 31															
22	CV=EV-AC															
23	SV=EV-PV															
24	CPI=EV/AC															
25	SPI=EV/PV															
26	Estimate at Completion (EAC)															
27	Estimated time to complete															
28																

REFERENCES

Christian, D.S. and Daniel, V.F. 1995. Using Earned Value for Performance Measurement on Software Development Projects, (Spring: 156) *Acquisition Review Quarterly.*

Christensen, D.S.1998. The Cost and Benefits of the Earned Value Management Process. http://www.dau.mil/pubs/arq/98arq/chrisevm.pdf accessed October 17, 2008.

Corovic, R. 2006. Why EVM Is Not Good for Schedule Performance Analyses, *The Measurable News,* Project Management Institute.

Department of Defense (DoD). 2006. Earned Value Management Implementation Guide, (October). www.osd.acq.mil/pm (accessed February 2, 2008).

Fleming, Q.W. and J.M. Koppelman. 1999. Earned Value Project Management, (July), *Crosstalk,* http://www.stsc.hill.af.mil/crosstalk/frames.asp?uri=1999/07/fleming.asp (accessed October 17, 2008).

Fleming, Q.W. and J.M. Koppelman. 2006. (June 16, 2006) Start with 'Simple' Earned Value on All Your Projects, *Crosstalk*, www,stsc.hill.af.mil/crosstalk (accessed October 17, 2008).

Gansler, J. 1999. New ANSI Standard on EVMS Guidelines. www.acq.osd.mil/pm/newpolicy/indus/ansi_announce.html (accessed October 17, 2008).

Lipke, W. and K. Henderson. 2006. Earned Schedule: An Emerging Enhancement to EVM, (November) *STSC Crosstalk*, www.stsc.hill.af.mil/crosstalk (accessed October 17, 2008).

Lipke, W. 2005. Connecting Earned Value to the Schedule, (June) *Crosstalk*, http://www.stsc.hill.af.mil/crosstalk/2005/06/0506Lipke.html (accessed October 17, 2008).

Lipke, W. 2003. Schedule Is Different, *The Measurable News,* Project Management Institute.

Lipke, W. 2006. Earned Schedule—An Extension to Earned Value Management, www.earnedschedule.com (accessed February 5, 2009).

Goodpasture, J.C. 2004. *Quantitative Methods in Project Management.* Boca Raton: J. Ross Publishing.

Haugan, G.T. 2003. The Work Breakdown Structure in Government Contracting, *Management Concepts.*

Henderson, K. 2003. Earned Schedule: A Breakthrough Extension to Earned Value Theory? A Retrospective Analysis of Real Project Data, (Summer) *The Measurable News,* Project Management Institute.

Henderson, K. 2004. Further Developments in Earned Schedule, (Spring) *The Measurable News,* Project Management Institute.

Henderson, K. 2005. Earned Schedule in Action, (Spring) *The Measurable News,* Project Management Institute.

Henderson, K. 2007. Earned Schedule: A Breakthrough Extension to Earned Value Theory? A Retrospective Analysis of Real Project Data. Originally published as part of 2007 *PMI Asia Pacific Global Congress Proceedings,* http://www.earnedschedule.com/Docs/Earned%20Schedule%20a%20%20Breakthrough%20Extension%20to%20EVM%20-%20Henderson.pdf (accessed December 30, 2008).

PMI, 2008. *A Guide to the Project Management Body of Knowledge (PMBOK® Guide)*, 4th Edition. Newtown Square, PA: Project Management Institute.

Schulte, R. 2005. What is the Health of my Project? The Use and Benefits of Earned Value," *A Welcom White Paper,* Retrieved on August 9, 2005 from www.welcom.com (accessed August 9, 2005).

Vandevoorde, S. and Vanhoucke, M. 2006. A Comparison of Different Project Duration Forecasting Methods Using Earned Value Metrics, *International Journal of Project Management.*

28 Tracking Project Progress

28.1 INTRODUCTION

This chapter will describe a model project status tracking framework in regard to the monitoring of certain common problems, progress tracking, and control problem situations. Basic examples for each of these will be offered. The intent here is to build a foundation on which a new or existing PM can develop his or her own methodology for managing project execution, tracking, and control in an effective manner. Since each industry and organization has its own nuances, no one source or approach can claim that it has the silver bullet or magic formula that will produce an improved control structure. The concepts and examples described here are intended to provide reasonable insight into general project control functions. In addition to this, examples of control parameters are discussed along with formal methods for dealing with each.

28.2 STATUS TRACKING

There are six project model status measurement categories: team efficiency, process efficiency, project efficiency, quality, value, and effectiveness. Each of these groups will be compared to the approved project plan and will be measured using both quantitative and qualitative techniques. The essence of control is measurement, comparison, and corrective action. A key management component of this is progress tracking.

Previous chapters have described the selection of specific metrics to track project status. The focus of this chapter is the use of these parameters in the management process. Specifically, measurement of defined project metrics is the base mechanic for project monitoring and control. As the execution process unfolds project tasks will deviate from planned values: tasks are executed, execution status is measured, performance results are reported, and appropriate management controls are applied. Through use of this process project status is used to aid in corrective decision making and to communicate status to the project team and other stakeholders.

Selecting an appropriate tracking method for a project depends on several factors including the following:

- *Size of the project:* If the project is very small, consisting of a dozen or so tasks, the manager may track task progress manually. Alternatively, if the project has more than 25 or 30 tasks a more automated tracking tool would be more desirable.
- *Tracking tool availability:* Email systems and other organizational communication tools are often used for progress tracking. Each of these tools provides some level of capability in providing methods to collect and disseminate task status information. The features offered by each tool add incremental value to the overall monitoring and control process.
- *Level of detail at which progress needs to be tracked:* Typically, the needs of the project determine the level of detail at which the manager needs to track progress. For example, if specific resources are assigned to tasks, a tight budget or deadline may require detailed

tracking of the work and costs associated. But, if this level of detail is not needed and the manager has little time to spend tracking, the PM may choose to track status at higher levels in the WBS (summary tasks).

Each organization offers a unique set of tools and project personalities. From a project management standpoint the monitoring and control process requires the following six basic elements:

1. An approved and baselined project plan.
2. A WBS that reflects the level of control appropriate for a particular project. The model rule of thumb is that this means control WPs of 80 h of effort.
3. A defined method for recording work performed using some form of an Earned Value model (see Chapter 27). Work accomplished for each WP will be monitored according to the defined measurement technique. Measurement of schedule deviation, work effort, cost expended, along with risk and HR issues incurred are metrics normally used to monitor project progress. In many cases, these metrics are presented in a planned (baseline) versus actual manner.
4. Estimation of work remaining on each active WP. This provides a cross estimate for the work performed.
5. An Issues log linked to WPs that contain information related to any that are not progressing according to plan, or that are forecast to become an issue before completion. These items should outline what needs to be resolved to improve future performance. A prioritized list of these Issues then becomes a work list for the PM to follow in removing roadblocks for the project team.
6. A project communications model needs to be established during the planning stage. Mature organizations will have this structure in place, but in some organizations this will have to be cobbled together by the project team. This model involves defining formats for stakeholder status reporting and the mechanism to deliver that status to specific stakeholders.

Figure 28.1 describes a model summary of the basic project monitoring process. This process is driven by a set of defined control parameters that are then measured as part of the execution process. Delivery of the control metrics is reflected by the "information distribution" process, which can take on a wide variety of technologies. The traditional format of status distribution was paper manually delivered to the recipient; however, the more modern approach for information delivery is to use Internet-based technologies to produce on demand status in flexible formats.

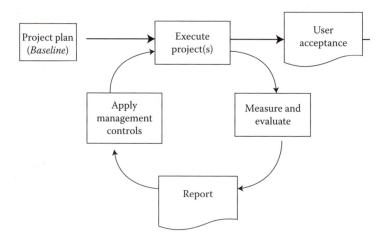

FIGURE 28.1   Project monitoring cycle.

28.3 TRACKING METRICS

As we have seen in earlier discussions, the project schedule consists of a detailed layout of WPs to be performed and the expected completion time for each. The aggregation of scheduled WPs then represents the tracking schedule targets for the reporting and overall project periods. Next, in order to determine if execution is within budget, work effort and accomplishment must be monitored. Planned hours of work and labor expenditures are measured against actual hours worked. For instance, the project may be running under labor budget because of lack of skilled labor at the point and time the status measurement is taken. In addition, actual hours accrued can be more than planned because team members may be working additional time in order to meet a project deadline. Deviations from planned resource expenditures from the original schedule will often lead to corresponding changes in the cost of the project. In order to collect actual project status, it is often necessary to consolidate inputs from various operational systems such as the organizational accounting system, procurement, and others.

The measurement of project status includes, but is not limited to, project team members, supporting team members, external contractual team resources, materials, facilities, training, defects, risk events, and team issues. The latter term, Issues, are often reported to provide a general qualitative indicator for the project. Severity and number of Issues can give insight into whether the project will be completed within the originally budgeted cost, time, and scope. Analysis of each status parameter can provide an overall mosaic of the project. Various stakeholder groups have unique views regarding what is most important to them and this broad perspective allows the stakeholder to judge the project from their point of view. The PM must sort through this myriad of status items to decide how to proceed and what items in his tool kit to apply to which problem. There is an old adage about the one-armed paperhanger that would seem to apply to this point. Many things popping at once can be a real management challenge to resolve. The one final point about the tracking problem is that the data received must occur in a time frame appropriate to effectively use that data. Oftentimes, accounting data are 6 weeks old when available. So, keep in mind, having accurate data too late is worse than not having data at all.

28.4 INFORMATION DISTRIBUTION

Distributing project status is an integral part of the execution phase. The goal for this activity is to focus on effectively relaying status metrics to team members and stakeholders. Formal status reports should be a regular means of communication throughout the life cycle of the project. These communications should contain appropriate information summarizing detailed information regarding previous activity status, current activity status, accomplishments, upcoming activities, nontechnical and technical project issues, actions to mitigate critical issues, resource usage, and risk status. PMs not only have an obligation to effectively manage a project, they must also be an honest broker of information, whether that information is what the stakeholders want to see or not.

A real example of status tracking is provided by America Online (AOL), which is a project intensive organization and recognized for their mature approach to the activity. The following case summarizes an internal AOL initiative to integrate Primavera and the enterprise-reporting package, SAP. Woodward reports the following scenario (Woodward, 2006):

AOL completed a yearlong implementation of SAP in July 2005, installing SAP 4.7 with nine modules for 90 core users and 1100 limited users.

A key scope decision was to continue using Primavera TeamPlay as the project and portfolio management solution. Primavera was in use prior to the SAP implementation and PMs liked its graphical display and interactivity. 400 PMs and 300 other stakeholders were using SAP in addition to 4300 timesheet users. AOL considered replacing Primavera with SAP's PS Project Systems module, but decided the switch was unnecessary and would have been too disruptive.

AOL's average project activity involves approximately 1000 projects, with a duration varying from a few weeks to more than a year. 20–30 new projects are initiated every week. There are four project categories: maintenance, support, R&D, and development.

The SAP implementation was kicked off in June 2004 and went live in July 2005. It was championed by AOL's CFO. IBM acted as the integration partner. The project team consisted of 100 people at peak staffing, some offshore. Very few customizations were allowed, for example, by making minimal use of the WBS user-defined fields. Most of the issues were culture related. For example, AOL's previous ERP system did not use the WBS concept. Overall, the implementation was considered successful.

The decision to retain Primavera created a need to synchronize the WBSs in SAP and Primavera. This is simplified a little by only synchronizing the financially relevant WBS levels. SAP is considered the system of record and is used for labor costing and settlement. Primavera is used for project labor estimation and forecasting, time sheet entry and approval, resource assignment, and scheduling including baselines.

Impress Software facilitates the transfer of master data and the nightly synchronization. The 1000 projects are arranged in four groups of approximately 250 each, with each group synchronized simultaneously. This allows the nightly synchronization to be completed in approximately two hours. The nightly synchronization is supplemented by a weekly run to report inconsistencies.[*]

AOL lists a number of benefits of this solution:

- PMs focus on their schedules in Primavera TeamPlay
- Financial analysts focus on costs in SAP
- Data consistency between the two applications is very tight
- The mapping rules identify nonstandard or "rogue" WBSs
- There is a significant time savings by importing WBS structures rather than rekeying
- It provides the capability to capitalize eligible project costs at an appropriate level of detail
- It puts PMs and accountants on the same page.

AOL also reports some lessons learned:

- Use a preconfigured integration application
- Get project management and accounting involved early, together
- Book on-site installation support early; put it in the SOW
- Document business requirements before technical requirements are mapped out.

As this scenario points out, status of a project is more than just defining and executing the scope requirements. In this example, we see various best practices regarding the whole life development cycle process and related decisions. A key point to recognize here is that status reporting by itself is passive. Getting effective metrics in a timely manner provides valuable insights to the management process. An effective PM can then take that information and use it to *guide and influence* future project direction. Measurement of planned versus actual variances is the visible part of this process, but is not the goal itself.

28.5 CONTROL

As described above there are several aspects of a project that must be measured, guided, and maintained. Consequently, effective tracking of project status is a key management step to its success. Control of the project is exercised through formal and informal processes exercised by the PM, project team, and stakeholders. Also, the process of conducting reviews and monitoring reports is a common strategy to exert a degree of control over the project.

---

[*] Note: SAP is a complex enterprise level resource planning software utility.

We summarized the basic project control model components earlier. Let us look a little deeper into what is needed for each of these. Specifically, the core components that are required to support the project control process are the Project plan, the Issue Management process, the Configuration Management processes, and the ICC process. A brief explanation of the role each of these plays is offered below.

28.5.1 PROJECT PLAN

Planning is the process of using detailed research to chart the step-by-step approach to solve a problem, take advantage of an opportunity, or meet a competitive challenge. The project plan establishes the target goal, sets the project schedule, work requirements, and resources needed to complete the project. This plan is then used to guide the execution of WPs; it also establishes procedures for dealing with quality, risk, communication, and change management. The ongoing planning process must be flexible and open to modification as communications within the team provides new insights. The more that planning is based on good research, the better the chances of ultimate success. The full life cycle plan considers everything in an organization and its long-term achievement of a desired objective. In this view, planning may consist of an orchestrated collection of subplans for specific short-term projects.

28.5.2 ISSUE MANAGEMENT PROCESS

Issues are unplanned supporting events that need to be resolved. PMs cannot predict what or when a specific issue will occur, but will agree that they will. Issues occur as a result of management failures or environmental forces: natural, technological, skewed management values, deception, misconduct, unplanned business trends, and economic fluctuations. Within this unwieldy domain an Issues Management process provides a mechanism for organizing, maintaining, and tracking the resolution of these ad hoc items that would not be resolved in the normal course of activity or in a timely manner.

The Issues Management process entails establishing appropriate identification and control mechanisms and defined processes to aid the project team in identifying, addressing, and prioritizing ongoing problems and issues. In order for this process to be effective, documentation and communication is key. Detailed understanding regarding the issue needing resolution triggers the process. From this base point, the PM makes sure that the item has appropriate documentation and priority for resolution and a named person responsible for tracking that resolution. Without a system of this type various daily issues get lost in the maze of work and can lay unresolved until they become major roadblocks. Proper management of this class of activity is a critical success factor for the PM (see Chapter 24 for more discussion on this topic).

28.5.3 CONFIGURATION MANAGEMENT

This is a key processes used to manage and control change in documents created throughout the life of the project. Various work activities result in continual modification for most product and management documents and these changes are a normal part of every project. Failure to monitor this activity and provide appropriate management oversight is considered by many to be the number one operational control requirement.

The configuration management process involves ensuring that both the process and product artifacts are properly managed. For example, in a durable manufacturing project this process would ensure that a drawing for a part was the latest version. A common source of problem in product processing is having two individuals work on a document at the same time, but only the last person saving the document would have their changes reflected. In the project management domain the control process would ensure that all management documentation is the latest version. Most configuration management solutions require a formal controlled repository for managing the flow of these project artifacts. Some organizations call this activity *Version Control*. In the contemporary environment computer document management software is often used to manage the flow of documents. Prior to this, manual processes struggled with how to ensure changes to descriptive documentation created in the project would reflect true status.

28.5.4 INTEGRATED CHANGE CONTROL

The ICC process allows the PM, sponsor, and clients to be aware of and manage changes made to the project during its life cycle. Considering that many projects have scope increases of 2% a month, projects can easily get out of hand. Change control is the process that is designed to manage these requests and ensure than any changes made are appropriate to the goals of the project. Approval for these changes should occur external to the project and a project board representing management usually handles the approval process. Without this type of control, a project can become a runaway effort at an alarming rate. In addition, this activity adds the required higher-level management oversight to the project. Key focus points of change management are project scope requirements, schedule, budget, risk, and quality.

28.6 KA CONTROLS

Basic status tracking processes exist for each of the KAs. A brief summary of scope, schedule, cost, and quality-tracking issues is provided below.

28.6.1 SCOPE CONTROL

Scope changes can be stimulated by either external requests or from activities within the project team. In either case the scope control process is designed to manage approval of any changes, manage the actual changes as they occur, and to ensure that the change is properly integrated into the overall project deliverables. Execution of this process requires a formal change procedure to document the request and then map the process through which such requests flow. This includes project board approval, tasking to the team for execution, checking results to ensure proper results, and monitoring of the activity through its life cycle. The term for changes in a project is "scope creep." Certainly one of the major goals of scope control is to minimize this activity based on its potential adverse impact on project success.

28.6.2 SCHEDULE CONTROL

This is managed at the project level by the PM and should be proactive in nature as is the case with all control strategies. When variations from the planned schedule are measured they must be evaluated as to root cause, then appropriate action must be taken to solve the issue with the least negative impact on the project.

28.6.3 COST CONTROL

The *PMBOK® Guide* defines cost control as follows: Influencing the factors that create changes to the Project Budget Estimates to ensure that the changes are beneficial, determining that the project budget estimates have changed, and managing the actual changes when as they occur (Commonwealth of Virginia, 2006). Effective project cost control includes the following.

- Monitoring expenditures to detect variances from the project spending plan
- Executing a proper change control plan to prevent incorrect, inappropriate, or unauthorized changes from being made
- Recording authorized changes accurately in appropriate source documents.

Budget issues should be considered strategically, not just taking into account cost but the department's responsibilities concerning payment and the details thereof.

28.6.4 QUALITY CONTROL

Emphasis on quality management is a process performed throughout the project life cycle in all stages. From a project management model standpoint a quality management plan will be defined during the planning phase in order to define the process for measuring the attributes of work performed at each stage, as well as to provide control specifications and guidelines for team members to ensure the quality of work required for the deliverable and overall workflow of the project.

There are differing schools of thought regarding the proper control metrics and processes for a project. However, the control examples outlined above should be considered a requirement for keeping the project on task. Do recognize that every approach to standardization brings with it a potential negative if misunderstood or mismanaged. Some feel that the style of control outlined by the model described here is very inefficient and causes more harm than good. For example, Koskeia and Howell state that "control as described in the *PMBOK® Guide* causes problems" (Koskeia and Howell, 2000). They cite the following five counterproductive issues that arise from this style of control:

- First, the standard guide control process is focused on stimulating explanations rather than corrections. In other words, more time is spent explaining why there is a problem instead of resolving the problem. In this mode, team members can be distracted from current real project-related tasks in order to create historical accounts of the previous day's work. Unless this activity is related to finding root cause solutions, such actions fall under the banner of sunk costs that are not productive to pursue.

- Second, Koskeia and Howell suggest that managers will find ways to manipulate tasks and schedules to give the illusion of good performance. "In order to make cost variance positive, managers try to decrease the actual cost of work performed as much as possible" (Koskeia and Howell, 2000). This can have the effect of making the metrics appear normal, but in fact there will be underlying negative implications in the actual output. Manipulating output measurements for the simple goal of improving status appearance is obviously against the goal of good project management.

- Third, a tight control strategy can limit the ability to manage the team's resources by taking away flexibility of the PM. Low-level external monitoring can lead to increased pressure to stay within planned parameters at a low level. This, in turn, hinders project performance because resources cannot be moved from one activity to the next throughout the project as dictated by the skill of the PM. In this situation, there is the implied suggestion that project schedules force WP managers to focus only on their portion of the project and show no concern for the overall project. Therefore, managers may hoard idle resources that could be actively used to complete other aspects of the project.

- Fourth, "control may give the wrong interpretation of performance" (Koskeia and Howell, 2000). Continuous measurement and comparison of each activity does not yield control of a project. Large projects cannot be effectively measured using such yardsticks.

- Fifth, control does not always succeed in revising the plan after variances have been detected. Regardless of how a plan or schedule has been laid out for a project, sticking to that plan without analysis of current issues or planned deviation may result in poor judgment calls because management may be overly focused on staying within the bounds of an already failed plan.

Implementation of any control process must include the awareness of over control.

Projects are complex undertakings. On the one hand it is important to keep key stakeholders in the communications loop regarding honest status of the effort. Proper selection of status metrics is important in this activity. So long as the project is moving along with minimal plan versus actual variances the control process tends to be orderly. However, when these variances start to become larger and the goal becomes one of moving the project back into some baseline structure the corrective actions become more radical. At that point, the need for a more enlightened management approach becomes critical and the PM needs to have more flexibility in how to approach the problem and make "appropriate" trade-off decisions. In this situation, the future outcome will likely be some major deviation from the baseline plan. For a PM hamstrung with minimal decision authority, this situation becomes untenable.

28.7 PROJECT STATUS TRACKING CASE STUDY

A good example of project control comes from the F/A-18 Advanced Weapons Lab (AWL) software upgrade. Bowers writes the following description of that effort which consisted of an extraordinary number of project requirements representing new technology in the specified capability, aircraft systems, and configuration parameters (Bowers, 2002):

> "The biggest challenge, by far, was providing for efficient use of critical mission computer resources to allow for successful implementation of all the requirements," says Brestal. "An MC resource team was formed to devise and implement risk mitigation plans for each affected resource."
>
> Truly this project was large and complex agrees Capers Jones, a Top 5 judge. "The combination of low rates of delivered defects and high levels of customer satisfaction indicates this project was very well planned and managed." Jones cites the processes as a key to their success. "The project was produced by a SEI CMM Level 4 organization, and demonstrates the value of the higher CMM levels."

To achieve this quality goal, the AWL team performed the following:

- Improved organizational maturity levels to achieve repeatability.
- Used a formal maturity measurement process to assess organizational maturity and process area capability. Established priorities for improvement and methods to implement these improvements.
- Published, updated, and distributed a strategic plan that defines basic core beliefs, visions, and mission.
- Tested jointly with the Operational T&E Squadron throughout the verification phase of 15C. This gave them an early look at the product and gave the AWL earlier insight into operational problems in the product.
- Published an F/A-18 AWL Management and Systems Engineering Process Manual to systematically identify and apply leverage to areas of weakness and expand on what they do right.
- Maintained and improved its system-configuration review board process to obtain a very solid, well-thought-out, and adequately funded set of requirements.
- Improved on and used a comprehensive set of metrics. An example of the numerous metrics used is the indicator used to indicate software maturity level. At 0.12 software anomaly reports per test hour, the software is ready for operational test.

Embedded in this case is a clear indication that success was achieved by formulating formal control processes linked to tracking and control. Also, note that even a mature organization such as Boeing undertook custom activities for this project. Custom-designed processes and metrics were used to monitor the outcome and there is evidence of a well-developed project plan.

Lessons learned from previous projects offer the best design guidance for new projects within the organizational culture. Companies such as Raytheon and NASA report how the analysis of past less than desirable results provided them the guidance for project tailoring. Raytheon went through several iterations before they finally developed a standardized process that was in line with the company's culture. NASA learned from eight previous failed projects by identifying common issues in their project failures and using this knowledge to foster a culture of project success.

Although it is somewhat difficult to map these specific examples across other project types, the underlying thought processes illustrated here are consistent with the model described in this chapter. This general approach has been used successfully by other projects and there are case studies of failed projects that have not followed this model.

Collectively, knowledge gained from experiences across many organizations have led to the conclusion that in order to have successful project execution a PM should have a detailed and flexible plan of attack. Progress tracking and control should be primary components of day-to-day management activities, but the mechanics regarding how to handle these measured results is the key to success. Regardless of how structured the organization can make the project management process, they ultimately conclude that humans defy complete structure. One should not view project status measure as a cookbook exercise for that reason.

28.8 CONCLUSION

Successful control of project execution is complex for any project. For this reason, tracking progress and control-related activities are key elements to success. The methods discussed here represent a cross section of core monitor and control concepts used in the project management arena. Beyond the raw mechanics of this activity, it is important to recognize that the collection of a planned versus actual status metric neither hinders nor enhances project success. These simply describe passively what is occurring. It is up to the management process to take this data and use it properly in influencing an improved outcome. Also, this discussion has highlighted that the design of the tactical control process must be reviewed for each project to decide which combination of methods and degree of granularity will work best in that set of project characteristics.

The concept of management control flexibility is an important ingredient to effective control and one that many organizations do not adhere to. In these situations the goal of control standardization is so strong that management flexibility is designed out of the process. For average skilled PMs and less technical projects this may be a suitable strategy, but for complex projects being managed by a highly skilled manager, this overly structured approach can result in a less effective outcome. It is important to not tie the hands of the PM to the point where he has no degrees of freedom remaining in resolving the deviant issues.

REFERENCES

Commonwealth of Virginia, 2006. Project Management Guideline, Section 4 Project Execution and Control Phase, ITRM Guideline CPM 110-01. http://www.vita.virginia.gov/uploadedFiles/Library/cpmg-appendix-c.pdf (accessed December 15, 2008).

Koskeia, L. and G. Howell, 2000. Reforming Project Management: The Role of Planning, Executing, and Control. http://cic.vtt.fi/lean/singapore/Koskela&HowellFinal.pdf (accessed December 15, 2008).

Woodward, H, 2006. Integrating SAP and Primavera at AOL (January), *Project Management World Today*.

Bowers, P. 2002. The F/A-18 Advanced Weapons Lab Successfully Delivers a $120-Million Software Block Upgrade. *Crosstalk*, http://www.stsc.hill.af.mil/crosstalk/2002/01/fa18.html (accessed December 15, 2008).

29 Enterprise Reporting Using the Balanced Scorecard

29.1 INTRODUCTION

As mentioned multiple times in previous chapters the project lives within the enterprise and exists within that environment to further the goals of its host organization. Up to this point the monitoring and control discussions have isolated their focus on the project. This chapter will open up that view to show that project reporting should be synchronized with enterprise level goals. The contemporary approach to enterprise level project and organizational status reporting was developed in the early 1990s by Harvard professors Robert Kaplan and David Norton. They named this system the *balanced scorecard* or BSC as it is now frequently known. The breadth of this model deals with some of the weaknesses and gaps of previous status presentations. By expanding the traditional narrow project status focus, the BSC approach provides a clearer framework prescription as to what companies should measure in order to balance their overall goal perspective.

The BSC model scorecard is not only a measurement system but also a *management system* that enables organizations to clarify their vision and strategy and then translate these into action; it is designed to provide feedback regarding both the internal business processes and associated external outcomes in order to drive continuous improvement results. When fully deployed, the BSC transforms strategic planning from an academic exercise into the nerve center of an enterprise and projects become key working elements for this transformation process.

BSC methodology contains a mechanism to translate strategy into action. Working through the elements and layers of the process enables management to define those key perspectives that will drive the business to success, as well as to define how to measure them. The BSC helps organizations align multiple strategies from various business units into the organizational operational strategy by linking their deliverables to those key goal elements that management has selected to drive the business. BSC outputs provide a clear communication of the company strategy and how it is supported by the commitment to defined objectives from various divisions and functional units of the organization.

Kaplan and Norton describe the BSC as follows:

> The balanced scorecard retains traditional financial measures. But financial measures tell the story of past events, an adequate story for industrial age companies for which investments in long-term capabilities and customer relationships were not critical for success. These financial measures are inadequate, however, for guiding and evaluating the journey that information age companies must make to create future value through investment in customers, suppliers, employees, processes, technology, and innovation (Kaplan and Norton, 1996).

29.2   SCORECARD IMPLEMENTATION

Implementing the scorecard typically includes four processes:

1. Translating the organizational vision into operational goals
2. Communicating the vision and linking it to individual unit performance
3. Business planning
4. Feedback and learning and adjusting the strategy accordingly.

BSC methodology builds on various key concepts of previous management ideas such as organizational maturity, quality management, employee empowerment, and so on. Simply stated, it is designed to move the organization toward formally defined goals and obtaining employee buy-in as part of the process.

29.2.1   COMMUNICATING STRATEGIC OBJECTIVES

It is common for company strategies to not be well communicated to all levels of the organization. This includes not just the vague motherhood mission statements, but more understandable details outlining how a particular strategy will result in reaching the vision of the company in the years to come. The underlying mechanics in developing the scorecard provides a mechanism for the organization to define strategy in terms of key business objectives with agreed upon targets for reaching such objectives in a set time period. These objectives are then communicated to all persons involved.

High-level scorecard objectives are cascaded from the enterprise level into divisional, business unit, and work group objectives. This process involves heavy employee input to derive the lower-level specifics. The translation process of high-level goals into operational objectives is designed to make all elements of the organization aware of the goals and then focus them on achieving their particular contribution to those goals, thus the BSC becomes a communication mechanism to align the work force to formal strategies and goals.

29.2.2   COMMUNICATING STRATEGY

Since strategy development and communication is a complicated issue, management can use the BSC methodology to map key drivers of the business and establish cause-and-effect relationships between those drivers and the desired outcomes. Once the cause-and-effect relationship is established, one can identify correct means to measure the lower-level business drivers. When implemented, the BSC contains understandable organizational goals and objectives that are then visible and communicated to the entire organization in a way that they can understand and participate in. Measurement of local drivers then provides management with a comprehensive picture regarding how various business entities are progressing toward desired objectives. Proper manipulation of these targets should then result in an organization moving in the direction designed by its senior management.

29.2.3   ASSIGNING RESPONSIBILITY

Every key business driver and subparameter used to measure status must have a person assigned to it who has responsibility for the driver performance. In this fashion, the BSC formally delegates the management of achieving the desired performance to operational levels who understand their internal processes best.

29.2.4   ALIGNING STRATEGY

Implementation of the BSC requires consensus building throughout the organization and this involves more than having senior management edict solutions. Also, it is incumbent on senior management to develop a coherent and visible strategy that can be understood at lower levels. Eventually, the WHAT of the model will have to be turned into the HOWs associated with the more abstract goals. At the lowest levels, the HOWs will have to be measured in some form—quantitatively or qualitatively. As all of these elements are integrated, it is important that the measured metrics represent desired outcomes. In theory, organizations and their employees will tend to move in these measured directions, so an improper metric can create the wrong result.

29.2.5 PROCESS INTEGRATION

The overarching goal of the BSC is to assist various organizational units in implementing management processes that support the defined enterprise strategy and align their internal processes with that strategy. Processes such as planning, forecasting, budgeting, and performance management must be linked and supported by the various monitoring and control processes. At the operational level, the resource allocation process will tend to follow the elements that are deriving the highest enterprise goal achievement. Gone are the days when the manager with the best presentation skills could claim the biggest share of the budget. Now, the BSC will chart actual performance versus goals. This linkage is not a simple numerical scale, but does allow management to see a broader overall picture of the business unit or project.

29.3 BSC IMPLEMENTATION PRINCIPLES

There are several design principles to follow in implementing BSC. The summary list below offers general design concepts to follow:

1. Gain senior management support.
2. Gain consensus on terminology and notation.
3. Find your internal sponsors and champion.
4. Define carefully the assessment framework.
5. Measure what matters—the organization will follow these factors.
6. Communicate and agree on objectives and targets.
7. Align local activities and initiatives to objectives and targets.
8. Implement supporting processes for measurement and communication.
9. Assess suitability of outputs periodically.
10. Focus on approaches to improve the overall process.

29.3.1 BEST PRACTICES

There are many best practices mechanics that need to be understood in creating this monitoring system. These can be summarized as follows:

1. Measure actual performance of all strategic goals.
2. Maintain a balanced set of measures.
3. Hold employees personally accountable for results.
4. Develop solid baseline data.
5. Match resource allocations to goals and objectives.

First, it is important to recognize that establishment of an enterprise level goal is important and not just a statement. Given this view, it is equally important to measure status of that goal as part of the operational activity.

Second, the "B" in BSC is important to keep in perspective. Balance is a concept that attempts to advance the organization across a broad front and not just a few isolated segments.

Third, the linkage of lower-level management to higher-level translated goals is designed to ensure that the linkage is maintained and pursued. Having operational level management focused on enterprise level goals is one of the main value elements of the BSC. This concept applies equally well at the project level as it gives the PM a perspective regarding how his local scope can impact the higher-level goal set.

Fourth, since the BSC model is strategically oriented, it is important to track status over a longer period of time than is typical for most control-oriented systems. One approach for this is to compare not only the progress of the enterprise, but also quite likely how this competitively measures against other similar organizations. External benchmarking could be oriented toward quality objectives, profit per employee, or a myriad of other industry-defined performance metrics.

Fifth, and finally, one of the key reasons for a system of this type is to aid management in accomplishment of their objectives. An effective BSC should provide insights into where resource allocation is most effective and requires careful attention to this aspect of management. This would involve removing resources from less profitable targets and reassigning them to higher producers.

In order for these best practices to work, one can see that management involvement in the process is a requirement, since their formal statement of goals is the starting point for the rest of the process. The defined goals must be clear and understandable to the layers below. But, goal definition alone is not a sufficient condition for success. Each of the five items above contributes to success and omission of any one will weaken the overall chain. These subprocesses must serve as the centerpiece of the overall management process for the enterprise.

29.3.2 Barriers to Success

Any integrated process of this sophistication will be a management challenge to implement successfully. All of the layered elements must be working in concert in order to achieve the desired outcome. Beyond this, there are several other more cultural items that can constrain the process. Some of the more notable of these are as follows:

- Inability of management to reach consensus on goals or related measures
- Insufficient involvement of end users in creating the measurement system
- Inability to break old habits, inflexible processes, ineffective legacy systems, and a static culture are all obstacles to successful measurement and implementation
- Fear or unwillingness of individuals to change
- Measuring what is easy or known, rather than identifying what needs to be measured.

There are no surprises in this list and it could be used for any organizational change objective. However, the BSC does offer one carrot that provides a potential advantage. That is, management reaction to the measured output of this system can easily be linked to a reward structure that is understood by all. If the culture of the organization begins to recognize this, there would be an added motivation to follow.

29.4 BSC MODEL

The most noticeable characteristic of the BSC model is its breadth of perspective. Traditional goal or control systems tended to be much less integrated and less broad in their view. In regard to the higher-level vision and goal drivers, there is a much stronger view of making these something more than platitudes of good performance. They are couched in organizational strategy terms from which the lower levels can link cause-and-effect drivers. It would be much easier to just state "jump tall buildings with a single bound" rather than being more prescriptive as to how one might do that, or how high is the building. In addition to these more clearly defined objectives, there is a corresponding need for an equally clear measure that reflects status of the objective or associated driver. Finally, there is recognition in the BSC model that organizational performance is more than a financial metric. The sections below will show more regarding how these pieces come together.

Financial performance measures alone can lead organizations to make short-term decisions at the expense of long-term value. Many organizations today are questioned by their quarterly stock market value approach to performance measurement. This short-term reactive goal tends to force tactical decision making and avoid addressing real-value creation mechanisms. Tactical project initiatives will be chosen over those with significant competitive value that might take 5 years to pay back, but have long investment cycles. Also, financial measures are often "lag indicators" meaning that they reflect outcomes of actions previously taken. One might argue that all they show is how good you were, not how good you are going to be. The BSC enhances the financial view by using lead indicators that tend to offer more of a prediction of future economic performance. The trick in this is obviously to identify such indicators. For example, if a strategy is to increase market share in a particular region, the challenge would be to find an indicator that reflected progress in that direction. Recognize that a growth goal typically requires an investment that will shrink current financial performance metrics.

All the measures in the BSC model are designed to serve as translations of the organization's strategy. This process allows the organization to translate its vision and strategies by providing a framework, one that reflects the strategy through the objectives and measures chosen. Rather than focusing singularly on tactical financial measures that provide little in the way of guidance for long-term employee decision making, the BSC uses measurement of various goal drivers as a link to key elements related to strategy. The BSC contains the following four focus perspectives:

- Learning and growth
- Business process
- Customer
- Financial.

Figure 29.1 illustrates this structure schematically.

The Financial and Internal Business Process perspective would be considered a more traditional view, but a Customer perspective is obviously a newer view and challenge. Also, the recognition that one of the objectives of an organization is to improve Learning and Growth of the employee has not been very visible in past measurement systems, especially at the enterprise level. Each of these macrolevel objectives contains the following four definitional components:

- *Objectives:* A goal-oriented statement that fits the view.
- *Measures:* Drivers that can be used to show status of the objective.

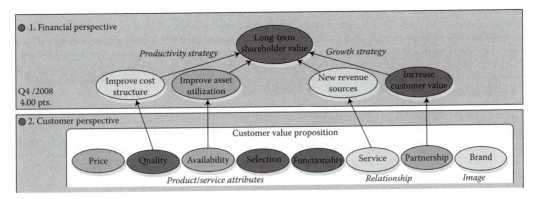

FIGURE 29.1 BSC structure. (From QPR Software, 2008. Balanced scorecard introduction. http://www. qpr.com/solutions/Balanced_Scorecard/?gclid=CJmivTr1IsCFRYNgQodEXB4VQ [accessed December 15, 2008]. With permission.)

- *Targets:* Time-phased values for achievement of a particular measure.
- *Initiatives:* Specific organizational actions that are designed to support accomplishment of the objective.

In each of these high-level perspective views, the challenge is to define methods by which the organization can improve its position through those mechanisms. Basically, the underlying theme of BSC says that the customer is a key part of success and in order to satisfy that customer the organization needs excellent processes and a highly skilled work force. The financial metrics are a result showing how well the organization has historically performed. Lastly, all of these component parts need to be integrated into one view.

29.4.1 FINANCIAL PERSPECTIVE

This perspective is designed to help the organization or project learn what they should do in order to appear to stakeholders that they are financially sound and successful in their current direction. Data in this category will always be a performance measurement priority and managers understand that they need to produce profitable results. In fact, quite often there is currently more than enough handling and processing of financial data. As organizations have invested in more sophisticated accounting-type systems, the level of financially oriented data detail appears to have grown without much concern over value added in the process. Over emphasis on historical financial data leads to an "unbalanced" perspective situation with regard to the other views. A desirable modification would be to add supplementary financial-related data, such as risk assessment and cost–benefit data. Simply measuring the consumption of resources is not something that easily translates to corrective management decisions. Comparison of planned versus actual resource consumption is a common performance metric that can be one of the more misleading metrics that one can create. Drivers selected in this category should strive to measure cost–benefit-type status.

29.4.2 INTERNAL BUSINESS PROCESS PERSPECTIVE

The Business Process perspective relates to underlying business processes that support output generation. Metrics based on this perspective help to evaluate how well the business is running and whether its products and services are being produced compared to defined objectives. Process-oriented metrics have to be carefully designed by those who know the processes most intimately. In addition to the strategic management processes, there are two other categories of business processes that should be identified:

- Mission-oriented processes
- Support processes.

Mission-oriented processes are the key processes that produce the competitive position of the organization, whereas support processes represent the more repetitive "keep the lights on"-type activities. Support processes are generally easier to measure and benchmark using generic metrics. The following list of questions needs to be answered in evaluating this category:

- What products or services will your customers value in the future?
- What processes best deliver the outcomes desired by the customers?
- Looking into the future, what are the new business processes that you must excel at?
- What will be valued in the future, and how will innovation deliver future values?

29.4.3 LEARNING AND GROWTH PERSPECTIVE

The Learning and Growth perspective includes employee training objectives and corporate cultural attitudes related to both individual and overall self-improvement. In knowledge-oriented organizations, people are the critical resource and in most organizations the HR is a key factor in their success. The rate of technological change in the organization makes it necessary for knowledge workers to be in a continuous learning mode. Many organizations have a difficult time finding the required levels of skilled workers, yet at the same time they allocate little to upgrade the current workforce. Kaplan and Norton emphasize that learning is more than training; it also includes things like mentors and tutors within the organization, as well as creating a culture of teamwork (Kaplan and Norton, 1996). Project teams in particular need an environment where all participants are working for the good of the team and are willing to help in whatever way they can. Some measure of coaching and mentoring is needed to track organizational direction of this BSC component.

29.4.4 CUSTOMER PERSPECTIVE

The Customer Perspective relates to the capability of an organization to provide quality goods and services, effective delivery, and overall customer satisfaction. Recent management philosophy has shown an increasing realization regarding the importance of customer focus and customer satisfaction in any business. These attributes are leading indicators since lack of customer satisfaction will eventually lead them to find other suppliers that will meet their needs. So declining trends from this perspective is a leading indicator of future sales decline, even though the current financial picture may look good. In developing metrics for customer satisfaction, attention should be paid to defining the kinds of customers and the kinds of processes for which we are providing a product or service to those groups. The following are the questions that need to be answered in this perspective:

- Who is your customer?
- What services or products do they expect from you?
- How do you listen to and learn from your customer?
- How do you retain and acquire new customers?
- How do you meet your customer's needs?
- How do you measure customer satisfaction and dissatisfaction?

29.5 BSC AS A STRATEGIC MANAGEMENT SYSTEM

For many organizations the BSC has evolved from a measurement tool into what Kaplan and Norton have described as a "Strategic Management System" (Kaplan and Norton, 1996). While the original intent of the model was to balance historical financial numbers with the drivers of future value for the firm, as more and more organizations experimented with the concept they found it to be a critical tool in aligning short-term actions with their strategy. Used in this way the scorecard alleviates many of the following issues of defining effective strategy options.

Figure 29.1 summarizes the reasons why organizations fail to effectively execute their strategy. Note that each of the shortcomings mentioned are included as part of the BSC model design and intent (Figure 29.2).

29.5.1 OVERCOMING THE VISION BARRIER

The scorecard is ideally created through a shared understanding and translation of the organization's strategy into objectives, measures, targets, and initiatives in each of the four scorecard components. The translation process of vision and strategy forces the executive team to specifically determine what is meant by often vague and nebulous terms contained in their vision and strategy statements. For example, terms such as "best in class," "superior service," and "targeted customers."

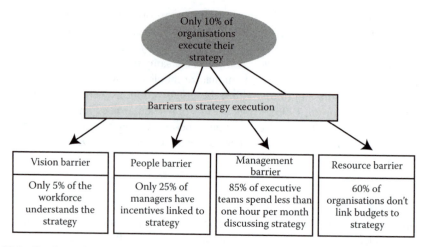

FIGURE 29.2   Barriers to implementing strategy (From QPR Software, 2009. Balanced scorecard introduction, With permission.)

Through the process of developing the scorecard model, it will be necessary to determine that "superior service" means 95% on-time delivery to customers. From this specification, all employees can now focus their energies and day-to-day activities toward the now clear goal of on-time delivery, rather than wondering and debating the definition of the higher-level abstract term. Translation of the model in this fashion creates a framework for turning strategy into actionable items. More importantly, this creates a new language of measurement that serves to guide all employees' actions toward the achievement of the stated direction.

29.5.2 OVERCOMING THE PEOPLE BARRIER

To successfully implement any strategy, it must be understood and acted upon by every level of the enterprise. Cascading the scorecard means driving it down into the organization and giving all employees the opportunity to demonstrate how their day-to-day activities contribute to the company's strategy. All organizational levels distinguish their value creation activities by developing scorecard measures that link to the high-level corporate objectives. The process of cascading creates a linkage from the employee on the shop floor back to the executive boardroom. Some organizations have taken cascading all the way down to the individual level with employees developing personal BSCs that define the contribution they will make to their team in helping it achieve overall objectives. Rather than linking incentives and rewards to the achievement of short-term financial targets, managers now have the opportunity to tie their teams, department, or business unit's rewards directly to the areas in which they exert influence. All employees can now focus on the performance drivers of future economic value along with the decisions and actions necessary to achieve those outcomes.

29.5.3 OVERCOMING THE RESOURCE BARRIER

Developing the BSC model provides an excellent opportunity to tie important processes together. Note that each element defined contains linkages for objectives, measures, and targets. From this core definition there is also the requirement to put in place initiatives or action plans for each along with consideration of timing. If long-term stretch targets are defined for a particular measure, then the initiatives can be used to produce incremental improvement steps along the path to their achievement. The human and financial resources necessary to achieve scorecard targets form the basis for the development of the annual budgeting process. No longer will departments and business units submit budget requests that simply take last year's amount and add an arbitrary 5%. Instead, the necessary costs (and profits) associated with defined targets can be more clearly articulated in the submission documents. This process enhances executive learning about the organizations capabilities as they are now forced, assuming that resources are finite, to make tough choices and trade-offs regarding which initiatives to fund and which to defer. Management of the BSC process also affords a great opportunity to critically examine and compare the current myriad initiatives taking place in an organization. Fundamentally, this class of decision is the primary operational role of senior management.

29.5.4   OVERCOMING THE MANAGEMENT BARRIER

Many management teams spend their precious time together discussing historical variances and looking for ways to correct these "defects." The BSC parameters provide a method to move away from this decision paradigm into a new model in which measured results become a starting point for review, questioning, and learning about strategy variances. In order to accomplish this role, the scorecard measures must link together to develop the full story related to their strategy. Instead of translating defects into root cause understanding the variances now link more directly to strategy deviations that can be probed more directly.

29.6   BSC AS A COMMUNICATION TOOL

When implemented properly the BSC is considered to be an effective tool in communicating organizational goals and monitoring the ongoing status of those goals. The model provides an elegant method of describing organizational strategy by precise statements related to organizational objectives. This process aids in goal communication and brings them alive through defined performance measures linked to these goals. Sharing scorecard results throughout the organization provides employees with the opportunity to understand the objectives and this stimulates discussion regarding the assumptions underlying the strategy. As the measurement process unfolds the organization has an improved capability to learn from the reported results. From this base, a more coherent dialogue can take place related to future modifications required to bring the overall process back in line. A clearer understanding of the firm's strategies can unlock many hidden organizational capacities as employees perhaps for the first time will be able to comprehend where the organization is heading and how they can personally contribute. This same view is afforded to the PM who is attempting to produce a new product or process as part of a planned initiative spawned from the scorecard parameters.

29.7   BALANCING THE BSC COMPONENTS

The BSC was originally conceived to overcome deficiencies related to a more isolated reliance on financial measures of performance. By design the model attempts to balance the reporting approach through the use of a broader set of objectives and linked drivers that are believed to produce improved performance. The concept of balance remains a central feature of the model. This is reflected in the following two ways:

- *Balance between internal and external constituents of the organization:* Shareholders and Customers represent the external components represented in the scorecard, while Employees and Internal Business Processes represent internal components. The scorecard attempts to recognize these sometimes-diverse goal structures by ensuing that both viewpoints are dealt with in the model description.
- *Balance between lag and lead indicators of performance:* Lag indicators generally represent past performance, whereas lead indicators are forward looking. Typical examples of lag

measures include customer satisfaction or current revenue. While these measures are usually quite objective and accessible they tend to lack any predictive power. Lead indicators on the other hand are the performance drivers that lead to the achievement future results. These measures include processes and their activities. For instance, on-time delivery might represent a leading indicator for future customer satisfaction. While these measures are normally thought to be predictive in nature, the correlations may prove subjective, and the data may be difficult to gather. Lag indicators without leading measures do not communicate how we are going to achieve our targets. Conversely, leading indicators without lag measures may demonstrate short-term improvements but do not indicate whether these improvements have led to improved results for customers and ultimately shareholders. Because of these interrelated characteristics a scorecard should include a mix of both type indicators.

29.8 ADVANTAGES AND DISADVANTAGES OF BSC

Any complex organization process such as the BSC will have both advantages and disadvantages associated with it.

29.8.1 ADVANTAGES

Taking the four BSC diverse reporting perspectives into one integrated whole helps to ensure that senior management will take a more balanced view regarding organizational performance. Within this broad perspective, the model supports the management of short, medium, and long-term views in an ongoing and cohesive manner. Previous discussion of the model mechanics has described how the higher-level objectives become linked through the layers of the organization. From this integrated structure and its associated drivers, the enterprise reporting system is much more likely to be focused on the specifics necessary to stay competitive in the long term and realize value for the stakeholders. In addition to these basic advantages, the following items are recognized as operational positives:

- Parameter measurement provides users with a rapid exception alerts indicator
- Access to the integrated data repository provides the structural basis for supplying drill-down details about various initiatives and how their performance impacts higher-level objectives
- Dependency paths are inherent in the model and these help show cause-and-effect relationships
- Graphical reporting of measurements and relationships are supported by the model architecture
- A centralized BSC model structure facilitates control over data access for security purposes
- Company vision and strategy is defined to action and shared with employees
- Organizational learning is supported by testing defined drivers relations against actual results
- Model metrics and actual measurement will tend to influence the employees to move in the defined direction
- Formal recognition of customers and employees in the goal structure has great positive potential in future organizational direction. The four-perspective view adds a measure of balance to the overall goal structure
- Top-level strategy and middle management level actions are clearly connected to the goal structure and appropriately focused
- The organization's performance reporting system (and the organization itself) is more likely to be focusing on the targets necessary to stay competitive and thus realize value for its stakeholders
- Analysis of scorecard measurements aligns key performance measures with high-level strategy at all levels of organization

- A BSC provides management with a comprehensive picture of business goals and strategies at all levels of an organization.

29.8.2 DISADVANTAGES

Implementation of this model is not a quick fix to some organizational problem. Successful operation requires careful thought and involvement by senior management. There is no canned template that can be used to hurry this process through design. The linked relationships described clearly indicate that the layering process will require time to wade through. Getting each level in the organization to understand the concept and reorganize their local processes in this direction make the implementation process complex. As with all high-level management concepts this is not a magic wand to be waved over the organization for instant results. There is formal training required to understand the overall concept and related methods of defining that various components. Use of outside consulting resources to accomplish this can make installation very costly, but that may be necessary.

The main driver for this approach has to come from very high levels in the organization. Management focus on structuring this new process may take attention away from operational issues, which could cause short-term problems. Some organizations may feel that this level of complexity is not worthy of the time commitment. *A halfhearted installation of BSC may well be worse than none at all.*

29.9 FUTURE OF THE BSC

BSC methodology continues to evolve into more organizations. In addition, the support tools that organizations need to create model solutions are becoming easier to use. The Coote Harvard website states that a BSC fundamentally complements financial measurement of past performance with measures of drivers of future performance, and thus enables effective decisions by the management of the organization (Coote Harvard, 2008). Regardless of a specific implementation strategy, the advent of BSCs has had a considerable impact on management reporting practices. Many organizations take the model directly and implement it as described here, while others modify the design to fit their vision. In both cases, the structures have similarity to the original Kaplan–Norton model description. What this broader recognition has done for organizations is to draw attention to the role that communicated formal objectives that are linked to measured initiatives plays in successful organizations. This recognition alone means that a multiparameter goal-oriented model similar to the BSC will be around for a long time to come.

29.10 CONCLUSION

In many ways the BSC model represents a management revolution. Not because it brings some brand new management theory to the organization, but more for the fact that it brings an understandable implementation of something that was not well understood prior. Implementation of this model in many organizations over the past few years has greatly impacted the management culture of those organizations.

Status reporting using a BSC model structure has proven to be an excellent means for an organization to assess its true performance and determine where it is headed in the future. The broad goal perspective of the BSC expands the focus away from historical financial measures and offers a truer status view of the organization and its various initiatives. Recognizing the customer perspective provides strong action drivers in directions that have long-term positive value for the organization that might well have been hidden by traditional financial metrics. Likewise, business process performance serves as a supporting role in various aspects of the operational performance.

The employee learning and growth perspective of BSC focuses on the most critical aspect of most organizations. That is, its HRs. Managing employee growth has great potential for the organization in many ways that are difficult to quantify; however, it is easy to observe an organization where the employee is well trained for their job and equally easy to see those that are not.

The BSC model offers a clear method to recognize the role that customers, processes, employees, and finance play in overall success. Guiding the organization toward selected targets is a dynamic process requiring both specification and measurement. Projects are common methods to achieve improved outputs and therefore they represent the bottom of the action chain in this structure. Every PM should be aware of his linkage to organizational objectives and how his project contributes to those objectives. Losing this perspective is a recipe for project cancellation.

REFERENCES

Coote Harvard, 2008. Introduction to balanced scorecard. http://www.cooteharvard.co.uk/introducing_the_balanced_scorecard.php (accessed December 15, 2008).

Kaplan, R.S. and D.P. Norton, 1996. *The Balanced Scorecard: Translating Strategy into Action*, Boston: Harvard Business School Press.

QPR Software, 2008. Balanced scorecard introduction. http://www.qpr.com/Solutions/Balanced_Scorecard/?gclid=CJjmivTr1IsCFRYNgQodEXB4VQ (accessed December 15, 2008).

QPR Software, 2009. Balanced scorecard introduction.

Part VIII

Closing the Project

LEARNING OBJECTIVES

Upon completion of this chapter the reader will understand the value of the following closing activities:

1. Obtain final acceptance of deliverables through formal approval procedures from appropriate stakeholders and customers
2. Document lessons learned by surveying project team members and other relevant stakeholders to use for the benefit of current and future projects
3. Facilitate administrative and financial closure in accordance with the project plan in order to comply with organization and stakeholder requirements
4. Preserve essential project records and required tools by archiving them for future use to comply with legal and other requirements
5. Release project resources by following appropriate organizational procedures in order to optimize resource utilization.

Source: Initiation Goals adapted from the PMI, *Handbook of Accreditation of Degree Programs in Project Management*, Project Management Institute, Newtown Square, PA, 2007, pp. 17–20. Permission granted by PMI for use in this context.

30 The Closing Process

The *PMBOK® Guide* defines the Close Project process as a formal set of activities at either project termination or completion of a major stage (PMI, 2008, p. 99). For multiphase projects, the Close Project process closes out that portion of the project scope and associated activities applicable to the phase. The closing process should address activities related to all knowledge areas. "This process also establishes the procedures to coordinate activities needed to verify and document the project deliverables, to coordinate and interact to formalize acceptance of those deliverables by the customer or sponsor, and to investigate and document the reasons for actions taken if a project is terminated before completion" (PMI, 2008, p. 99). These activities include actions related to project administrative and contractual issues.

Administrative closure procedure: This procedure details all the activities, interactions, and related roles and responsibilities of the project team members and other stakeholders involved in executing the administrative closure procedure for the project (PMI, 2008, p. 64).

Contract closure procedure: This procedure includes all activities and interactions needed to settle and close any contract agreement established for the project, as well as define those related activities supporting the formal administrative closure of the project. This procedure involves both product verification (all work completed correctly and satisfactorily) and administrative closure (updating of contract records to reflect final results and archiving that information for future use) (PMI, 2008, p. 344).

Seningen describes the closing process as (Seningen, 2008):

> Sharing knowledge in a systematic format, documenting lessons-learned, and ensuring frequent communication will maximize project success factors.

Whether managing one project or multiple projects, the value of lessons learned sharing and communication should not be undervalued. It can be the difference between total project success and missing key issues. When managing a team of PMs who work on similar and ongoing projects, the value of lessons learned documentation and communication should be evident, although polls would probably show that this discipline is limited in practice (Seningen, 2008).

30.1 PROJECT IMPLEMENTATION REVIEW

When conducting a Project Implementation Review (PIR), updating the lessons learned documentation, as well as holding a lessons learned review meeting, is critical for at least the following:

1. Not repeating the same mistakes
2. Improving the probability of balancing the triple constraint so as to not have cost or schedule overruns on future projects.

Some of the mistakes and problems that typically occur and that could be described as part of lessons learned documentation occur in these are as follows:

- Vendor management
- Equipment

- Project approval
- Budget approval
- Communication (lateral and vertical)
- Testing
- Technical support
- Training, and so on.

All projects commence with an optimistic forecast for their completion. Unfortunately, experience suggests that the actual outcome is often different. In fact, these variances can have implications for future project outcome. Some may say that a project team just does the best they can under uncertain conditions, so why worry about documenting these results after the project is completed. In fact, there are multiple reasons why a formal closing process is not only prudent, but a required activity. The sections below will explore this point in greater detail.

Cleland and Ireland (2002, p. 435) describe that all projects end, even though it may not be because the planned deliverables have been achieved. Whatever may be the final state of a project, all projects need to be closed in a formal fashion as outlined in this chapter. Cleland (2004, p. 503) describes two kinds of a project termination:

"Positive termination occurs when the project comes to closure with a positive outcome and an upbeat relationship with the customer and stakeholders." Negative termination occurs when the project is terminated but with less-than-positive sentiments between the project and client organizations.

Meredith and Mantel (2003, pp. 644–648) offer a broader view of reasons for project termination and an adaptation of that list follows:

- The project is terminated when the required deliverables have been completed. Planned schedule and budget may or may not be as planned.
- Termination by extinction—original reason for the project has changed.
- Project is terminated because the time, cost, and functionality balance is no longer deemed worthy of continuation. As an example, although a new product is developed, it does not show any remarkable result.
- Affiliation between two companies or organizational groups makes the project redundant. One of the projects would be terminated as a result.

Cleland (2004) describes five specific objectives for a formal closing process:

1. When the project is over, what do you expect to see changed in the organization?
2. What documentation and physical materials do you expect to have at your disposal after the project is disbanded?
3. What form do you expect these preferred changes to take?
4. What communication media will be used for sharing lessons learned data (e-mail, telephone, instant messaging, etc.)?
5. What specific protocol is to be followed to communicate closure results to the organization?

30.1.1 NORMAL PROJECT TERMINATION

In this normal termination situation the project produces a successful outcome, and a new product or process is now installed and working. For this case the lessons learned format and protocol should follow a standard set of activities. One of the basic management activities required is to formally transfer the project dedicated property, equipment, human resources, and material to appropriate organizational entities. The deliverables from the project now are operational entities in the organization, and so a new support infrastructure is required. Meredith and Mantel (2003, p. 647) provide a scenario example for this situation:

> The project team that installed a new piece of software, instructed the client in its operation and maintenance, and then departed, probably left only minor problems behind it, problems familiar to experienced managers. If the installation was an entire flexible manufacturing system, however, or a minicomputer complete with multiple terminals and many different pieces of software, then the complexities of integration are apt to be much more severe.

As a general rule, the more mature global organizational processes are, the more likely that project closing activities will be handled in a consistent manner. Organizations that do not formally recognize this activity would be categorized as process immature.

30.2   ABNORMAL TERMINATION

When projects are terminated for reasons other than successful completion, the closing process will tend to be more customized in nature. Contractual relationships will need to be carefully negotiated and subclauses related to early termination will have to be reviewed. This will involve more than just checking payment and delivery status for a completed contract. If termination occurred because of a management action, those conditions will have to be addressed. In these early termination situations, the skill levels required may be higher than simply closing out accounts.

30.3   TERMINATION MODEL

Figure 30.1 outlines a WBS structure of activities structured around project, scope, contract, and site-related activities required to close the project (Taylor, 2001).

When a project is finished or terminated, some organizations employ specially trained managers and technicians to close the project. One of the key activities for this team is to review the status of work packages by budget, schedule, and technical performance parameter. Cleland and Ireland (2002, pp. 443–444) summarize their view of these activities as follows:

1. Ensure that all project deliverable end products are properly transferred to the new asset owners, along with appropriate standard records
2. Review that all contractual requirements have been met and properly record any variations along with their resolution conditions
3. Define the list of stakeholders related to new environment (i.e., product or process support)
4. Help the project team members find other project assignments
5. Prepare "lessons learned" to assist future project teams in assessing similar situations
6. Analyze the weakness and strengths of the project, and explain how the project team dealt with the problems and what is necessary for future project teams to avoid negative situations and utilize the positive items identified.

30.4   PROJECT TERMINATION CHECKLIST

One of the typical organizational approaches to implementing a formal closing process is to produce a checklist that the project team must execute for official project ending. Horine (2005, pp. 282–284) outlines the following 13 important topic areas to be covered in the project checklist:

1. *Gain client acceptance:* This stage has to be accomplished before the team attempts to close the project. In this activity, the most important result is that the client formally verifies and accepts the project deliverable and that this event is formally documented.

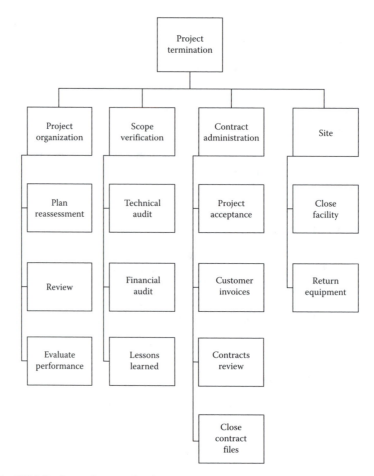

FIGURE 30.1   WBS for the project termination.

2. *Transition deliverables to owner:* In this activity, the team formally hands off the project deliverables to the new owner. This includes not only possession of the item, but also ability to support the item long term.

3. *Close out contract obligations:* The project team will coordinate with procurement personnel to document status of all contractual relationships. This should include not only the status, but resolution plans for all variances to the agreements.

4. *Capture lessons learned:* Documenting the project team's experience-related activity enables future projects to avoid some mistakes and challenges faced by the existing team. Horine explains that "lessons learned should be documented throughout the project lifecycle and include both positive and negative aspects of the project."

5. *Update organization's central information repository:* This activity involves documenting project records and deliverables as a formal archive for the organization. Horine states that "it is a powerful way to reduce learning curves and gain efficiency on future projects."

6. *Document final project financials:* This activity involves documenting the final project financial reports such as a budget status summary and variance analysis.

7. *Close various accounts and charge codes:* This activity involves the process of closing team member accounts and codes related to financials, infrastructure, and security.

8. *Update resource schedules:* Work to ensure that team members have appropriate job opportunities following the closure.

9. *Conduct performance evaluations:* The PM needs to ensure that appropriate performance feedback is performed and documented for all team members.
10. *Update team resumes:* The team members should update their resumes to reflect the new project activity.
11. *Market project accomplishments:* Formally recognize team member accomplishments and overall project positive experiences.
12. *Review project performance with clients:* Horine says that "the best testament to evaluate client satisfaction is to see whether the sponsoring or [user] individuals (organization) will officially endorse team's work." This is a process to know whether the team really achieves the desired goal.
13. *Celebrate:* From at least a morale standpoint, it is important to find something to celebrate at the conclusion of a project. Almost all projects have had some good experience, and reminiscing about these is a good way to leave the project team feeling about their experience. These experiences have long-lasting motivational value to the individuals involved and help build a positive culture.

The steps above have value not only for the termination phase but can also be of intelligence value as each life cycle stage is completed.

30.5 PROJECT TEAM AND CLIENT RELATIONSHIP

Basically, the role of the project team is to gain client acceptance for the items delivered. In many cases this will not be exactly what the original plan defined. Wysocki (2007, p. 358) states that "... acceptance can be very informal and ceremonial, or it can be very formal, involving extensive acceptance testing against the client's performance specifications." Client acceptance can be in the form of ceremonial acceptance or formal acceptance.

Ceremonial acceptance represents the opposite of formal acceptance, and two situations fall under this heading. The ceremonial form might be orchestrated as a scheduled demonstration of the product capabilities. This is usually a management level session that occurs after the technical acceptance has been completed. Wysocki describes formal acceptance as a predefined process through which the project team proves conformity to planned results that the client wanted. In this case, a checklist has important meaning, and "... is used and requires a feature-by-feature sign-off based on performance tests. These tests are conducted jointly and administered by the client and appropriate members of the project team." Both forms of acceptance are important client relationship processes.

30.6 CREATING LESSONS LEARNED DOCUMENTATION

Creating lessons learned documentation is a key evaluation methodology for the project. Two of the key guiding philosophies of project management are planning and measurement. The lessons learned documentation serves the role of describing the project environment beyond what has previously been communicated in formal reporting documents related primarily to scope, time, and cost. This exercise gives the team an open format opportunity to analyze for all to see what has gone on during a phase or for the entire project. Organizations should have a goal of continuous improvement and this is one of the key processes that can support this goal. Morris and Pinto (2007, p. 253) comment that an effective internal evaluation is necessary for at least three reasons:

1. Evaluation defines what the unfinished project needs to accomplish and helps in classifying future project objectives
2. Evaluation provides formal feedback to the project team and their peers in regard to measurable accomplishments

3. Evaluations are a core element of organizational learning and should be carried out with a view of providing guidance on future ventures.

Baca (2007, p. 447) states that the lessons learned process helps in evaluating processes, tools, and techniques that worked either well or poorly on the project. The lessons learned process can be produced either by the project team or by an external audit model. Regardless of the data collection process used, the following ground rules should be followed:

- No topic area is off-limits
- Speak in terms of process, not people problems
- When successful events are documented, it is appropriate to credit specific individuals.

30.7   LESSONS LEARNED REPORT

Wysocki (2007, p. 363) describes the final lessons learned report as a formal representation of the project's memory and history. This document serves as a valuable source for others to learn and research situations faced by the project team. During the project life cycle the project team can use this document for their own analytical and review purposes; however, the more likely value will be for future project efforts. This means that the document should be readily accessible for other teams to use as a reference document. There are several contemporary computerized document management tools that can be used for this purpose. In the operational mode, this archive should be a shared, searchable online database. One issue that may be encountered when implementing shared knowledge comes from the behavioral trait that some people tend to be territorial about the knowledge they have gained.

One sample format for the lessons learned document is as follows:

- *Overall success of the project:* The project team should summarize the areas they feel the project succeeded in.
- *Organization of the project:* The life cycle used by the project team is an important item for others to see and this can be an important factor in how the project progressed. Issues such as major stages, milestones, project authority details, project board organization, and the like are key items.
- *Techniques used to get results:* The document should record a list of major problems faced and the solution strategies used to resolve them.
- *Project strengths and weaknesses:* Each project is unique with their strengths and weaknesses. These should be expressed in terms of features, practices, and processes utilized by the team. The goal would be to avoid less successful items and take advantage of the more successful ones. Weaknesses that lead to individuals should be avoided, but positive attributes linked to an individual are appropriate for the document.
- *Project team recommendations:* The project team should translate the project-related items outlined above into an organizational improvement recommendation format. Peer recommendations are one of the strongest support sources for subsequent teams. These recommendations are not formulated by remote staff, but by individuals who have fought through the real problems.

30.8   PROJECT TEAM CELEBRATION

Regardless of the reason for a project's termination, the project team should celebrate prior to disbanding. Essentially all projects have both high and low events and the celebration process helps to build a positive morale boost for the time spent together. Rationale for this is that it is easy to just walk away with the feeling that the entire effort was a failure and thankless. The fact is that there is always something positive in the experience and the team members need to have this reinforced for both professional and morale reasons. For projects that have successfully delivered a complex product or process the celebrations are easy and long lasting. However, in cases where the project produces less than planned results, it may be necessary to be more creative in finding a celebration theme, but the effort is a necessary management activity.

Team performance bonuses are a nice way to make the celebration process easier to execute for sure. The low end of a celebration can be just a team meeting to discuss the experience and thank the individuals for their hard work. At the other end of the spectrum, a catered party for the families with a live band certainly makes a positive impression. The final advice for this activity is to be sure that lessons learned are documented and the appropriate celebration is orchestrated. Do not underestimate the positive value of this activity.

30.9   CONCLUSION

One of the major closure activities for the project team is to produce a lessons learned document and transfer their acquired knowledge to others.

Project termination is one of the most mishandled tasks in the life cycle of a project. It means the end of the relationship between the project team and their client. At this point, final resolution of the effort is no longer in doubt. Hopefully, the effort has been successful and the team morale is high. Regardless, this is the time to dissolve the team and help them move on to another job location.

REFERENCES

Baca, C.M., 2007. *Project Management for Mere Mortals: The Tools, Techniques, Teaching, and Politics of Project Management*. Boston: Pearson Education Inc.

Cleland, D.I., 2004. *Field to Project Management*, Second Edition. Hoboken, NJ: Wiley.

Cleland, D.I. and L.R. Ireland, 2002. *Project Management: Strategic Design and Implementation*, Fourth Edition. New York: McGraw-Hill.

Horine, G.M., 2005. *Absolute Beginner's Guide to Project Management*. Boston: Que Publishing.

Meredith, J.R. and S.J. Mantel, 2003. *Project Management: A Managerial Approach*, Fifth Edition. Hoboken, NJ: Wiley.

Morris, P.W.G. and J.K. Pinto, 2007. *The Wiley Guide to the Management of Projects*. Hoboken, NJ: Wiley.

PMI, 2008. *A Guide to the Project Management Body of Knowledge (PMBOK® Guide)*, Fourth Edition. Newtown Square: Project Management Institute.

Seningen, S., 2008. Learn the value of lessons-learned, Retrieved April 1, 2008 from www.projectperfect.com.au

Taylor, J., 2001. The Project Management Workshop, Amacom.

Wysocki, R.K., 2007. *Effective Project Management: Traditional, Adaptive, Extreme*, Fourth Edition. Hoboken, NJ: Wiley.

Part IX

Contemporary Topics

LEARNING OBJECTIVES

Upon completion of this part the reader should understand the role of these selected contemporary topics in the overall project infrastructure. Specifically, the goals are as follows.

1. Expand the reader's view of the project environment with selected contemporary topics
2. Understand the concept of organizational maturity and its impact on the project environment
3. Understand the operational issues of Project Portfolio Management (PPM)
4. Understand the theory and challenges in the PMO structure
5. Review the logic and challenges of outsourcing project work to third parties
6. Understand the current thinking regarding techniques to develop high-productivity project teams
7. Review the concepts of effective project governance.

The concepts discussed in this part of the book are devoted to a selective set of topics that have not yet found a clear or mature direction, yet represent important concepts for the PM to understand. The learning objectives summarized above provide insight into the basic topics included and the sections below will provide a short rationalization for each.

ORGANIZATIONAL MATURITY

Projects must exist inside their support organizations. The logic behind maturity is that the project will be more successful and effective with the support of a mature organization. Defining the specifics of maturity is the goal of Chapter 31.

PPM

In the past few years, there has been a growing recognition that many projects are underway at any one time, each consuming critical organizational resources. A proper organizational approach to this is to deal with the overall project world in much the same fashion as one would invest in a diverse set of stocks. The basic goal of PPM is to maximize the return on resources invested in project ventures. Chapter 32 will provide an overview of this strategic view of project management.

Enterprise Project Management Office (EPM)

As a companion activity to PPM the EPM concept represents a strategy to centralize various project process and decision making at the enterprise level. This concept is extracted from the PMO model, but extended in scope to all projects. There is not a single standard definition for roles and responsibilities of a PMO-like structure and within the multiple options have potential advantages and disadvantages. One simple way of looking at an EPM organization is to charge it with providing some form of leadership role in the selection, development, and implementation of a variety of project management processes. Chapter 33 will outline other options for this emerging organizational approach to project selection and management.

Outsourcing

The popular press has made this topic familiar to all and many individuals have been personally "touched" by its impact as their job moved offshore. This strategy remains open for review as to whether it is a long-term positive process for an organization, but it is clear that the use of third-party vendors to provide many organizational services will continue for the foreseeable time period. The impact of this decision on the project is far reaching and in effect turns the project organization into a geographically dispersed virtual collection of work units. Regardless of the value obtained by outsourcing, there is little debate that it complicates the job for the PM. The overall evaluation of outsourcing is now and will likely remain a controversial strategy for organizations. Chapter 34 will dwell further into this topic.

High-Productivity Teams

Behaviorists have long talked about employee motivation and morale, but techniques to put these theories into actual practice in a project environment have been recognized. Motivation and morale are potentially positive in generating higher productivity, but they are not the sole drivers. Some would argue that they are not even significant factors. In any case, the SEI has developed a teachable and somewhat mechanical technique that has proven to produce higher output in multiple-type projects. Even though this theory is still in the formulation stage, the basic message is deemed worthy of review by the modern PM. Chapter 35 will translate the SEI team management model into what is described as a Team Process that is amenable to all project types.

Project Governance

Every project requires a properly working decision-making structure. Surprisingly, this is not always understood and the roles of the PM, sponsor, stakeholders, team members, and senior management are often misunderstood as a result. Project governance is the decision model by which a project is managed within an enterprise so that it aligns with business needs. Hence, it can be viewed as the decision coupling of project's activities with business vision, strategy, and objectives. This process has always been difficult, given the many different expectations and aims that it attempts to embrace. Chapter 36 will provide a general theoretical overview of this topic from a project perspective.

It is important to recognize that the contemporary label of this part title can be translated to mean that any of the ideas defined here are not yet mature or well established in the project environment. So, anticipate that each of the topics discussed will still need to mature and it is important for the PM to stay up to date with these ideas. The PMI requires that the certified PMP dedicate a minimum of 20 h per year in continuing education. It is areas such as these that justify such a requirement.

31 Organizational Maturity

31.1 INTRODUCTION

The goal of this chapter is to review three major enterprise level maturity models and their underlying concepts. The first of these is the classic CMM (Capability Maturity Matrix) from SEI and its follow-on extension CMMI (Capability Maturity Model Integrated). The second model is PMI's Organizational Project Management (OPM3) model. The CMM model was originally designed to describe the software development environment, but since inception the concept has broadened to include similar attributes for a general project environment.

Traditional thinking in regard to project management has focused on the internals of executing the overall life cycle efficiently. As understanding of the more global issues has grown, this perspective has broadened in two key ways. First, there is now a recognition that projects live within their host enterprise level processes and culture. Secondly, organizations are pursuing a wide array of projects at any one time, all of which are competing for scarce resources. Both of these situations impact the management actions necessary to produce successful outcomes. One implication of this view is that a project can be more effective if the organizational processes support its needs. The project will be negatively impacted if the organization is not equipped to deal with these and other similar issues. We will translate this supporting organizational cocoon concept as "maturity." The basic theory of this idea is that the higher the maturity the better support for the project environment.

The following metaphor was used earlier in Chapter 4, but is worthy of repeating here. This deals with the relationship of organizational culture to the project. It can be described as follows:

> Imagine the organization as a flowerbed and the project as a seed that needs to grow in that flowerbed. If the flowerbed does not have good soil, water and other nutrients then the seed (project) won't blossom to its full potential.

Translating this abstraction into the project world is meant to show that there are many organizational processes that are needed to optimally support a successful project. If these are not in place and working well, the project (seed) will suffer and never blossom to its full potential. In the absence of an appropriate enterprise support structure, it will be up to each project to build its own flowerbed for support and this extra effort fragments the project resources, adds expense, and consumes additional time. Each such remedial effort is subtle, but clearly has a negative influence on the final results.

If our flowerbed metaphor represents organizational maturity, what is maturity? Cooke-Davis defines this as "the extent to which an organization has explicitly and consistently deployed processes that are documented, managed, measured, controlled, and continually improved" (Cooke-Davies, 2004). Key to this view is that maturity is related to attributes such as efficiency, consistency, and continual improvement in operational processes. Paulk et al. (1993a) describe maturity as the "potential for growth in capability" and continuous improvement "through focused and sustained effort towards building processes and management practices." Figure 31.1 illustrates a visual picture of the maturity concept. This indicates that the ability of an organization to have a stable platform (stool), its legs (technology, processes, and people) must all be in place and balanced.

FIGURE 31.1 Balanced organization elements.

To add substance to this idea Table 31.1 contains a brief summary of characteristics of mature and immature organizations as described by Paulk.

31.2 OVERVIEW OF FORMAL CMMS

Assessing the maturity of any organization is a very complex undertaking and can only be done empirically by observing various discrete components within the organization. The capabilities of these components are measured using some form of performance indicator. So, the concept of a CMM involves a review of certain defined processes representing low maturity through high maturity. "It also describes an evolutionary improvement path from ad hoc, immature processes to disciplined, mature processes with improved quality and effectiveness" (CMMI Product Team, 2002).

TABLE 31.1
Characteristic Differences between Mature and Immature Organizations

A Mature Organization	An Immature Organization
• Ensures organizational processes are accurately communicated to all members	• Lacks proper communication between levels and departments in the organization
• Ensures that all activities are carried out according to the developed processes	• Is reactionary, and managers are usually focused on solving immediate crises
• Established processes are continually reviewed, updated and tested as required	• Have no defined process for ensuring high and consistent product quality
• Ensures that the established processes complement the organizations area of competence and business model	• Will tend to overlook failing processes to meet deadlines
• Establishes clear and defined roles and responsibilities for all members	• Lacks clear and defined roles for its members
• Ensures that product quality and customer satisfaction are priority	

Source: Paulk, M.C. et al., 1993a. Capability Maturity Model for Software, Version 1.1, Technical Report: CMU/SEI-93-TR-024 ESC-TR-93-177: 1-5. With permission.

31.3 SEI'S CMM

Although the original CMM developed by the SEI of Carnegie-Mellon University focused on software-oriented projects, its concepts have been applied across many other industries and project types. Paulk et al. (1993b) describe the CMM objectives as containing the following attributes (Ref. SEI-93-TR-025 ESC-TR-93-178):

- Based on actual practices
- Reflects the best of the state of the practice
- Reflects the needs of individuals performing software process
- Improves software process assessments or software; capability evaluations
- Technically documented
- Publicly available.

The CMM high-level structure as shown in Figure 31.2 shows that each maturity level is made up of broader process capabilities and key process areas (KPAs) (with the exception of the first level that has no defined structure). Each KPA achieves some defined operational goal that is deemed to improve management and control of the project.

31.4 CMM STRUCTURE

Maturity levels: There are five maturity levels defined in the CMM model. A maturity value is a grade level representing the "degree of process improvement across a predefined set of process areas" (CMMI Product Team). Process capabilities are measured to develop the maturity level grade for the organization.

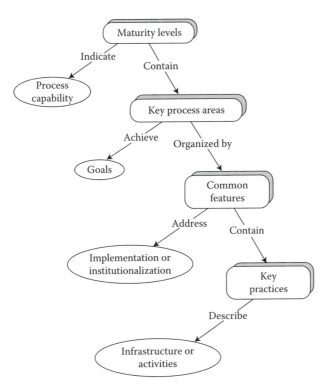

FIGURE 31.2 Maturity structure [From CMMI Product Team, 2002. *Capability Maturity Model Integration (CMMI) Version 1.1.* Pittsburgh: Carnegie Mellon Software Engineering Institute. With permission.]

KPAs: A KPA identifies a cluster of related activities that, when performed collectively, achieve a set of goals considered important (Paulk et al., 1993b).

Goals: This represents controls that identify "scope, boundaries and intent of the KPA."

Common features: These address the implementation and/or institutionalization of KPAs. There are five common features of each KPA: Commitment to perform, ability to perform, activities performed, measurement and analysis, and verifying implementation.

Key practices: Describe the elements of infrastructure and practice that contribute most effectively to the implementation and institutionalization of the KPAs.

31.5   CMM MATURITY LEVELS

CMM maturity levels are viewed as a numeric measure of organizational capability. Figure 31.3 describes five defined levels of maturity and a basic descriptor for each level (Paulk et al., 1993b).

31.5.1   INITIAL LEVEL (LEVEL 1)

This level is termed the chaos stage since organizations operating within this realm lack proper procedures, defined roles for its members, and its departments are typically disconnected from a goal alignment view. Operating at the initial level of maturity means that the organization would heavily depend on competency of its employees, rather than on defined and tested processes (Paulk et al., 1993a). A survey conducted from 1996 to 2000 by SEI involving 1012 organizations indicated that a significant number (32.2%) of organizations operate within the realm of the initial level (SEI, 2001, p. 10).

31.5.2   REPEATABILITY LEVEL (LEVEL 2)

As the name implies, organizations operating in the repeatability level have developed some processes that can be applied to multiple projects within the organization. While these processes are not always enforced, they exist within the processes of the organization. Projects operating in

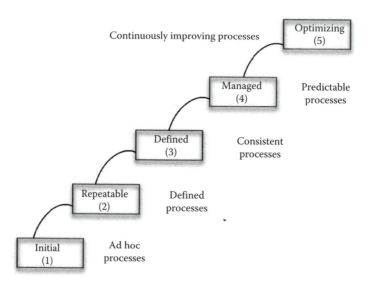

FIGURE 31.3 Organizational maturity levels.

level 2 organizations have data to help track cost, schedule, and apply best practices. Also, these organizations are somewhat similar to those at level 1 in that they are still prone to higher operating cost and overdrawn schedules.

31.5.3   Defined Level (Level 3)

Level 3 organizations use developed and tested processes. Roles of members are defined and tailored toward proven organizational processes. Processes are enforced throughout the organization as standards. Operational processes are frequently reviewed and appropriate revisions are issued when required. This maturity level organization tends to have formal training programs to continually improve understanding of processes within the organization (Paulk et al., 1993b, pp. O-9-16).

31.5.4   Managed Level (Level 4)

One of the key differential characteristics of operating at level 4 is the ability to measure and possibly quantify organizational progress and status within a project. This distinction also involves the predictability of process performance. Level 4 organizations use competitive benchmarking best practices to assess all aspects of their processes. Organizational processes are followed systematically to execute all projects, and changes are enacted through planned procedures. These organizations also have the ability to *predict* outcomes for future projects.

31.5.5   Optimizing Level (Level 5)

Operating at level 5 signifies that the organization is fully matured according to the model definition. There are only a few organizations that can effectively claim to operate at this level. Effective customer-oriented QA and continual improvement of the organizations processes and products become the cultural norms at this point.

As a final note on CMM, research evidence shows positive value produced by achieving each of the higher levels. However, it must be recognized that the time to achieve each level is typically in the 2-year range and requires significant support by senior management. In some cases there is a diminishing return at level 5 compared to the cost of maintaining it, but all organizations should evaluate the cost versus value of achieving each of these levels.

31.6   CAPABILITY MATURITY MODEL INTEGRATION (CMMI)

CMMI was created by SEI in association with the Office of the Secretary of Defense (OSD) and the National Defense Industrial Association (NDIA) as an evolutionary extension of CMM. The CMMI Product Team describes this model as a means of improving an "organization's processes for development, acquisition, and maintenance of products or services" (CMMI Product Team, 2002). Similar to the architecture of CMM, CMMI is made up of components including specific practices, generic practices, specific goals, generic goals, process areas, capability levels, and maturity levels that collectively establish the framework. Figure 31.4 illustrates the CMMI component structure.

CMM and CMMI are well-thought-out and excellent concepts to describe the operational role of our flowerbed concept. These models provide an enhanced understanding regarding the processes required to effectively produce project results. Thousands of organizations have and are continuing to pursue through the sponsorship of SEI the CMM model and its new extensions to these concepts. The interested reader should review the SEI web site for significant technical documentation of these two models (Ref. www.sei.cmu.edu).

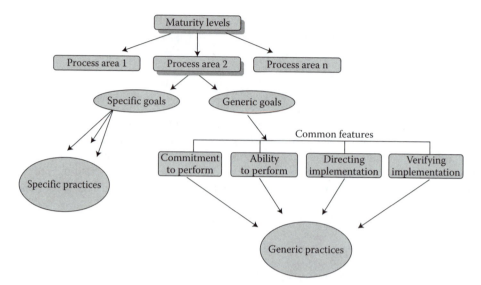

FIGURE 31.4 CMMI component structure [From CMMI Product Team, 2002. *Capability Maturity Model Integration (CMMI) Version 1.1*. Pittsburgh: Carnegie Mellon Software Engineering Institute. With permission.]

31.7 VALUE OF ORGANIZATIONAL MATURITY

Intuitively, it seems logical that improved processes should lead to better productivity and there is industry evidence that this is indeed the case. Dr. William Ibbs, University of California at Berkley, has developed a project management maturity model based on a maturity grading structure similar to CMM (Ibbs, 2007). The relationship between organizational maturity and project schedule performance as defined by SPI is derived from this research and is shown in Figure 31.5.

The correlation of project schedule performance improvement versus organizational maturity is a strong motivator to embrace this concept. Note that as organizations approach level 5 maturity, they are actually able to beat scheduled completion dates, whereas the lower maturity organizations tend to overrun schedules in the range of 200% of plan (i.e., Schedule Performance around 0.5). Similar performance results have been observed in other organizations, so this is not an isolated observation.

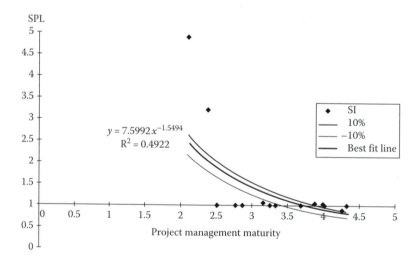

FIGURE 31.5 Project schedule performance versus maturity level. (Courtesy of Dr. William Ibbs, private communication, 2007. With permission.)

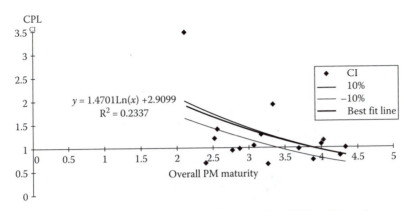

FIGURE 31.6 Cost performance versus maturity level (Courtesy of Dr. William Ibbs, private communication, 2007. With permission.)

Although not as extreme as the schedule case, a similar result is shown in Figure 31.6 for cost performance. Typical industry cost overruns from broader surveys indicate that average project cost overruns are closer to the 75–100% range, but the Ibbs research shows that organizations in the level 3 maturity category have average cost overruns closer to 16%, whereas level 5 organizations experience cost overruns in the 9% range.

Two conclusions are reflected by the schedule and cost survey data. First, cost seems to be a more carefully controlled variable than schedule, and maturity level has a positive impact on the project budget performance. The Ibbs survey results represent a significant quantification of the maturity phenomenon. In addition to the cost and schedule implications, this survey also indicates that higher maturity organizations spend more time planning projects and closing them. The higher level maturity organizations also spent more time communicating that the lower levels. One of the classic arguments against formal project management is the belief that the effort is not worth the time spent given all of the vagaries associated with projects. This research data would support an argument that improved processes, developing better requirements, and more project postmortem evaluation all add value to the outcome. If all organizations shared this belief, the job of implementing formal management methods would be easier.

31.8 ORGANIZATIONAL PROJECT MANAGEMENT (OPM3)

OPM3 is an evolutionary step from the CMM model in that it was explicitly defined to fit all organizational types. Around 1998, various members of PMI became sensitive to the fact that the *PMBOK® Guide* was focused heavily on a single project view and basically ignored the overall organizational impact on the project environment (Schlichter et al., 2003). Perspective gained from the CMM experience suggested that they should look into a way of integrating organizational maturity into the PMI project model structure and that became the initial stimulus for the development of OPM3. PMI defines this model as "the systematic management of projects, programs, and portfolios in alignment with the achievement of strategic goals" (PMI OPM3, 2003). The design concept behind OPM3 is that all initiatives of an organization can be divided into projects that can be grouped into any of the three sectors mentioned in the definition—projects, portfolios, and programs. The OPM3 model mechanics are based on industry best practices that are used to assess the organization or subcomponent maturity level (PMI OPM3, 2003, pp. xi–xv). The OPM3 model is structured into the following three main elements:

- *Knowledge:* The main body of the model that explains organizational maturity with respect to Best Practices

FIGURE 31.7   Elements of OPM3. (From PMI, 2003, *OPM3 Knowledge Foundation*. Newtown Square, PA: Project Management Institute, p. 8. With permission.)

- *Assessment:* Describes the evaluation methods on the basis of Best Practices and Capabilities
- *Improvement:* Describes the practices for implementing changes

Figure 31.7 illustrates the elements of OPM3 (PMI OPM3, 2003, p. 8). Note that the interrelationship between these three items involves a comparison of current organization to best practices, then an assessment of the gaps followed by an action plan to move the organization to a higher level of maturity.

31.9   OVERVIEW OF OPM3

During the early model development project phases, the design team struggled to find the best way to define specifics regarding what elements comprised maturity. Eventually, this process resulted in the identification of 10 individual elements required to provide those capabilities. The defined components were grouped into the following 10 categories (Schlichter et al., 2003):

1. Standardization and Integration of Processes
2. Performance Metrics
3. Commitment to the Project Management Process
4. Alignment and Prioritization of Projects
5. Continuous Improvement
6. Using Success Criteria to Cull or Continue Projects
7. People and Competence
8. Allocation of Resources to Projects
9. Organizational Fit
10. Teamwork

Embedded in each of these 10 groups are more detailed subprocesses, called *design cells*, which were then elaborated and transformed into a directory of 600 operational best practices.

   PMI defines OPM3 as "the application of knowledge, skills, tools, and techniques to organizational and project activities to achieve the aims of an organization through projects" (PMI OPM3, 2003 p. 5). The design concept of OPM3 was to apply the general structural guidelines established by the *PMBOK® Guide* to organizational processes at all levels. This concept describes the management structure of the organization in terms of project, programs, and portfolio and links the organizational capabilities to these KAs. Evaluating the degree to which an organization practices, the

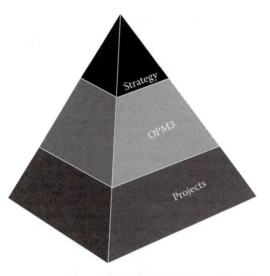

FIGURE 31.8 OPM3 strategy bridge. (From PMI, 2003, *OPM3 Knowledge Foundation*. Newtown Square, PA: Project Management Institute,. p. 5. With permission.)

defined levels of maturity provides a grading mechanism for improvement (Paulk et al., 1993b). In addition, "OPM3 provides an added advantage to organizations by allowing them to bridge long standing gaps between an organization's strategic plans and its ability to achieve those plans through the execution of projects" (Haeck, 2004). Figure 31.8 shows schematically the relationship of organizational strategy to projects through OPM3.

Knowledge, assessment, and improvement are three components of OPM3 that encompass the key KAs or processes as defined by PMI. Figure 31.9 illustrates the operational process for implementing the model.

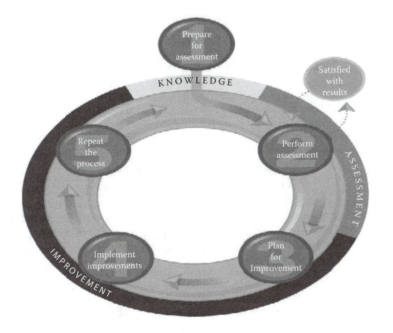

FIGURE 31.9 The OPM3 operational cycle. (From PMI, 2003, *OPM3 Knowledge Foundation*. Newtown Square, PA: Project Management Institute, p. 9. With permission.)

During the early model definition stage, there were many design disagreements among the PMI volunteer-driven teams. Because of this the design and implementation efforts basically floundered through the first two rounds of volunteer teams. Finally, in 2003, in the third round the team produced a beta version that after user testing was issued by PMI with the title OPM3 (Schlichter et al., 2003).

31.10 OPM3 COMPONENTS

31.10.1 KNOWLEDGE

Attempting to implement OPM3 standards to any organization requires a firm understanding of the model knowledge base. The first step in this process is to become familiar with the Best Practices and assessment guides published by PMI in the *OPM3 Knowledge Foundation* (Haeck, 2004).

31.10.2 ASSESSMENT

The second step in the implementation cycle is to assess the current degree of maturity of the organization with respect to existing processes compared to the defined Best Practices. This step is carried out in two stages; maturity assessment with respect to Best Practices and assessment of specific capabilities. In this stage, the organization would use the Self-Assessment procedure provided by *OPM3 Knowledge Foundation* to determine maturity levels of specific processes with respect to Best Practices. From this assessment, process capability gaps are identified.

The second stage requires the organization to examine capabilities related to the gap practices identified in the first stage (PMI OPM3, 2003, p. 39). Essentially, the goal of moving toward a Best Practices environment is to consider this as the optimum state for the organization to execute project management processes and deliver project management services and products (PMI OPM3, 2003, p. 171).

31.10.3 EVALUATION PROCESS

Once an organization has determined its capability gap areas, an improvement strategy is developed. However, the improvement cycle should be looked at as a four-step approach.

1. Plan for improvement
2. Implement improvement
3. Assess improvement
4. Repeat the process.

The mechanics for these steps are briefly explained below.

Plan for improvement: At this stage, the organization must review their current capabilities in relation to benefits that would occur at the higher maturity level. This must be evaluated in regard to both cost and productivity impact to the organization. Finally, the issue of change management complexity for the new capability is a consideration to ensure a smooth and easy transition between capabilities.

Implement improvement: After the completion of the planning step, the next step is to execute the improvement. Organizations must however keep in mind that these implementations could affect factors such as organizational structure and strategy.

Assess the improvement: As with all organization change activities, a post assessment is required to evaluate results. Deviations from the planned result must be evaluated and lessons learned captured.

Repeat the process: This final step implies that the model is a continuous improvement activity. Quality management says that the organization must understand that no state is optimum forever.

TABLE 31.2
Example of an OPM3 Directory of Best Practices

BP ID	Title	Description	Project	Program	Portfolio	Standardize	Measure	Control	Improve
1000	Establish OPM3 Policies	The organization has policies describing the standardization, measurement, control, and continuous improvement of OPM3 processes	X	X	X	X	X	X	X
1010	Project Initiation Process Standardization	Project Initiation Process standards are established	X			X			
1020	Project Plan Development Process Standardization	Project Plan Development Process standards are established	X			X			
1030	Project Scope Planning Process Standardization	Project Scope Planning Process standards are established	X			X			
1040	Project Scope Definition Process standardization	Project Scope Definition Process standards are established	X			X			
1050	Project Activity Definition Process Standardization	Project Activity Definition Process standards are established	X			X			

Source: PMI, 2003. *OPM3 Knowledge Foundation.* Newtown Square, PA: Project Management Institute. With permission.

31.11 BEST PRACTICES

A Best Practice is the currently recognized best method to achieve a stated goal or objective (PMI, 2008). The OPM3 directory defines the model's best practices, indicates which project phase they deal with, and provides a brief explanation of the practice. Collectively, this set of processes provides the organization with a standards checklist to review for operational status. Table 31.2 contains an example subset of an OPM3 best practices definitions.

31.12 CAPABILITIES DIRECTORY

The Capabilities directory contains detailed data regarding Best Practices. This directory is used in the Assessment activity to aid in defining the state of capabilities existing in the organization. There are also unique identifiers that define the relationship of the capabilities to various best practices. Table 31.3 contains a sample subset of the Capabilities directory.

31.13 IMPROVEMENT PLANNING DIRECTORY

The Improvement Planning directory provides the dependencies between capabilities. This directory (Table 31.4) shows the capabilities that are necessary for the best practices along with the dependent capabilities. This also shows defined dependencies between best practices.

TABLE 31.3

An Example of a Capabilities Directory

Capability ID 1410.010	**Capability Name** Know the Importance of Competent Resource Pool		**PPP** Project	**SMCI** Standardize	**IPECC** Planning
	Capability Description The organization is aware of the processes needed to provide qualified people to projects				
	Outcome ID 1410.010.10	**Outcome Name** Organizational Process Analysis	**Outcome Description** The organization is aware of its current state with respect to the processes that provide qualified people	**KPI Name** Results of the Current State Process Analysis	**Metrics Name** Exists
Capability ID 1410.020	**Capability Name** Identify Process Requirements for Resource Pool		**PPP** Project	**SMCI** Standardize	**IPECC** Other
	Capability Description The organization identifies the process requirements for ensuring a competent resource pool				
	Outcome ID 1410.020.10	**Outcome Name** Process Requirements for Managing Resource Pool	**Outcome Description** The organization defines the requirements for managing a competent project resource pool	**KPI Name** Requirements for the Process	**Metrics Name** Exists
Capability ID 1410.030	**Capability Name** Develop a Skills Database		**PPP** Project	**SMCI** Standardize	**IPECC** Other
	Capability Description The organization has a skills database				
	Outcome ID 1410.030.10	**Outcome Name** Skills of Individuals	**Outcome Description** The organizational skills database includes skills of individual staff members	**KPI Name** Skills Gap Analysis Results	**Metrics Name** Exists

Source: PMI, 2003. *OPM3 Knowledge Foundation.* Newtown Square, PA: Project Management Institute. With permission.

The relationship between Best Practice and Capability is represented in Figure 31.10. A Best Practice is structured to generate outcomes that can be measured or quantified through Key Performance Indicators (KPIs). The organization must therefore be able to link Capabilities to a Best Practice (Figure 31.11). A firm understanding of Best Practices can assist the organization in standardizing their organizational processes. This will include training programs and formal organizational procedures and strategies.

TABLE 31.4
Sample of Improvement Planning Directory

Best Practice	1410	**Name** Manage Project Resource Pool		Project	Program	Portfolio	Standard	Measure	Control	Improve
		Description The organization has the mechanisms, systems, and processes that provide projects with professional project managers and competent, committed project team members								
				X			X			

Capability	Name	Outcome Checklist	
1410.010	**Know the Importance of Competent Resource Pool**	☐	
1410.020	**Identify Process Requirements for Resource Pool**	☐	
5220.030	Implement Staff Acquisition Policies and Procedures	☐	
1410.030	**Develop a Skills Database**	☐	
1400.040	Review Human Resource Plan	☐	
3100.030	Staff Technical and Administrative Resources	☐	
5630.010	Assign Professional Project Managers	☐	
1410040	**Determine Training Requirements**	☐	☐
1410.050	**Match Project Resource Requirements**	☐	☐

Source: PMI, 2003. *OPM3 Knowledge Foundation*. Newtown Square, PA: Project Management Institute. With permission.

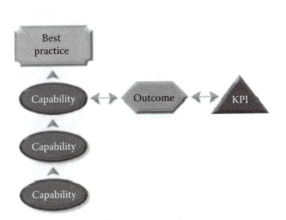

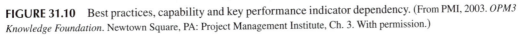

FIGURE 31.10 Best practices, capability and key performance indicator dependency. (From PMI, 2003. *OPM3 Knowledge Foundation*. Newtown Square, PA: Project Management Institute, Ch. 3. With permission.)

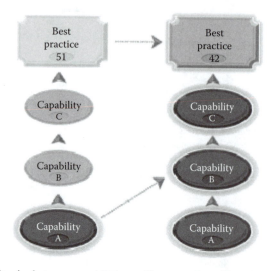

FIGURE 31.11 Dependencies between capabilities and best practices. (From PMI, 2003. *OPM3 Knowledge Foundation*. Newtown Square, PA: Project Management Institute, Ch. 3. With permission.)

31.14 OPM3 PROCESSES

The OPM3 model adopts its basic concepts and guidelines from the *PMBOK® Guide* and divides its views into three process groups:

- Project Management
- Program Management
- Portfolio Management.

From this structural perspective, a strategic planning model is constructed. Based on the internal evaluation cycle, the organization can derive action plans ranging from the project level to departmental and to organizational levels. A schematic depiction of these components is described in Figure 31.10 and shows that each plan has to be integrated with other plans and then moved through the required process improvement stages (Figure 31.12).

Finally, Figure 31.13 illustrates the evolution of OPM3 as it evolves through the organization. Each of the horizontal and vertical aspects of this evolution must be kept in synchronization.

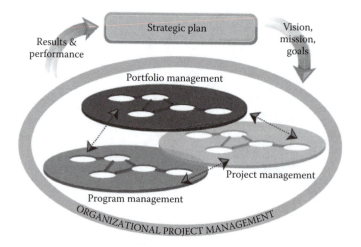

FIGURE 31.12 OPM3 Planning Model. (From PMI, 2003. *OPM3 Knowledge Foundation*. Newtown Square, PA: Project Management, p. 21. With permission.)

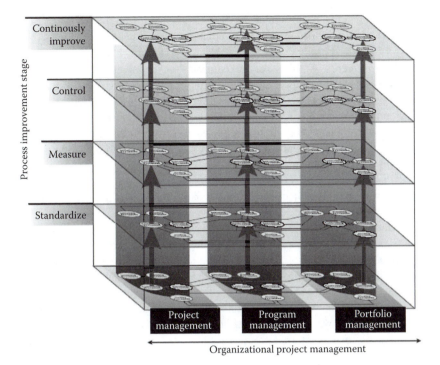

FIGURE 31.13 OPM3 process construct within the organization. (From PMI, 2003. *OPM3 Knowledge Foundation*. Newtown Square, PA: Project Management Institute, p. 23. With permission.)

31.15 APPLYING OPM3 IN AN ORGANIZATION

There are different ways for an organization to apply OPM3. The two basic steps outlined in the "OPM3 knowledge foundation" are summarized below (PMI OPM3, 2003, p. 25).

1. *Prepare for assessment:* The members who are going to be in the assessment team must be familiar with the OPM3 model, the directories, and the OPM3 assessment cycle. A Self-Assessment CD is used to guide this process.
2. *Perform assessment:* The team uses a worksheet template to perform the assessment survey. This process produces two lists: one of these is the list of best practices that the organization demonstrates and the other is the set of practices that are not at the best practices level (gaps). Various reports are also created to show the current maturity, domain specific status, and process views. From this evaluation, the team will decide where to focus further attention.

The Best Practices directory provides sample solutions for the gap target areas. Corresponding Best Practice and Capability information in the respective directories can then be used to review target processes. For example, one of the target capabilities could be "Develop awareness of Project Management Activities" (PMI OPM3, 2003, pp. 23–35). This capability is associated with the Best Practice "Establish Internal Project Management Activities." One of the defined outcomes for this capability is "Organization supports and utilizes local initiatives." The survey assessment showed that this is not the case. A second defined capability is "The Organization has intelligence on important Issues and activities in the project management community." In this case, the survey verified the capability. From this type of analysis capability gaps are evaluated as potential targets for possible upgrade in subsequent steps.

The team repeats the process of evaluating capabilities to see how they stand in regard to Best Practices. Once completed, this results in an organizational overview of its status relative to Best

Practices. The question at this point is to decide how to proceed toward higher levels of maturity. All gaps do not have the same priority or severity, so that assessment has to be made as follows:

1. *Plan for improvements:* At this stage the team uses the Improvement Planning directory to decide how to proceed based on their capabilities gaps as compared to Best Practices. At this point, the selection issue for improving performance involves finding the capability gaps that can be improved at least cost while making the greatest improvement. Technical and organizational knowledge is also required to review feasibility and political issues surrounding this decision. Once defined the result is a portfolio of action items that the organization needs to pursue.

2. *Implement improvements:* Each identified action item evolves to a project and is managed in the same way that all projects are. Some look at this type of activity as the "cobbler's children fixing their own shoes." Historically, many organizations have focused on developing products and services for external customers and ignoring their own needs. OPM3 represents an example of cleaning your own house so that the organization can better execute the external needs. This is a simple concept, but one not often practiced.

3. *Repeat the process:* As improvements are implemented through this process, the Self-Assessment tool is used to evaluate whether the planned results have been achieved. This data are used to update the improvements plan for all best practices action plans. The process described above is ongoing and becomes the mechanism for an organization to continuously improve its internal processes.

31.16   OPM3 BENEFITS AND CASE STUDIES

OPM3 formalizes the study, understanding, and application of organizational maturity models for a broad range of organizations. It provides working definitions of key processes from not only the project management perspective but also from program and portfolio management aspects. Through its definition of best practices, OPM3 offers organizations an opportunity to link their maturity improvement strategies to corresponding operational performance and thereby help guide a path toward continuous improvement of those processes. Improving portfolio, project, and program management outcomes plays an important role in achieving strategic objectives; therefore increasing organizational maturity has a corresponding strategic importance.

OPM3 implementation documentation outlines its three key benefits to the organization as follows:

- It helps the organization eliminate the gap between strategy and individual project selection as signified in Figure 31.4

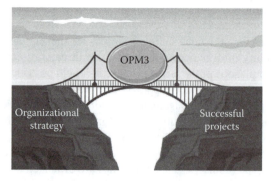

FIGURE 31.14   OPM3 linkage role. (From PMI, 2003. *OPM3 Knowledge Foundation.* Newtown Square, PA: Project Management Institute. With permission.)

- It provides a mechanism for organizations to analyze their maturity gaps and from this guide, the development of focused action plans for continuous improvement
- It defines a library of model best practices that identifies the characteristics of specific process targets. This will help organizations implement improvements without requiring extensive requirements analysis time.

Figure 31.14 illustrates in cartoon format the role of OPM3 in creating an organizational set of strategies that lead to successful projects.

From OPM3's formalized definitions, organizations can evolve their adoption of predefined Best Practices to help guide process improvement changes within the organization. "The OPM3 methodology offers companies a rare opportunity to introspectively look at the holistic link between their strategies and how well they are able to systematically *translate* those lofty ambitions into tactical reality" (Haeck, 2008). "OPM3 provides an added advantage to organizations by allowing them to bridge gaps between an organization's strategic plans and its ability to achieve those plans through the execution of projects" (Haeck, 2008).

Although the experience with using OPM3 is fairly new, a couple of published actual case study results can be cited.

Case 1: Pinellas County IT Turns Around Performance and Customer Care (PMI OPM3 Case Study, n.d.)

Background: The Pinellas County Florida IT department was in a crisis. The department had built a reputation for budget overruns on projects and needed to take immediate action. It had lost business relationship with other county agencies and as such lost viable sources of income.

Case Conclusion: The county IT department was able to institute Best Practices that aided them in predicting outcomes of projects more accurately. From this, they were able to change customer perception through implementation of measures and controls that allowed better on time, within budget projects. The department was also able to motivate individuals at the project level by synergizing projects with business model strategies.

Case2: *OPM* ProductSuite in Action: Savannah River Site (PMI OPM3 Case Study 2, n.d.)

Background: The Washington Savannah River Co. (WSRC) was used as one of the initial test locations for the OPM3 model suite. Although the company had embraced formal methodologies they continued to face various issues that ranged from scope creep with stagnating budget to scheduling constraints.

Case Conclusion: WSRC was assessed against Best Practices for its Project and Program management processes. Although the assessment was able to conclude that WSRC was operating at a relatively high maturity level (95%), the assessment team was able to conclude that additional maturity opportunities can be explored, even in mature organizations. OPM3 was able to provide "WSRC with knowledge of specific areas in which project and program management could be improved as part of an internal project improvement process."

31.17 CONCLUSION

Model-driven evaluation of organizational processes is a relatively new concept. In order for this type of high-level analysis to be accepted widely, it will take more tangible evidence of its value. However, there is little doubt as to the logical approach embedded in the basic concepts of maturity models such as CMM, CMMI, and OPM3. Organizations that understand the idea of process maturity, formal internal evaluation, and continuous improvement are on the right trail to overall organizational productivity and success. One of the key operational questions regarding maturity assessment involves how to continually maintain a Best Practices structure and then evaluate whether PMI or SEI will best serve the role of standards supporter.

Some organizations have still not accepted the notion that improved maturity as defined by a model does in fact lead to improved operational performance. So, before this concept will move forward these two issues will have to be dealt with—(1) the reference source and (2) philosophical acceptance of the assessment process. Only time will tell on these issues, aggressive organizations who can follow this general direction are likely to move upward in competitive position. Regardless of how the host organization pursues this strategy, the modern PM must understand how his project is impacted positively or negatively by the maturity of his organizational environment. Insight into some of the key support processes can be gained through an understanding of these models, so even that level of involvement has merit.

REFERENCES

CMMI Product Team, 2002. *Capability Maturity Model Integration (CMMI) Version 1.1*. Pittsburgh: Carnegie Mellon Software Engineering Institute.

Cooke-Davies, T.J., 2004. Innovations—Project Management Research 2004. http://www.humansystems.net/papers/measuring_organizational_maturity.pdf (accessed March 17, 2008).

Haeck B. 2004. Organizational Project Management Maturity—The Next Wave in Excellence. http://www.onetooneinteractive.com/resource/whitepapers/0029.html (accessed March 17, 2008).

Ibbs, W., 2007. Ibbs Consulting Group, private communication with Dr. William Ibbs.

LLC, Grail. 2006. Grail FAQ. Grailllc. [Online]. http://www.grailllc.com/FAQsforOPM3.htm (accessed April 20, 2008).

Paulk, M.C., B. Curtis, M.B. Chrissis, and C.V. Weber, 1993a. Capability Maturity Model for Software, Version 1.1, Technical Report: CMU/SEI-93-TR-024 ESC-TR-93-177: 1-5.

Paulk, M.C., C.V. Weber, S.M. Garcia, M.B. Chrissis, and M. Bush, 1993b. Key Practices of the Capability Maturity Model, Version 1.1, Technical Report: CMU/SEI-93-TR-025 ESC-TR-93-178: O-9-16.

PMI, 2003. *OPM3 Knowledge Foundation*. Newtown Square, PA: Project Management Institute.

PMI, 2008. *A Guide to the Project Management Body of Knowledge* (*PMBOK*® *Guide*), Fourth Edition. Newtown Square, PA: Project Management Institute.

Schlichter, J., R. Friedrich, and B. Haeck, 2003. *The History of OPM3*. The Netherlands: PMI's Global Congress.

Software Engineering Institute, 2001. Process Maturity Profile of the Software Community 2000 Year End Update.

32 Project Portfolio Management

32.1 INTRODUCTION

Companies today are awash with projects. Whether creating and launching new products, implementing new technologies, integrating new acquisitions, or upgrading existing products or technologies, organizations spend billions of dollars and millions of work hours on countless thousands of projects. Whether they realize it or not, each business has some method of selecting and managing multiple projects already. The real issue is to ensure that the right projects are being selected at the right time, and does this process continue to assess that selection through the development life cycle. Also, decision processes are needed to ensure that there is comparison made between other proposals, existing installed processes, other ongoing projects, and the newly proposed potential new one. The goal is not to select a good project; it is to select the best one from all competitors.

Project portfolio Management (PPM) represents a centralized management process for this area of activity. It includes "identifying, prioritizing, authorizing, managing, and controlling projects, programs, and other related work, to achieve specific business objectives" (PMI, 2008, p. 433).

Included in this process is a single, organized view of all proposed and active projects and programs including data regarding objectives, costs, timelines, status, resources, risks, and other critical factors.

Just as any investor must select the right balance of investments to meet their "value" goals, each company must select the right mix of projects to meet its goals. PPM is a strategic view of this valuation process and it attempts to find the optimum mix of projects in pursuit of those valuation objectives. Recognize that the wrong project well executed is still the wrong project. Also, organizations have countless choices on which to spend their time, money, and HRs, which means that making these decisions in a highly competitive environment is critically important and will continue to grow more so in the future. In contrasting the goal of portfolio management to project management, the former is the strategic arm of organizational continuous improvement actions, whereas project management is the tactical arm focused on ensuring that individual projects are completed according to plan. These two processes must be linked together to achieve the proper outcome.

32.2 ROLE OF PPM

The logic of looking at project activity as a collected whole represents the theory of portfolio management, but what is its role in the organization? Is this just another gimmick management technique with little substance that will die away when the next one comes along? The answer to these basic questions is the theme of this chapter. It is important to point out that this topic is still maturing and that is the rationale for placing it among the contemporary topics rather than as a core management process. Likewise, its organization framework is evolving and in many organizations this is closely associated with their Project Management Organization (PMO) initiatives that will be discussed in the next chapter. In fact, there are three evolving roles intertwined—project management, PPM and PMO. In this chapter, we will focus on the portfolio aspects and deal with the PMO role Chapter 33.

One of the major values for having a portfolio level view of projects is to get away from the more isolated departmental (stovepipe) view of funding these initiatives. The portfolio approach places all projects on the same evaluation plane and allocates resources centrally to those selected. The vocabulary term for the array of ranked projects is the *efficient frontier*, meaning essentially to choose the highest value option first, then second berst and so on. Thus, the output goal of portfolio management is to define that aggregation of options.

There is no simple method for defining the value of a working portfolio management process, which effectively quantifies valuation of individual project proposals, but think of the following example as a justification starting point. Some have said that having an effective project management process can improve overal project value by 25% (time and cost). We have touted the merits of this idea through many previous sections of the text. Now, what if a good portfolio selection process could do essentially the same for the entire slate of projects? This suggests that the two organizational processes working together could produce a major operational improvement in terms of resources and competitive aspects. Pennypacker and Cabinis report industry examples where project lead times to market have been reduced by as much as 60%, development costs significantly declined, quality improved, and forecasting accuracy increased (Pennypacker and Cabanis, 2003). Reports of these type of results are causing organizations to look carefully at this process. In response to this interest, there is a growing vendor market for support tools, consulting, and process documentation.

As we have said previously , *doing the wrong project right does not make the right project*! Selecting the proper targets is the piece of the management puzzle that we are tackling here. By making better decisions regarding which projects are funded, how they are governed, and which projects are allowed to continue to completion supports the goal of being better able to deliver competitive value to the marketplace.

32.3 IMPROVING PROJECT SELECTION DECISIONS

PPM processes and tools provide real-time information to help executives make better decisions about their overall project environment. In order to support the PPM each initiative (existing and planned) must have timely valuation and status metrics collected and analyzed. These data are then continually matched against organizational goals to help define which initiatives best align the entire portfolio with the organization's business strategies and objectives, thereby maximizing the competitive value.

32.4 IMPROVING VISIBILITY OF PROJECT PERFORMANCE

The improved visibility provided by PPM processes and software tools gives an unprecedented view into the overall "health" of the organizational projects. This process produces project status in the form of graphical summaries, project scorecards, bubble charts, and communication portals utilized by a wide range of stakeholders. Individual project health can be evaluated and analyzed through the use of predefined metrics related to cost, schedule, scope, quality, risk, and customer satisfaction. As a result of this improved overall visibility, the project decision-making process can also be improved.

32.5 BETTER UNDERSTANDING OF PROJECT VALUE

Thomas and Mullaly have documented the complexity of assessing project value (Thomas and Mullaly, 2008). Project value as we now understand it is more than simple, measurable financial metrics. This multivariable view complicates all project assessment techniques and that remains an unresolved issue for PPM. However, the PPM processes and tools have expanded the ability to heuristically (rules of thumb) and financially evaluate the portfolio. Also, the ability to group and score the portfolio allows greatly improved insight into the overall picture. For example, projects could be grouped into categories such as new applications, maintenance, or upgrades. Scoring methodologies based on ratings such as competitive value, improvement rating, risk assessment, and financial calculations can be specified to better understand the return on investment from each project within or across the different types of projects. All of these capabilities support improved project selection and oversight.

32.6 CONDUCTING "WHAT IF" ANALYSIS

Just as in financial situations, business trends change continuously. A primary tool in managing dynamic and uncertain situations of this type is simulation, or conducting "what if?"-type analysis. Analyzing these characteristics can enable a better understanding of how uncertainty can impact the outcome of a project. This knowledge can be used to evaluate how sensitive the outcome is to various assumptions or changes in the project. A sample of this type analysis was discussed previously in Chapter 16. More use of this same type of technology is relevant at the portfolio level. Simulation is a critical tool for situations that have more variables and complexity than can be reduced to deterministic mathematical equations. Use of "what if"-type analysis allows various assumptions to be reviewed that would not be possible with traditional tools.

32.7 PROJECT INVESTMENT MANAGEMENT

The traditional view of project success has been defined by the delivery of defined requirements on time and on budget. The PPM view of project success is to deliver an optimum set of projects that contribute the most collective value relative to their cost and to other potential project investments. If all projects are aligned with the organizational goals, then executing them to optimize their performance and value will deliver the greatest returns.

Fundamentally, PPM is a process discipline used to ensure that a correct mix of investment activity is initiated, grouped, funded, and managed. In order to execute this activity, the following five components must be in place:

- Investment organization to manage the overall process
- Prioritization techniques based on organizational value metrics
- Evaluation process that includes senior management
- Decision insight and support—understanding the organization and the value that technology can bring
- Balancing current needs and future requirements—tactical versus strategic perspectives.

32.8 WHO NEEDS A PPM?

Given the theoretical logic presented thus far, one might conclude that every organization needs a PPM. If that were true all would have one installed and working. That is not the case in reality. Summarized below are visible indicators that the existing internal de facto project selection process is not working effectively (Greer, 2008):

- Frequent difficulties in finding enough qualified HRs for selected projects
- Excessive project overruns from "not enough resources"
- High personnel turnover due to "burn out" of key project contributors because they are working on too many projects and spending too many overtime hours
- Frequent changes in project status (i.e., moving from "active" to "on hold" to "top priority" and back)

- Completion of projects that do not meet organizational goals—too much tactical and not enough strategic views
- Intense competition among departments rather than cooperation. Tendencies to fund local efforts at the expense of more global projects
- Local projects are favored over global projects.

At the minimum, a PPM process aids in the following ways:

- Improving fiscal management
- Improving communication between the project team and business management
- Quantifying the benefits of a project
- Deciding how best to assign resources
- Setting priorities
- Identifying and managing risks
- Assessing the impact of adverse events
- Selling a project vision concept
- Obtaining funds for a project.

Effective portfolio management ensures that projects continue to support the organizational mission as defined by formal goals. Successful project management by definition means that projects are completed on schedule and budget.

As organizations develop a strategy for achieving their mission, executives have to consider the following business drivers as they make their investment decisions:

- How do we ensure that we are working on the right initiatives to support business objectives?
- Do we have a consistent method to measure the key performance indicators on active initiatives?
- Can the organization define standards that improve the consistency of business initiative outcomes?
- Do we have enough people and dollars to deliver our commitments?
- Can the organization respond adequately to changing business conditions and realign the initiatives required to respond to the identified changes?

There is often a linkage between a project, or portfolio of projects, and an organizational initiative. Once an organizational level initiative is established, one or more projects may be envisioned to move the organization in that direction. From that view a portfolio entry would follow to quantify the value of that initiative. At this point the portfolio entry represents a conceptual project. As these conceptual projects are identified, various resource and performance metrics are added through an analysis process. This includes a summary of the dollars, people, and time requirements for the initiative. Also, a corresponding valuation score is defined to correspond to the various resource requirements and risk characteristics. Other decision considerations might be added to indicate items such as a timing constraint. As all of these decision elements are combined, it becomes possible to assess how the individual project visions best fit into the overall portfolio scheme given aggregate constraints for the organization (usually people and dollars).

32.9 PPM GOAL STRUCTURE

At the highest level we have described the goal of PPM as matching projects to organizational goals and through this make related decisions to improve organizational competitiveness. Associated with this high-level goal are more definable subgoals that support this. They are described below.

32.9.1   SUBGOAL 1: STRATEGIC GOAL ALIGNMENT

Why do we want strategically aligned projects? Basically, a project should exist to achieve a company goal or a set of goals. Projects are initiated to improve an existing state for a product, a service, produce a result, or some variation of these three scenarios. A project should not exist if it does not support organizational objectives. The solution to this requirement does not "just happen." Many times, the selection process is disjointed and fragmented throughout the organization. The PPM process helps to ensure that all organizational projects are in alignment and contributing to company goals. One area that it deals with is departmental bias as a selection criterion regarding which project to fund (i.e., no "pet" projects unless they specifically bring appropriate value to the organization). Regardless of all other considerations, the final portfolio of projects should truly reflect the business' strategy (Cooper et al., 2001). Additionally, aligning the portfolio selection to the business strategy helps companies to adapt quickly to meet new business challenges as well as leverage current investments, and make better investment decisions (Spizzuco, 2005).

32.9.2   SUBGOAL 2: RESOURCE INVESTMENT FOCUS

Why is resource investment focus important? Resources are neither unlimited nor infinite. Consequently, they must be allocated to targets wisely in order to deliver maximum return where expended. Careful investment focus helps to optimize the value returned. An effective PPM process brings an investment focus to the analysis of projects and allocates resources to obtain maximum gain. Without investment focus the organization can easily undertake projects that essentially drain resources and deliver less value. It is a common practice to over allocate resources to too many projects. In some cases this is done because the organization does not have a valid information resource that matches the allocations to the available resource pool, while in other situations the planners are just not concerned about the supply side issues. When either of these situations occur, the result is a project gridlock (Cooper et al., 2001). The problem of too many projects and too few resources can be partly resolved by having a formal process that matches estimated resources to a defined resource pool. In order to be successful with the project management aspects of this process, it is vital that some meaningful measure of resources for all projects is used as part of the ultimate project selection process. Overstretching HRs by avoiding capacity management leads to lower productivity and unsuccessful project outcomes.

32.9.3   SUBGOAL 3: BETTER PROJECT CONTROL/GOVERNANCE

Projects are dynamic entities that if not controlled can quickly spiral out of control on multiple fronts. The governance process affects many decision domains including setting business priorities, budgeting, project selection, resource allocation, application portfolio strategy, and performance measurement (Gruai, 2005). An effective control and governance process is required to provide continuous oversight of projects and portfolios ensuring that they remain within the organization's defined performance framework. However, an appropriate balance has to be struck between too much control (nothing significant gets done) and too little control (things get done that are not of value). PPM processes and status data can provide much of the informational framework required for this activity. The organizational question involved here involves how to energize that data into an effective management and control environment. In order to accomplish this, it is necessary to define an effective governance framework for both the project and portfolio levels. Putting portfolio management in place can force companies with weak governance structures to improve them (Datz, 2003).

A PPM structure serves as a driver of greater engagement in project activity by senior management through the selection process and greater project visibility. PPM governance involves providing oversight, control, and decision making for all ongoing initiatives (Hanford, 2005). Improved project governance also leads to improved cost control, which is often a weakness for many projects. As a part of continuously managing the portfolio, project costs are also carefully analyzed. Areas where costs can be reduced across the portfolio or within particular projects are more visible. In addition, the increased level of attention to overall status helps to answer the key question of whether the project is still viable given its current status. If the answer to this question is "no," then there is an opportunity to restructure the portfolio and save resources. If this type of ongoing analysis is also not done, poor value initiatives are allowed to continue to completion, even though the result is not viable on a cost–benefit basis.

32.9.4 SUBGOAL 4: EFFICIENCY

Closely tied to better governance and cost control is the concept of efficiency. A company's goal is to have projects perform well and achieve desired results without wasted resources or effort. At the extreme the goal is to achieve an optimal level of performance for the portfolio. This process begins at project selection by ensuring that the best project option is selected for funding and continues through the project life cycle with dynamic monitoring and control. Also, the efficiency concept involves providing continuous oversight of the portfolio and to seek out areas which can be corrected and/or improved. This process ensures a more efficient overall use of organizational resources—humans, materials, time, and financial. PPM also helps to achieve efficiency through its process of measuring efficiency and performance as projects are tracked through their life cycles.

Some of the issues that PPM can surface include the following:

- Projects not having clear critical success criteria
- Project management processes and tools that are not working well
- PMs not having clear mandates, sufficient competence, or capacity to achieve satisfactory project outcomes.

By bringing these issues to the surface and ensuring that they are dealt with, portfolio management methods help create greater efficiency in project completion (Wideman, 2005).

32.9.5 SUBGOAL 5: BALANCE

Modern organizations increasingly have multiple and changing goals that they desire to attain over varying time periods. Consequently, it is necessary to have an appropriate mix of projects matched to those goals in order to achieve these diverse objectives. A balanced portfolio takes into account the timing factor as well as the goal structure. A large number of projects in one goal area while neglecting other goals will violate the concept of balance. Effective balancing through the PPM evaluation process helps to achieve a desired balance of projects in terms of a number of parameters; for example, the right balance in terms of long-term projects versus short ones; or high-risk projects versus lower-risk projects; and across various markets, technologies, product categories, and project types (Cooper et al., 2001). Obviously, the driver for the balancing process is a set of articulated organizational goals and strategies that can be translated into specific portfolio initiatives. From this base, PPM will help match viable projects across the various goals.

32.9.6 SUBGOAL 6: VALUE OPTIMIZATION

Projects are envisioned and designed to bring value to the organization, and while there are many that can be funded and do bring value, not all can be pursued because of various constraints. So, a critical subgoal of PPM is to maximize the value produced from the selected projects given the constraint set.

Having all the projects overseen from "under one roof" and with a standardized set of evaluation criteria is a prerequisite for the PPM process. In reality not all projects will subsequently bring value despite the best attempts and good intentions of the management processes. However, this reality should not deter the PPM process, but should stimulate interest in improving the evaluation process over time. Recognition that an optimum solution was not being achieved is a strong motivator for improvement.

The final point on value optimization is to recognize that better project visibility obtained at the executive level through PPM methods enables a quicker termination of projects that turn out to be "nonvalue added," thus saving resources for the organization over time. From a pragmatic viewpoint, this is one of the quickest value paybacks for a PPM process.

Financial portfolio analysis: Portfolio balance and risk mitigation are achieved by spreading investments over a number of different initiatives. In this fashion, projects are balanced across a number of categories that can include strategic or tactical business objectives, compliance, required maintenance, operational efficiency, and research and development. Depending on the organization's objectives these factors would be used in the selection process to achieve the best match regarding overall goals and market dynamics.

Top-down/bottom-up approach: There are two basic methods to develop the portfolio—either by looking at top-level goals and decomposing downward to projects that will develop those goals, or by sorting through individual project proposals and lower-level business unit objectives to aggregate a draft portfolio. Both of these approaches are popular with immature project organizations (Pennypacker and Cabanis, 2003). If the portfolio design process is conducted separately with either option it will produce disconnected view of the interrelationships of projects to the organization goals. A combined top-down/bottom-up approach is the desired solution in order to link the two layers and that is essentially what the PPM mechanics are designed to do.

32.10   MODELS OF PPM

There are two traditional organizational models for implementing a PPM system (INNOTAS, 2007, p. 1):

- The *engagement profitability* model
- The *budget alignment* model.

In the *engagement profitability* model, "projects" and "programs" are vehicles for managing revenue-generating engagements with customers that produce profit margins. Decisions and behavior are driven by the profitability of customer engagements. Examples of this model include IT services firms and professional services departments within product companies. The focus at each level includes project, resource, and customer performance evaluation.

As Figure 32.1 implies, the PPM initial focus is typically on project and resource performance. As the process matures over time, automation and expertise evolves, the management focus often turns more toward the higher-order benefits of improved performance based on increasing focus toward customers, products, and lines of business.

The *budget alignment* model corresponds with an operational environment where the value of a projects is harder to assess (and these projects typically do not generate revenue directly), and project costs are considered overhead. In this type of environment decisions and behavior are driven by the need to squeeze value out of the available budget. Examples of the budget alignment business model include Enterprise IT and product development organizations.

Once the project portfolio is identified and managed as a coordinated portfolio of investments, the management challenge becomes a question of prioritization among the viable options. At this stage the key issues involve an assessment of existing projects and dealing with the resource allocation activities needed to support those initiatives. A final parallel concern is the ongoing maturation of standardized processes and metric development. If we look at this evolutionary

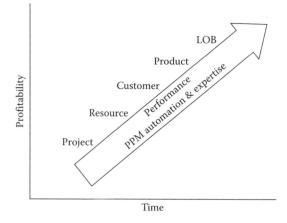

FIGURE 32.1 PPM profitability ladder.

process in stages, the sequence focuses first on demand related to identification and valuation of the target portfolio. Stage two involves methodologies for planning and scheduling the portfolio, which in turn deals with matching resources and constraints to the target portfolio. This results in an identification of the efficient frontier and scheduling considerations. Finally, at stage three the concerns are executing the plan, monitoring status, and implementing the solutions.

In both of the models described, achieving *customer satisfaction* is a critical driver, but the customers are different in each case. In the engagement profitability model, the customer is the target market being served, whereas in the budget alignment model, the customer is usually internal. In both of these models the PPM process has the potential to create improved value.

32.11   THE HYBRID MODEL

Some environments structure a combination of the two models described above, taking the best benefits from both. An example of this would be to produce a development organization that uses an in-house professional services organization to implement their products. To drive development decisions and deliver new products, the product development team will use the supply–demand delivery framework, while the professional services team will use the performance stack for their segment.

32.12   EFFICIENT FRONTIER

The efficient frontier maps specific projects against their value in rank order. So, the highest value items would start at the origin. Each succeeding project would have a lesser value and the shape of the curve would be as shown in Figure 32.2. This curve defines the maximum value that can be achieved from the optimum portfolio. The second consideration is the resource constraint. This level defines the maximum value that can be achieved for the optimum portfolio.

In the operational mode, it will be necessary to allocate some projects that are not on the efficient frontier. Regulatory requirements are a typical example. When these are included in the portfolio the overall value is decreased. In concept, this curve represents the portfolio, target and helps to assess the penalty taken by choosing lesser alternatives. Also, it helps to assess the penalty taken by a resource constraint being at some particular point. Various vendors know how PPM tools that can produce these figures and manipulate decisions impact for alternative portfolio options.

32.12.1   PROJECT ASSESSMENT

As the PPM database is populated with various data regarding project proposals and status the question becomes how to display meaningful information for decision making. There are various types of

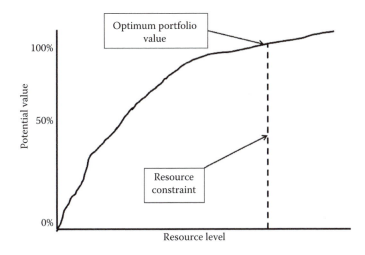

FIGURE 32.2   Efficient frontier.

reports that can be created to show status grouped in any imaginable form. The efficient frontier is one new assessment tool and a second one is called the *bubble chart*. This is a visual representation of projects based on risk versus value. Figure 32.3 provides a hypothetical sample view of this format.

In this example, the highest value and lowest-risk projects would be reflected by bubbles in the top-right quadrant, whereas lowest value and highest-risk initiatives would be closer to the origin. Also, the size of the bubble is a visual measure of the size of the project. A display of this type is useful for envisioning the overall portfolio. Multidimensional bubble diagrams offer an interesting perspective and visibility into the portfolio. The portfolio database supports different views and options with graphics and other decision analytics.

Information capabilities as briefly summarized here help facilitate an enhanced dialogue between the project team and business unit executives, allowing a better balance between project demand and the optimum use of available capital and HRs. These new informational capabilities represent just one of the ways in which PPM processes provide a new perspective for an organization to achieve more with less.

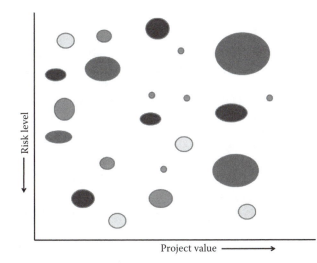

FIGURE 32.3   Bubble chart.

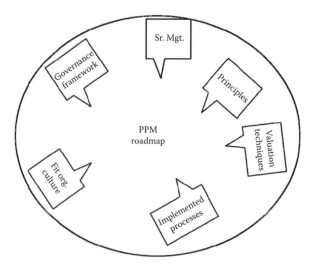

FIGURE 32.4 Implementing PPM.

32.13 KEYS TO IMPLEMENTING PPM

Portfolio management does not just consist of selecting the right project investments; it also means ensuring proper execution and regularly measuring effective performance, notably in the delivery of the anticipated benefits and the correct usage of the invested budgets. Figure 32.4 summarizes the basic components of the overall PPM structure. This is not a prescriptive view in that selection of the particular item to deal with first depends on the particular organizations needs. We have discussed these elements along the way, so this diagram is considered a summary of the basic implementation roadmap.

One of the difficulties in PPM implementation is the need to review ongoing status and add that perspective to the overall process. This concept does not show well in the schematic above, but failure to recognize it would mean that the process just continued to throw new projects into the pool and not consider status of the ongoing items. Business changes, variables in ongoing project performance, risk events, and a host of other factors can well mean that an optimal decision 6 months ago is not of zero value to the organization. The status system results must be frequently evaluated to ensure that previous decisions are still valid in relation to other projects.

32.14 PPM PRINCIPLES

PPM is not just another project management process. PPM is also a philosophy—one that, in accordance with the analogy based on financial portfolio management, is focused on value creation. Getting the most from PPM requires that an organization fully embrace the following principles:

- Projects will be managed as a portfolio of resource investments
- Projects will be identified based on their match to organizational goals and related value
- Resource availability and risk will be considered in the decision process
- Projects will be defined and selected to include the full scope of activities necessary to generate value
- Value delivery practices will recognize that there are different types of projects that will be evaluated and managed differently
- The delivery process will be managed throughout the life cycle to ensure that the value equation is intact
- All relevant stakeholders will be engaged in the process and assigned appropriate accountability for the delivery of capabilities and the realization of value
- Projects that fall out of the optimum mix will be terminated in place.

32.15 FINDING THE APPROACH THAT FITS

PPM is not a "one-size fits all" solution. Despite the general applicability of common principles, there is not a single roadmap approach for implementation or organizational structure. The alternative approaches reflect different views regarding how best to accomplish PPM goals in light of the individual culture, maturity, and other realities. These different approaches reflect different assumptions, methodologies, models, structures, roles and responsibilities, reporting lines, resource demands, and levels of authority. The implementation challenge involves designing an approach that will work well within a specific organizational culture.

The first step is to define the PPM processes that will initially best support the business need and from this obtain buy-in and general consensus for the approach. Without this basic rationale and consensus, the subsequent steps will fail. Once the appropriate initial approach has been designed, use the desired output requirements of that approach as a checklist for choosing the right operational support processes.

32.16 EXECUTIVE SUPPORT

According to surveys, the biggest challenge for implementing PPM is lack of adequate executive support. Introducing PPM into an organization requires a significant investment of time and money. It requires learning new concepts and skills, instituting new processes, and achieving a significant cultural change. Realistically, the deployment of PPM within the organization will not be popular with everyone. Support from the top is needed to lend credibility and authority and to guide the right behavior in the organization.

32.17 GOVERNANCE FRAMEWORK

Effective governance also starts with executive leadership, commitment, and support. However, this by itself is not sufficient to achieve the final result. An overall hierarchical governance framework must be defined with roles and responsibilities for all participants. A sample summary of roles and responsibilities for various stakeholder groups is outlined in Table 32.1.

Lee Merkhofer offers a model organization to illustrate roles and responsibilities for the PPM function. A sample organizational structure is shown in Figure 32.5, although there are obviously

TABLE 32.1
PPM Roles and Responsibilities

Role	Responsibilities
Executive team	This group establishes portfolio funding constraint levels, approves project recommendations, and provides policy guidance
Portfolio management team	The organizational unit is a delegated responsibility for carrying out PPM duties
Portfolio manager	Head of the Portfolio Management Team, responsibilities include making project recommendations and working with the Executive Team
Portfolio administration	Individual or group responsible for collecting project information, applying tools, and coordinating the day-to-day steps of the portfolio management process
Business managers	Persons responsible for managing outputs of projects in the business environment. Responsibilities include verifying project cost, value, and risk estimates for projects within their respective business areas
Project managers	Persons responsible for day-to-day management of individual projects. Responsibilities include providing project proposal data and communicating project status to Program Managers and the Portfolio Manager

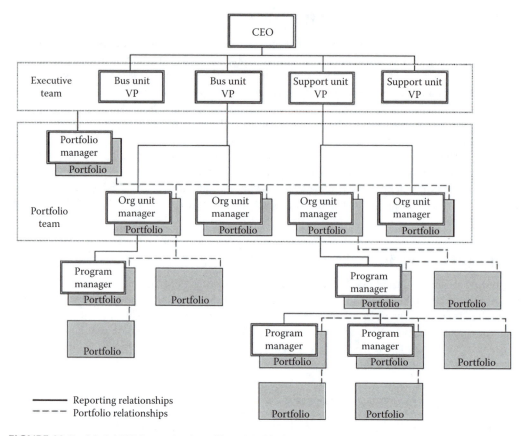

FIGURE 32.5 Model PPM organization. (From Merkhofer, L., 2007. Seven Keys to Implementing Project Portfolio Management, p. 22. Retrieved from http://www.prioritysystem.com/PDF/implementingppm.pdf. With permission.)

other options as well. Note that some coherent grouping of program managers in an organizational structure implies their control of the lower levels of this structure. In many cases this might be viewed as a departmental grouping more than a program grouping, but the responsibilities would be similar for both roles.

32.18   VALUE-MEASUREMENT FRAMEWORK

A value-measurement framework defines how the organization will create a value expression for a particular initiative. There are various methods used for this process; however, the fact that different organizations create value in different ways means that the models for measuring project value are necessarily somewhat unique for different organizations. Among other things, creating a value-measurement framework requires that the organization decide for whom value is to be created (i.e., shareholders, external customers, internal customers, etc.), and how to measure the different kinds of value that are being created (i.e., financial, customer satisfaction, quality level, speed of service, etc.). A value framework will need to help answer the following questions:

- What is the value of conducting this project (probably in a numeric score format)?
- What are the sources of value (e.g., reduced costs, increased revenue, increased customer satisfaction, new learning, process capability, etc.)?
- What are the risks compared to the organization's risk tolerance and what is the risk-adjusted value of the project?

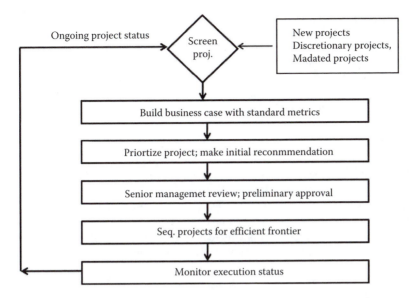

FIGURE 32.6   PPM process steps.

32.19   INSTITUTE EFFECTIVE PROCESSES

An organizational PPM process must be established as a formal, documented, and repeatable activity. By definition, these process steps cannot be *ad hoc*. The rigor by which the process is executed will drive the resulting impact on the success of organization. Excellence can only be achieved when standardized procedures, tools, training and support functions are established, implemented, and continuously improved upon. In sum, PPM should be looked at as an ongoing improvement process. Each step along the implementation trail will uncover a weakness that will need to be resolved. The basic process steps for PPM are summarized in Figure 32.6.

32.20   PPM IMPLEMENTATION ROADMAP

The basic process components of the PPM approach have now been described. The key at this stage is how to move these components into the organization successfully. Each implementation step should build off of the previous one. This requires a formal assessment of each step, then package the next step design from that foundation. The following nine steps should be considered as a basic implementation process (INNOTAS, 2007, p. 3):

- Assess the current organizational status as to management gaps in project selection and execution.
- Review the gap analysis with key stakeholders.
- Develop a formal vision for the future PPM process.
- Create a project charter for the first stage process; senior management approval required for this charter.
- Design stage one PPM tools and processes.
- Test stage one concepts in pilot organization.
- Design production version tools and processes from pilot lessons learned. Production scale tools required for next stage. Plan roll out program.
- Execute enterprise roll out.
- Monitor ongoing status and evolve process.

32.21 EXTERNAL EXPERTISE

The PPM process involves specialized expertise that will likely be in short supply within an organization. Oftentimes a consulting organization provides the fastest access to best practices and brings an outsider's perspective. Consultants also generally get better access to senior executives and from that perspective help serve as catalysts for change. There are many training and mentoring activities that can also be supported by a specialized consulting firm.

32.22 IMPLEMENTATION GOALS

PPM can be a significant enabler for enhanced efficiency, effectiveness, and productivity in the project domain, while reducing exposure to risks related to project failures. However, it would be a mistake to think that this is a cookie-cutter solution that is easy to implement, or that there is one defined solution. Any new idea that involves large segments of the organization brings with it a complex change management process. For this reason, choosing the right scope and approach can dramatically improve the initial success. Choosing the wrong approach can harm the organization by increasing costs, wasting valuable time, and generating useless and inaccurate information. For this reason, it is important to first invest the effort it takes to identify and understand alternatives and to make the right choices for the specific organization.

32.23 KEY PPM INTERFACES

There are five key organizational interface components to PPM. These are

- Strategic planning
- Business stakeholders
- Executive management
- Project management
- Project management office.

Strategic planning: Strategic planning is a high-level function to guide organizations focus on long range, critical business areas. One of the key goals of this function is to define, assess, and adjust the organization's direction in response to a changing environment. Executives use this information in their management role in leading the organization toward strategic objectives and the strategic planning function is a major support element for this activity. Decisions made in the strategic planning process are a prerequisite for PPM. Without a clear strategy, there will be no defined criteria for choosing among the many competing project requests.

Business stakeholders: Stakeholders are individuals or groups affected by the projects in the portfolio. These individuals are not only affected by the projects' performance, but often will be the future users of the product or process being developed. By making project performance more visible, PPM allows not only executives more insight and better decision making, but also stakeholders benefit from increased information. This group needs to be involved in the process and they offer key knowledge regarding how to use the initiatives effectively. Failure to involve key stakeholders is a major source of project failure. By facilitating project communication with stakeholders early and often, PPM ensures that stakeholders know what is going on and fully understands the project benefits. This means that stakeholders can provide better feedback and actively support the development process. The PPM communication mechanism also means that executives get improved feedback and better know when to kill a project that stakeholders have indicated will not work as envisioned.

Executive commitment: A critical prerequisite to executing PPM methods and tools is the commitment of senior executives. Just as with any other major initiative, the organizational implementation of PPM is a megaproject that requires commitment, focus, and dedication from senior leadership, otherwise it is lost in the sea of competing priorities. The entire executive team must be committed to the principles described here and an appropriate senior executive should serve as the project sponsor. Implementation of PPM processes will require changing how functional managers, PMs, project workers, and stakeholders deal with the project culture. Creating this cultural change will require strong, consistent, focused leadership from the highest level, otherwise the project will be doomed to be another "band aid" or "management fad" that will be announced with great fanfare, practiced half heartedly for a few months, forgotten as soon as the next crisis surfaces, and then the organization returns to business as normal.

Project management: As indicated earlier, it is the role of the project teams to execute the defined initiatives. PPM processes drive the high-level project requirements and project management then supplies status back to the PPM for overall assessment activities.

PMO: A formal PMO provides a commonly used organizational focal point for managing the portfolio. As such, it is the companion process to PPM (or vice versa) and may in fact be the organization entity that runs the PPM process. The PMO is often formally delegated to oversee strategic PPM processes and tools as well as individual project management practices, plus selecting and utilizing tools and methods that help stakeholders understand the status and progress of projects. In many organizations that already have a PMO established structure the PPM function and all PMs may be housed there. The PMO also provides support to sponsors and senior management in establishing cross-project priorities and to PMs in organizing work to help keep projects moving forward.

32.24   PPM IMPLEMENTATION CHALLENGES

Typically, projects are driven and defined by customers who set project and goal milestones, schedules, and other requirements. This approach leads to difficulty in defining what a project should be from a goal value point of view. Quite likely, the major challenge in implementing a PPM process is the change it will require in the organizational project selection process and the resource allocation process associated with that. PPM processes clearly have the flavor of centralization and that characteristic has long been correlated with implementation complexity and resistance by lower levels of the organization. As with all centralization strategies, lower-level decision freedoms will be reduced by this process. Also, the PPM analysis process makes it difficult to hide project mistakes and much of the organization will not appreciate this uncovering activity.

32.25   ADVANTAGES OF IMPLEMENTING PPM

In many cases one can judge the merit of an idea by how many organizations are moving in that direction. A survey by the META Group showed that nearly 60% of the 219 CIO's questioned had implemented portfolio management or had an official plan to do so. Also, the level of product and consulting activity among the commercial vendors suggests an active market.

It is easy to see that having a comprehensive and shared view of all ongoing or planned projects and initiatives and associated key indicators have value from a management perspective and should lead to better project decision making. The basic principle of selecting the best global project slate is obviously logical as is the notion of matching resources to projects. These are nothing more than basic management best practices. For these reasons the rationale for doing PPM-type processes is essentially irrefutable. It would also be difficult to say that ideas such as the efficient frontier did not make sense. That communicating overall status was a bad idea. Similarly, matching projects against organizational goals, assessing risk, and continuous monitoring of status all seem very difficult to reject. The simple step of data collection and consolidation enables organizations to realize how much work is in progress and the related status of that work. All components of the concept seem worthy for any organization using projects to achieve change.

32.26 SUMMARY

PPM is now one of the most popular project management concepts. Its obvious logical reason to exist has created an industry in itself. Some organizations have found the concepts to be highly effective if managed efficiently, whereas others have failed in their attempts to change their existing culture. Implementing a PPM structure to support project selection is an essential element in achieving organizational goal alignment, investment focus, governance, cost control, and valuation assessment.

Two standard models for implementing a PPM structure were introduced; namely, the engagement profitability model and the budget alignment model. A third hybrid model is a new approach stemming from the combination of the two models. Regardless of the structure selected for an organization, it needs to be carefully thought out, as there is no one single best choice for this process.

PPM implementation success will be fueled by embracing the key principles, selecting an approach that fits the organization, securing executive support, establishing an appropriate governance structure, developing a value measurement framework, instituting effective processes, and following a well-planned incremental roadmap for implementation. All large organizational process changes have challenges and so does the implementation of PPM. Once in place, project variances become public and this makes it difficult to hide mistakes. For that and many other reasons the project culture will be slow in accepting PPM. As a matter of fact, the real value of PPM may be more high level than low level if one just looks at the day-to-day work. In spite of the inherent challenges, PPM has a variety of advantages. It provides the advantage of having a comprehensive and shared view of all ongoing or planned projects and initiatives and associated key value and cost indicators. It also offers various information distribution features that help the stakeholder community understand what is going on.

PPM and project management together represent two of the most significant strategies that an organization can follow to produce more goal achievement with less resources. Accomplishment of both represents a major undertaking for the organization and requires significant involvement of senior management. Chapter 33 will follow this discussion with more details regarding the PMO model, which is also a popular management strategy dealing with this same topic area.

REFERENCES

Cooper, R.G., S.J. Edgett, and E.J. Kleinschmidt, 2001. Portfolio Management: Fundamental for New Product Success. http://www.stage-gate.com/downloads/working_papers/wp_12.pdf (accessed October 24, 2008).

Datz, T., 2003. Portfolio Management Done Right. http://www.cio.com/article/31864/Portfolio_Management_Done_Right?page=1 (accessed October 24, 2008).

Greer, M., 2008. What's Portfolio Management (PPM) and Why Should Project Managers Care About It? http://www.michaelgreer.com/ppm.htm (accessed January 11, 2009).

Gruia, M., 2005. Measure What Matters: New Perspectives on IT Portfolio Selection. http://www.umt.com (accessed October 24, 2008).

Hanford, M., 2005. Portfolio Management: An Introduction. A white paper published by IBM. http://www.ibm.com/developerworks/rational/library/oct05/hanford/index.html (accessed January 11, 2009).

INNOTAS, 2007. The Two Models for Implementing Project Portfolio Management. http://hosteddocs.ittoolbox.com/Innotas102406b.pdf

Merkhofer, L., 2007. Seven Keys to Implementing Project Portfolio Management. Retrieved from http://www.prioritysystem.com/PDF/implementingppm.pdf

Pennypacker, J. and J. Cabanis, 2003. Why corporate leaders should make project portfolio management a priority. http://www.primavera-aus.com/pdf/npd/PPM.pdf

PMI, 2008. *A Guide to the Project Management Body of Knowledge (PMBOK® Guide)*, 4th Edition, Newtown Square, PA: Project Management Institute.

Spizzuco, T., 2005. Optimize Your Vision: Aligning Your IT Portfolio & Business Strategy. http://hosteddocs. ittoolbox.com/TS072407.pdf

Thomas, J. and M. Mullaly, 2008. *Researching the Value of Project Management*. Newtown Square, PA: Project Management Institute.

Wideman, R.M., 2005. Project Portfolio Governance Guidelines (But are they complete?) http://www. maxwideman.com/papers/governance/governance.pdf

33 Enterprise Project Management Office

33.1 INTRODUCTION

As a companion activity to portfolio management the Enterprise Project Management Office (EPMO) concept represents a strategy to centralize various project-related decision-making processes at the enterprise level. This concept is adapted from the more traditional PMO model, but extended in scope to all enterprise projects.

Definition: *A "Project Management Office (PMO) is an organizational body or entity assigned various responsibilities related to the centralized and coordinated management of those projects under its domain"* (PMI, 2008, p. 1). The formal scope of this function can vary from simple project coordination to full ownership of the project portfolio and related resources. Other potential roles involve defining and maintaining the project process standards, documentation, and metrics related to the OPM3 activities. In its normal organizational role the PMO is the IT-oriented operational component of portfolio management charged with high-level oversight of the project management activities in that segment of the enterprise. As discussed here the scope is enterprise-wide, therefore the acronym EPMO is used.

33.2 PMO FUNCTIONS

There is not a singular functional description for a PMO and working versions of this organization unit are quite varied. The scope and depth of a particular PMO will be based on the organizational philosophy of the host. As one of the contemporary processes this concept has had both high praise and failure. Mature organizations that have good project governance can have high success with a centralized PMO with board functional authority. Less mature organizations still struggling with formalizing project management will not be able to achieve the same level of success with centralizing these functions. There are a variety of potential responsibilities that can be grouped under a PMO, depending on a particular organizational philosophy. These functional responsibilities will have a strong influence on the operational project management. The following is the list of operational functions that might be assigned to the PMO (Richardson and Butler, 2006, p. 407):

- Standardized processes and methods for project development
- A formal archiving system to capture lessons learned
- Administrative support for project teams
- Assistance or management activities in staffing projects
- Training programs for project teams
- Consulting and mentoring of project teams
- Evaluating and managing the resource capacity issues related to overall project requirements
- Centralized tracking and communication of project status to appropriate stakeholders

- Aiding or managing the alignment of project activity to business goals
- Performing project quality reviews and audits
- Performing postimplementation reviews
- Managing technical resource capacity for project efforts
- Assessing status of current project WIP.

33.3 ENTERPRISE PROJECT MANAGEMENT OFFICE (EPMO)

The organizational label PMO has been most typically linked to information technology project structures; however, the basic principles are valid for the organizational project portfolio. Because of this view we are calling the function Enterprise Project Management Office (EPMO) to denote that it is organization-wide in scope. This function would be similar in concept to a PMO and involves the integration of processes, technology, organizational structure, and HR to align overall enterprise strategy with the execution of its projects. It includes very contemporary project management approach that includes a framework to deal with the challenges of organizing, managing, and tracking projects throughout the organization. Figure 33.1 contains an schematic overview of this concept and clearly shows the related layers of management and key processes.

Note that there are three tiers of management in the EPMO structure that collectively form the link between lower-level project resources and organizational strategy. At the lowest level, project management focuses on the efficient execution of selected projects. The second layer of portfolio management (PPM) coordinates projects selection and resource allocation. EPMO represents the highest level in the management structure and serves the general role of connecting the strategic vision of the enterprise to the selection of specific projects and programs that enable this vision. In order for these layers to work together, sophisticated collaborative technology is required among the players. A physical organizational structure for EPMO has the same potential variability as that for a PMO. This can be focused on strategic planning or buried in various functional departments. Ideally, the EPMO and PPM functions would be housed together to facilitate coordination between them.

One issue that is visible from Figure 33.1 is the organizational layering issue involved in creating a decision-oriented management level above the project layer. PMs tend to be an independent lot and many will feel that they do not need this level of "help." There is an adage in organizations that says "I'm from headquarters and I'm here to help you." Sometimes, a centralized group is viewed as a bureaucracy that adds no value. This is the sticky human side of the PMO- or EPMO-type organizations.

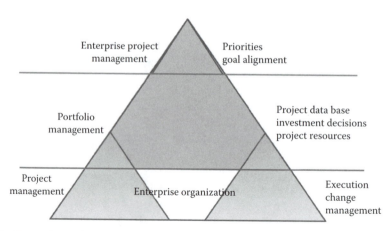

FIGURE 33.1 Enterprise project management model.

33.4 COMMUNICATION

The EPMO function must have an effective communication system between the project teams, key stakeholders, and senior management in order to carry out their required functions. The basic information flows involve the following:

- Project selection activities
- Metrics collection for ongoing projects
- Information distribution
- Project status assessment.

The level of authority attached to these activities is the undefined variable. An EPMO organization can alternatively be an information collector, observer, and transfer agent, or they can have significant authority over the entire pyramid shown in Figure 33.1. Regardless of the balance between these two extremes, an active communication system is required if this concept is to be successful.

33.5 PERFORMANCE METRICS

The EPMO decision environment area is rich in its use of various metrics. Some of the early metrics relate to the valuation of the project proposal, whereas the later metrics become more status oriented. In all of these roles, it is important to identify the success criteria that will be used for decision making. Also, some aggregate level metrics are needed to evaluate the EPMO function itself. This activity is overhead when one looks simply at the work being accomplished. Some of the key evaluation questions related to an EPMO follow:

- Is project throughput better than it was?
- Is the organization improving competitively as a result of better project selection?
- Are resources being managed better than they were previously?
- Is overall organizational maturity improving as a result of EPMO functions?
- Are negative attitudes from lower-level groups compromising the EPMO organization?

The specific metrics to evaluate this level of assessment are complex, but some formal assessment of EPMO value is needed. If a value perspective cannot be developed and sold to senior management, the natural inclination will be to scrap the structure as not adding value.

When the EPMO structure is created and its formal charter is set, its operational role within the organization is also determined. Based on these defined roles (which are highly dependent on the level of authority the organization has over project teams), the performance metrics would also be defined. Performance metrics defined and collected should be chosen such that they motivate the desired behavior in project teams. Use of metrics is one of the primary ways an organization can communicate its values and expectations to project teams. It is up to senior management to validate the metrics they want to use for this.

As a result of implementation complexity, the initial metrics collection process will be somewhat crude, but is a required component for the start-up. The initial data requirement at this stage is to obtain valid project status information from which to make GO/NO GO decisions for ongoing projects. One of the major value equations for the EPMO is to curtail projects that either no longer fit the organizational goal structure, or those whose value has declined because of poor team performance. In both cases metrics plays a major role in this assessment process. We point out once again that this role for the EPMO is not going to be popular with the project teams if their project is cancelled. It is probably best to keep this decision away from the EPMO and place it in the hands of a senior steering group who would be supplied metrics for that purpose.

The basic theory of metrics is to use a relatively small number so that the user population does not get confused with volumes; however, EPMO function would have so many stakeholders with different perspectives and motivations that this rule may be suspect. For each target group the basic rule is still valid, but given the breadth of target groups, it is easy to envision that there will be a significant set of data collected.

For all of the reasons outlined here, the role of metrics will increase in an EPMO environment and this requires a mature operational project management framework to support the lower layers of the communication structure. EPMO will fail because of either poor goal definition from the top or poor metrics reporting from the bottom.

Once the EPMO function has gathered all the relevant project metrics, they need to present the information to senior management. This section of communication is important because it helps the management team make informed decisions regarding the allocation of resources to various projects.

33.6 STATUS REPORTING

As the volume of raw data grows the EPMO organization has a significant communications system design consideration to make. Basically, it has to opt for either a "pull" or "push" delivery model. Depending on the needs and maturity of the organization, one of these strategies may be more effective than the other. In a "pull" model, the target user has the ability to actively requests status information and progress on projects in a wide variety of formats. The benefits of this approach are that the responsibility of collecting the information lies with the user and they are free to choose which items to concentrate on. In order for this approach to be successful, the data must be clearly defined as to meaning and the user has to be analytically oriented.

In other cases, the user may have defined a "canned" view of metrics that they are interested and do not want to browse the data. In this case, the "push" model is more appropriate. Here, the process is to push out a formatted view of the data on a preset schedule.

In the contemporary environment the pull strategy is becoming more popular as the user community information literacy level increases and delivery techniques become more intuitive to use. At the top of the EPMO information structure, a much more analytic process will be needed by the analyst in order to mine the data in multiple formats and define trends and issue.

The information distribution characteristics outlined above signal a need for a much more sophisticated communication system than exists in most organizations today. This includes both the structure of the data and the delivery technology. The most notable item is that each project should utilize a consistent status-reporting format so that all projects can be combined into a single comparative status view. This involves much more than a schedule and budget report, especially if we add the analytic analysis capability to the equation.

Creation of the formal information distribution process to deliver this class of information is yet another example of the overhead required to drive the EPMO capabilities. This requirement can be a schedule constraint for the implementation process. Personal observation suggests that many organizations are still struggling to collect their portfolio information. Until that is successfully accomplished, the portfolio selection decision-making framework above that level is hampered. Also, if an organization does not know what projects are being currently pursued and what resources are being allocated to those projects, then the EPMO level decision functionality is hamstrung. The breadth of this problem suggests that the implementation of the communication system will be a layered process and likely require significant time and resources to accomplish.

The EPMO charter should describe the standard metrics and forms of communications that the project team is expected to deliver regarding status reports. One of the implementation tasks will be to define the data collection process and report formats for at least the initial operational stage. One goal of the reporting process is to make it as transparent as possible, meaning that the data collection is not intrusive to normal work activities. Conflicts between the project teams and the EPMO implementation need to be minimized.

Project status reports need to be standardized. This has the twofold benefit of reducing the amount of work required to generate the report and making it easier to understand. When a report is standardized, the inputs and the content of the report are well known by the producers and consumers of the report. This standard report format implicitly defines an agreement between the EPMO and their stakeholders regarding both format and data content. Getting buy-in on this item will be a challenge, but an important one to accomplish.

33.7 EPMO COMMUNICATION LINKAGES

Much of the formal communication channels between the various project teams and the EPMO strategic layer are supplied by the standard data collection process. After the project is selected and moved into formal development, subsequent analysis may show that it is no longer a viable effort and should be cancelled. When that situation occurs, it would be best to have a high-level management steering committee deliver that decision in order to keep the project team and EPMO relationship harmonious. In the case where the EPMO organization has a high-level of formal authority over the project teams there is the additional responsibility for more direct communication with the PMs. In this formal structure, the EPMO organization is chartered to influence necessary changes to achieve the overall project goals. In any case, the EPMO function would analyze status of ongoing projects and develop conclusions related to the health of the project. Their level of authority would dictate actions beyond that point.

33.8 EPMO ORGANIZATIONAL MODELS

The models presented in this section are derived from PMO organization experiences and extrapolated here to the enterprise level. They collectively provide a vision regarding how the EPMO organization might be structured based on the design philosophy and current maturity level. The models shown below also help to understand how the design philosophy would lead to a physical organization for EPMO.

As indicated earlier, there are many different ways of packaging these functions into an organizational structure. Some of the structures focus on a desired maturity level for the function, some simply describe the authority structure, some are structured to concentrate solely on the operational roles, and so on. Fundamentally, an organizational structure reflects the goals that the host has for the function.

Mark Mullaly describes four alternative organizational role categories for a PMO and each of these would also fit an EPMO structure (Mullaly, 2008):

Scorekeeper

- Monitor and report progress of project portfolio
- Program and project information conduit
- Clearing house for consolidated status updates

Facilitator

- Enable improvement efforts
- Source of best practices

Quarterback

- Focus on project delivery
- PMs report to the PMO
- Central point of accountability

Perfectionist

- Control focused improvement
- PMO is the "center of excellence"
- Agent for change regarding how the organization does projects

In addition to these four, there are three other common groupings as described in more detail below.

33.8.1 Weather Station

33.8.1.1 Organization Driver

When senior management gets nervous about all the money they are spending on projects and do not have a good feeling about the value derived, they are looking for information. The current status is a collection of different reporting formats coming from various projects with different varieties of metrics, jargon, and formats. To end this confusion, they set up a "Weather Station" organization to standardize the reporting, but essentially leave the projects alone. In this mode, the EPMO organization is a clearing house for project information. More robust versions of this may perform the following roles:

- Maintaining a database of action items, project archives and lessons learned
- Developing a formal enterprise reporting system for executives and key stakeholders
- Defining advanced standard metrics for reporting such as EV
- Track postproject actual results and communicate this status to management.

33.8.1.2 Formal Authority

Although the Weather Station's functions sound minor, almost clerical, they require that the Weather Station set the frequency, format, method of delivery, and associated tools for reporting and planning. Table 33.1 lists the key advantages and pitfalls for this type of model. Weather Station is probably the best choice when frequent project failures are noted and there are insufficient data to ascertain a corrective strategy. It also is the easiest to implement without conflict and negative political reaction. This type of organization does not apply management focus if the goal is to fix the project situation.

If the Weather Station model is not given enough authority to ensure cooperation with the project teams, it will have to resort to "nag authority." This approach typically leads to inefficiencies, embittered relationships, and being disliked by the project teams. If the function is to have any accountability, it must be given commensurate authority. Although the rationale for creating this structure is valid, it should not be viewed as the long-term format for the reasons described.

TABLE 33.1
Advantages and Pitfalls of the Weather Station Model

Advantages	Pitfalls
Easy start up	Authority to ensure cooperation
Creates consistency in terms, PM methodology, and progress reports	Participation viewed as optional
Standardization of tools used for PM	Viewed as "red tape"
	Not authorized to tell PM's and clients how to do things

Source: Executive Leadership Group (ELG), Choosing the Right PMO Setup. 2008. http://www.elg.net/index.php?option=com_content&view=article&id=32&Itemid=5 (accessed December 28, 2008). With permission.

33.8.2 Control Tower

33.8.2.1 Organization Driver

This organization form comes from a term coined by project management consultant Jan Renerts. In this model the project management function is recognized as a valuable process is creating new strategic value for the organization; however, management believes that the overall function is not working as well as it could. A few visible indications for this conclusion follow:

- Project team training is haphazardly done or not at all
- Multiple expensive and voluminous "methodologies," have been purchased but are essentially unused. Projects seem to follow whatever structure the team decides
- Business stakeholders do not know how to support their respective projects and conflict results between them and the PM over roles and relationships
- Lessons learned on one project are not applied to other projects. The same issues seem to recur across projects
- Management rewards heroic efforts of projects teams rather than projects that are well run. The culture seems to be nurturing the heroic model.

33.8.2.2 Formal Authority

The basic goal of the control tower is to treat projects like business processes that need to be protected and nurtured—an envelope around them if you will. This view follows W. Edwards Deming's classic quality dictate to reduce variability. In practice, this model has four general functions:

- Establishing standards for managing projects
- Consulting project teams regarding how to follow those standards
- Enforcing the use of those standards
- Improving the standards through experience.

So, this organization would be somewhat viewed as an in-house consulting organization with some degree of conformance clout. Note that the focus is still only downward to the project and there is not visible role in project selection or portfolio management. Table 33.2 highlights the major pros and cons of this type of structure.

Effective control tower-oriented organizations are rare. They belong only in organizations that have solved the authority problems of cross-functional projects and developed a cadre of skilled PMs who apply a consistent protocol of planning, budgeting, and tracking their projects (ELG, 2008).

One of the other potential pitfalls of this model is that standards writing can be infinite. There has to be a review of standards imposed and not just have standards for standard sake. If project teams are required to meet a set of standards, then they need to understand those standards and the reason they exist. The internal consultants become the key to success in that they can carry the message as

---

TABLE 33.2
Key Advantages and Pitfalls for the Control Tower Model

Advantages	Pitfalls
Detailed Standards	Temptation to "over-control"
Projects are prioritized based on need and resource availability	Underestimate additional workload for PM's
Not viewed as optional	Audit mentality versus coaching
PM is recognized as center of excellence or as a business asset	

Source: Executive Leadership Group (ELG), Choosing the Right PMO Setup. 2008. http://www.elg.net/index.php?option=com_content&view=article&id=32&Itemid=5 (accessed December 28, 2008). With permission.

experts and gain firsthand experience in their value. An intrinsic authority level would come from accountability of both sides. The project team would profit from improved performance and the control tower organization would share. This has the elements of a partnership, if done well. Without this element the whole EPMO organization will be treated as a bureaucratic nuisance—or worse, a joke.

One of the pitfalls of all standards-oriented organizations is that the philosophy of "build it and they will come" does not work. One might think that PMs would naturally seek out standard, proven processes, and that they would embrace the internal consultant who aims to help them install those standards. That is not true! Intrusion of a staff function into the project is almost always viewed as meddling. A technique for adding some level of authority and responsibility to both groups is needed and this is where the organizational design becomes difficult. Quite possibly, the "trick" to getting this form of organization working is to first educate the PMs on techniques for success. Assuming that some level of this can be achieved, the next step would be to test the standards on a project with a good PM. Let the success be advertised at the grass roots level, then move the successful PM to either the consulting role or the EPMO manager role. At this point, the respectability of the EPMO is established, the standards have been shown to work and migration toward a standard overall use is begun. This is no small task, but the steps seem appropriate from experience. Once organizational standards have been embedded in the culture to some reasonable degree, it then becomes possible to start improving those standards. This organizational change management scenario provides a glimpse of the complexity related to implementing macroprocess such as an EPMO.

33.8.3   RESOURCE POOL

33.8.3.1   Organizational Driver

Often, the organizational function that hires, assigns, and manages PMs knows less about project management than the PM does. Therefore, project management talent tends not to be managed as an asset, which then contributes to other related issues in the long term. The organizational driver in this case is an increased recognition that something needs to be done to improve the project management talent within the organization.

33.8.3.2   Formal Authority

The perceived solution in this case is to deal with a centralized "resource pool" method for PMs. Project teams would then "hire" a PM from this repository of expertise. This role would make logical sense to combine with other EPMO-like functions, but given its complexity is probably not the first management issue to be dealt with.

With a resource pool properly in place, the EPMO functions would be involved with the following.

- Hiring and preparing the resource pool for allocation
- Developing high-quality skills in the resource pool
- Managing career paths for the resource pool members.

In this model, the resource pool resource is a highly skilled "hired hand" brought in to deliver the stated requirements. The basic function of the resource pool member is to ensure that projects are done correctly, not that the correct projects are done. There is some hanging question as to what would happen if the sponsoring organization would not follow good practices. It may be necessary to have some type of high-level audit function by the EPMO organization that is then delivered to management for their action.

The major advantage of this model is that it helps to protect and improve the skill level of each PM in the resource pool, which in turn improves and ensures the quality of the project's deliverables.

TABLE 33.3
Key Advantages and Pitfalls for a Resource Pool Model

Advantages	Pitfalls
Helps to ensure that projects are effectively done	How do we ensure the right projects get done?
Skilled PMs	Who governs the tools and methodology?
PMs have authority	How do we capture lessons learned?
PMs have tools needed	

Source: Executive Leadership Group (ELG), 2008. Choosing the Right PMO Setup. http://www.elg.net/ index.php?option=com_content&view=article&id=32&Itemid=5 (accessed December 28, 2008). With permission.

When comparing a Weather Station and a Control Tower EPMO with a Resource Pool EPMO, the main pitfall is that consideration is not given to the correct projects being completed based on priorities. Instead, the focus is more on the growth and skills of each individual PM and less attention in developing or improving the EPMO methodology. Table 33.3 lists some of the major advantages and pitfalls of the resource pool model.

The resource pool model is another example of a structure in which "build it and they will come" does not apply. Executive leadership must agree on some basic authority regarding how this process would work in regard to the project process. Described below are two authority-based options for this process:

First, the resource pool manager would be given formal and exclusive authority to supply PMs to approved projects. There would need to be some form of negotiation process used for this because each project would be after the PM with the best reputation.

A second option would be to give the pool manager formal authority only over the pool resources, but not so much over the assignment of the resources to projects. Project sponsors might be able to select whomever they wish to run their project. In this case the pool manager would handle the following:

- Hiring members of the pool
- Supervising pool member's performance
- Coaching pool members to help them upgrade their effectiveness
- Managing various career development actions such as professional certification
- Disciplining pool members who do not prove their mettle.

On an informal basis, the pool manager should also help remove organizational barriers to the practice of good project management by working to persuade and educate executives and on the overall value of this process.

33.8.4 WHICH MODEL IS THE RIGHT ONE?

In the discussion above, we have summarized seven different ways of looking at this function. Admittedly, there is overlap between the various views. Each option represents a distinctively different goal for the EPMO function. From this, we are still left with the basic question "what's the right choice?" This is a good question and one might guess that a response of "It depends" would not be satisfying. However, that is the real answer. Organizations should be constructed in a manner to focus best on solving their problems.

One of the keys to selection of the right EPMO strategy is the general maturity level of the overall host organization (remember the metaphor of the organizational flowerbed supporting the project seeds). In some cases, the organization itself is not capable of supporting the concepts underlying the EPMO. So, the organizational choice requires matching the EPMO charter to the corresponding organizational level. In the section below, we will take a look at how EPMO maturity might be matched with a corresponding organizational level maturity.

33.9 EPMO MATURATION STAGES

Figure 33.2 shows a logical progression of EPMO solutions that generally matches an equivalent CMM-type maturity score. If an organization recognized the merit of the full EPMO functional scope and decided to implement this scope as described, it would require a significant time period, probably measured in years more than months. The normal implementation strategy for a macro-initiative of this type is to pick out the most pressing need for the organization and initially focus on that target as phase one. As each stage is proven, more functions are added in priority order. Over time, the organizational EPMO maturity would increase to the point where adding more functionality does not deliver equivalent value, so every organization may choose to stop short of level five. At that point the EPMO-related processes would stabilize.

As we have seen in previous discussions, the typical maturity grading scale is based on five levels as originally defined by the CMM research. Following that general approach a five-stage EPMO maturity process is shown in Figure 33.2.

There are other functional packaging options that could be shown, but the one outlined here fits a general functional evolutionary approach. Regardless of the defined stages, the key concept of EPMO maturity is that the lower levels focus on identified critical, start-up project problems, whereas the higher levels move the organization toward achieving the broader-valued aspects of the concept. At the top maturity level, the organization would be fully integrated in its EPMO, project management, and PPM processes. Project portfolios would be mapped to the business goals in such a manner as to optimize their value to the enterprise and project teams would be developing their products in a standardized manner. The EPMO organization would be the focal point for orchestrating these various processes and would be accountable for their outcomes. Significant incremental value would be anticipated from each of these maturity levels. Moving through these stages should not be anticipated as an overnight journey, but more like a 3–5 year effort with good management sponsorship, dedicated resources, and overall support. The major difficulty during the evolution will be the insertion of a new and powerful layer in the existing organizational culture. Project teams will have to understand that they are truly global organizational team players and not just a project team focused on a local goal.

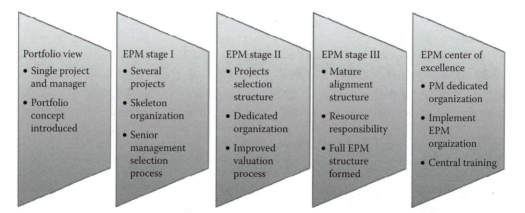

FIGURE 33.2 EPMO maturity stage model.

33.9.1   EPMO Tools and Technology

In today's business world information is a critical asset. PMs must be able to gather, store, and analyze large amounts of data to properly manage project processes. As a result of this, tools and technology are needed to support the requirement. Addition of an EPMO organization increases this requirement. Multiple projects must be tracked at the same time and the initial selection will be much more complex than that exists today. In order to deal with the data collection, analysis, and information distribution processes, a sophisticated set of tools is needed.

Before attempting to buy a particular tool, the first consideration must be to have a workable process and then attempt to match the tool to the process. Not vice versa.

In most cases EPMO tools will not exist in the organization, or at least not fully. This means that the tool issue will be one of reviewing what is available both internally and commercially. Hill (2008) suggests an approach to dealing with the tool question. His process is summarized in Figure 33.3. The first phase of the selection process is to define the high-level steps that will be taken, which are very similar to all projects—requirements, execution, and implementation.

Once these tools have been decided upon, the proper technology can be chosen. The software chosen can then be used throughout the remainder of the tools function model. The second phase of the model outlines the steps for tool implementation. This includes the training of users. Once the implementation of the tool package is complete, the process moves into the third phase to evaluate the chosen tools. This model follows the basic idea of quality management in that it can be repeated as often as necessary to ensure constant improvement.

Hill further states there are three important keys in the evaluation of project management tools. First, an organization must analyze the existing environment. The target EPMO process should have been defined at this point and from this the starting point for tools is to evaluate availability with the goal being to buy rather than build internally. Technicians have a bias toward custom building and then get caught having to maintain a tool. The general rule of thumb is that the organization is not in business to build tools. Commercial vendors are and, if selected properly, will do it better and cheaper. In some cases an existing tool already in use might make a good starting place and that would be first choice in the selection.

Step one: Once the assessment is done to evaluate the internal tool usage and external vendors, the technical and economic analysis would lead to a strategy to either acquire, upgrade, or custom develop the required tool.

Step two: The normal implementation cycle for a new process or tool is to first pilot test it in an appropriate role. Very seldom should a "big bang" all at once new tool be used. This does not seem to be one of those requirements, so the pilot option is recommended. Training is an important element is getting the new user comfortable with the tool. Done properly,

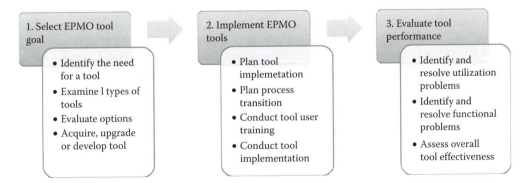

FIGURE 33.3   EPMO tool selection steps.

this should be a motivational process. At this stage, there should be a back-out plan in case the pilot is not successful. After the pilot, it will be very hard to change directions.

Step three: All along the implementation process, a formal postinstallation review should be undertaken. Wherever possible, react to the reported negative situations.

As the EPMO organization evolves, more standardized tool usage will be required. Some examples of common tools follow:

- Project proposal and analysis
- A life-cycle stage-gate project management process
- Common templates for the various stages of projects (charter, requirements document, project plans, deployment plans, support plans, etc.)
- Financial analysis models
- Risk analysis templates
- Project planning templates
- Status report formats
- Project close procedures and audits.

Beyond these base level tools, there is a requirement for other tools related to information distribution and archiving. These would fall into the category of infrastructure, but are vital to the overall communication process. At least a crude version of these would have to be defined in the early organizational stages. Email is not a suitable method to accomplish this class of communication management.

Until recently, there has been a lack of comprehensive software packages with the ability to support this class of activity; however, in recent years several products have been introduced that focus on the portfolio aspects of management. Eric Verzuh states that an EPMO package needs to have broad capabilities in that the software must track numerous individual projects while integrating the status of those projects into a coherent set of metrics that can judge overall status and adherence to organizational goals (Verzuh, 2008). The following broad capabilities are needed for the enterprise level information views:

- Individual project management status
- General project team communications and collaboration
- Visibility of project interdependencies
- Visibility of resource use across all projects
- Project portfolio status summary
- Project status reporting
- Cost information (current and projected)
- Interfaces to other support (i.e., HR, accounting).

Standalone project management software on the personal computer has existed for the management of single projects since the 1980s. These packages have taken different approaches, but as the PC matured the robustness of these applications matured. By the late 1990s and the early 2000s the need for a broader view of this area was recognized and tools with this characteristic were introduced. At this point, several vendors are attempting to capture this market. Gartner Research tracks this vendor class and reports periodically on there tool capabilities (Light and Stang, 2008). Their prediction is that tools in this class will begin to mirror the *PMBOK®* Guide KAs in their role, with risk being one of the fastest growing areas of concern.

The Gartner research organization summarizes the grading of vendors in each major usage category by strategic vision and capability dimensions. Their two-dimensional view is to produce a scatter diagram with vendors shown in one of four quadrants as follows (Light and Stang, 2008):

- Niche player with low capability
- Niche play with high capability

- Visionary with low capability
- Visionary with high capability.

Vendors are thus allocated to one of these four quadrants, which Gartner names the *magic quadrant* and this is a well-known term in industry. Obviously, all vendors would like to be graded as both visionary and high capability. A niche vendor can survive if they do their smaller scope very well but long term they are in danger of a broader competitor duplicating their capability. Research vendors such as Gartner are given great credibility in their opinions and would likely be a source for at least initial tool vendor screening evaluation.

Interestingly in 2000, Gartner evaluated 32 different portfolio class applications, while in 2008 there were 21, yet the market was growing significantly. During this period many of the smaller companies were bought out by competitors or went out of business. This is one of the selection concerns that are important in an evolving area such as PMO, PPM, or EPMO. Currently, because of the immaturity of the EPMO concept, its tool set would probably have been designed for an IT-oriented PMO.

The following major vendors are now starting to appear as leaders in this area and seem focused on capturing market share: Microsoft, IBM, Oracle, Computer Associates, and Hewlett Packard. Microsoft currently owns the desktop standalone project tool and they are making inroads into the enterprise version of that tool. Associated with that is a heavy push into document management using Sharepoint and SQLServer (database utility). Add to that the already captured Office tool set and one would have to consider that they will be a major competitor for some time to come. IBM has been slower to move in this direction, but obviously has the internal expertise and various advertised initiatives that could make them a market leader. Oracle has become a major player through recent acquisitions of Primavera (project) and Pertmaster (risk). Computer Associates and Hewlett Packard are currently rated as leaders in the space and are obviously funded well enough to compete. Given that these tools are still dynamic, there is no attempt here to rank or rate or even attempt to define these products. Given the complexity and dynamics of emerging tools, an organization must enter the selection process carefully and work with vendors and experts in choosing the best fit for their needs. The investment cost and time associated with major tool adoption projects is too great to be taken lightly.

Tool selection and EPMO adoption should be managed just like a project. The process necessary to make the right decision in choosing the future of an organization's PMO is vitally important. This process can consume large amounts of time and capital, so it should be planned, executed, and managed properly. Future users should be involved in the selection process and not just forced on them after the fact.

33.10 EVALUATING AND PRIORITIZING NEW PROJECTS

The most common approach for creating a new project is to build a business case describing the value and cost for the vision that is then presented to executive management for approval. The proposal would deal with the characteristics of the project in regard to its objectives, benefits, impacts, and deliverables. In concept, the new project assessment should be linked to both the organizational strategic plan and other new proposals or existing work-in-process projects (although often not done other than casually).

One common approach for comparing project value or merit is to define certain evaluation criteria and weigh the project against each of these either qualitatively or quantitatively. The use of weighted criteria can help eliminate much of the political flavor in project selection and it helps provide more emphasis on the value equation related to the proposal.

33.11 WEIGHTED CRITERIA EXAMPLE

The sample weighted project ranking matrix example identifies six key criteria areas for assessment. Each criteria item is assigned a score (1 for low to 5 for high). In the example calculation two

TABLE 33.4
Weighted Project Ranking Matrix

		Weighted Project Ranking Matrix						
		New Request		Funded Projects		Your Request	
	Weight %	Criteria	Weighted			Criteria	Weighted
Criteria	(Total = 100%)	1 = Low 5 = High	Weight % Times Criteria	Weighted Score		1 = Low 5 = High	Weight % Times Criteria
				A	Z		
Supports company's strategy mission/goals	25	3	0.75	1.25	1.00		
Increase revenue	20	2	0.40	0.80	0.60		
Likelihood of success	20	4	0.80	0.80	0.80		
Readiness-skill sets/ geographical/culture/HW/SW	15	3	0.45	0.75	0.45		
Urgency-completive advantage or legal requirements	10	2	0.20	0.50	0.30		
Decrease costs	10	3	0.30	0.30	0.20		
Weighted total score			2.90	4.40	3.35		
Project ranking (1 = highest weighted total score)			3	1	2		

Source: This worksheet was authored by Ron Smith who gave permission to use it in this context. It was originally published in Baseline Magazine, March 2008. With permission.

projects are graded (A and Z) and compared with a funded (or existing) project. Note the final grade for the two new projects is 4.40 and 3.35, respectively, compared to 2.90 for the funded project.

As one gains experience with weighted evaluation tools such as this, new criteria or weight values can be added so that the resulting evaluation score best fits the organization. Performing an assessment such as this also sensitizes the evaluation team to these defined multiple criteria. Use of a standardized multicriteria grading approach helps to ensure that a particular set of criteria are reviewed for each project proposal.

Various other methods are used to identify initiatives for the project portfolio. These range from highly strategic to tactical in their perspective. Two example approaches for this are summarized below.

33.12   SUMMARY

This chapter has introduced the concept of a contemporary organization unit designed to integrate project selection, portfolio management, and project management processes. This concept was named EPMO and design concepts were illustrated based models from on the lower-level IT-oriented PMO. The rationale for managing projects globally is similar for both of these organizations. Both consume organizational resources and both have the same basic investment objective of getting maximum efficiency out of organizationally aligned projects.

Whether one looks at this as simply an IT initiative or an enterprise level initiative, the discussion should fit equally well. In the case of PMO installations the early history suggests that they are frequent failures, yet the theory seems ironclad. Much of the difficulty in implementing this class of process comes in the cultural change required by the organization and realignment of responsibility and authority to the new organization. The fact is that a PMO or an EPMO type organization represents a centralization image and lower-level units often resist such initiatives. Keep in mind, then, that the real implementation issue is not in deciding what appears to be a very logical solution to a universal problem but more one of significant change of the organization culture. A senior manager once described his concept of delegation and centralization this way—"you can call it either term so long as I get to make the decision." The implication is that if it is delegated below that level they lose control and above that level they also lose control. There is a simple message in this statement that needs to be understood.

REFERENCES

Executive Leadership Group (ELG), 2008. Choosing the Right PMO Setup. http://www.elg.net/index. php?option=com_content&view=article&id=32&Itemid=5 (accessed December 28, 2008).

Hill, G., 2008. *The Complete Project Management Office Handbook*, Second Edition. Auerbach Publications: New York.

Light, M. and D.B. Stang, 2008. Magic Quadrant for IT Project and Portfolio Management, Gartner Research. http://mediaproducts.gartner.com/reprints/ca/157924.html (accessed January 13, 2009).

Mullaly, M., 2008. Four Archetypes of the PMO. www.gantthead.com (accessed November 15, 2008).

PMI, 2008. *A Guide to the Project Management Body of Knowledge* (*PMBOK*® Guide), Fourth Edition. Newtown Square, PA: Project Management Institute.

Richardson, G. and C. Butler, 2006. *Readings in Information Technology project Management*. Boston: Course Technology.

Verzuh, E., 2008. *The Fast Forward MBA in Project Management*. Wiley: Hoboken, NJ.

34 HR Outsourcing

34.1 INTRODUCTION

Over the past 20 years there has been significant growth in moving project work to external third parties. This action has been called various things—outsourcing, offshoring, contracting, virtualizing, and so on. Chapter 20 has earlier described the general process of procuring products and services from third parties and this chapter is a specialized case of the general procurement action. In this chapter, the main focus is on the procurement of human-related services rather than a durable product. This has been the fastest growing area of procurement over the past decade or longer. The primary rationale for pursuing this class of relationship comes from one of the three general motivations:

1. The external labor cost is lower than equivalent internal costs
2. The external vendor has some specialized skill that is better to acquire via contract than to create that skill internally
3. Internal labor capacity is inadequate to handle requirements and the action is viewed as acquiring extra short-term capacity.

Regardless of the motivation, the project management issue becomes one of managing a portion of the project using resources that are often housed external to the project team and owned by an external organization. This form of staffing adds increased complexity to the management of the project.

Many outsourcing arrangements involve moving the work to a foreign country, which can further complicate the communication and coordination actions for the PM. The most frequently stated reason for this decision is the low cost of labor in these locations. An International Data Corporation (IDC) survey identified that approximately 23% of IT budgets were being allocated to offshore outsourcers in 2007 (Konrad, 2003). Growth trends over the past few years have been significant and this indicates a bias toward this strategy; however, there is also evidence that many of these decisions are made without a strategic view or detailed understanding of the issues. In some cases the decision to outsource is appropriate, whereas in other cases it brings more negatives than positives. In this chapter we will take a critical overview of the outsourcing decision process and offer some appropriate criteria to evaluate this class of decision.

Morstead describes the requirement for a formal business model as a prerequisite for making outsourcing decisions (Morstead and Blount, 2004). Closely related to this customer-centric view is the concept of an organizational "value chain" as described by Michael Porter (1985, p. 37). Collectively, these two views provide high-level guidance into the cost and revenue aspects of the enterprise resources. One of the critical issues in regard to what processes can be moved external is the impact that this has on customer relationships. Second, where in the value chain can cost efficiencies be obtained through the use of third parties. Third, within the value chain a critical question is to determine what strategic value is derived from internal worker intellectual value. Fundamentally, the answers to these three questions should help drive many of the decisions related to the use of third-party providers.

The lure for using international resources historically has been the extremely low wages compared to U.S. workers doing the same job. That issue still remains but has weakened in more recent times as a result of labor rate inflation in prime outsourcing areas. This, in turn, has stimulated further geographic dispersion of the target vendors and even further complicated the sourcing selection and management process.

Large numbers of case studies reviewing outsourcing success or failure of these ventures seem to be polarized. Some will argue vehemently that this activity is both productive and required; whereas others would say that these decisions are short sighted and in the longer term the cost advantages will disappear. Let us just say that both sides of this argument represent the complexity of this strategy. The real question for the PM is when to take which side of the argument. There are certainly factors in this decision that involve much more than a person-to-person labor rate, yet that is the level of analysis that many organizations use for the decision. If nothing more is done to justify the outsourcing, then all decisions will be to follow that path. The fact is that many outsourcing decisions that looked like they were sure cost savers were later found to be disastrous, so take that history lesson to heart. Another attitude to beware of is the "throw it over the fence one." What this means is to get rid of the problem internally by giving it to another. Unfortunately, this often does not work either.

The outsourcing decision is likely never a sure thing and is probably also not correct for 100% of the areas reviewed. However, once a viable technical vendor is identified there is reasonable probability that they may add some value to the resource question, but this needs to be carefully defined and managed.

With these two vague positions on this topic let us see if we can outline the decision structure of the outsourcing issue and comment on some of the myths of outsourcing.

34.2   MANAGEMENT DRIVERS

As stated above, the most visible driver for this decision is a perceived cost saving. However, it is important to examine both the cost and the creative value derived from various other components within the organization. It is also important to recognize that outsourcing of a function generally strips that knowledge capability from the enterprise and moving it back later will be very difficult since the internal skills will often have left the organization. Retooling those skills can be expensive and time consuming. One of the most difficult aspects of an outsourcing arrangement is the changing business requirement related to that venture. This dynamic feature affects how well the outsourcing relationship can be contractually defined and whether the target vendor can evolve along with that. If vendor A has skill B and the business evolves to skill C, it may well be that the vendor is no longer appropriate. Changing vendors along this line can be very complex. As an example, during the 1990s it seemed clear that outsourcing mainframe computer operations was an obvious good decision. Surprisingly, these contractual relationships did not fare well as the underlying business systems tried to evolve and the preferred technology was evolving at the same time. Many of these outsourcing arrangements came completely unglued as the smaller system architecture replaced the mainframe and new skill requirements emerged.

There are multiple lessons to be learned from these previous experiences. First, many organizations are not very good at strategic planning and forecasting trends over 5-year periods. Certainly, this statement is true in higher technology areas. Second, the impact of an outsourcing decision permeates throughout the organization and can be significantly impacted longer term by technological discontinuities, which are difficult to fix because of lost internal skills from the previous outsourcing decision. As business systems evolve, organizations need to be able to migrate through the new waves of change in an orderly fashion. Outsourcing does not naturally help with this in that the loss of internal skills leaves much of the work to be done by two vendors—one of which is not highly motivated because they are losing business.

One of the subtle aspects of business over the past 25 years is its significant metamorphosis of embedded operational technologies, which are prime targets for outsourcing. This includes hardware, software, networks, and the related business systems application environment. Would a vendor be able to be the preferred provider of these support services over the next 10–15 years? Likely not! So, built into the decision must be a plan to migrate. This means that if an organization is going to stay competitive it has to have the wherewithal to re-engineer itself and that requires reasonable internal technical skills. It is one thing to say that an outsourcer can provide a current service cheaper, but quite another to say that they can re-engineer your organization without access to internal skills and business knowledge. Many outsourcing vendors have expertise in dealing with current technology but they have less motivation to help a customer organization move to a different technology model.

34.3 FIRST WAVE OUTSOURCING

The first wave of outsourcing moved the work to local vendors. In some cases, these vendors even moved into the organization and took some of the former employees to transition the work. Many of those employees were "transitioned" in a year or so, so outsourcing was often viewed as a way to downsize the organization. Following this, came a group of specialized vendors who had significant expertise in doing some niche service. Typical areas for this were mainframe operations, call centers, manufacturing, maintenance of some large common application, specialized applications that were licensed to the buyer, and so on. In this mode, the outsourcing trends became fragmented around specialized vendors. As this trend continued to evolve, the scope expanded into highly specialized areas. The most radical version was announced recently when a hospital advertised that they were looking into robotic surgery where the doctor was located elsewhere and through robotic technology and networks could do specialized surgeries. In the author's view this is pretty far down the niche specialization trail and might be a good personal test of just how biased you are for or against this topic. If the price of robotic surgery was 50% of the traditional option would you choose the robot over the internal resource?

34.4 OFFSHORING WAVE

As the outsourcing trend continued, vendors began to emerge in far-off locations where labor rates were extremely low. Moving work outside the U.S. continental boundary would be defined as *offshoring*. Manufacturing and IT activities were the initial targets for this new set of vendors. Initiatives involving manufacturing appear to have been more successful in regard to producing lower cost results. One of the major differences in manufacturing versus IT is the communication model. The science of "blueprinting" and now CAD systems is very advanced in that a manufacturing design model can be transmitted to a remote outsourcer with very little additional collaboration required. On the other hand, IT does not have that formalized design capability at present, so the translation process is cumbersome at best.

One interesting side note to show the multiple dimensions of this issue is to look at technical manufacturing ventures. In its early form, these decisions were made to simply take advantage of low-cost labor. However, in some parts of the world such as Japan and Taiwan the organizations absorbed sufficient technical capability to essentially "bleed" the technology out of the United States into those locations. Manufacturing of computer-related hardware and chip manufacturing activities have diluted the U.S. capabilities in this area. Most electronic and computer devices are now produced in these areas of the world and the U.S. organizations have been weakened in their ability as a result. So, we clearly observed lower cost production in the beginning, but one might debate what the longer-term societal implications are for the work force and future innovation in these industries.

34.5 ISSUES WITH OUTSOURCING RELATIONSHIPS

Relationships with a third-party vendor are different from having the traditional project team located in one physical location. Let us review the more significant outsourcing-related issues from the PM's perspective.

Communication issues: Physical distribution of the project team increases the complexity of the communication process, but even more so with international relationships, staggered time zones, different languages, and varying work processes. In order for an outsourcing arrangement to be successful, good communications are mandatory. Many outsourcing projects fail because of poor collaboration methods resulting in slow decision cycles. To help combat these issues, organizations must develop good collaboration procedures and make the subject of communications even more a priority than normal.

Quality certification issues: Many outsourcing companies tout their quality or maturity certifications. These certifications range from ISO 9000 quality (Chapter 21) certification or SEI's CMM Level 5 (Chapter 31) development process certifications. The key question for the buyer is whether these certifications actually improve performance in the relationship. In any case, quality assurance and quality control requirements need to be carefully documented in the formal contract language.

Project control: Many outsourced projects fail in the same ways as internal projects. Certainly one of the key reasons for this is inadequate controls, particularly in the scope change arena. Strongly related to this is the availability of appropriate collaboration tools across the broad geographic boundaries. Effective control requires the timely distribution of technical information.

Poorly defined requirements and deliverables: If the offshore entity is viewed solely as a resource to perform some defined unit of work the relationship it will suffer, then the stated requirements are not accurate. In order for these relationships to work effectively, the two parties need to view their arrangement as a win–win partnership. This means that both are dedicated to resolving issues and not fighting over the details and whose fault a particular issue might be. One of the critical aspects of that is to develop high-quality requirements, define roles and relationships, and have well-defined management processes as defined throughout this book. The higher the maturity levels of the two organizations, the more likely these issues can be kept in bounds.

Intellectual property issues: There are many complex issues related to sharing Intellectual Property with both local and foreign vendors. Some countries legal systems do not offer effective Intellectual Property protection. This can allow the contractor to extract the buyer's intellectual property and become a competitor much like the history of the U.S. computer and electronics industry. In addition, U.S. export laws now require that companies' document equipment and Intellectual Property shared with foreign nationals. So, this problem is both a regulatory issue as well as a potential loss of competitive advantage if the knowledge is leaked to others. Potential lost of both Intellectual Property and internal technical capabilities are two factors that must be carefully managed in the outsourcing process.

34.6   OUTSOURCING SUCCESS AND FAILURES

Duke University conducted a study of firms that outsourced engineering work and found the following (Wadhwa et al., 2006):

1. Only 1% believed that their outsourced engineering employees did better quality work than their U.S. counterparts, however, 40% felt that the quality of work was equivalent.
2. 50% of respondents noted that Intellectual Property concerns are a barrier to expanding their outsourcing efforts, whereas 27% noted technical expertise as a barrier to expansion and 40% noted wage inflation was a negative issue.
3. Cultural differences and engineering capability are concerns with firms that outsource. Over 50% of companies noted that cultural differences are a barrier to expanding their outsourcing efforts. 44% of respondents noted that their U.S. engineering jobs are more technical in nature than the work done by their outsourced counterparts.

In an *InformationWeek* survey of 420 IT professionals who were asked the question "How successful has your company's outsourcing experience been so far," 50% of respondents rated their outsourcing efforts a success, while 33% were rated neutral, and 17% rated their efforts as disaster (McDougall, 2006). The following list summarizes the top six reasons for negative outsourcing performance:

1. Poor customer service, vendor responsiveness, or flexibility—45%
2. Hidden vendor costs—39%
3. Insufficient up-front planning by the buyer company—37%
4. Insufficient vendor technical expertise—33%
5. Not enough contract or vendor management by the buyer organization—28%
6. Insufficient vendor or industry expertise—25%.

34.7   BEST OUTSOURCING PRACTICES

The success scorecard for outsourcing is mixed as the discussion above has summarized. This suggests that there are decision criteria and processes that could be characterized as best practices. Organizations that have successfully used third-party vendors have followed many of the best practices outlined below. These provide at the very least a checklist to consider before undertaking an outsourcing strategy. Just looking at an outsourcing arrangement as a low-cost option is asking for failure without understanding what the future relationship should entail. The following nine principles correlate with improved results and should be understood and defined in an outsourcing relationship:

1. Solid WBS
2. Well-written contracts or statement of work
3. Formalized communications
4. Formal and informal reviews
5. Standardized development process
6. Intellectual property information control
7. Project status tracking
8. Selecting qualified vendors
9. Developing a long-term relationship.

Solid WBS: The foundation of a successful outsourcing project is a good WBS definition. Too often companies simply "outsource development" without carefully defining the actual work. Often missing is the elaboration on issues such as who will be responsible for what? Use of a fully populated WBS and companion dictionary as described earlier in the book provides a method of defining work units in sufficiently small size to ensure better understanding of the required results. In the process of elaborating work units in this manner, the various parties will feel more engaged and future status tracking of the work is improved.

Well-written contracts or statement of work: Too often the initial approach to an outsourcing relationship is based on the idea that the company is simply adding internal resources rather than treating the outsource resources like the independents they are. The fact is that the relationship should be formalized and contractual in nature. When conflict occurs the contract is the last document that will be used to resolve the outcome. In a typical contractual relationship, the contract is designed to support the agreed upon framework of work activities. According to the American Management Association Do's and Don'ts in Subcontractor Management, good subcontract management includes the following (Sammet and Kelly, 1980):

1. Adequately defined requirements from well-prepared statements of work, specifications, formal schedules and data requirement lists
2. Include all necessary milestones, performance reports, cost reports, follow-up, and expediting agreements and surveillance requirements in the contract prior to execution

3. The prime contractor must specify the output goal than how that goal is to be met
4. The selected method of funding a subcontract can be a valuable tool to exercise the leverage necessary to maintain control over manpower and dollars. It is essential that all elements of a subcontract be negotiated and agreed upon by both parties as early as possible.

Formalized communications: The keystone of successful outsourced projects is good communications. These processes should be well thought out and formalized, but must also include the ability to have informal communications. Even the simple practice of email communication must be done correctly to account for time cultural differences. Given the loss of frequent face-to-face interaction, the communication must be very sensitive to the way in which the wording will be interpreted. Protocols should be in place for various message types. For example, any email to a contractor should always be copied to a key list of contacts and placed in a communications file.

Good communication practices should also include formal periodic status meetings. The content of these meetings should be documented and archived. In addition, formal tools should be in place for tracking issues and action items. The existence of web-based information distribution systems can help with availability of needed items and should be explored. The formal communication system represents a major component of project governance tools and processes. Mark J. Power outlines the value of such tools in his *Outsourcing Handbook*:

> Why use governance tools? The advantages are many. For example, they can reduce the cost and over-head associated with the governing process. Also, they can help the organization achieve continuous improvement in quality of performance and in satisfaction to end-users. When used effectively, these tools can help maintain the outsourcing vision and mission alignment with business goals. In addition, they can be instrumental in identifying issues like performance, quality, and resource utilization. (Power et al., 2006)

Holding effective periodic status meetings across a wide geographic boundary means that face-to-face sessions will not be the norm. To replace this loss, implementation of newer collaboration tools such as Google, WebEx, Microsoft Meeting Center, or Citrix GoToMeeting will provide a more feature-rich communications environment. Time zone differences remain a scheduling issue as one party may be up in the middle of the night. This evolving class of communication tools allows the sharing of personal desktop views for graphs and documents over the internet along with an audio connection.

Even though modern collaboration tools are employed, there is no substitute for face-to-face communication. Every effort should also be made to provide some personal contact and recognize that the lack of this can result in misunderstandings and possibly morale issues.

Project status tracking: Use of computerized project tracking is essential with and international project team. The rationale for this is more than the tool technology. This provides the necessary timeliness and data access in various operational aspects of project communication. A well-engineered technical communication system aids in transmitting the work plan to the necessary parties, especially in regard to the daily dynamics of that plan. Less sophisticated tools such as spreadsheets do not serve that role as well.

Selecting qualified vendors: Obviously, one of the key outsourcing issues in successful outsourcing is selecting a qualified contractor in the first place. Organizations that are looking for a relationship with an outsourcing company should look for a company that has core competency in the target area. The vendor selection process should follow a formal RFP format as discussed in the procurement section of the book (Chapter 20). Formal selection techniques such as developing a vendor scorecard and comparing multiple vendors are important. Debbie Friedman outlines in her book *Demystifying Outsourcing* that establishing a selection criterion is "the most important factor in choosing an outsourcing partner is the talent level of the staff" (Friedman, 2006, p. 75). In other words, cost may initially make companies consider outsourcing, but in the end outsourcing vendors should be selected by the potential quality of their results over the long term.

Formal and informal reviews: As with any third-party activity, there should be periodic reviews tied to performance milestones. These include both schedule and technical sessions and may include budgets if the relationship passes cost through to the customer. Collectively, these reviews are crucial to the success of project and the outsourcing relationship. According to Stuart Morstead in his book *Offshore Ready,* "execution requires commitment and, above all, accountability … You must build a collaborative environment in which individuals are held accountable for their participation" (Morstead and Blount, 2003).

Standardized development process: Another factor that can lead to less than desirable results is that lack of a common development process across the two participants. There are many flavors of process standards. ISO 9000 is the most recognized one, but others such as OPM3 from *PMBOK® Guide,* TenStep, Method123, Prince2 from the UK, and CMM from Software Engineering Institute are also widely recognized. All of these, to one degree or the other, establish procedures and standards for project management. Any standard only certifies that a process is in place, not that it is being used. Nor does it guarantee quality work or on-time delivery. The major advantage comes from the standardization of procedures, common vocabulary, and having better process compatibility across the organizations. Since no two organizations will ever be in perfect accord, this will be a normal area of issue for the relationship.

Intellectual property information control: With U.S. vendors there is a reasonable level of control through contractual terms, however, the same is not necessarily true with offshore vendors. When dealing with a foreign vendor it may be necessary to segregate the Intellectual Property work into pieces so that the overall view is more difficult to extract. If a company has particularly sensitive information, it may want to take into consideration the laws and culture of the vendor's country before moving sensitive work to that area. In any case, concern with the potential loss of intellectual property is a major issue to deal with in the outsourcing environment and the risk of this should not be overlooked.

Developing a long-term relationship: Outsourcing is often viewed more than a single project, it is a long-term relationship between the parties. A successful arrangement can be created for a single project; however, most outsourcing success stories come from the fruits of a long-term relationship between the parties. Morstead states that you must "first establish whether you view the other party as a vendor or a partner" (Morestead and Blount 2003). This distinction has important implications. In the long-term view the vendor is likely to build up knowledge about the buyer's proprietary work product or process and is an entity whose long-term financial health is important to both parties. This makes them a partner" (Morestead and Blount 2003, p. 225). If organizations recognize that outsourcing is a long-term partnership, both can reap the benefits of improved innovation and reduced development costs.

Outsourcing started with companies who recognized the opportunity for cost arbitrage, but experience has shown that true profitability from the experience involves recognizing that it involves more than just cutting costs (Prahalad and Krisnan, 2004). Only when there is mutual process compatibility and capabilities for resource leverage will there emerge a fundamental level of innovation between the parties. A singular focus on cost to the exclusion of the managerial innovation potential is a risk few companies should take.

34.8   OUTSOURCING VENDOR EVALUATION WORKSHEET

It is important to perform a formal evaluation of each vendor being considered and the decision should not be made on cost alone. Other issues such as quality programs, knowledge capability, skill levels, and others are equally important to success of the relationship. As with all sample worksheets described in the book, this one is meant to show the breadth of review necessary in the decision process. Use this calculator to help ensure that the related decision considers the proper breadth of vendor analysis. For a particular procurement situation, it may be necessary to modify or add criteria in order to focus on required issues. Also, it may be necessary to change weights for the criteria used for the same reason.

Instructions: Use Table 34.1 to fill in your own rating numbers in the column labeled "Your Project," then add them up to get your total.

TABLE 34.1

Outsourcing Vendor Evaluation Worksheet

		Example Oracle Upgrade	Your Project
1	Do we have a methodology for evaluating the business case for outsourcing	Got it covered?	1 = Yes 0 = No
	Identify your strategic reasons (improving business focus, gaining access to capabilities you do not have) and tactical reason (reducing operating costs, making capital funds available)	1	
2	Have we identified the hidden costs?		
	Hidden costs could include underestimating the price of hardware, software, or the length of the project losing key resources and scope creep	1	
3	Do we have other projects or events that will hinder the outsourcing project?		
	Overlap might include duplicating efforts with other departments, unknown dependent activities and working on tasks out of sequence	0	
4	Will our stakeholder and project sponsor be committed?		
	If the stakeholders and/or sponsor are not committed, your project's progress will be hindered and your chances of success will decrease	1	
5	What are our conditions of satisfaction (COS)?		
	Define your COS; ensure that your expectations are met or exceeded and knowledge transfer will be completed successfully before your project starts Make sure your outsourcer understands the expectations	1	
6	How will we manage our relationship with the outsourcer?		
	At the start of the project, the outsourcer must develop a communication plan that has to be approved by you. The plan should include how stakeholders will be kept informed on a consistent and finely basis	1	
7	What are our responsibilities before the outsourcing project begins?		
	A good outsourcer will have your responsibilities—the facilities you are expected to provide, for example—outlined in the final agreement. Remember that the outsourcer's performance depends on you following through on your responsibilities	1	
8	How do we meet our responsibilities once the outsourcing project begins?		
	The degree of discipline your leadership exercises can determine project success. You must have discipline to meet schedule commitments, such as equipment and available people, to be successful	1	
9	Can we free up internal resources to work on the outsourcing project?		
	Your resource considerations should include matching skills to what is needed, staff ramp-up/roll-off, training time and learning curve, holiday, vacations, and part-time resources	0	
10	Have we selected an outsourcer that understands our business?		
	You must spend the time and energy to select the best outsourcer. You will have to live with the decision during and after the project. Always get two or three bids and consult your legal representatives	1	
	Total score	8	

What your score means?

    8–10: High probability of project success! Work on weak areas

    0–7: Re-evaluate your project and responsibilities

    Failure to do so may impede project success

Source: This worksheet was authored by Ron Smith who gave permission to use it in this context. It was Originally published in Baseline Magazine, March 2008. With permission.

34.9 CONCLUSION

Outsourcing relationships can be executed successfully provided an organization takes the time to manage the relationship with best practices. There are many issues and pitfalls awaiting an organization that decides to outsource, but there are also many potential benefits as well. If executed correctly through long-term relationships, companies can improve their level of innovation, decrease their development costs, and improve their time to market for new products.

REFERENCES

Friedman, D., 2006. *Demystifying Outsourcing*. San Francisco: Wiley.

Konrad, R., 2003. *Start-Ups Take Outsourcing to Extremes*. Associated Press.

McDougall, P., 2006. In depth: When outsourcing goes bad, *InformationWeek* (December 15). http://www. informationweek.com/showArticle.jhtml;jsessionid=CLSNBSIGYC31EQSNDLRSKH0CJUNN2JVN? articleID=189500043&queryText=outsource+best+practices (accessed June 19, 2008).

Morstead, S. and G. Blount, 2003. *Offshore Ready, Strategies to Plan and Profit from Offshore IT-Enabled Services*. Houston: ISANI Press.

Porter, M., 1985. *Competitive Strategy: Creating and Sustaining Superior Performance*, New York: Simon & Schuster.

Prahalad, C.K. and M.S. Krisnan, 2004. The Building Blocks of Global Competitiveness, *Information Week* (September 9). http://www.informationweek.com/news/showArticle.jhtml?articleID=46800203 (accessed January 13, 2009).

Power, M.J., K.C. Desouza, and C. Bonifazi, 2006. *The Outsourcing Handbook, How to Implement a Successful Outsourcing Process*. Philadelphia: Kogan Page Limited.

Sammet, G. and C. Kelly, 1980. *Do's and Don'ts in Subcontract Management*. New York: AMACOM.

Wadhwa, V., B. Rissing, and G. Gereffi, 2006 (October 24). *Industry Trends in Engineering*, Offshoring. Available at SSRN: http://ssrn.com/abstract=1015839.

35 High Productivity Teams

This chapter focuses on techniques to produce high productivity in project teams. Much of the material is adapted from various SEI's literature on Team Software Process (TSP) and Personal Software Process (PSP). SEI is a federally sponsored research center dedicated to improving quality of system development, primarily in the IT arena. Many of their findings have found application outside of IT as is the case with this topic.

This chapter uses the base notation described for TSP and translates it into a more generic description that is amenable to all project types. Watts Humphrey was the person who originally described this process and since that time a significant amount of research has occurred to expand and verify the process. The reader can refer to voluminous SEI research results and technical literature related to this topic for further background (www.sei.cum.edu).

35.1 BACKGROUND AND OVERVIEW

Beyond describing the SEI model, the basic goal of the chapter is to provide a more general mechanism for high team productivity beyond behavioral management techniques. The methods described are based on mechanics pioneered by SEI in a method called TSP. This approach was selected for inclusion because it has been tested in a variety of real projects and fits the operational theme of this book, even though the TSP design was originally conceived for use in software development. The approach used here will translate the original TSP fundamental process into a general project team productivity model. Underpinning this original approach are basic quality-oriented methods for improving productivity. The TSP concepts will be distilled here into a general process that will fit any project team.

An examination of failed projects shows that they often fail for nontechnical reasons (Humphrey, 1998a). The essence of the TSP model is that proper training of the project team will produce a higher-quality output regardless of other improvement strategies employed. This idea has many similarities to the organizational maturity theme, but in this case is focused on the human element and the project team.

Over the past several years various SEI projects explored various aspects of the project and organizational environment. Their most notable contribution has been in introducing the technical project world to the concept of organizational maturity (Chapter 31) with CMM and CMMI models. Both of these are now widely recognized and the five point maturity scales are used by various other researchers in their niche areas. Humphrey's work expanded this view downward into the project team structure with two operational models titled TSP and PSP. These techniques evolved from research efforts over the past several years. As with all of the topics outlined in Part VIII of the book, these would be classified as contemporary. However, the importance of workable techniques that go beyond behavior theory is very important to the technical PM and for that reason alone these concepts deserve a place in this set of material.

The initial idea behind the development of better team management processes was to help software engineers use better process improvement principles in their development work and to find a workable method to put team members in charge of their work and to make them feel personally responsible for the quality of the products they produced (Humphrey, 1998b). This work led to the formulation of PSP and TSP concepts companion processes to guide project teams to consistently produce quality products on aggressive schedules and within cost constraints. The design objective of the TSP model was to create a highly productive, self-directed team environment. This chapter will focus on the team model mechanics required to produce that result.

Experience gained from the use of these project team management tools have shown them to be valid for all team environments and not just those dedicated to software. As we have stressed throughout the text, all projects have more similarity than is generally recognized. Based on that view, this discussion will explore these two model approaches from the view of managing high-performance teams in a general project environment regardless of output objective. To keep that objective clearer, the SEI acronyms will be translated to *PP for Personal Process* and *TP for Team Process* as the concepts are generalized. When the terms PSP and TSP are used here, they are referring to the SEI literature. All of the basic concepts discussed map back to the Humphrey and SEI research on the topic.

35.2   INTRODUCTION TO TSP CONCEPTS

SEI research results document that the TSP methodology helps to reduce defects, improve productivity, and reduce test time (Humphrey, 2000a). In essence, use of CMM processes provides an overall *operational context* for effective project engineering, whereas the TSP process guides engineers (team member) in actually doing the work. TSP is the process element focused on the project team activities.

One of the complex process issues within a project team is management of activities such as requirements management, enterprise goal alignment, management constraints, technical capability, and meeting customer expectations. To balance these often conflicting forces, a project team must understand the complete context of technical and business issues in order to produce a consensus approach that minimizes the ongoing conflict. This means that the team must be familiar with and have the capability for dealing with the following attributes (Davis and Mullaney, 2003):

- Understand business and product goals.
- Produce their individual work plans to address those goals.
- Make personal work commitments.
- Direct their own tasks.
- Consistently use the methods and processes that they select.
- Manage output quality.

In order to achieve these capabilities there must be a defined relationship among management, project team, and members. Figure 35.1 shows how these relationships are distributed in the PSP and TSP models.

Chick (2006) and de Oca and Serrano (2002, 2004) have evaluated the use of PSP and TSP techniques in various types of multidiscipline projects. These experiences suggest that the same concepts can be used as the operational and management foundation for any project team. With this as the historical introduction, the rest of this section will introduce the key ideas of PSP and then focus on operationalizing the TSP concepts into a general project team environment. In this discussion, it will be assumed that the external organizational maturity is sufficient to support the ideas discussed here.

35.3   PP CONCEPTS

This section will translate the SEI PSP model into a general PP model. This model's general principles are based on the following five basic planning and quality principles (Humphrey, 2000a):

1. Every team member is different; to be most effective each must plan their own work and base their plans on personal capabilities.
2. To consistently improve individual performance, actual results must be measured and then used as a guide for improvement.

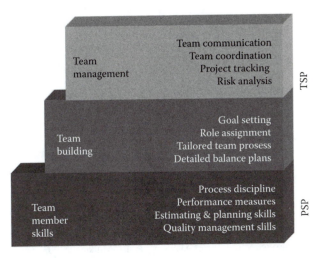

FIGURE 35.1 Elements of the PSP and the TSP. [From Davis, N. and J. Mullaney, 2003. *The Team Software Process^SM (TSP^SM) in Practice: A Summary of Recent Results Technical Report.* Pittsburgh: Software Engineering Institute. With permission.]

3. To produce quality products, individual team members must feel personally responsible for the quality of their products. Superior products are not produced by accident, but through a conscious individual striving to do quality work.
4. The team culture must clearly understand that it is more efficient to prevent defects than to find and fix them.
5. The team culture must clearly understand that the right way is always the fastest and cheapest way to do a job.

35.3.1 PP EXAMPLE

One of the primary tenets of the PP model is that team members must plan their work before committing to or starting on a job. In addition, they must use a defined process to plan that work. To understand personal performance results, team members must measure the time that they spend on each job step, the defects that they inject and remove, and the sizes (scope) of the products they produce. The goal of these steps is to consistently produce quality products, plan their work, measure results, and track product quality throughout the project life cycle. Finally, the individual must analyze the results of each work activity and use these findings to improve their PPs. It is not surprising that if team members are not trained in these disciplines, the resulting culture will be chaotic. The management dilemma for this level involves methods to get teams to try to use disciplined methods, since most team members do not believe that this level of rigor is justified or efficient. The PP model addresses this by putting the team through a rigorous training course to learn the methods and see results in simulated case activities. From this experiential approach, team members are more likely to buy into the method as they see visible results.

The PP training design structure is composed of several incremental components. Figure 35.2 outlines one example of this showing how the SEI PSP courses are introduced in six upwardly compatible steps titled PSP0 through PSP2.1 (Humphrey, 2000a). These courses are used to train software engineers in coding and QC, but a similar approach can be used for any project team. Note that the structure of the training fundamentally deals with techniques for scope definition, estimating based on requirements, and then design of work units. In this training mode, team members plan their work, produce a sample output, and then gather and analyze quantitative data from their work. From this process the team member is taught to analyze their results to improve the next iteration. If one examines this process, it is very similar to classic Deming quality management approach (i.e., PDCA).

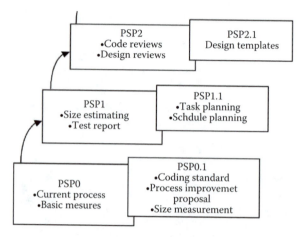

FIGURE 35.2 The PSP course structure. [From Davis, N. and J. Mullaney, 2003. *The Team Software Process^{SM} (TSP^{SM}) in Practice: A Summary of Recent Results Technical Report*. Pittsburgh: Software Engineering Institute. With permission.]

PSP0 and PSP0.1: At this initial training stage, team members will develop three assignments using a standard methodology. The objective of this phase is for the members to learn how to follow a defined process and to gather basic size, time, and defect data for that class of work. In the more general case, this could involve techniques to collect requirements from users, WBS development, and collecting standard performance metrics as previously described in Part VI of the book. From these focused training sessions, the team member is sensitized to a defined work process and performance data collection.

PSP1 and PSP1.1: Once team members have learned to manipulate the standard work process and compile historical data, the focus moves to estimating and control. In this step, statistical methods are taught and used to produce size and resource estimates for individual work units. Participants are taught to use EV techniques for schedule planning and tracking.

PSP2 and PSP2.1: At this third stage, team members have learned the basic model techniques for developing project plans and measuring output results. This stage focuses attention toward a more quality-centric management view. In this phase, team members are taught techniques to identify defects in design and perform root cause prevention at the design level.

After these first six assignments, the team members begin writing technical reports related to the class assignments. At the end of the training, an overall report summarizing the results is produced. The report documents the evolving state of their performance based on measured results improvement. From this analysis the team members are charged with defining challenging yet realistic improvement goals and to identify the specific changes that they will make to achieve those goals. The final training phase will implement those changes.

By the end of the training course, team members are able to plan and control their personal work, define processes that best suit them, and consistently produce quality products on time and for planned costs.

Obviously, translating this process to a particular project environment would require some work based on the characteristic of the planned deliverables, but conceptually this process is simply oriented toward teaching standard methods that have strong similarity to the *PMBOK® Guide*-type life cycle model processes. Standard templates could be used for several of these activities. General training modules and techniques would have to be designed to deal with the following activities:

1. Requirements facilitation and documentation—user group sessions designed to collect and catalog requirements
2. Scope definition based on WBS decomposition techniques

3. Size estimating based on local or industry standard tools and models for the application area
4. Design reviews—this type approach is already documented in the professional literature and deals with techniques for technical peers to review designs
5. Collecting actual performance metrics and then using them to analyze status. The host organization would have to define appropriate metrics for that class of project
6. Root cause analysis sessions could use the Ishakawa model to teach problem analysis-type techniques for the local case.

Each of the processes outlined above could be sequenced through the same course structure as outlined for the PSP model approach.

35.3.2 INTRODUCING PP TO THE TEAM

There are several important management points to understand in introducing the PP components to a team. As indicated above, team members should be trained by a qualified PSP instructor using customized curricula. Although many of these concepts can be introduced quickly, they must also be done properly. Potential trainer resources for this could initially come from a growing number of SEI-trained PSP instructors who offer commercial training courses (see www.sei.cmu.edu). These instructors would have to understand the local model design approach as well as the PSP model and this would require support from an internal technical resource.

The second important step in PP introduction is to perform the training in coherent groups or teams who should be ready to use the methods in the not too distant future. When organizations ask for volunteers for PP training, they could get a broad sprinkling of skills that will be hard to adapt to a specific training target project. If this were to occur, the productivity impact is diminished since the concepts taught would be harder to utilize. Ideally, core project teams should go through the process together. Alternatively, homogeneous skill groups could have basic material tailored for their particular skill segments. Selection of the participants and course materials are critical issues.

Third, effective PP introduction requires strong management support as we have pointed out frequently for other change management activities. This means that management must believe in the approach, know how to support their staff once they are trained, and regularly monitor project performance to see how the new model is working. Without proper management attention, many team members gradually slip back into their old habits. The general problem is that most technical professionals find that it is difficult to consistently follow a disciplined work process if nobody notices or cares. Team members need regular coaching and support to sustain high levels of personal performance.

The final implementation issue is that even when all team members are PP trained and properly supported, they still have to consciously figure out how to combine their PPs into an overall TP. The mechanics for this is the interface point between the two models.

35.4 TP PROCESS

Humphrey's initial goal in developing TSP was to design a teachable approach to building and sustaining effective project teams. As described in this section, TP is an team-level extension and refinement of the higher-level CMM and lower-level PSP methodologies as defined by SEI.

TP represents a prescriptive approach for building a productive self-directed team and it outlines how individual members should perform in their project organization. It also defines how management should guide and support their teams and how to maintain an environment that fosters high team performance. The principal benefit of TP is that it shows team members how to produce quality products for planned costs and on aggressive schedules. It does this by teaching team members how to manage their personal work in a team environment and by making each individual the owner of their plans, processes, and results.

35.5   TP WORK OBJECTS AND PRINCIPLES

As team members start applying their PP skills on the job, it will soon be discovered that they need a supportive team environment that recognizes and rewards sound methods. In many organizations, the projects in crisis receive all the attention. Projects and individuals who meet commitments and do not have quality problems often go unnoticed. If managers do not provide a supportive environment and do not ask for and constructively use planning and results data, the team members soon stop using the process as well.

An adaption of the five design objectives of TP are as follows: (Adapted from Humphrey, *TSP Design Objectives*, 1998c.)

1. Build self-directed teams that plan and track their work, establish goals, and own their processes and plans. Team sizes should be from 3 to 20
2. Teach PMs how to coach and motivate their teams and how to help them sustain peak performance
3. Accelerate process improvement by installing a culture similar to CMM level 5-type behavior to be normal and expected
4. Provide improvement guidance to the external organization
5. Facilitate teaching of industrial-grade team skills through appropriate training programs.

Linked to these objectives are six underlying principles or beliefs (Davis and Mullaney, 2003):

1. Team members know the most about the job and can best define the related plans
2. Team members who plan their own work are more committed to the plan
3. Precise project tracking requires detailed plans and accurate data
4. Only the people doing the work can collect precise and accurate data
5. To minimize cycle time, the team members must balance their workload across competing activities and understand the relative priority of these activities
6. A focus on quality will lead to maximum productivity.

Within this structure TP has two primary components: a team-building component and a team-working or management component. The team-building component of the TP is called the *TP launch*, which challenges the team to follow the model development process and should produce some evidence to motivate that behavior. The management component focuses on ensuring that the process is followed through appropriate management behavior.

35.5.1   TP LAUNCH STRUCTURE

TP mechanics provides the project organization with explicit guidance regarding operational techniques for accomplishing their objectives. For example, a WBS oriented, top-down decomposition approach is used to define the overall technical scope of the effort. From this base, estimating metrics are used to determine an overall schedule. These methods are then used to develop an aggregate schedule that is broken into manageable phases with detailed estimates done only for the current phase or segment. Each defined work unit has a named individual who is responsible for the work and for reporting status of their individual pieces. Each time a new project phase begins, whether at the start of the project or at a later transition from one phase to the next, a formal project launch is held. Figures 35.3 and 35.4 illustrate how TP guides the team through four defined phases of a project. It is assumed that projects may start or end on any phase, or they can run the complete life cycle from beginning to end. Regardless, before each a phase, the team performs a complete launch or relaunch activity where they plan and organize their work for the next phase. Generally, once team members are trained in the personal work model, a 3–4-day launch workshop provides sufficient

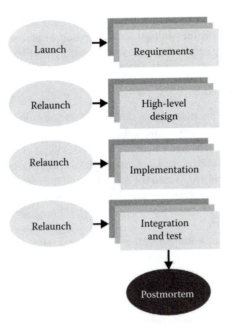

FIGURE 35.3 TSP launch stages. (From Humphrey, W.S., 1998c. *Crosstalk, The Journal of Defense Software Engineering*, 4. http://stsc.hill.af.mil/crosstalk/1998/04/index.html, [accessed October 15, 2008]. With permission.)

guidance to complete a full project phase plan. After this, teams would hold a 2-day relaunch workshop to kick off each of the second and subsequent phases. These launches are not considered training; they are part of the regular development process.

Figure 35.3 illustrates the periodic relaunching process. This approach follows an iterative and evolving development strategy; therefore periodic relaunches are necessary so that each phase or cycle can be planned based on the knowledge gained in the previous cycle. The relaunch process also requires team members to update detailed plans, which are usually accurate for only a few months. Primary output of the TP launch is an overall aggregate plan and a detailed plan for about the next 3–4 months. After team members have completed all or most of the next project phase or cycle,

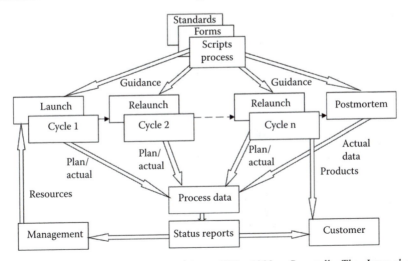

FIGURE 35.4 TSP process flow. (From Humphrey, W.S., 1998c. *Crosstalk, The Journal of Defense Software Engineering*, 4. http://stsc.hill.af.mil/crosstalk/1998/04/index.html, [accessed October 15, 2008]. With permission.)

they revise the overall plan as needed and make a new detailed plan to cover the next 3–4 months. This process is generally called the *rolling wave* approach to planning. This relaunch process is taught as part of team training and is illustrated schematically in Figure 35.4.

35.6   TP LAUNCH DETAILS

The TP launch represents an important step in developing a team culture and common understanding. A *consensus* detailed plan is a primary output from the launch process. This artifact provides the communication vehicle to help the team reach a common understanding of the work and the approach to be taken. Some formal indication that management supports this plan is also an important launch event.

The TP launch script is designed to lead the team through the required planning steps. This script is customized to the size and characteristics of the project. Humphrey outlines the following items that need to be defined and resolved as part of the launch (Humphrey, 1998c):

1. Review project objectives with management and agree on and document team goals.
2. Establish team roles.
3. Define the team's development process.
4. Produce a quality plan and set quality targets.
5. Plan the needed support facilities.
6. Produce a general plan for the entire project.
7. Develop detailed work unit plans for each for the next phase. (The TSP model says that this should be for the individual task level.)
8. Balance team workload to achieve a minimum (viable) overall schedule.
9. Verify that the individual plans will produce the team plan requirements.
10. Assess project risks and assign tracking responsibility for each key risk.

In the final launch step, the team reviews their plans and the project's key risks with management. Once the project starts, the team conducts weekly team meetings and periodically reports their status to management and to the customer.

Figure 35.5 illustrates a TP launch process composed of nine team meetings over a 4-day period. The output objective for each meeting is listed in each cell.

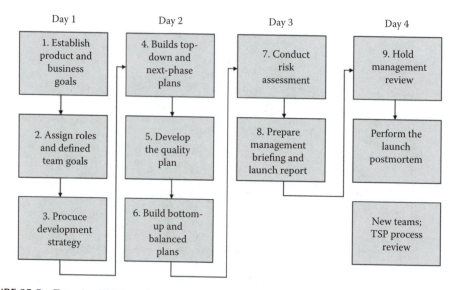

FIGURE 35.5   Four-day TSP launch process. [From Humphrey, W.S., 2000b. *The Team Software Process*ˢᴹ *(TSP*ˢᴹ*), Team Software Process Initiative.* Pittsburgh: Software Engineering Institute. With permission.]

By the end of the launch training, the team should have formed into a cohesive unit and created a plan that balances business, technical, and customer aspects. As a result of this activity there is now an agreed upon technical solution for the effort and the whole team understands how the planned product will satisfy business and customer needs. The underlying work processes have also been reviewed and agreed upon. These are fundamental project team decisions that should be in place for any project regardless of the methodology. As a result of this activity, the team now has learned how to produce a detailed plan that they can use to guide and track the work. Less obvious in the above process is that the following items are communicated to the whole group. Specifically, team members:

- Know who is responsible for which tasks and areas.
- Understand and agree with the quality goal.
- Have a common approach for monitoring progress against the plan.
- Have explored the project risks and performed reasonable mitigation of those risks.

Once again, each of these should be team objectives regardless of the approach taken.

35.7   TEAMWORK PROCESS

Once the launch process training is completed, the principal need is to ensure that all team members follow the plan through execution. This includes the following key operational activities (Humphrey, 2000b):

- Leading the team
- Process discipline
- Issue tracking
- Communication
- Management reporting
- Maintaining the plan
- Estimating project completion
- Balancing team workload
- Relaunching the project (phase)
- Quality management.

The team leader's primary responsibilities are to provide high-level guidance and motivation to the team members, handle customer issues, deal with external management, and maintain the process discipline. Another important team leader responsibility is to ensure that all of the issues that the team members identify are managed and tracked. Through all of these activities the team leader is responsible for maintaining open and effective team communication. Proper communication practices are a key part of maintaining the team's energy and drive necessary to keep the effort moving in a positive direction.

Within a team organizational structure, the project plan provides formal guidance in regard to work timing and sequence. Work performance is tracked against the plan using EV and other performance metrics (Part VI of the book covers these topics). Tracking activities are designed to help team members evaluate project status and provide timely information to help various stakeholders understand current status and future completion projections.

35.8   QUALITY MANAGEMENT

All projects have variances from the plan and it is common to find product output not meeting planned standards. The TP model places principal quality emphasis on defect management. In order to manage quality, teams must establish quality goals and associated measures to define the status of those goals.

From this base they then establish plans to meet the goals. During the team launch training, team members are taught how to produce a quality management plan. This plan structure is based on the estimated size of the product and historical data on defect rates. Estimates are then made on defect patterns through the process. Subsequent progress tracking will monitor performance against this profile.

As defects are identified, the training program should have provided guidance regarding methods to correct or deal with these problems. In this mode, much of the quality management process is housed inside the PP or TP work activity. Quality management techniques taught as part of the TP and PP training follow the theoretical processes outlined in Chapter 21, Quality Management.

Defect management is more than just measuring results and repairing the problem. The theme of PP and TP processes is to identify ways to prevent problems before they occur. By improving individual work processes, team members typically learn how to reduce their defect rates by 40–50%. Improved design methods can further reduce defect rates as the team members become more proficient with the method. Existence of a formal quality plan and defect tracking processes make the team members more sensitive to quality issues so that they are more careful, which reduces defects even further. As with other project management processes, the quality management focus continues through the life cycle. Defect status is reviewed at the end of each phase and the lessons learned process is used to improve downstream activities.

35.9 EXPERIENCE EXAMPLES AND EVALUATION

Various SEI Technical Reports outline IT project case data from 13 organizations and over 20 projects. These data compare results from typical non-TSP projects with those managed using the process outlined here (Davis and Mullaney, 2003; Humphrey, 2000b). Product sizes ranged from small 600 lines of code (LOC) to 110,000 LOC projects. Corresponding team sizes ranged from 4 to 47 team members, and project duration ranged from a few months to multiple years. Application types include real-time software, embedded software, utility software, client-server applications, and financial software, among others. In these reviews, TSP teams delivered products that were more than two orders of magnitude better in quality than typical projects. Also, TSP projects completed in less than 50% of the time and resources spent in the typical project. Certainly, quantified results of this type are sufficient to get any organization's attention.

One specific set of sample data from a large Boeing avionics project team is shown in Figures 35.6 and 35.7. After Release #9 of PSP/TSP training, the number of defects detected reduced 75%, and the system cost reduced 94%. The final project delivered a high-quality product ahead of schedule (Humphrey, 2000b).

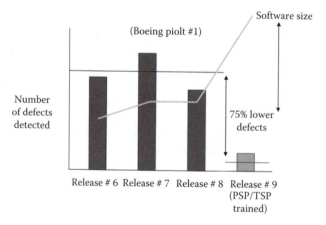

FIGURE 35.6 TSP test defects. [From Humphrey, W.S., 2000b. *The Team Software Process^SM (TSP^SM), Team Software Process Initiative.* Pittsburgh: Software Engineering Institute. With permission.]

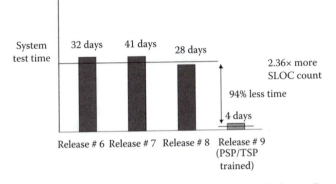

FIGURE 35.7 TSP test time. [From Humphrey, W.S., 2000b. *The Team Software Process*^{SM} *(TSP*^{SM}*), Team Software Process Initiative.* Pittsburgh: Software Engineering Institute.]

Two conclusions can be gained from this example. Leaving a project team alone to follow whatever path they choose is not the most productive strategy. Second, a disciplined process, once learned, can have significant positive impact on project deliverables.

35.10   TSP QUALITATIVE FEEDBACK RESULTS

Besides the raw data analysis of team performance, qualitative results are equally important to review. Davis summarizes both positive and negative comments from team members in his SEI technical report. These responses form a valuable source for lessons learned. Some typical positive comments from the Davis and Mullaney survey are listed below (Davis and Mullaney, 2003):

- "The best part about PSP/TSP is that collecting the metrics is for my benefit, not for some-one else. I found that collecting the data proved to me that using a better process really does help my quality and productivity."
- "Gives you incredible insight into personal performance."
- "... [TSP is a] transparent project management paradigm—everybody has a common understanding of the plan and everyone knows what is going on in the project and where we are in the project at any time."
- "The first TSP team I coached was surprised when unit test was completed in half a day. They said they had done a prototype of this code before the project started and it took 1.5 weeks to get it to work well enough to see any results. They have found only two defects since the code has been integrated with the rest of the software."

Some negative comments were also expressed and these points are also valuable for the future development of the process (Davis and Mullaney, 2003):

- "I am a very creative person. I liken doing software to an artist painting a picture, and so I still worry about the PSP structure taking some of the fun and creativity out of the software process. PSP tends to distill the repetitive measurable tasks out of the creative and innova-tive ones that occur early in the design phase. The purpose of design is to provide an early analysis that leads to products with fewer of the more costly defects later. You have to have a good design to get good code."
- "No tool support. SEI's TSP tool is not sufficient at all."

It is important to recognize that a team development process cannot be completely distilled to a set of mechanics related only to planning and data collection. A successful project requires open human interactions that allow individuals and teams to use their intellectual skills. Also, the conflict between disciplined and creative work must be dealt with from a management perspective. Getting all humans in a team to follow a standard approach will always result in resistance, so motivation of the mechanics will be an important aspect.

35.11   FUTURE TRENDS

Documentation of different TSP experiences in various organizations and projects, multidiscipline systems, and different industries shows that the TSP model approach has resulted in mostly successful outcomes. However, there are clear areas for improvement in aspects such as the process for introducing the concepts to the team, extension to large teams, and combining TSP and other traditional project management methods in a complicated project system. TSP was originally designed as a software engineering technique, so more work is needed to translate the original process model into general concepts that can be readily adopted for a broader array of project types. Techniques for customization have been outlined here but this issue will remain a challenge for specific situations.

35.12   LARGE, MULTIDISCIPLINED PROJECTS

A typical large project consists of multiple team skills such as engineers, IT support, business process SMEs, quality and risk management, financial, and other support personnel. This diversity and size of skill mix creates additional management issues in maintaining the type of process discipline required for a TSP model effort. Also, this diversity of background raises a few additional complexities to the implementation process.

The most obvious management issue in a large project is simply managing the flow of HR with their multiple skills in and out of the work plan. Basic TSP theory suggests that work plans should be developed for each individual. This level of detail would appear onerous for the larger project and in fact may not be feasible given that the specific individual may not be identified at the planning stage. Most likely, work plans in this environment would be limited to a WP level, which would then involve a more skill group or subteam level focus. From this base, the day-to-day work definition would typically be managed by the owning work group manager. This means that the model process would have to be translated into a format suitable for that small subteam size group, with an overall project management focus embedded in that view. Much of the project management theory described in this book fits that definition, but is a modification of the TSP process. Basically, this means that all project participants would have to be trained in methods that were compatible with the TSP model, but also focused and segmented into their skill groups. In other words, a business process participant would not likely need to be trained in technical design review procedures, but would need to understand the concepts of requirements definition for the type of work they were allocated and produce status reporting for that segment.

Second, the general problem of project team training is now much more complex than for a homogenous team. The variety of skills involved suggests that these training examples would have to be concocted in such a way that the audience would use the needed training in their specific segmented work environment. Examples were provided earlier showing general tools for useful for general management activities (i.e., requirements definition, WBS techniques, change management, etc.). As an example, a mixed skill group of users, engineers, operations, manufacturing, and other support groups would normally perform scope definition at the WBS level in the large team model. A skilled facilitator could guide a group of this type through the process without all members of the group being experts in the technique. What must be understood in the larger team environment is how a particular work unit interfaces with others and why it is important to produce in some particular form.

After the basic planning activities are completed, different mixtures of participants would likely perform future design reviews as lower-level technical details are debated. So, in this area, training would be more skill group oriented to deal with the specifics of that project phase. Regardless of the team size and complexity, basic literacy training is needed so that all team participants will understand the general process and the overall logic of the processes being employed. The theme of this training would be to tout the test value of the methods. Otherwise, it may well be perceived as bureaucracy without value.

Project tracking and control concepts should be basically the same for all projects and based on WP performance and a project level performance measurement methodology. Status reporting would be handled at the work group level, but individuals would need to follow a standard reporting format.

35.13   SUMMARY

The basic management concepts described in the PSP and TSP models are not new. Their value to the management discussion is that they were developed and verified in a real project environment by a world-class organization with highly skilled researchers. Also, the results were tested across multiple organizations and publically available. All of these factors add legitimacy to this approach. Clearly, the organizational reaction to any form of process standardization will be resistance so long as reasonable proof of value is not communicated. This means that the training program must be sensitive to this goal and changes should be introduced carefully and with support of top management. With each step there must be strong evidence that the techniques used brought desired results.

The PSP and TSP team management models represent powerful techniques to facilitate project execution through tested team management processes. They provide clear methods for producing improved team skills, discipline, and commitment required for successful project execution. Research shows that in most TSP project experiences the process results in reduced cost, time, and quality defects in the project. For these basic reasons, it is important that the contemporary PM understand and utilize these basic concepts. More research is needed to broaden the understanding and implementation of the PP and TP concepts outlined here.

REFERENCES

Chick, T., 2006. Using TSP With a Multi-Disciplined Project Management System, *Crosstalk, The Journal of Defense Software Engineering*, 3. http://stsc.hill.af.mil/crosstalk/2006/03/0603Chick.html (accessed October 15, 2008).

Davis, N. and J. Mullaney, 2003. *The Team Software Process^SM (TSP^SM) in Practice: A Summary of Recent Results Technical Report*. Pittsburgh: Software Engineering Institute.

de Oca, C. and M. Serrano, 2002. Managing a Company Using TSP Techniques, *Crosstalk, The Journal of Defense Software Engineering*, 9. http://stsc.hill.af.mil/crosstalk/2002/09/index.html (accessed October 15, 2008).

Humphrey, W.S., 1998a. Three Dimensions of Process Improvement Part I: Process Maturity, *Crosstalk, The Journal of Defense Software Engineering*, 2. http://stsc.hill.af.mil/crosstalk/1998/02/index.html (accessed October 15, 2008).

Humphrey, W.S., 1998b. Three Dimensions of Process Improvement Part II: The Personal Process, *Crosstalk, The Journal of Defense Software Engineering*, 3. http://stsc.hill.af.mil/crosstalk/1998/03/index.html (accessed October 15, 2008).

Humphrey, W.S., 1998c. Three Dimensions of Process Improvement Part III: The Team Process, *Crosstalk, The Journal of Defense Software Engineering*, 4. http://stsc.hill.af.mil/crosstalk/1998/04/index.html (accessed October 15, 2008).

Humphrey, W.S., 2000a. *The Personal Software Process^SM (PSP^SM), Team Software Process Initiative*. Pittsburgh: Software Engineering Institute.

Humphrey, W.S., 2000b. *The Team Software Process^SM (TSP^SM), Team Software Process Initiative*. Pittsburgh: Software Engineering Institute.

Serrano, M. and C. de Oca, 2004. Using the Team Software Process in an Outsourcing Environment, *Crosstalk, The Journal of Defense Software Engineering*, 3. http://stsc.hill.af.mil/crosstalk/2004/09/index.html (accessed October 15, 2008).

36 Project Governance

36.1 INTRODUCTION

Merriam Webster dictionary defines governance as "the organization, machinery, or agency through which a political unit exercises authority and performs functions and which is usually classified according to the distribution of power within it." This definition can be applied to both total organizational structure and project management wherein governance is described as a set of management and control relationships. At the project level this involves the principles, decision-making processes related to project resources, performance metrics, changes, budget, schedule, and other related actions to ensure successful completion of the venture.

Project governance is the model by which a project is managed within an enterprise so that it aligns with business needs. Hence it can be viewed as the tight coupling of project's activities with business vision, strategy, and objectives. This process has always been difficult given the many different stakeholders, expectations, and aims that a project embraces. Conceptually, governance provides a decision framework that can help achieve the objectives of a project. It relates to accountabilities and responsibilities for the management of the underlying work processes.

Figure 36.1 shows a schematic illustrating the nature of governance, which is an envelope wrapped around the project structure to ensure that all of the underlying processes are effectively managed.

36.2 NEED FOR PROJECT GOVERNANCE

Project governance extends the general principle of enterprise governance downward into the management of individual projects. Many organizations today have recognized their shortcomings in this aspect of project management methodology and are developing formal project level governance structures. This level structure differs from an organizational equivalent structure in that the project view has to recognize the fragmented, virtual, and transient nature of projects. In both cases, the basic intent is to clarify how decisions are made and accountability established. This discussion focuses on that role in project decision making. More specifically, the governance activity is particularly focused on processes such as change management and strategic (project) decision making that lie outside of the general PM's decision domain.

In recent years, corporate level governance has placed additional responsibilities on their board of directors to monitor enterprise performance and this requirement has caused a cascading downward into lower levels of the organization into the project environment. Board members are being held responsible for low-level issues; therefore they want better visibility into those levels. This higher level management requirement encompasses the need for control and to assure that

- Projects are being managed well and in accordance with the requirements of defined governance procedures
- PPM is optimizing the return from corporate resources and maintaining alignment with strategic objectives
- Strategic projects are being properly managed.

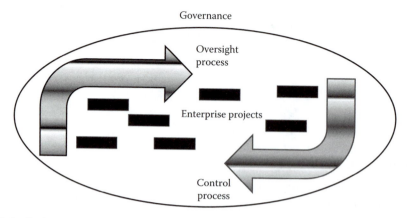

FIGURE 36.1   Project governance life cycle.

36.3 PROJECT GOVERNANCE DEFINITION

As used here the term *project governance* implies external controls linked to enterprise management regarding project activities and designed to ensure that the project serves its intended goals. Lack of senior management commitment is a consistent cause of project failure and the governance processes are meant to better deal with this aspect of a project's management. The need for increased project governance is at least partially stimulated by this recognition. Today, adequate governance processes are not in place in many organizations and even less well executed overall. Formal project management methodologies provide various templates and processes that support the notion of external management requirements, but these alone do not accomplish the required decision support requirements needed to make a project successful. When implemented properly governance forms an integrated decision cycle that forms a closed-loop feedback system between strategic planning and budgeting, and execution and delivery controls (Jennings, 2005). The basic requirements for a project governance system consist of the following eleven key roles:

1. How does the organization formally identify opportunities? (Competitive analysis, track and asses employee suggestions and customer feedback, etc.)
2. Select/authorize/fund the go ahead of projects? (e.g., Only the Strategic Executive Committee has the authority to update/re prioritize the Enterprise Portfolio!)
3. Establish the basic approval and measurement processes including defining roles and accountabilities, policies and standards, and associated processes.
4. *Evaluate* project proposals using a defined methodology to select those that represent the best enterprise investment of funds and scarce resources and those that are within the firm's capability and capacity to deliver.
5. *Enable* staffing of projects through the allocation of internal and external HR, along with business support. If the scope is sufficient, this should include an experienced PM, business knowledgeable resources, and technical resources.
6. *Define* the desired business outcomes (end states), benefits, and value for approved projects, along with business measures of success and an overall value proposition.
7. *Control* the scope, contingency funds, overall project value, and other business attributes of approved projects.
8. *Monitor* approved project's progress, stakeholder's commitment, results achieved, and leading status indicators.
9. *Measure* the outputs, outcomes, benefits, and value of project performance against both the plan and ongoing expectations.

10. *Management action* defined to steer the project into goal alignment with the organization, remove obstacles, manage the critical success factors, and provide guidance on benefit-realization shortfalls.
11. *Develop* the organization's process maturity delivery capability by continually building and enhancing its ability to deliver more complex and challenging projects in less time and for less cost while generating the maximum value.

36.4 ORGANIZATIONAL LEVEL PROJECT GOVERNANCE PRINCIPLES

To support the project level governance processes, the following governance capabilities must be defined at the enterprise level. Ultimately, senior management up to the board of directors needs to recognize their project governance responsibilities in the following four areas:

1. Portfolio selection and effectiveness
2. Project sponsorship effectiveness and efficiency
3. Project management effectiveness and efficiency
4. Disclosure compliance.

In order to satisfy these requirements, all senior management should take an active interest in the following project-related areas and activities:

- Recognize that they have ultimate responsibility for governance of projects
- Be actively involved in defining the roles, responsibilities, and performance criteria for the governance of projects
- Ensure that disciplined governance arrangements, supported by appropriate methods and controls, are in place within the organization and are being properly applied throughout the project life cycle
- Maintain a coherent and supportive relationship between the overall business strategy and the activities related to project selection. This is a responsibility that a PMO or EPMO (Chapter 33) organization might be utilized for
- Ensure that all projects have an approved plan containing management milestone authorization points at which the status is formally reviewed and approved. Decisions made at authorization points are formally recorded and communicated
- Ensure that the project resource allocation process is guided by a formally delegated authorization body that has sufficient representation, competence, authority, and resources to enable them to make appropriate decisions
- Oversee project selection processes using business case data supported by relevant and realistic information that provides a reliable basis for making authorization decisions
- Ensure that appropriate independent scrutiny of projects and project management systems is performed and properly communicated to relevant management groups
- Ensure that clearly defined criteria for reporting project status including risks and major issues is being performed for all projects
- Support a positive organizational culture of improvement and frank internal disclosure of project information
- Support processes to ensure that project stakeholders are engaged at a level that is commensurate with their importance to the organization and in a manner that fosters successful completion of the requirements.

In reviewing the responsibility list above, one would quickly conclude that most senior management groups are more removed from the tactical project working level that this implies and this is one motivation for including an overview of this topic. Government regulations such as Sarbanes-Oxley are heightening recognition and consequences of these high-level responsibilities. Given this some-what theoretical view of governance, the PM needs to understand the logic inherent in these requirements and must take a supportive responsibility in achieving these management principles.

36.5   TACTICAL LEVEL PROJECT GOVERNANCE

The starting point for the establishment of appropriate project level governance process is to define what elements need to be managed and those that will make a positive contribution to the project outcome. The trade-offs in this selection are management on one side and overhead bureaucracy on the negative side. Governance must be geared toward helping the project be successful and not slowing it down for control sake. One example of this help would be to assist in resolving internal conflict over whether a large change request should be approved or not. In this mode, the governance process serves a Supreme Court role when the lower-level processes are unable to resolve the issue, or the issue is larger than their delegated authority levels. This same type role would be appropriate for issues related to resources, budgets, and cross-departmental issues. Beyond the examples above, there are many other situations that are in the domain of project governance that create management gaps in the project governance environment. The list below represents governance-related events or processes that need to be delegated to either the PM, project board, or other organizational entity. Failure to successfully assign responsibility for any one of these, either initially or during the project, represents a governance gap that will then have to be resolved ad hoc with no clear management process. A basic project-oriented checklist of these items includes the following:

- Existence of a documented business case stating the objectives of the project and specifying the in-scope and out-of-scope items
- An acceptance mechanism to assess the compliance of the completed project to its required objectives
- Formal identification of all stakeholders with an interest in the project
- A defined method of information distribution to each defined stakeholder
- A defined set of project requirements formally agreed to by all appropriate stakeholders
- An agreed upon scope specification outlining the project work to be performed and associated deliverables
- The formal appointment of a PM (before or soon after the project Charter is approved).
- Clear assignment of project roles and responsibilities
- A current, published project plan that spans all project stages from project initiation through to operational status
- A system of accurate status and progress reporting including resources consumed and accomplishments
- A formal central document repository for the project archives
- A centrally-held glossary of project terms
- A defined process for the management and resolution of issues that arise during the project
- A defined process for processing change requests that involves appropriate management. Supporting elements for this process are an ICC procedure and an authority level Project Board for review and approval
- A process for the recording and communication of risks identified during the project.

A review of this summary list raises the point that these are all basic project management processes, so why go through this? The answer is that one of more of these will be absent from most projects with some associated governance issue result from the absence. Mature project management environments will have each of these items and corresponding organization responsibilities defined.

36.6 OPERATIONAL GOVERNANCE MODEL

There is yet a middle tier to the concept of governance. This tier involves the method used by the organization to map project requirements to business needs. We described these processes in Chapters 32 and 33 in regard to project portfolio management (PPM) and EPMO. In order for this layer of management to be effective, this activity should not be a haphazard process and requires some of the same support processes as outlined for the project and senior management levels. Capgemini defines four key component layers involved in this process. These include both the project and normal operational environments (Capgemini, 2006). A fifth layer, "innovation," also exists as an overriding driver to guide the overall adoption of new technical and business capabilities that emerge with new technology options. Experience has shown that high levels of innovation will be disruptive to existing activities and cannot therefore be expected to fit neatly into one of the other four categories without some additional imaginative thought as to its application. A brief summary of the five layers follows (Capgemini, 2006, p. 13):

1. *Innovation.* Understanding new technologies, products, and practices to build proposals regarding how to improve a technology or business area. The organization needs the ability to make decisions as to the best time to adopt a technology and to ensure a persistent rate of improvement.
2. *Information.* This area involves the form, content, and context of data management processes to actively support and record business decisions. Current information related to key business processes is increasingly important with faster moving markets and the demands for compliance.
3. *Integration.* The definition of all standards, naming conventions, practices, and architecture reference models required to support cost-effective integration technology aspects. This supports the ability to be adaptive and collaborative in terms of internal and external business flows.
4. *Infrastructure.* Technical architecture capability to support common project elements; networks, processors, directories, security, and storage. This activity supports low operational flexibility and high reliability.
5. *Industrialization.* The awareness of the methods, best practices, and suppliers that can be used to improve operational effectiveness support of a market competitive position.

Accomplishment of these attributes aids in overall goal alignment between the organization and the project.

The matrix shown in Figure 36.2 structures this view for a combined business and project perspective. Capgemini refers to this as the "5i & 3c" (5 rows and 3 columns) matrix and it forms the high-level structural first step in identifying the necessary goals for governance by delineating technology in terms of what it provides to the overall enterprise, as opposed to a single-point evaluation view (Capgemini, 2006, p. 14). Grading of the adequacy of governance in each cell of this matrix would be done by assessing the status of each element for a particular project with its specified technology.

The 15 cells shown in Figure 36.2 represent an assessment categorization model for evaluating future technology roles in the organization with a governance focus. In this particular example, the governance process is tightly linked to the project selection activity. The major difference is that in this case the high-level view is more governance focused than just looking for improved processes. Also this view forces considerations at the enterprise level. Goal visions identified by this process would then lead to more specific proposals that in turn would be structured into specific project targets through the normal project selection processes. The tactical governance process would then guide and support further definition into a deliverable product specification. Level of detail required for this overall process is variable based on individual enterprise needs, but the framework is still

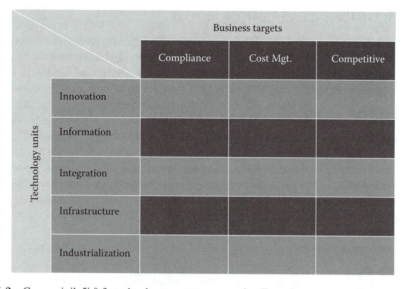

FIGURE 36.2 Capgemini's 5i & 3c technology governance matrix. (From Capgemini, Governance—Improving Governance between Business and IT Services, www.capgemini.com, 2006, p. 14, 16, 20, 21. With permission.)

useful for establishing high-level targets and responsibility assignments at the matrix cell or row level. Also, the matrix can help guide resource allocations and other guiding aspects such as constraints, issues to consider, and the like. During a technical assessment review, it is also possible to target specific technology vendors' products by using the cells to partition how a particular product supports that goal segment.

One should view the Capgemini matrix as an enterprise level guide to governance-oriented project formulation. In this perspective, we see a completely different view than that which would be visible from a typical departmental organization approach. In this view, the selection process becomes one of mining strategies to improve the overall organizational governance and then formally assigning roles and responsibilities to accomplish the identified targets. Even though our focus here is on the project role, the same activities could be mapped to other organizational tiers without altering the process described. Any organization looking at its environment in this manner should not be looking at low-level processes to repair. It also seems reasonable to suggest that the column headings for the matrix shown in Figure 36.2 could also be expanded to include other specific goals.

36.7 DEFINING RESPONSIBILITIES

One point that is not intuitively obvious from this description is the distinction between governance versus project selection. Basically, governance is more of a principle-based view, whereas the selection process is more nuts and bolts rigid mechanics based on data collection and analysis. The best results of a well-defined governance culture will come from the intelligent application of its principles combined with clear delegation of responsibility and managerial monitoring of internal control systems (Reid and Bourn, 2004).

The decision-making element of governance—who makes what decision, when and based on what information—although not easy to implement is relatively straightforward in concept. Basically, some organization entity needs to be charged with each of the elements outlined here as they will not naturally occur well without that. The procedures outlined in this section can be applied to all projects and the corresponding governance management principles outlined here should be reviewed to ensure that appropriate governance exists. Gaps in governance processes will create nagging problems for the PM, as his levels of authority are breached. It is important to note that in most cases the PM can not fix a governance gap, but realizing that the gap exists can help in mitigating its

TABLE 36.1
Governance Roles and Responsibilities Implementation Options

Group	Characteristics
Business monarchy	Individuals, or group of senior business executives, up to and including C × O level, but not including project senior executives other than the CIO
Project monarchy	Individuals, or a group of senior project executives including the CTO
Feudal	Business unit leaders, key function or process owners, and their delegates are independent groups
Project duopoly	Project senior executives work with only one business function or department
Anarchy	Individual users and/or project staff make decisions to suit their individual needs

Source: Capgemini, 2006. Governance—Improving Governance between Business and IT Services, p. 15. www.capgemini. com. With permission.

impact. Also, recognize that governance issues cover the entire organization and there needs to be continuity regarding how the layers of the enterprise decision layers interact.

The roles outlined below in Table 36.1 are certainly adequate to start with in order for the basic "who" aspects to be clarified, although in most cases there would need to be more detail regarding naming individuals to the roles. The challenge now is to define what types of decisions should be made by who and who should be responsible for preparing the information to support the decision-making process. The MIT Sloan Business School collected data from 100 top-performing organizations to categorize how they made governance decisions. Table 36.1 summarizes the roles and responsibilities options for these organizations. The one striking thing about these options is their variability. One might conclude that many different approaches will work so long as they are well defined.

Figure 36.3 illustrates how a custom governance matrix can be constructed based on the various organizational governance options described here. Each organization will have to design its own

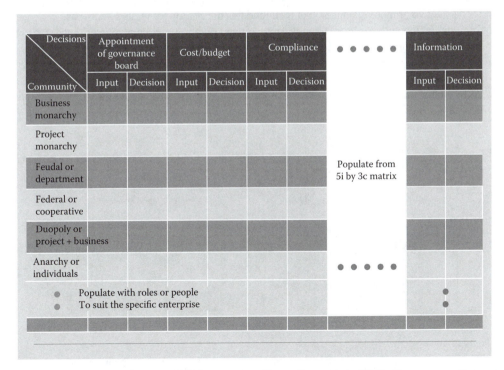

FIGURE 36.3 Roles and responsibilities matrix. (From Capgemini, 2006. Governance—Improving Governance between Business and IT Services, p. 14, 16, 20, 21. www.capgemini.com. With permission.)

unique specifications and further clarify the roles. Even if the first version of this effort is just a working draft, it would provide the basis for further clarifying general decision categories.

36.8 POPULATING THE PROJECT GOVERNANCE FRAMEWORK

In order to populate the Capgemini framework, defined core values need to be established for each decision area. In this view, we see a slightly different perspective. That is, we now recognize that there is a wide variety of goals to examine. Surprisingly, in some cases the goal for a decision area is not always to improve. There are situations where the goal is to follow a more static view. As can be seen in Figure 36.4 of the four-cell matrix, three of the cells are more static. Only one fits the highly competitive view initially described.

There are several reasons why for some areas of the organization it might be best to select one of the static modes. Examples to illustrate this point are as follows:

1. That area is not a core objective and by holding cost down there more resources can be applied to another higher priority area
2. There is no great operational value added by improving this area beyond what it is currently
3. The target area is being considered for spin off and sale.

Certainly there are many more examples to justify a mixed goal strategy for various organizational units, but the key thought is that continuous improvement is not always the operative goal. It is important to note that one of the fundamental reasons for an effective governance structure is to be able to properly prioritize and orchestrate this type of decision granularity. All organizations attempt to do this through their planning processes; so the point here is that the governance process helps the organization focus on more than just a singular global improvement view. By selecting the right growth targets at the expense of others a more efficient allocation of resources can occur. It is also possible for governance processes to vary by, project given the diversity of goals as described above. Experience has shown that if an individual project initiative cannot be defined into one of the four quadrants, the governance requirement overlaps across two or more quadrants that then muddies the criteria for related decisions.

The model framework outlined for governance formalization can be also used for determining whether there is a benefit in moving individual new project objectives, or existing systems and services, from one quadrant to another in order to better define their value. Realize that there is a significant management decision style difference in moving from the top right "Innovation/Lead" quadrant counter clockwise to the bottom right "Reduce Exposure" quadrant. The variations in

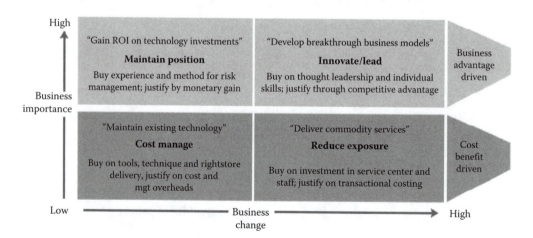

FIGURE 36.4 Business importance versus business change. (From Capgemini, 2006. Governance—Improving Governance between Business and IT Services, p. 14, 16, 20, 21. www.capgemini.com, With permission.)

approach indicated through the cells of Figure 36.4 also represent variability in governance strategies and likely any related array of project types related to those cells.

There are other uses for assigning projects into one of the four quadrants. For instance, the business planners should question each target area as to its desired quadrant position of Figure 36.4. This would lead to a decision for an excellent proposal in a "Cost Manage" category of rejecting since the goal is to not invest. In another case, if moving a process from "Maintain Position" to an "Innovate/Lead" would add particular value, that proposal would be accepted. Similarly, a "Cost Manage" situation might be improved in value by some selected technology initiative. Capgemini describes the boldest strategy with the biggest payoff as one which links the top right to bottom right in a transformational outsourcing to view the area as a commodity rather than a valued leadership position (Capgemini, 2006, p. 20). This strategy can free resources and money for reallocation to other initiatives. The desire to pursue a new wave of technologies may be the driver for a new approach to governance, but a roadmap summary showing what the technology aspects of this change may bring and will also help define a governance process.

The diagram in Figure 36.4 charts the evolution of an IT organization as it moves its information environment through a series of steps involving integration, information, industrialization, and innovation. Although this specific example may be hard to extrapolate to other environments, it is a good example of what a high-level governance process can yield in terms of high-level benefits. In this example, the result was lower costs and increased capability through what could be mapped as strategic governance. This process would not occur with traditional planning techniques to look at each lower-level item for improvement. If you look at the key steps that drove this result, it required sharing, high-level process redesign, and process integration (implementation). None of these would occur naturally without a governance process that drove the project necessary to create this result (Figure 36.5).

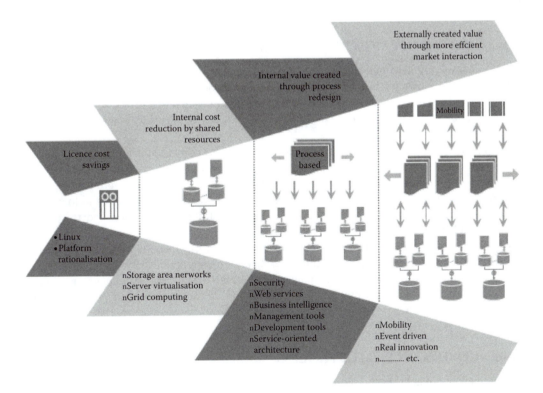

FIGURE 36.5 Evolution changes with respect to CapGemini's 5i model. (From Capgemini, 2006. Governance—Improving Governance between Business and IT Services, p. 14, 16, 20, 21. www.capgemini.com. With permission.)

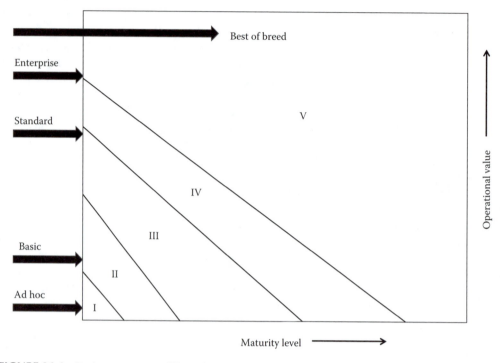

FIGURE 36.6   Project governance life cycle maturity model.

36.9   GOVERNANCE LIFE CYCLE MATURITY MODEL

Figure 36.6 shows a conceptual model illustrating the maturity levels of organizational governance. As with most maturity models, there are five separate levels outlined here. The capability labels on the left of the diagram represent the general characteristics of each level.

A translation of this maturity scale into a more general governance view is summarized below:

Level-1: The first level of project governance is basically the admission that there is a requirement for some form of control, but it is an ad hoc process.

Level-2: The second level shows the development of a basic governance structure with regular reviews and implemented standards, trust between business and the project begins to occur.

Level-3: The third, or intermediate level, reveals a portfolio prioritization and increased respect between the business and project layers.

Level-4: At level four, the maturity level is graded "advanced," which indicates above average capabilities in portfolio selection and governance. An enterprise level view of projects is recognized with common project management methods that deal with change management. This also includes program and PPM reporting mechanisms that provide senior management with metrics and overall project status visibility at an appropriate level.

36.10   GOVERNANCE VALUE PROCESS

Measurement of project value is complex, multifaceted, context-specific, and dynamic. Actual value received occurs from successful execution of well chosen and focused project strategies.

Making that happen is the fundamental goal of governance. The key word here is focus. No one strategy or approach will accomplish everything, every time. One underlying cause of the difficulty

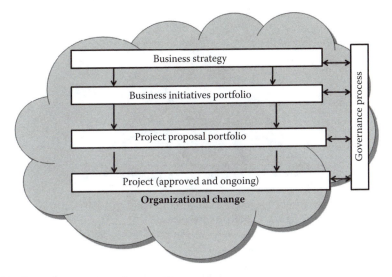

FIGURE 36.7   Strategic governance framework.

in realizing business value lies in the increasing rapidity and complexity of the business environment and in the changing sources of value creation. Organizations today increasingly generate value from intangible assets, such as brand, knowledge, information systems, e-business, improved processes, re-engineered organizational structures, and the enveloping management represented by enterprise governance.

The project level governance implementation strategy challenge is in understanding and managing a project-oriented value-creation process that is dynamic and complex and is characterized by these kinds of intangible assets. This process must be anchored to an explicit, clear, and focused business strategy. Without a clearly articulated and understood strategy, it is difficult to align investment decisions with strategic direction, to select the right things to do, and to decide what you will not do (see Figure 36.7).

If organizations are to be successful in tackling the question of value, they must recognize that the basic challenge is managing change and refocusing organizational initiatives in the proper direction. They must shift their focus beyond improving what they see now to looking outward at desired future states. This involves the following:

- Defining comprehensive programs of business change—programs that include all aspects of the business, processes, people, technology, and the subsequent change efforts that are both necessary and sufficient to deliver the desired business outcome
- Developing complete and comparable business cases to aid in proper target selection
- Selecting investments based on the overall value to an enterprise, not to just a functional or geographic unit
- Recognizing that the decision to select and proceed with an investment is only the beginning of an ongoing governance process. Successful execution requires clear accountability and relevant measurements of realized value

Most organizations have elements of strategic governance in place; however, project failure rates indicate that management remains one of the top causal failure factors. Rarely do all of the governance model components that have been described in Chapters 33, 34, and in this chapter exist in a organization. This would require at least a level four maturity. The driving theme of governance is to provide value to the business through alignment of organizational objectives and priorities with project and other initiatives.

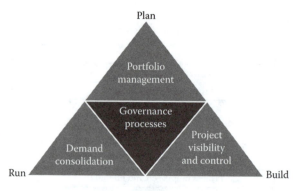

FIGURE 36.8 Mercury governance model. (Adapted from Lobba, A., 2004. *IT Governance: The Value Engine*, White paper, Mountain View, CA: Mercury Interactive.)

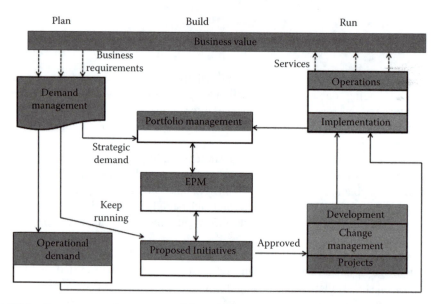

FIGURE 36.9 Governance value process. (Adapted from Lobba, A., 2004. *IT Governance: The Value Engine*. White paper, Mountain View, CA: Mercury Interactive.)

One of the key management questions involves where to start maturing the organizational governance processes. The schematic model outlined in Figure 36.8 shows how the governance process is the central component that drives a series of linked high-level processes of planning, running, and building the organization process environment.

A more detailed view of this overall process that shows more of the decision flows is shown in Figure 36.9, which is adapted from Lobba (2004).

36.11   CORPORATE GOVERNANCE AND PROJECT TEAMWORK (HALAS)*

The web site www.maxwideman.com contains a wealth of project management information and sometimes humor. This allegory authored by Peter Halas was originally published on the web site

---

* Note: This section supplied courtesy of Peter Halas and Max Wideman.

www.maxwideman.com as a somewhat abstract metaphor for corporate governance. Organizations work in much the same way that the human body does.

The following amusing little allegory by Peter Halas clearly illustrates how confrontation across "elements" can quickly cause a situation to get out of control. Underlying the story line, however, is a clear message for PMs and project practitioners generally—as we shall discuss later.

Scenario: Final Report from the Body Management Governance Team

In January, the Liver said to the Lungs, "Justify yourself. I do not see your value." The Lungs were so shocked at this "power play" that they started hyper ventilating. By February, the brain had to issue a general alert (bio email) saying that lack of oxygen is starting to affect the entire body, and started orchestrating a coordinated release of hormones, adrenalin, blood supply constriction activators, and so on.

In March, the Brain proceeded to monitor the levels of hundreds of actions to see, in real time, that they are again doing the job—and issued reports to the Heart, Kidneys, and Skin, advising them to make adjustments as needed. During April and May, the entire Body had benefitted from this response. Flows of work orders were being processed efficiently by all stakeholders, even traditionally less than cooperative areas such as the Thyroid, Pancreas, and Adrenalin Glands were now if not happy, at least not throwing stones.

By June the crisis was over; but in July the Liver that had been relatively unaffected by all this, and had not really seen (nor fully understood) what had just unfolded, issued a message to the Brain. Evidently, the Liver had determined that too large a percentage of the Body's blood supply was residing in the head. But not only that, the blood was too rich. It knew that other species can live without so many red blood cells and white blood cells, so it recommended the Brain cut back its blood supply. Further, it recommended "thinning out the blood" itself by 14%. (This was a magical number it had learned at a Gartner Executive briefing on "Offshoring, and Cutback Best Practices.")

By August, the Brain, having made these adjustments, was then just in subliminal mode and the autonomic Nervous System was by then the one really running the Body. The Brain was no longer functioning as before, but was just handling the vital "needs of the business."

From September onward some of the results of the thinning out of the Blood (i.e., outsourcing, and resource cutbacks) and the fact that there was now less blood in the cranium began to show (i.e., Willy-nilly project initiatives, no clear "alignment," no tools to assist in "governance" structures).

The effects were as follows:

- The Eyes did not see as well and as far down the road as they used to. So any truck coming in the opposite direction could "blind side the Body" at any time.
- The Hands did not work as fast and as nimbly, so many items picked up slipped out of them, were dropped and broken.
- When the Body consumed alcohol or smoked, it was so overwhelmed that it practically shut down. Similarly, it could hardly withstand any competition at all.
- Since many other cells did not like the climate that the liver had unintentionally caused, they started leaving inconspicuously through skin pores, by being exhaled through the lungs, and yes, whenever the opportunity arose, through the gastro and urinary tracts. Since metrics were no longer kept of this turnover rate, no one noticed.

By October, the Stomach was no longer working efficiently because of the new restrictive (cost cutting) measures, bile that was spewing out the liver was starting to accumulate. As a result, the bilirubin, biliverdin, and haematoidin metrics were off the charts, and no one seemed to know what to do about it anymore.

In November the Liver started fighting with the Stomach, not realizing that the Brain was the real reason for the way things being as they were. The fact is that whether one likes it or not, one cannot ignore the Brain, for it orchestrates the entire health of the organization and its ability to cope with its environment. Finally, the Body died (went into liquidation) in December.

The Cast:

Brain	Played by PM
Heart	Played by HR (Human Resources)
Liver	Played by Finance
Lungs	Played by Sales/Marketing
Hands, stomach	Played by Manufacturing and Inventory Logistics

36.12 COMMENTARY

Especially for those in the project world, does this scenario look familiar? It is mainly a result of how the various players perceive each other. For example, there are significant differences between two prevailing viewpoints, namely:

- "PM" as a "commodity provider—just do what I tell you"
- "PM" as a "strategic partner—help me make my business better."

This is *not* an "either/or" comparison. The project team and its leader need to be respected as a value added provider, and for the most part, are looked at as falling somewhere on a continuum between these two perspectives.

But when it comes to *projects* and their management, we can gather from this allegory the sort of complexity involved. This is where it is essential to distinguish the difference between managing the technology and managing the project. You need specialists in each bodily area (i.e., technology) to maintain the health of each (i.e., tactical management). Then you also need a wellness campaign (i.e., strategic project management) to ensure that the whole collection runs smoothly.

A "wellness campaign" (i.e., project management) can be as intensive or casual as you wish (i.e., level of project management ceremony), but without it, things will quickly unravel. But one thing is clear; it is quite a different exercise from the brain's day-to-day control (governance) of the body (technology) as a whole. In project work, project management and technology management cannot be divorced but they do need to be understood as different disciplines and, where necessary, managed separately often by different people working together.

36.13 CONCLUSION

In this chapter, we looked at the breadth of the governance process from the enterprise level down to the project. The final section left us with hopefully a little serious humor on this topic. As we saw, if all of the components are not working together the organism will eventually fail. A governance process represents the driver for the enterprise level integration.

Organizations that recognize and start the governance process with portfolio management are those that are sensitive to controlling their "front door"—that is, projects being pursued, but they may now feel that their biggest management gap is the project approval process and not having "apples to apples" data that allows meaningful comparisons. In this environment, there is often a disorganized collection of funding that does not deliver optimum value to the business.

Secondly, organizations that have tactical project management problems in staying on schedule, on budget, and in using resources efficiently, tend to focus their governance initiatives initially on improving project visibility and control.

Third, organizations determined to minimize how much they spend in the day-to-day "keeping the lights on" activities will focus their initial governance on better demand management, better prioritization, and operational efficiency—all aspects of running the organization. This will lead to various initiatives for demand consolidation efficiency.

Regardless of which side of the governance triangle (figure 36.8) is viewed as the appropriate initial starting point, once value from this activity is demonstrated, there will be a tendency to expand into the other sides of the triangle on a priority basis. Typically, this evolution happens gradually, folding in more processes and bringing more value into the organization. The key to starting is in finding where the organization will get the biggest initial gain. Typical strategies for this include eliminating/combining project, defining resource allocations, reviewing decision-making processes for project approval, make versus buy decisions, and so on.

Governance of project management activities involves those areas that are specifically related to project activities. Effective governance of project management ensures that an organization's project portfolio is dynamically aligned to its goals and objectives, is delivered efficiently, and is sustainable. This activity also supports the means by which senior management and other major strategic project stakeholders are provided with timely, relevant and reliable information (i.e., visibility into the tactical realms). An associated project governance role is to balance the risk of the organization's investment against the opportunities and benefits that those outcomes will provide the business. It addresses the risks to ensure that the value to the organization is received in balance with properly mitigated risks.

Current organizational governance processes tend to be inadequate in managing what is, in most cases, an uncertain project journey to a fuzzy destination. To support such an environment there needs to be an evolutionary process directed toward a full project investment life cycle from goal setting to fruition; one that senses and responds to changes in the internal and external environment and improves internal understanding regarding what is working and not working as expected. Without such a process, the risk of ending up in the wrong place, with the associated undesirable business consequences, is significantly increased.

To be effective in delivering value, the governance process must ensure that organizations understand their desired business outcomes, understand their sources of value, and develop value-focused strategies. From this view, they can then take a structured approach to developing comprehensive, value-based business change programs to execute those desired business strategies, and manage the realization of value.

REFERENCES

Capgemini, 2006. Governance—Improving Governance between Business and IT Services. www.capgemini. com, p. 14, 16, 20, 21.

Halas, P., 2008. This section was adapted from http://www.maxwideman.com/musings/governance.htm with permission of Peter Halas, QuinneTech LLC.

Handler, R., 2004. *IT Governance—The Natural Evolution of IT*. Stanford, CN: MetaGroup, Inc.

Jennings, T., 2005. Planning for Value: Artemis Solutions for IT Management. www.konsultex.com.br/artemis/ arquivos-e-imagens/artemis-butler-report.pdf (accessed on February 2008).

Lobba, A., 2004. *IT Governance: The Value Engine*. White paper, Mountain View, CA: Mercury Interactive.

Meta, 2004. IT Governance: The Value Engine, A webinar presented November 16, 2004.

Reid, B. and J. Bourn, 2004. Directing Change A Guide to Governance of Project Management. www.apm.org. uk/download.asp?fileID=319 (accessed January 15, 2008).

Part X

Professional Ethics and Responsibility

This final chapter describes various concepts regarding the project management code of professional ethics and social responsibility as outlined by PMI. Specifically, the professional project manager should exhibit the following characteristics and skills:

1. Ensure individual integrity and professionalism by adhering to legal requirements and ethical standards in order to protect the community and all stakeholders
2. Contribute to the project management knowledge base by sharing lessons learned, best practices, research, and other information within appropriate communities in order to improve the quality of project management services, build the capabilities of colleagues, and advance the profession
3. Enhance individual competence by increasing and applying professional knowledge to improve services
4. Balance stakeholders' interests by recommending approaches that strive for fair resolution in order to satisfy competing needs and objectives
5. Interact with team and stakeholders in a professional and cooperative manner by respecting personal, ethnic, and cultural differences in order to ensure a collaborative project management environment.

Note: This list of learning objectives is adapted from the Project Management Institute Code of Ethics and Professional Conduct, http://www.pmi.org/pdf/ap_pmicodeofethics.pdf

37 Ethical Project Management Practices

37.1 LEARNING OBJECTIVES

The purpose of this chapter is to outline fundamental practices that the professional PM must understand in the course of executing his or her job responsibilities. It is no longer sufficient to just get the job done. Daily news articles reinforce the negatives that can occur when a manager of any activity decides to short cut either ethical or legal limits for which they are responsible. This occurs whether the individual is president of the United States, CEO, or PM. A professional without these traits will eventually crumble. Surprisingly, the act of being ethical and honest is not as easy as one might think as this chapter will attempt to illustrate. From a PMI perspective the PM must understand the basic tenets of the topic and from an operational view one must be able to translate these into workable job traits.

37.2 INTRODUCTION

The positive value of ethical behavior has been increasingly recognized in recent years. Improper actions by middle and upper management can destroy promising carets and even the organizations that work for. In all such cases a poor choice of decisions lay at the root of the failure. We categorize these under the category of professional ethics.

PMI has developed a formal Code of Conduct for Professional Responsibility. The basic structure of this code deals with the following (PMI):

- Values (as defined in the *PMBOK® Guide*)
- Responsibility
- Respect responsibility
- Fairness
- Honesty.

Although each of these terms is reasonably familiar to most, implementation of the underlying concepts behind the term is not necessarily easy. Figure 37.1 provides a good overview of the basic actions related to each of the key structure components.

It is essential that PMs not only be good employees for the organization that they work for, but they also must behave as a broader representative of the project stakeholder population that may consist of external parties outside of their organization. Since some of these behavioral relationships may in fact be external to the host organization behavioral rule, interpretation can become complex in the real world. The external influences come from the various needs and interests of groups with various stakes in project. The key theme of this code is that the PM must follow a

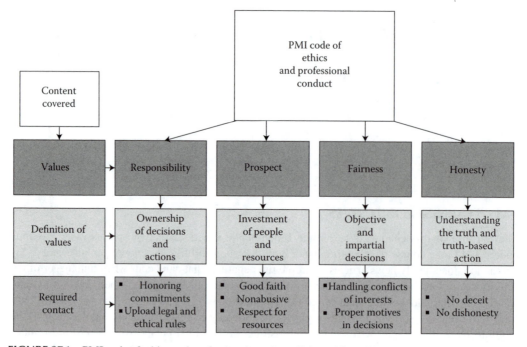

FIGURE 37.1 PMI code of ethics and professional conduct. (Adapted from PMI, 2006. Project Management Institute Code of Ethics and Professional Responsibility. http://www.pmi.org/pdf/ap_pmicodeofethics.pdf, [accessed October 17, 2008].)

prescribed set of professional responsibility and ethical principles. These principles include the following areas:

- Ensuring individual integrity
- Contributing to the base of project management knowledge
- Enhancing individual performance
- Balancing stakeholders' interests
- Interacting the project team and stakeholders in a professional and cooperative manner.

Ensuring one's individual integrity requires taking ethically based responses to a number of common project scenarios. Generally speaking, the PM must do what is right, but that term is difficult to define in every case. Specifically, written and oral communications with project stakeholders and governmental authorities must be truthful. The PM must also adhere to the approved processes for project management activities. Lastly, any violations of applicable laws and ethical standards must be immediately reported to the appropriate authorities.

To contribute to the project management knowledge base, a PM must do a number of things. First, any lessons learned from their personal project experiences must be shared with others who would profit from that knowledge. Following this principle also requires that the PM contributes to the education of and mentoring to less experienced PMs. In addition, he needs to engage in research to determine how to improve the profession and its processes, and then the findings of this research. Finally, he should strive to find techniques for improved measurement of project performance and work to continuously improve those outcomes.

In order to improve individual competence, the PM must take a number of steps. Initially, it is important to appraise and understand his personal strengths and weaknesses. Next, the PM needs to take advantage of learning opportunities to address these weaknesses. Furthermore, he should prepare and execute a personal development plan in much the same manner as he pursues a project objective. Finally, the individual must continue to improve knowledge related to relevant professional topics. This entails seeking out new information about project management and the industry.

Balancing project stakeholders' interests requires that a PM consider the interests of all players in the project. Initially, the PM needs to examine the interests and needs of these individuals and groups, and then seek to understand ways in which these diverse interests can be best met. Following this, he must also work to resolve conflict with the understanding that the customer's needs must usually take precedence over the other groups. One strategy to minimize future conflict of interest is to obtain the clearest and most complete requirements possible prior to undertaking execution of the project. Using this philosophy as a base, the PM needs to address scope change issues and other problems when they arise, rather than waiting until time has passed and they have become more serious.

Interacting with the project team and stakeholders in a professional and cooperative manner requires a number of actions. Cultural differences can impact the smooth functioning of a project team. The PM needs to understand these and take them into account in his dealings with the various stakeholders. In addition, he must uncover any differences in communications preferences, work ethics, and work practices among these groups. All future stakeholder dealings should recognize and respect these differences. Cultural and language differences across country boundaries are particularly critical. When projects contain multinational stakeholders, the PM needs to follow local practices and customs so long as doing so does not violate laws.

Significant differences in interests can also occur within a single country environment. These include customers, government agencies, other business functions involved in the project, sponsors providing financial resources for the project, the internal project team, parties from inside or outside the organization, end users of the project's product, members of society who will be affected by the project, and others. Clearly, this diversity of interest requires that PMs give adequate attention to numerous potential problem areas during the course of a project. These dealings will almost assuredly create ethical concerns in regard to information handling and project decision making. It is only by giving this area proper attention that the manager can ensure that his ethical responsibilities to these stakeholders can be effectively and properly discharged. The list below summarizes five basic rules of thumb to guide PMs in their behavior regarding honesty and ethics. Following these rules do not guarantee appropriate behavior, but should help guide managers in conducting themselves in an ethical and professionally responsible manner. In dealing with stakeholders a PM should

1. Not misuse access to, or control over financial resources that stakeholders have given them for legitimate use in the project, for example, engage in illegal manipulation of organizational resources
2. Not mislead stakeholders in regard to the status of the project by providing them with inaccurate information or failing to provide them with timely information relevant to the project
3. Inform the proper authorities regarding legal or professional violations by other stakeholders taking place in the context of the project
4. Not reveal trade secrets provided to them in confidence, unless holding such information in confidence would violate a law, contractual provisions, or professional responsibility/ethical rules
5. Not use information obtained in the context of the project for the purpose of gaining an unfair advantage over the stakeholder, or that would be harmful to the stakeholder if revealed.

Each of the items outlined above seem straightforward and easy to follow; however, situations will occur in the project environment that will challenge individual interpretation of a specific action. PMs are frequently faced with opportunities to obtain additional rewards for themselves or their organizations if they are willing to take a professionally or ethically marginal or inappropriate actions, such as conducting one less audit or inspection than they know they should or looking the other way when faced with clear evidence of others' wrong doing, and so on. Unfortunately, such actions can have far-reaching negative impacts, as well as having the potential to ruin the manager's career if the acts are discovered. Even the simple acts of accepting a meal or a football ticket are judged unethical in some organizations. For all of these reasons it is important to know what each organization defines as its ethics and behave according to those rules. Avoiding such situations can be a veritable minefield if a PM does not clearly understand which actions constitute inappropriate professional or ethically irresponsible behavior.

37.3   PMI'S CODE OF PROFESSIONAL CONDUCT

Given recent experiences in American industry regarding misplaced ethics among senior executives, PMI has elevated the level of interest and focus on this topic. The PMP certification exam will ask numerous questions to verify that the candidate understands the breadth of this topic. Basically, the PMI code deals with the following (PMI, 2006):

1. Organizational rule and policy compliance
2. Personal ethics in terms of reporting qualifications and representations
3. Respect and honesty toward the profession
4. Honesty in reporting facts to stakeholders
5. Maintaining proper confidentiality of data and other information
6. Care in avoiding conflict of interest
7. Care in avoiding receipt of payment from outside sources for questionable reasons.

As we have emphasized previously, ethical behavior can be a much more complex topic than first appears. Generally speaking, understanding the local policies and following basic rules of courtesy and honesty will suffice, but not thinking about these issues can create situations that can sabotage one's professional career and reputation. Think of this requirement as being ethical and honest in all dealings involving both internal and external relationships. The PM must think of themselves as *honest brokers of process and information.*

REVIEW QUESTIONS

Although the idea of honesty and ethics would seem like a simple issue, the real world often presents some tough interpretations for the PM. The sample questions shown below will provide an opportunity to review this theoretical material in simulated real-world settings and see how to react according to the formal policy definitions. Also, discussing these questions with your peers and changing the scenarios slightly will offer good discussion material.

1. You are working for a U.S. organization and have just finished a very large and successful project in a foreign country. It is common in this culture to reward people for good work. The sponsor is extremely happy with the outcome of the project and wants to hold a formal ceremony and give expensive presents to you and key team members to show his appreciation. He has told others that it will hurt his feelings if you do not take these gifts. What do you do?
2. You are a member of a PMP certification study group. One of the members says that he has a friend who is wired into the exam question authors and can get significant input on key topics being addressed this round. What do you do? Do you report this situation to the certifying organization?

3. Success of your foreign project depends on receiving materials in a timely manner; however, the goods are being held up in local customs for long periods of time. A nephew of the local *Grand Po Bar* says that he can provide an expediting service for you and get the goods moved through customs quickly. In checking around, you find that this seems to be in fact true and also seems to be the only way that the goods get moved. What do you do? Who would you coordinate this decision with?

4. You are told by your boss to cut your project budget estimate by 20% in order to get your project plan approved by senior management. All other cost "cutting" strategies have been exhausted and you now feel that the current estimate is accurate for the work defined. What do you do? What if your boss says "just cut the budget and get the project done with the cut specified?"

5. Your boss asks you to write an invited article for a national industry publication for him. You do this and the boss does not include your name on the article. What do you do?

6. While working on an external (contracted) project that has extra budget funds your customer asks you to perform some additional tasks that are not included in the formal contract. You should
   A. Honor the customer's request as a sign of cooperation to ensure future business
   B. Refuse the request and report the customer to your sponsor
   C. Acknowledge the request and advise the customer to submit a formal change request
   D. Convene a meeting of the project team and rewrite the scope statement

7. You are managing an internal project. The initial product test results are very poor and do not meet the minimum customer requirements. If these results are made available to your customer you are afraid that they might cancel the project and this could reflect poorly upon you. Rerunning the product test can be done quickly and inexpensively. Based on this set of circumstances you should
   A. Be the first to recommend canceling the project
   B. Inform your external sponsor about the results and wait for a response
   C. Inform your management immediately and recommend retesting for verification
   D. Withhold the information from management until you perform additional tests to verify the initial results

8. Your project is running out of budget allocation and significant work remains. You are directed by senior management to instruct your team to charge their work time to another project's account. Given these instructions you should
   A. Follow instructions
   B. Inform the corporate auditors
   C. Understand the background of management's instructions before taking any action
   D. Shut down the project, if possible

9. You are working in a country where it is customary to exchange gifts between contractor and customer. Your company code of conduct clearly states that you cannot accept any gifts from a client. Failure to accept the gift from this client may result in termination of the contract. The action to take in this case would be
   A. Provide the customer with a copy of your company code of conduct and refuse the gifts
   B. Exchange gifts with the customer and keep the exchange confidential
   C. Contact your project sponsor and/or your legal or public relations group for assistance
   D. Ask the project sponsor or project executive to handle the gift exchange question with the client

10. You are a PM working on a time and material contract. The target price for the project is $2,000,000 and the project schedule is 12 months. The most recent completion estimate indicates that the project will finish 2 months early and if this happens your company will lose about $250,000 in billings. What should you do?

    A. Bill for the entire planned amount since this was the approved budget

    B. Bill for the target amount by adding nice to have features to the design at the end of the project so that the schedule and budget are met

    C. Report the project status and completion date to the customer

    D. Report the project status and completion date to the customer and ask if they would like to add any additional features to account for the monies not spent

11. In order to balance the needs of the many stakeholders involved in your project, the most desirable method to achieve resolution of conflicts would be

    A. Compromise

    B. Forcing

    C. Controlling

    D. Confrontation

12. You receive a contract to perform testing for an external client. After contract award, the customer provides you with the test plan to use for the acceptance process. The vice president for QA says that the customer's test plan is flawed and he will correct the plan that will be used and it will be more in line with the organizational quality program. The contract says that the customer will supply the acceptance test plan. In this case you should

    A. Use the customer's test plan

    B. Use the QA manager's revised test plan without telling the customer

    C. Use the QA manager's revised test plan and inform the customer

    D. Tell your sponsor that you want to set up a meeting with the customer to resolve the issue

13. You have just been assigned as the PM for an ongoing project and discovered that your project team is routinely violating OSHA, EPA, and affirmative action regulations. You should

    A. Do nothing; it is not your problem

    B. Start by asking management if they are aware that regulations are being violated

    C. Talk to the corporate legal department

    D. Inform the appropriate government agencies about the violations

14. One of your employees has an opportunity for promotion in another area. If this promotion is granted, the employee will be reassigned elsewhere causing a resource problem for the project. You have the authority to delay the promotion until your project is completed. You should

    A. Support the promotion but work with the employee and the employee's new management to develop a good transition plan

    B. Ask the employee to refuse the promotion until your project is completed

    C. Arrange to delay the promotion until the project is completed

    D. Tell the employee that it is his responsibility to find a suitable replacement so that the project will not suffer

15. In accordance with the compensation agreement for your project you have been given a $70,000 bonus to be distributed to your seven-person team as you see fit. One of the team members has not performed particularly well and another of the team is in your car pool. Based on this situation you should

    A. Allocate an equal share to each team member to avoid the image of favoritism

    B. Provide everyone a share based on your personal assessment of their performance

    C. Give the decision to the team and follow their advice

    D. Ask the sponsor to make the decision

16. You are the PM and your customer has requested that you inflate your planned cost estimates by 25%. His logic is that management always reduces the cost of project estimates by about this amount so this strategy would balance out the required budget. Which of the following is the best response to this situation?

A. Do as the customer asked to ensure that the project requirements can be met by adding the increase as a contingency reserve

B. Do as the customer asked to ensure that the project requirements can be met by adding the increase across each task

C. Do as the customer asked by creating an estimate for the customer's management and another for the actual project implementation

D. Complete an accurate estimate of the project. In addition, create a risk assessment showing why the decreased project budget would be inadequate.

17. You are the PM working in a foreign country. Your local support person from the client organization presents you with a list of local team candidates for you to hire and you find that several of these are related to him. What is your reaction?

A. Reject the team leader's recommendations and assemble your own project team

B. Review the résumé and qualifications of the proposed project team before approving the team

C. Determine if the country's traditions include hiring from the immediate family before deciding on how to address the family member situation

D. Replace the project leader with an impartial project leader

18. You discover that one of your project team members has sold pieces of equipment that were allocated to the project. Upon further investigation you find that his rationale was a need for cash to pay for his son's college tuition. He says that he considers this remuneration for overtime hours worked without pay and he asks for your support in this view. You also find that his claim of unclaimed and unpaid overtime is true and that he has been a hard worker on project. What should you do?

A. Fire the project team member

B. Report the team member to his manager

C. Suggest that the team member report his actions to the HR department

D. Tell the team member that you are disappointed in what he did, and advise him that you will consider this a fair trade for the unpaid overtime. You also inform him that this will be grounds for dismissal if it occurs again.

19. You are a PM working within your functional organization and you do not get along well with the departmental manager. There is a serious disagreement regarding how the project should be conducted. This disagreement involves schedules, sequence of tasks, quality objectives, and other aspects of the project. While this disagreement is still unresolved the department manager tells you to start work on what he considers critical activities. Which of the following choices is the best for you?

A. Go to higher-level senior management and voice your concerns

B. Complete the activities as requested

C. Ask to be taken off of the project

D. Refuse to begin activities on the project until the conflict issues are resolved

20. PMI has contacted you regarding a PMP candidate's experience claim. The individual involved is a friend and he says that he worked as a PM in your organization. He did work there, but not in the capacity individuated. This is a violation of which of the following?

A. The PMP code to cooperate on ethics violations investigations

B. The PMP code to report accurate information

C. The PMP code to report any PMP violations

D. Law concerning ethical practices.

REFERENCE

PMI, 2006. Project Management Institute Code of Ethics and Professional Responsibility. http://www.pmi.org/pdf/ap_pmicodeofethics.pdf (accessed October 17, 2008).

Appendix A: Financial Metrics

Financial metrics represent one of the core techniques used to evaluate the business value of a project proposal. Organizations use various forms of these metrics and three are illustrated here. The role of financial measures is to evaluate a life-cycle view of benefits and costs related to a particular project proposal. In this view, the goal is to assess the relative measure of benefits versus the associated costs over some expected useful life.

Three classic financial measures are demonstrated here: payback (PB), net present value (NPV), and internal rate of return (IRR). Each of these has inherent advantages and disadvantages in their constructs and these should be understood before making decisions based on any one of them.

Step one in the financial measurement process is to estimate the net benefits and costs of the project proposal over some useful life cycle. In many cases, this is done for 5 years given the inaccuracy of estimates longer than this. As an example, let us assume that the project had the following cost-benefit estimates:

0	– $50,000
1	$15,000
2	$30,000
3	$10,000
4	$50,000
Total	$55,000

The interpretation of the forecast data shown above is that the initial project would cost $50,000 and would be installed 1 year later at which point positive returns would occur. During its expected lifetime (years 1 through 4), the project would generate net business benefits (benefits minus cost) as shown above. Summing the total list of values shows that the project would generate $45,000 in net benefits over the 4-year period. A basic financial question related to this set of data is "Do we consider this a good financial project?" The answer to this depends on how we look at the data using the basic financial measures. Let us explain the three classic measures before going further.

PB method is the simplest and most used financial measure. This calculation simply looks at the cash flow estimates and defines the time required to recoup the initial investment. The goal in using this measure would be to favor projects that pay back their investment the quickest.

NPV is the calculated present value of the benefit stream. This measure is sensitive to the time value of money, and the analysis is based on some *hurdle rate*, that is, it has to be measured positive compared to say 15%. This would favor any projects that returned values greater than this amount.

IRR represents the rate of return calculated for the estimated benefit stream. This is analogous to investing money and getting more in return, and then calculating how much return you received on that investment. This is the most difficult financial calculation, but the one that has the best comparative value across the project portfolio.

A.1 CALCULATION MECHANICS

PB—time required to pay back the initial investment.

NPV—this value is calculated by the formula $(1 + i)^n$, where i is the interest rate and n is the number of years in the future. In the NPV calculation, each out year value would be translated back to the current time using the formula above, and then all values are added together for the NPV value.

IRR—this uses the same concept as the NPV, except that the value of i is unknown; therefore it is necessary to repeat the calculation until a value of NPV is zero. So, IRR is the value of i where the NPV is zero.

Timing Issues Related to Estimates

In the use of these measures there is a timing issue to consider, that is, do we assume in the estimate that all of the cost are incurred at the beginning of the year, middle of the year, or spread equally through the year? For these examples, we will follow the Excel assumptions, but for a particular analysis, the timing question should be reviewed to make sure that the calculated values properly reflect the estimates used.

A.2 COMMENTS ON EXCEL INTERNAL ASSUMPTIONS

In calculating NPV, Excel assumes that the cash flows occur at the end of the indicated period.

A.3 SAMPLE CALCULATIONS

The ABC company is considering investing in three projects. The projects are expected to cost the company $500,000 and the future cash inflows for the three projects being considered are as follows:

	Project X		Project Y		Project Z
1	$50,000	1	$325,000	1	$130,000
2	$150,000	2	$175,000	2	$130,000
3	$300,000	3	$75,000	3	$130,000
4	$100,000	4	$50,000	4	$130,000
5	$50,000	5	$25,000	5	$130,000
Total	$650,000	Total	$650,000	Total	$650,000

Note that all three projects have the same total benefit stream; however, timing of these benefits is different for each. We will see the impact this has on the financial measures that each produce.

- One more important thing about the Excel's NPV function is that if we include the initial project cost or the initial investment into the Excel NPV function, we get incorrect results that are indicated in the worksheet below as *NPV (Incorrect)*. This is an idiosyncrasy of the Excel's NPV function.

A.3.1 OTHER INFORMATION

The company wants to make 10% or more on any project that they accept. Each of these projects is only expected to produce cash inflows for 5 years.

The worksheets above illustrate the calculations both manually as well as using Excel's functions for the given examples.

Project X Excel Calculations

Project X

Year		Initial Cost
0		$ (500,000)
1	2008	$ 50,000
2	2009	$ 150,000
3	2010	$ 300,000
4	2011	$ 100,000
5	2012	$ 50,000
Total Returns		$ 650,000

NPV = ($5,836.66)   **Calculations Below**

NPV is the value from IRR calculations table below
It is the value closest to zero for the assumed R.O.I - r

IRR = 10%

Payback = 3 Years
Just add the # of years required to get back the initial invested cost

1) NPV Calculations

Using: $PV = FV / (1 + r)^n$

PV = Present Value
FV = Future Value - Return in given year
r = Rate of Interest
n = Period - No. of years considered

Ex.: PV Calculations

Year 1: $PV = (50000)/(1+0.01)^1$
Year 2: $PV = (150000)/(1+0.01)^2$
Year 3: $PV = (300000)/(1+0.10)^3$
Year 3: $PV = (100000)/(1+0.10)^4$
Year 3: $PV = (50000)/(1+0.10)^5$

Year (n)	Present Value (PV)
1	$45,454.55
2	$123,966.94
3	$225,394.44
4	$68,301.35
5	$31,046.07
0	($500,000)
NPV	-$5,836.66

Project X Hand Calculations

2) IRR Calculations

For different values of "r" calculate NPV, the closest value of NPV to zero will have the r = IRR

Calculated NPV = Initial Cost + PV (Year 1 + Year 2 + Year 3 + Year 4 + Year 5)

Calculated NPV	Assumed R.OI - r	Year 1 - PV	Year 2 - PV	Year 3 - PV	Year 4 - PV	Year 5 - PV	Initial Cost
$ 64,271.30	0.05	47619.04762	136054.42	259151.28	82270.247	39176.308	$ (500,000.00)
$ 49,127.34	0.06	47169.81132	133499.47	251885.78	79209.366	37362.909	$ (500,000.00)
$ 34,572.97	0.07	46728.97196	131015.81	244889.36	76289.521	35649.309	$ (500,000.00)
$ 20,578.94	0.08	46296.2963	128600.82	238149.67	73502.985	34029.16	$ (500,000.00)
$ 7,117.69	0.09	45871.55963	126252	231655.04	70842.521	32496.569	$ (500,000.00)
$ (5,836.66)	0.10	45454.54545	123966.94	225394.44	68301.346	31046.066	$ (500,000.00)
$ (18,308.51)	0.11	45045.04505	121743.36	219357.41	65873.097	29672.566	$ (500,000.00)
$ (30,320.84)	0.12	44642.85714	119579.08	213534.07	63551.808	28371.343	$ (500,000.00)
$ (41,895.29)	0.13	44247.78761	117472	207915.05	61331.873	27137.997	$ (500,000.00)
$ (53,052.31)	0.14	43859.64912	115420.13	202491.45	59208.028	25968.433	$ (500,000.00)
$ (63,811.16)	0.15	43478.26087	113421.55	197254.87	57175.325	24858.837	$ (500,000.00)
$ (74,190.05)	0.16	43103.44828	111474.44	192197.3	55229.11	23805.651	$ (500,000.00)
$ (84,206.20)	0.17	42735.04274	109577.03	187311.17	53365.005	22805.558	$ (500,000.00)

	IRR	0.10

For r = 0.10 the NPV value is closest to 0, hence we choose IRR = 0.10

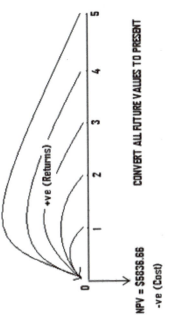

+ve (Returns)

NPV = $5836.66   CONVERT ALL FUTURE VALUES TO PRESENT

-ve (Cost)

Project Y Excel Calculations

NPV = $46,104.96
NPV is the value from IRR calculations table below
It is the value closest to zero for the assumed R.O.I.- r

IRR = 16%
Payback = 2 Years

Year		Project Y
0	Initial Cost	$ (500,000)
1	2008	$ 325,000
2	2009	$ 175,000
3	2010	$ 75,000
4	2011	$ 50,000
5	2012	$ 25,000
	Total	$ 650,000

Project Y Hand Calculations

1) NPV Calculations

Using: $PV = FV / (1 + r)^n$

PV = Present Value
FV = Future Value - Return in given year
r = Rate of Interest
n = Period - No. of years considered

Project Y   Initial Cost= $500000

Year (n)	Present Value (PV)
1	$295,454.55
2	$144,628.10
3	$56,348.61
4	$34,150.67
5	$15,523.03
0	($500,000)
NPV	$46,104.96

2) IRR Calculations

For different values of "r" calculate NPV, the closest value of NPV to zero will have the r = IRR

Calculated NPV = Initial Cost + PV (Year 1 + Year 2 + Year 3 + Year 4 + Year 5)

Calculated NPV	r	Year 1 - PV	Year 2 - PV	Year 3 - PV	Year 4 - PV	Year 5 - PV	Initial Cost
$ 93,765.07	0.05	309523.8095	158730.16	64787.82	41135.124	19588.154	$ (500,000.00)
$ 83,610.73	0.06	306603.7736	155749.38	62971.446	39604.683	18681.454	$ (500,000.00)
$ 73,781.85	0.07	303738.3178	152851.78	61222.341	38144.761	17824.654	$ (500,000.00)
$ 64,263.71	0.08	300925.9259	150034.29	59537.418	36751.493	17014.58	$ (500,000.00)
$ 55,042.44	0.09	298165.1376	147294	57913.761	35421.261	16248.285	$ (500,000.00)
$ 46,104.96	0.10	295454.5455	144628.1	56348.61	34150.673	15523.033	$ (500,000.00)
$ 37,438.90	0.11	292792.7928	142033.93	54839.354	32936.549	14836.283	$ (500,000.00)
$ 29,032.59	0.12	290178.5714	139508.93	53383.519	31775.904	14185.671	$ (500,000.00)
$ 20,874.99	0.13	287610.6195	137050.67	51978.762	30665.936	13568.998	$ (500,000.00)
$ 12,955.63	0.14	285087.7193	134656.82	50622.864	29604.014	12984.217	$ (500,000.00)
$ 5,264.64	0.15	282608.6957	132325.14	49313.717	28587.662	12429.418	$ (500,000.00)
$ (2,207.37)	0.16	280172.4138	130053.51	48049.326	27614.555	11902.825	$ (500,000.00)
$ (9,469.28)	0.17	277777.7778	127839.87	46827.792	26682.502	11402.779	$ (500,000.00)

IRR = 0.16

For r = 0.16 the NPV value is closest to 0, hence we choose IRR = 0.16

NPV =	($7,197.72)
IRR =	9%
Payback =	4 Years

Project Z Excel Calculations

Project Z

Year	Cost	
0	2008	$ (500,000.00)
1	2009	$ 130,000.00
2	2010	$ 130,000.00
3	2011	$ 130,000.00
4	2011	$ 130,000.00
5	2012	$ 130,000.00
	Total	$ 650,000.00

Project Z Hand Calculations

Using:

1) NPV Calculations

$$PV = FV / (1 + r)^n$$

PV = Present Value
FV = Future Value - Return in given year
r = Rate of Interest
n = Period - No. of years considered

Project Z	Initial Cost= $500000

Year (n)	Present Value (PV)
1	$118,181.82
2	$107,438.02
3	$97,670.92
4	$88,791.75
5	$80,719.77
0	($500,000)
NPV	-$7,197.72

2) IRR Calculations

For different values of "r" calculate NPV, the closest value of NPV to zero will have the r = IRR

Calculated NPV = Initial Cost + PV (Year 1 + Year 2 + Year 3 + Year 4 + Year 5)

Calculated NPV	r	Year 1 - PV	Year 2 - PV	Year 3 - PV	Year 4 - PV	Year 5 - PV	Initial Cost
$ 62,831.97	0.05	123809.5238	117913.83	112298.89	106951.32	101858.4	$ (500,000.00)
$ 47,607.29	0.06	122641.5094	115699.54	109150.51	102972.18	97143.562	$ (500,000.00)
$ 33,025.67	0.07	121495.3271	113547.03	106118.72	99176.378	92688.203	$ (500,000.00)
$ 19,052.30	0.08	120370.3704	111454.05	103198.19	95553.881	88475.816	$ (500,000.00)
$ 5,654.66	0.09	119266.055	109418.4	100383.85	92095.277	84491.08	$ (500,000.00)
$ (7,197.72)	0.10	118181.8182	107438.02	97670.924	88791.749	80719.772	$ (500,000.00)
$ (19,533.39)	0.11	117117.1171	105510.92	95054.88	85635.027	77148.673	$ (500,000.00)
$ (31,379.09)	0.12	116071.4286	103635.2	92531.432	82617.35	73765.491	$ (500,000.00)
$ (42,759.94)	0.13	115044.2478	101809.07	90096.521	79731.435	70558.792	$ (500,000.00)
$ (53,699.47)	0.14	114035.0877	100030.78	87746.297	76970.436	67517.926	$ (500,000.00)
$ (64,219.84)	0.15	113043.4783	98298.677	85477.11	74327.922	64632.976	$ (500,000.00)
$ (74,341.83)	0.16	112068.9655	96611.177	83285.498	71797.843	61894.692	$ (500,000.00)
$ (84,085.00)	0.17	111111.1111	94966.762	81168.172	69374.506	59294.45	$ (500,000.00)

IRR	0.09

For r = 0.09 the NPV value is closest to 0, hence we choose IRR = 0.09

A.4   PROJECT SUMMARY COMPARISON

Project	NPV	IRR	PB
X	− 5836.66	0.10	3
Y	46,104.96	0.16	2
Z	− 7197.72	0.10	4

From the values calculated, we can see that project Y is the best financial option in all three metrics. The PB method though very simple often favors short projects with quick returns and penalizes excellent longer-term projects where large benefits do not start until years 3 and beyond. Also, the PB method completely ignores the time value money concept. In this example, all three methods favor the same project. This would not be the case if the flows were changed to showing benefits later in the stream.

There are many different reasons for a company to select a project that has benefits other than financial and therefore would not be reflected in the measurement techniques discussed above. For example, the following additional criteria need to be assessed along with the basic financial measures:

- Strategic fit
- Availability of resources
- Likelihood of success
- Increases revenue
- Process efficiency impact.

The example below adds some additional assessment criteria to the evaluation and assigns weights to each of these. Note that some of the attributes are pure numeric and others are qualitative assessments translated into an integer score. Let us see what summary values emerge from this type analysis.

		Project X		Project Y		Project Z	
		Criteria Rating	Weighted	Criteria Rating	Weighted	Criteria Rating	Weighted
Criteria	Weight (%)	1 = Low, 5 = High		1 = Low, 5 = High		1 = Low, 5 = High	
Strategic fit	20	4	0.8	3	0.6	5	1
Availability of resources	10	3	0.3	4	0.4	4	0.4
Probability of success	5	4	0.2	3	0.15	5	0.25
Revenue impact	10	2	0.2	4	0.4	2	0.2
Operation cost impact	10	4	0.4	2	0.2	4	0.4
IRR	40	1	0.3	4	0.8	1	0.2
PB	5	3	0.15	4	0.2	2	0.1
	Score		2.65		3.55		2.75
	Ranking		3		1		2

As we can see from the weighted average shown above, Project Y is still rated the best option based on the highest weighted score of 3.55. The weights assigned for each category are arbitrary, and it is often wise to compare other weightings to see if the answer is sensitive to some particular value.

DISCUSSION QUESTIONS

1. What other evaluation criteria might be added to the matrix?
2. What might be the value in selecting one of the other projects?
3. Do you see any flaws in the manner in which weights are assigned to the various criteria?
4. Is there a situation where PB might be the best financial metric?

Appendix B: Templates

One of the irritants of project management is the creation of formalized documents. As a result of this many projects shun the documentation process and produce only the bare minimum. For those external stakeholders dealing with such projects the lack of professional looking documentation can make the management process appear amateurish, whether that is a valid assessment or not. Use of templates can be a great asset in making the overall project look more professional and save a significant amount of time in the process. Throughout the book various project management artifacts have been described as to their role, but not so much in detail as to what they might look like in format. If one were to undertake developing a format for all of the documents, there would be significant time spent in that activity. That is really not a productive use of time and samples of these are readily available through many sources. The Internet has opened access to these and there is now a wide array of online sources where tested template formats can be found. Some of these have moderate costs and some are free. This appendix will summarize a few such sites that are worthy of reviewing.

A sample listing of template types would include the following:

- Project plan
- Change request
- Scope management plan
- Scope change register
- Change tracking
- Requirements definition
- Estimating worksheet
- Standard development network plan
- Life cycle development activity list
- Issue register
- Project status reporting
- Project standards.

The web locations listed below are well known and respected sources for various project management information, including methodologies, articles, training, and templates.

B.1 WEB SOURCES

- TenStep is a free web site offering technical source material in project management. One of the unique aspects of this site is its partnership with Cordin8. The pair offers an inexpensive approach to implementing a tested methodology along with the Cordin8 tool for information sharing and collaboration. Both of these options offer good potential for improving project selection, delivery, and collaboration. Their URLs are www.tenstep.com and www.cordin8.com
- Method123 is an online web reference for project management. They sponsor an excellent methodology and templates for various project management functions. Their URL is www.method123.com

- Gantthead.com is a good online source for various project management articles, training, webinars, and tools related to current project management. Their URL is www.gantthead. com
- The International Association of Project and Program Management (IAPPM), formed in 2003 through volunteers, is established as a global PMP organization and Association providing knowledge and useful content to PMs and program managers. IAPPM is the publisher of the CPPMBoK, currently in first draft format. It also provides certified project manager (CPM) certification to individuals meeting project experience and eligibility criteria. As an independent project organization, IAPPM is dedicated to helping individuals achieving success in the global project community.
- The Matt H. Evans web site contains an extensive listing of spreadsheets that collectively have potential value in various aspects of project management. The URL for this list is http://www.exinfm.com/free_spreadsheets.html. On particularly interesting general purpose, spreadsheet that should be reviewed is The Project Management one which is sponsored by IAPPM (reference # 84 on the list).

B.2 GOVERNMENTAL WEB SITES

- The State of Texas Department of Information Resources sponsors a large set of templates. Further details can be found at www.dir.state.tx.us.
- The State of Michigan hosts a very mature web site dedicated to project management (www.michigan.gov/dit). Use the search option to browse through their library of methodology and templates.
- The government of Tasmania sponsors a mature project management web site dedicated to various aspects of project management. This site includes a knowledge base, various resources including templates, and services. Further details can be found at http://www. egovernment.tas.gov.au/themes/project_management.

B.3 TEXTS

Several texts publish project management templates. One notable example can be found at Garton, C. and E. McCulloch, 2007. *Fundamentals of Project Management*. Lewisville, TX: MC Press.

In addition to the various sources shown here Google will uncover many more.

Appendix C: Project Repository Architecture

This appendix describes a project document architecture and content designed to be an information repository template for the project team and other stakeholders. The value of such a repository comes from having easy access to technical and status documentation regarding the project and its work products. These data groups have broad communication value to the various stakeholder groups involved. They also have value for management activities related to team collaboration, decision making, and information sharing. Organizations that successfully accomplish appropriate levels of information sharing report significant improvements in operational efficiency and improved decision making. In addition to this, the cross-communications value produced through the lessons learned process is greatly enhanced by having ready access to previous project's archival data.

An appropriate enterprise project documentation repository strategy consists of the following 14 basic component parts:

1. Product artifacts produced by the project (physical design and related artifacts dealing with the technical aspects of the project)
2. Operational metrics (raw performance data collected for the project)
3. Project formal status reports (complete history)
4. KA artifacts (filed by KA)
5. Quality Management artifacts (test plans, testing results, etc.)
6. Configuration management libraries (formal documentation repository filed by document type with version control references)
7. Initiation repository (original vision, project Charter, business case, etc.)
8. Operational documentation (User-related documentation)
9. Communication repository (meeting minutes, team communications, and external communications)
10. Lessons learned repository)
11. Reusable standard project management templates
12. Scope Management (ICC)
13. Risk Management (Risk database)
14. Issues log.

Note: Items 10 through 14 may be managed separately in centralized enterprise level repositories.

Each of these repository subsets has value in executing and managing the overall work flow of the project as well as the enterprise project portfolio. In addition to the project level items outlined in the first 10 document groups the last four groups have value to external parties as well as the project team. For that reason these repositories may be organized as central repositories to contain all projects in the enterprise. Finally, there is a top tier of related operational systems that need data from a project. If the various project data structures were standardized this could be an automated link. For example, data related to enterprise PPM, central accounting, human relations, corporate policies, and others would be managed by their respective custodians but linked to the project as needed.

Ideally, all of the data groupings described above need to be logically interrelated; however, practically speaking, that level of integration is probably not a reasonable tactical goal for most organizations. The core discussion here is to describe a subset of those items that are most critical to serve the basic project needs and that are more under the control of the PM.

In modern terms this repository should be viewed as automated. If that is not practical, a similar manual filing structure should be organized as indicated. Migration from manual files to automation should proceed in a priority order based on value to the team. Step one of that approach should be to develop an overall repository structure and work toward full automation.

In operation, value of a structured repository will be achieved through information sharing and improved decision making. Additional value will be recognized in supporting audit requirements and other requests for project data. One of the common complaints of project team members is the time wasted looking for various project documentation. One other aspect often not recognized is using an out-of-date paper document. The project repository is intended to be the "one source of the truth" place to retrieve data and one could view this as a "Google" database for all project artifacts. In order to achieve these goals, both the storage and the retrieval technical aspects of the repository need to be considered in the design. The view shown here is more logical and less technical. Access to the data store will require appropriate security access control through a common portal or collaboration tool that simplifies the access process.

Computer-based storage and retrieval of text and graphics documents has been one of the last frontiers for automation, but relatively recent technology developments in document management software has made this topic one that can be effectively dealt with. Basically, large volumes of documents can now be stored cost effectively, and retrieval speed for such documents is impressive. Based on this current state of technology the structure described here represents nothing more than an internal project to improve the overall working of projects. A manager once described this environment as a *project digital workbook*. Rather than having to pass physical documents around they can be retrieved electronically with more accuracy and timeliness than previous paper approaches.

A brief comment about each of the document subgroups is included below.

1. *Product artifacts.* This segment of the repository contains many of the technical objects needed by team members to execute their daily. Items related to the WBS and WBS Dictionary would be cataloged here. In addition, any logical or physical design notes would be stored under their respective WBS codes. Finally, any other related artifacts dealing with the technical aspects of the project are appropriate for this group.
2. *Operational metrics.* This subgroup is designed to store all project performance metrics, both planned and actual. Data required to report project status externally would be extracted from this source for both internal and external status information; however, other metrics stored in this area would be used for internal analysis of performance.
3. *Project status reports.* This subgroup would contain the chronology of all formal project status reports and would be the source used for transmission to appropriate stakeholders. In this manner, the person accessing these reports would be sure that they had the latest version.
4. *KA artifacts* (filed by KA). Each of the project KAs produce various documents related to that area. Initially, these are the KA subsidiary management plan for the project. Later, more detailed documents will be created. These groupings would contain operational items for these areas. *Note*: QC and QA could be separated from this group (see below).
5. *QC and QA artifacts.* This section would contain various operational quality artifacts such as test plans, testing results, user acceptance reports, quality audits, and so on.
6. *Technical design libraries.* Most projects have some form of technical design definition of the product. Physical projects would have CAD drawings and IT projects would have various logical design documents. These documents would be filed in some logical structure.

7. *Project initiation.* Many historical artifacts are created during the early initiation phase of the project. Items such as the original business case, vision statements, preliminary scope statement, and the official Charter authorizing the project should be saved here for future reference and possibly reuse on another project.

8. *Operational documentation.* In many cases, the user of the project output will need documentation regarding operational instructions for the device or system. Future maintenance support personnel will need this information related to their work. All documents in this category should be kept under version control and available to authorized personnel.

9. *Communications repository.* This group is intended to include items such as meeting agendas and minutes, internal team communications, and possibly other unstructured documents sent to or received from external sources. Portions of this group would be open to all and other portions would have limited access.

10. *Lessons learned repository.* The value found in this process has been recognized in recent times and all projects should be tasked with completing their lessons learned for each phase. This should be an enterprise level system, but might have to start as a project level effort and then evolve outward.

11. *Standard project management template library.* The concept of reuse has been around for quite some time in various industries and project types. Over the past few years one of the strategies to make the project management process more time efficient is to reuse templates for various required items (i.e., project plans, status reports, WBSs, etc.). This also should be an enterprise level effort and shared by all projects. The PMO/EPM organization is typically charged with supporting this effort.

12. *Scope Management* (ICC). This process will either be viewed as internal to the project structure or managed at a central level. The organizational option for change control will answer how the system data will be stored. However, the formal capture, tracking, and management of scope change requests represent the most important formal configuration management and tracking goals for the project and the organization. Failure to properly manage this aspect of the process can create severe cost and schedule problems.

13. *Risk Management.* This management area represents the least mature management process for most organizations. Assuming that the process is recognized as being part of formal project management, there is a need to identify and track risk elements in the same basic manner that scope is handled. That is, identify risk elements, decide how to handle that element, and then monitor the process. More detailed discussions on risk management processes are included in Chapters 22 and 24 for further thoughts on this data group. The key data would reside in a formal Risk Register database.

14. *Issues log.* Project issues are items that come up in the course of planning, execution, control, and closing. They simply mean that someone needs to deal with the "issue." For instance, an item of hardware needed by the project team is broken and needs to be fixed, or a decision is needed on a particular question. The issues log would identify this need, criticality, and person responsible. Since many of these events require support from external project team personnel this should be an enterprise system, but the starting point could be to formalize it for the project team.

Each of repository subgroups outlined above represents important data artifacts for the project. Mature organizations seeking to improve their project environment will support the evolution of this concept, but tactically the PM should follow these guidelines in whatever physical form is available—manual files, crude project file systems, enterprise formal systems, and the like.

C.1   IMPLEMENTATION STRATEGY

Data repositories of the breadth and complexity outlined here typically evolve over time through several steps. The initial development phase often is triggered by the recognition that a defined data group has value. From that initial starting point data are collected and stored, usually in a crude form such as a spreadsheet file, or a manual paper file. Similar repository groups often migrate in fragmented fashion horizontally across the organization. Finally, at some critical decision point high-level recognition and support will surface for formalizing and centralizing the data store in order to improve the overall operational value of the data.

Each of the repository groups outlined here often follows such an evolutionary path, but ideally this subject should be looked at in total and a formal solution sponsored by higher organizational levels. It is important to recognize the value of these repositories as part of the standard project work processes and to identify an official owner charged with care and feeding of each data group. As these groups become operational, there needs to be a mechanism for the cross-connections between these systems to be recognized and dealt with. Standardization of issues such as master data schema and date element definitions is needed for all subgroups. One development hazard is to allow stove-pipe, isolated developments, that later do not fit together. Likewise, data element ownership needs formal recognition (e.g., where does the data come from and who owns it). Philosophically, effort needs to be made to instantiate a single data version of the truth through these data stores by avoiding redundancy and having version control.

Step two in the evolution is the organizational push to embed the data into the operational culture. This is typically accomplished by using these data items for formal status reporting and operational management actions. Storage and publication of approved metrics fits into this theme as well. The major support facility for this is some type of retrieval strategy that opens up the data store to a wide variety of users with little pretraining requirement—an intuitive usage approach.

Step three in the evolution involves using the data store for analytical purposes to improve understanding of the project activities—that is, source of errors, productivity variables, resource issues, resource management across the organization, and the like. Data mining utilities and search capabilities represent two types of tools to aid in ad hoc exploration through the data. These capabilities support efforts to improve performance and understand the processes better than traditional tools or ad hoc judgment provide.

Step four represents data reusability across projects. The management templates will support quick methods of producing high-quality documents efficiently. Much of the project management process requires formatting and boilerplate wording. All of this can be supplied from the template library and other structures defined in the repository. Once the entire structure matures, the repository will provide the opportunity to bring significant changes to the traditional documentation and associated development paradigms.

As with all procedural changes to the organizational culture this process requires high-level management support and an understanding of the value. Historically, this has not been the case and project documentation methods have been characterized as being similar to the cobbler's children with no shoes. Process improvement can only come through a strategic focus on the topic.

Index